Probability, Statistics, and Stochastic Processes for Engineers and Scientists

Mathematical Engineering, Manufacturing, and Management Sciences

Series Editor:
Mangey Ram, Professor, Assistant Dean (International Affairs)
Department of Mathematics, Graphic Era University, Dehradun, India

The aim of this new book series is to publish the research studies and articles that bring up the latest development and research applied to mathematics and its applications in the manufacturing and management sciences areas. Mathematical tools and techniques are the strength of engineering sciences. They form the common foundation of all novel disciplines as engineering evolves and develops. The series will include a comprehensive range of applied mathematics and its application in engineering areas such as optimization techniques, mathematical modeling and simulation, stochastic processes and systems engineering, safety-critical system performance, system safety, system security, high-assurance software architecture and design, mathematical modeling in environmental safety sciences, finite element methods, differential equations, and reliability engineering.

Mathematics Applied to Engineering and Management
Edited by Mangey Ram and S.B. Singh

Mathematics in Engineering Sciences
Novel Theories, Technologies, and Applications
Edited by Mangey Ram

Advances in Management Research
Innovation and Technology
Edited by Avinash K. Shrivastava, Sudhir Rana,
Amiya Kumar Mohapatra, and Mangey Ram

Market Assessment with OR Applications
Adarsh Anand, Deepti Aggrawal, and Mohini Agarwal

Recent Advances in Mathematics for Engineering
Edited by Mangey Ram

Probability, Statistics, and Stochastic Processes for Engineers and Scientists
Aliakbar Montazer Haghighi and Indika Wickramasinghe

For more information about this series, please visit: https://www.crcpress.com/ Mathematical-Engineering-Manufacturing-and-Management-Sciences/book-series/ CRCMEMMS

Probability, Statistics, and Stochastic Processes for Engineers and Scientists

Aliakbar Montazer Haghighi and
Indika Wickramasinghe

CRC Press
Taylor & Francis Group
Boca Raton London New York

CRC Press is an imprint of the
Taylor & Francis Group, an **informa** business

First edition published 2021
by CRC Press
6000 Broken Sound Parkway NW, Suite 300, Boca Raton, FL 33487-2742

and by CRC Press
2 Park Square, Milton Park, Abingdon, Oxon, OX14 4RN

© 2021 Taylor & Francis Group, LLC

CRC Press is an imprint of Taylor & Francis Group, LLC

Library of Congress Cataloging-in-Publication Data
Names: Haghighi, Aliakbar Montazer, author. | Wickramasinghe, Indika, author.
Title: Probability, statistics, and stochastic processes for engineers and
scientists / Aliakbar Montazer Haghighi, Indika Wickramasinghe.
Description: First edition. | Boca Raton, FL : CRC Press, 2020. | Series:
Mathematical engineering, manufacturing, and management sciences |
Includes bibliographical references and index.
Identifiers: LCCN 2020005438 (print) | LCCN 2020005439 (ebook) |
ISBN 9780815375906 (hardback) | ISBN 9781351238403 (ebook)
Subjects: LCSH: Engineering—Statistical methods. | Stochastic analysis.
Classification: LCC TA340 .H27 2020 (print) | LCC TA340 (ebook) |
DDC 519.5—dc23
LC record available at https://lccn.loc.gov/2020005438
LC ebook record available at https://lccn.loc.gov/2020005439

ISBN: 978-0-8153-7590-6 (hbk)
ISBN: 978-0-367-50086-3 (pbk)
ISBN: 978-1-351-23840-3 (ebk)

Typeset in Times
by codeMantra

Visit the CRC Press website for figure slides: www.crcpress.com/9780815375906

بِسْمِ اللهِ الرَّحْمنِ الرَّحِيمِ

IN THE NAME OF GOD, THE ALL-MERCIFUL, THE ALL-COMPASSIONATE

I have always appreciated my lovely wife, Shahin Hamidi, for putting up with me for about 48 years of marriage. A significant part of what was supposed to be our "quality time" together, I have squandered in research and publication. Fully aware of her involuntary sacrifice, I have always devoted my scholarly work and achievement to her, along with others.

With all due respect to all other authors, writing mathematics books is extremely time-consuming, breathtaking, and sometimes frustrating. Nonetheless, it is very rewarding experience because it serves students, faculty, and researchers, internationally. For that reason, my part of this volume (my fifth in the English language that might likely be the last book I will publish) is dedicated to the following conceptual ideas: humanity, peace, honesty, mercifulness, kindness, and compassion.

Aliakbar Montazer Haghighi
Houston, Texas, USA
December 1, 2019

I would like to dedicate this book to my wife and two kids. Without their continuous help and understanding, this book wouldn't have completed on time. I also like to dedicate this book to my parents for their immense contribution to my success.

Indika Wickramasinghe
Houston, Texas, USA
December 1, 2019

Contents

Preface

There are textbooks in probability and statistics available in the market—some old and new prints or new editions—that contain basic ideas and are more suitable as undergraduate textbooks. There are also textbooks written particularly for graduate programs and study. Featuring recent advances in probability, statistics, and stochastic processes, *Probability, Statistics, and Stochastic Processes for Engineers and Scientists* presents a volume that includes up-to-date topics for engineers and scientists, like fuzzy probability as presented in *Probability and Statistics* or *An Introduction to the Stochastic Processes* with applications from various areas of stochastic processes and queueing models.

Often, for science, technology, engineering, and mathematics (STEM) majors, the concepts behind probability, statistics, and stochastic processes courses are taught in two or three separate courses with just as many textbooks. However, this book may be used for all three subject areas, with all necessary topics contained in one volume. If adapted for just probability and statistics, the Chapter 7 on stochastic processes offers students an opportunity to see real-life applications of the topics they are studying.

Although this book offers rigor to some engineering concepts, most engineers, after taking at least two semesters of calculus and a semester of ordinary differential equations, will be able to grasp the necessary mathematical concepts from probability, statistics, and stochastic processes as presented, and will inevitably engage a deeper understanding of the entire book. However, only the first two calculus courses are sufficient for the subject matter relating to probability and statistics.

As written, this book balances both theory and practical applications of probability, statistics, and stochastic processes, keeping practitioners in engineering and the sciences in mind. Figure slides are also available for downloading at www.crcpress.com/9780815375906. Additionally, this book offers descriptive statistics and basic probability materials for those needing descriptive and calculus-based probability and statistics courses at universities and colleges.

Notably, this book presents key information for understanding the essential aspects of basic probability theory and its applications. Beginning with the rich history of probability and statistics, this book presents a selection of both subject's standard terms and properties. From there, we move on to the three well-known types of statistics, namely, descriptive, inferential, and nonparametric. This book then offers Chapter 7 with a detailed coverage of stochastic processes and its applications in queueing models, random walk, and birth-and-death models with a variety of examples. Additionally, this book will illustrate how to engage different software options for calculation and data analysis, such as the Scientific Calculator, Minitab, MS Excel, MATLAB®, and the R software environment.

We hope this book can also be used by researchers and working professionals in the fields of probability, statistics, and stochastic processes. We anticipate the information as an excellent resource for public and academic libraries.

Lastly, we would like the opportunity to express our gratitude to Taylor & Francis Group, LLC. (CRC Press), Boca Raton, Florida, the United States, for offering us through Professor Mangey Ram and Ms. Cindy Renee Carelli and awarding us the contract to write this book, as well as those who helped us in this work. In particular, we express our most appreciation and thanks to Drs. Arouna and Kalatu Davies for performing excellently with their rigorous editing of this book manuscript they accepted. Last but not least, the first author is extremely appreciative of Dr. Danny R. Kelley, Dean of Brailsford College of Arts and Sciences at Prairie View A&M University (PVAMU), for his continuous support for him to complete writing this book.

Aliakbar Montazer Haghighi and
Indika Wickramasinghe
Houston, Texas, USA
December 1, 2019

MATLAB® is a registered trademark of The MathWorks, Inc. For product information, please contact:
  The MathWorks, Inc.
  3 Apple Hill Drive
  Natick, MA 01760-2098 USA
  Tel: 508-647-7000
  Fax: 508-647-7001
  E-mail: info@mathworks.com
  Web: www.mathworks.com

Authors

Dr. Aliakbar Montazer Haghighi is a professor and the head of the Mathematics Department at Prairie View A&M University, Texas, the United States. He received his PhD in probability and statistics from Case Western Reserve University, Cleveland, Ohio, the United States, under supervision of Lajos Takács, and his BA and MA in mathematics from San Francisco State University, California. He has over 50 years of teaching, research, and academic and administrative experiences at various universities globally. His research publications are extensive; they include mathematics books and lecture notes written in English and Farsi. His latest book with D.P. Mishev as co-author, titled *Delayed and Network Queues*, appeared in September 2016, which was published by John Wiley & Sons Inc., New Jersey, the United States; his last book chapter with D.P. Mishev as co-author, *Stochastic Modeling in Industry and Management*, Chapter 7 of *A Modeling and Simulation in Industrial Engineering*, Mangey Ram and J.P. Davim, as editors, appeared in 2018, which was published by Springer. He is a lifetime member of American Mathematical Society (AMS) and Society of Industrial and Applied Mathematics (SIAM). He is the co-founder and is the editor-in-chief of application and applied mathematics: *An International Journal* (*AAM*), http://www.pvamu.edu/aam. Additional information about Dr. Haghighi is available at https://www.pvamu.edu/bcas/departments/mathematics/faculty-and-staff/amhaghighi/.

Dr. Indika Wickramasinghe is an assistant professor in the Department of Mathematics at Prairie View A&M University, Texas, the United States. He received his PhD in mathematical statistics from Texas Tech University, Lubbock, Texas, the United States, under supervision of Dr. Alex Trindade; MS in statistics from Texas Tech University, Lubbock, Texas, the United States; MSc in operations research from Moratuwa University, Sri Lanka; and BSc in mathematics from University of Kelaniya, Sri Lanka. He has teaching and research experience at University of Kelaniya (Sri Lanka), Texas Tech University, Eastern New Mexico University, and Prairie View A&M University. Dr. Wickramasinghe has individually and collaboratively published a number of research publications and successfully submitted several grant proposals. More information about Dr. Wickramasinghe is available at https://www.pvamu.edu/bcas/departments/mathematics/faculty-and-staff/iprathnathungalage/.

1 Preliminaries

1.1 INTRODUCTION

Games of chance or concepts of probability are known to have been in existence for over a thousand years. For instance, the game of "knucklebones", ankle bone of animals or heel bone that can land in any of four different ways, has been played for time immemorial. Because of the irregular shape of the bone, called the **talus**, causing it to land in any of four possible ways, the landing considered random (in general, a **talus** is a randomizer similar to the die). This early brush with probability, however, was not systematically handled even though the solutions of the games assumed equally likely outcomes. As mentioned in David (1962), mathematicians in Europe were hardly familiar with the calculations of probabilities. Indeed, the first printed work on the subject seems to have been made in 1494 by Fra Luca Pacioli, while the basic game theory was written in 1550 by Geronimo Cardano. It was titled *Liber de Ludo Aleae* (*The Book on Games of Chance*).

The theory of probability once appeared in a correspondence between Pascal and Fermat in 1654, regarding a problem posed by a French gambler Chevalier de Méré. As stated in the literature, the seventeenth-century gambler was losing more than he expected. He then turned to his friend Blaise Pascal for an explanation. Fermat and Pascal are the founders of the theory of probabilities. Fermat's views on the fundamental principles of probability became the foundation of the probability theory. His theory of probability grew out of his early research into the theory of numbers. Most of his work, however, was published after his death. The theory of probability became a branch of mathematics in the mid-seventeenth century, much, much later than the calculations that first surfaced in the fifteenth century (Figure 1.1).

Fermat, on the one hand, tried to solve the problem of outcomes of a game by listing all possible cases such as tossing two distinct fair coins four times and listing the possible outcomes as *HH*, *HT*, *TH*, and *TT*. The list was referred to as **two players play four games**. The winner is the one with the first three win games. The outcomes will be win–win, win–loss, loss–win, and loss–loss. The score was sometimes measured in the "bit". At the end of the play, each player gains a proportional amount of the total bit (equally set from each player) of the number of wins. Hence, the first player, in this case, wins three-quarters of the total bit, while the other wins one-quarter.

Pascal, on the other hand, offered a different method called the **method of expectations**. This method enables the play for an extended number of games. He used mathematical induction and the recursive property of the Pascal–Khayyam triangle in the form, as shown in Figure 1.2.

1

(a)

(b)

FIGURE 1.1   (a) Blaise Pascal 1623–1662. (b) Pierre de Fermat 1601–1665.

```
1    1    1    1    1    1    1
1    2    3    4    5    6
1    3    6    10   15
1    4    10   20
1    5    15
1    6
1
```

FIGURE 1.2   Pascal–Khayyam triangle.

Game Played by Chevalier de Méré

Two dice are rolled. Each die having six sides with numbers 1 through 6 on each side, the possible sums are 2, 3, 4, 5, 6, 7, 8, 9, 10, 11, and 12. In other words, the maximum sum is a double 6. Chevalier de Méré first used to bet that on a roll of a die, he would get at least a 6 in 4 rolls of the die. However, he changed his strategy and played by betting to have a double 6 (a total of 12) in 24 rolls of the two dice. But he soon noticed that his previous strategy resulted in more gains. The cause of the difference was the problem posed to Pascal. Based on Pascal's calculation, the chance of winning with one die was 51.8%, while it with a double dice was 49.1% (see Smith (1966) for detailed calculations).

Problem posed by Chevalier de Méré

The idea is to compare the following two cases (assuming the dice are fair so that each side has the same chance to land):

Case 1. What is the chance of having at least one 6 in four rolls of a die?

Case 2. What is the chance of having at least one double 6 in 24 rolls of two dice?

Answer to Case 1

The chance of a 6 in four rolls of a die would be one from 6^4 possible combinations of numbers to occur, leaving 5^4 unfavorable combinations without a 6. Thus, there are $6^4 - 5^4$ favorable combinations to bet. Based on the equal chance assumption, the chance of obtaining a 6 is:

$$\frac{6^4 - 5^4}{6^4} = \frac{671}{1,296} > 50\%.$$

Answer to Case 2

The chance of a double 6 in 24 rolls of 2 dice would be one from 36^{24} possible combinations of numbers to occur, leaving 35^{24} unfavorable combinations without a double 6. Thus, there are $36^{24} - 35^{24}$ favorable combinations to bet. Based on the equal chance assumption, the chance of obtaining a double 6 is:

$$\frac{36^{24} - 35^{24}}{36^{24}} = 0.4914 < 50\%.$$

As noted in the entire discussion above, in each game played, concerns are focused on the outcomes of a game. Hence, we are led to consider the properties of a collection of object that is referred to as a **set**. Thus, to better understand the probability, it is necessary to briefly remind the readers of the basics of set theory.

1.2   SET AND ITS BASIC PROPERTIES

Sets are recognized as the most basic of structures in mathematics. They are of substantial mathematical significance, in various mathematical directions. By a **set**, we mean a **well-defined** (in mathematical sense) collection of objects. The objects may be referred to as the **members** or **elements**. In other words, a set is a notion of group objects. To denote a set by its elements, we simply list the elements within a pair of curly brackets, { }, separated by commas. For instance, $A = \{a, b\}$. Symbolically, if a is an element of a set A, as in this example, we write:

$$a \in A, \tag{1.1}$$

where the symbol "$\in$" is read as "belongs to".

Note 1.1

Elements of a set do not have to be numbers, they may be set by themselves, and they do not have to be related, other than all belonging to the same set.

Note 1.2

The term "well defined", in mathematical sense, means that an element is decisively a member or not a member of the set, but not both that are for a given set A, "for all x, either x is a member of A or it is not a member of A".

Definition 1.1

Two sets are **equal** if both have exactly the same elements, and vice versa.

Note 1.3

Based on Definition 1.1, arrangement of elements of a set is unimportant. For example, $A = \{1, 2, 3\}$ is the same as $A = \{2, 1, 3\}$.

Note 1.4

When elements of a set are real numbers, the set is denoted by (or expressed by) an **interval**. For instance, $[7; 9]$ is the set of all real numbers between 7 and 9, inclusive.

Definition 1.2

In the set of real numbers, $\mathbb{R}$, a **neighborhood** of an element (a real number), say c, is a set containing an open interval (a, b) such that $a < c < b$. By an **open set**, it is meant a set, which is a neighborhood of each of its points. A **complement** of an open set is referred to as a **closed set**. Equivalently, a closed set is a set, say A, containing all its limit (or boundary) points (such points are those that, within all of their neighborhoods, are elements both from A and outside of A, too).

Note 1.5

In $\mathbb{R}$, an open interval (a, b) is an open set.

Elements of a set may also be **descriptively** defined. In other words, a set may be defined by some "property" or "formula", say $p(x)$, where "x" could be any variable. In such a case, a set A can be symbolically written as:

$$A = \left\{ x \middle| p(x) \text{ is true} \right\}, \tag{1.2}$$

where "|" stands for "such that". Note that "such that" may also be symbolized as ":" or "∋".

Example 1.1

To describe a set of numbers, say the natural numbers between 1 and 9, we could write:

$$A = \left\{ n \mid n \in \mathbb{N}, 1 \leq n \leq 9 \right\},$$

where $\mathbb{N}$ refers to the set of natural numbers $1, 2, \ldots$.

Example 1.2

The set of all indigents in the United States in census 2010 can be written as

$$A = \left\{ x \mid x \text{ is an indigent in the United States in the census of } 2010 \right\}.$$

By definition, the **size of a set A**, the "number" of elements of A, is referred to as the **cardinal** or **cardinality** of A, and sometimes it is denoted by $|A|$. If the set has only one element, say a, it is called a **singleton** and it is denoted by $\{a\}$.

A set may be **finite** or **infinite**, depending upon its number of elements being finite or infinite, respectively. More precisely, we have the following definition:

Definition 1.3

A set A is finite (or, of cardinality, say n, where n is any natural number) if there is a bijection (i.e., a one-to-one correspondence) between its elements and those of a finite set of natural numbers $\{1, 2, \ldots, n\}$; otherwise, A is **infinite** (or, **of infinite cardinality**).

Example 1.3

The set $A = \{1, 2, 3\}$ is a finite set with three elements (i.e., of finite cardinality 3), while the set $B = \{1, 2, 3, 4, \ldots\}$ is an infinite set (i.e., of infinite cardinality) with infinitely many elements (i.e., as many as the natural numbers).

Note 1.6

The concept of **infinity** (symbolically, ∞) that may make a set "infinite" was a concern of Indian mathematicians in the East earlier than the fifth century BC and during the fifth century BC initiated by the Greek mathematician Zeno of Elea (490 BC–430 BC) in the West. But the infinity, in the modern era, was formally defined by Cantor during 1867–1871.

In general, one may ask how many natural numbers are there? In other words, how far can one count? The answer, intuitively, is "the natural numbers are endless". More formally, the "number" of natural numbers is **infinite** (cardinality), and this cardinality, denoted by ω, has the set of natural numbers, $\mathbb{N}$. If a set has the same number of elements as (i.e., if it is bijective with) the set of natural numbers, we say the set is **denumerable** or **countably infinite**, being also assigned cardinality ω. Also, a set is called **countable** if it is either finite or denumerable. If a set is not countable, it is **uncountable** with assigned cardinality one larger than ω.

The **classic set theory** (study of sets) was initiated by George Cantor and Richard Dedekind during 1870s, starting with their meeting in 1872 that leads to the paper by Cantor in 1874. Its complete development took almost a century (Figure 1.3).

The sum of two sets is referred to as the **union of the two sets**, denoted by $\cup$, and it is a set whose elements are in either of the two sets with no duplicate counting of elements. Symbolically, if A and B are two sets, their union is denoted by $A \cup B$.

(a)

(b)

FIGURE 1.3 (a) Richard Dedekind, 1831–1916. (b) Georg Ferdinand Ludwig Philipp Cantor, 1845–1918.

Example 1.4

If $A = \{1,2,3\}$ and $B = \{3,4,5\}$, then $A \cup B = \{1,2,3,4,5\}$.

The **intersection of two sets** A and B, denoted by $A \cap B$, is a set containing elements that are in both sets. In case there is no common element, the intersection would be an **empty or null set**, denoted by $\{\ \}$ or $\emptyset$.

Example 1.5

For $A = \{1,2,3\}$ and $B = \{3,4,5\}$, we have $A \cap B = \{3\}$, since 3 is the only common element in A and B.

Example 1.6

If $C = \{1,2,3,4\}$ and $D = \{2,3,4,5\}$, then $C \cap D = \{2,3,4\}$.

Example 1.7

For $A = \{1,2,3\}$ and $B = \{4,5\}$, we have $A \cap B = \emptyset$ since there is no common element in A and B.

Example 1.8

If $A = \{1,2,3,4,5\}$ and $B = \{3,4\}$, then $A \cap B = \{3,4\}$ and $A \cup B = \{1,2,3,4,5\}$.

In terms of elements, the union and intersection of two sets are defined as $A \cup B = \{x : x \in A \text{ or } x \in B\}$ and $C \cap D = \{x : x \in C \text{ and } x \in D\}$, respectively. Both definitions of union and intersection of two sets may be extended for any finite or infinite number of sets in the same fashion. For instance, for three sets A, B, and C, we have: $A \cup B \cup C = A \cup (B \cup C) = (A \cup B) \cup C$ and $A \cap B \cap C = A \cap (B \cap C) = (A \cap B) \cap C$.

For the union and intersection of infinite number of sets $A_1, A_2, \ldots$, we use the notations:

$$\bigcup_{i=1}^{\infty} A_i = A_1 \cup A_2 \cup A_3 \cdots \text{ and } \bigcap_{i=1}^{\infty} A_i = A_1 \cap A_2 \cap A_3 \cap \cdots, \qquad (1.3)$$

respectively.

The **universal set**, denoted by U, is a set assumed to be containing all sets under consideration. The universal set is used to define some definitions that are stated below.

For a set, say A, its **complement** is the set of elements in U not in A, and it is denoted by A^c, A', or A^-. However, the **relative complement** of a set, say A, with respect to another set, say B (subset of A), is the set of elements in B that are not in A. It is called **the difference of A and B**. More generally, for any two sets A and B, the **difference of the two sets**, denoted by B/A, is the set of elements in B but not in A. In other words,

$$B \backslash A = \{x \in B | x \notin A\}. \qquad (1.4)$$

Example 1.9

If $A = \{1,2,3\}$ and $B = \{3,4,5\}$, then $B \backslash A = \{4,5\}$.

If U denotes a universal set, then we list the following properties of compliment:

$$(A \cup B)^c = A^c \cap B^c \text{ and } (A \cap B)^c = A^c \cup B^c : \textbf{DeMorgan's laws} \qquad (1.5)$$

$$A \cup A^c = U \text{ and } A \cap A^c = \varnothing. \qquad (1.6)$$

$$\varnothing^c = U, U^c = \varnothing \text{ and } \left(A^c\right)^c = A. \qquad (1.7)$$

Note 1.7

As a result of De Morgan's laws (1.5), we have:

$$(A \cup B) = (A^c \cap B^c)^c \text{ and } (A \cap B) = (A^c \cup B^c)^c. \qquad (1.8)$$

Roughly speaking, a **Borel set** (named after the French mathematician Émile Borel, 1871–1956) is a set that can be constructed from open or closed sets through repeatedly taking countable union and intersections. It is denoted by $\mathscr{B}$.

Here are some examples:

1. Any subset of the interval [0,1] is a Borel set.
2. The set of all rational numbers (also irrational numbers) in the interval [0,1] is a Borel subset of [0,1].
3. Any countable subset of the interval [0,1] is a Borel subset of [0,1].
4. The complement in [0,1] of any Borel subset of the interval [0,1] is a Borel subset of [0,1].

Presenting sets through diagrams, there is a well-known graphic presentation of sets called **Venn diagram**. For instance, the intersection of two sets A and B and that of three sets A, B, and C are represented by Figures 1.2 and 1.4, respectively. The rectangle is referred to as the **universal set U**, that is, the set of all possible elements, under consideration, in some specific circumstances (Figure 1.5).

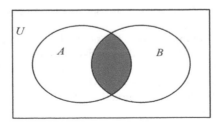

FIGURE 1.4   Intersection of two sets, Venn diagram.

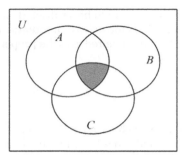

FIGURE 1.5 Intersection of three sets, Venn diagram.

Having defined a universal set, we can define a set in the following three ways:

1. As we have seen, when only a finite number of elements are available, a set may be defined **by listing all its elements**. For instance, if there are n elements in a set A, we might define A as:

$$A = \{a_1, a_2, \ldots, a_n\}. \tag{1.9}$$

2. A set may be defined by its elements' **property**. In this case, a set A is denoted as

$$A = \{a \mid P(a)\}, \tag{1.10}$$

where $P(a)$ denotes "the property of a". In other words, for each element a of A, the property P is either true or false.

3. Let I be a set and A a subset of I. Then, a function I_A defined on I is called a **characteristic function** or **indicator function** of A, which identifies an element of I, say i, as either in A or not in A. Such a set is called a **crisp set**. Symbolically, I_A is defined as:

$$I_A(i) = \begin{cases} 1, & \text{for} \quad i \in A, \\ 0, & \text{for} \quad i \notin A. \end{cases} \tag{1.11}$$

In other words, I_A states which elements of I are in A and which are not. Thus, (1.11) states that for each $i \in I$, i is in A if $I_A(i) = 1$, and for each $i \in I$, i is not in A if $I_A(i) = 0$. That is, in particular, the characteristic function I_A maps elements of I to elements of the set $\{0,1\}$, or

$$I_A : I \to \{0,1\}. \tag{1.12}$$

Thus, the characteristic function of a crisp set A is a function that assigns a value 1 or 0 to each element of the universal set: 1 if the element is in A and 0, otherwise. A universal set is always a crisp set. The name "crisp set" is usually used when one is in a fuzzy environment, as we will see later.

If A is a part of the set B, A is said to be a **subset** (denoted by $\subseteq$) of B; that is, $A \subseteq B$ if all elements of A are in B, too. In case the subset does not contain all elements of the original set, then it is referred to as the **proper subset** and is denoted by $\subset$.

Example 1.10

The set $A = \{1,2,3\}$ is a proper subset of $B = \{1,2,3,4,5\}$, because elements of A, that is, 1, 2, and 3, are in B and B has further more elements, too. Thus, we can write $A \subset B$.

Example 1.11

If $C = \{2,3,4\}$ and $D = \{2,3,4\}$, then $C \subset D$ or $D \subset C$. In this case, we could write $C = D$ since both sets have the same elements.

Note 1.8

The empty set is a subset of any set by assumption.

Note 1.9

With listing elements of sets, we can use Venn diagram to represent the union and intersection of sets.

Example 1.12

Let $A = \{1,2,3,a,b,c\}$ and $B = \{2,4,a,b,d\}$. Then, Figure 1.6a and b represent Venn diagrams for the union and intersection of A and B, respectively.

(a)
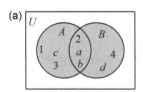

(b)
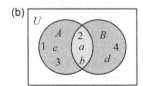

FIGURE 1.6   (a) $A \cup B = \{1,2,3,4,a,b,c,d\}$. (b) $A \cap B = \{2,a,b\}$.

Remark 1.1

Knowing the notion of "subset", we can redefine the characteristic function as follows: Let A be a subset of Y. Then, the characteristic of A is a function, say Y_A, such that $Y \rightarrow \{0,1\}$ that satisfies (1.11).

By the **Cartesian product** of two sets A and B, denoted by $A \times B$, it is meant the set of ordered pairs (a,b) such that a belongs to A and b belongs to B. In symbols,

$$A \times B = \left\{(a;b) \middle| a \in A \text{ and } b \in B\right\}. \tag{1.13}$$

Note 1.10

$$A \times B \neq B \times A. \tag{1.14}$$

Example 1.13

If $A = \{1,2,3\}$ and $B = \{a,b\}$, then

$$A \times B = \{(1,a),(2,a),(3,a),(1,b),(2,b),(3,b)\},$$

while

$$B \times A = \{(a,1),(a,2),(a,3),(b,1),(b,2),(b,3)\},$$

which are evidently not the same.

Cartesian product of two sets could be extended for n sets. Thus, if $A_1, A_2, \ldots, A_n$ are n sets, the Cartesian product is the set of all ordered n-tuples $(a_1, a_2, \ldots, a_n)$, where $a_i \in A_i$, for $i = 1, 2, \ldots, n$. In symbols,

$$A_1 \times A_2 \times \ldots, \times A_i = \left\{(a_1, a_2, \ldots, a_i) \middle| a_i \in A_i, \quad \text{for } i = 1, 2, \ldots, n\right\}. \tag{1.15}$$

Later, we would need the term set function, which we will define it here.

Definition 1.4

By a **set function**, it is meant a function whose domain is a set of sets and its range is usually a set of real numbers.

Example 1.14

A function that maps each set of the set of sets to its cardinality is a set function.

Definition 1.5

Let Ω be an arbitrary nonempty space (or set) of points. A **class** of subsets of Ω, denoted by $\mathcal{F}$, is called a **field** or an **algebra** if it contains Ω itself and it is **closed** under the formation of finite union and complements, that is,

$$\Omega \in \mathcal{F} \tag{1.16}$$

$$E \in \mathcal{F} \Rightarrow E^c \in \mathcal{F}. \tag{1.17}$$

$$E_1 \text{ and } E_2 \in \mathcal{F} \Rightarrow E_1 \cup E_2 \in \mathcal{F}. \tag{1.18}$$

Note 1.11

    i. Since Ω and $\varnothing$ are complement of each other, (1) and (2) imply that $\varnothing \in \mathcal{F}$.
    ii. From De Morgan's law, we will have:

$$E_1 \text{ and } E_2 \in \mathcal{F} \Rightarrow E_1 \cap E_2 \in \mathcal{F}. \tag{1.19}$$

We leave the details of proofs of the notes (i) and (ii) as exercises.

Definition 1.6

A class of subsets of an arbitrary space Ω, say $\mathcal{F}$, is a σ**-field** (read as "sigma field") or a σ**-algebra** if $\mathcal{F}$ is a field that is closed under countable unions. In other words,

$$E_1, E_2, \ldots \in \mathcal{F} \Rightarrow E_1 \cup E_2 \cup \ldots \in \mathcal{F}. \tag{1.20}$$

Note 1.12

Once again, from De Morgan's law, we will have:

$$E_1, E_2, \ldots \in \mathcal{F} \Rightarrow E_1 \cap E_2 \cap \ldots \in \mathcal{F}. \tag{1.21}$$

We can combine Definition 1.2, 1.3, and some Notes in between in the following definition.

Definition 1.6a

Let Ω be an arbitrary nonempty space (or set) of points. Also, let $\mathcal{F}$ be a class of subsets of the space Ω. Then, $\mathcal{F}$ is a σ**-algebra** if

$$\Omega \in \mathcal{F} \tag{1.22}$$

$$E \in \mathcal{F} \Rightarrow E^c \in \mathcal{F}. \tag{1.23}$$

If $E_1, E_2, \ldots$ is a countable sequence of sets in $\mathcal{F}$, then

$$E_1 \cup E_2 \cup \ldots \in \mathcal{F}. \tag{1.24}$$

Example 1.15

$\mathcal{F} = \{\emptyset, \Omega\}$ is a σ-algebra.

Example 1.16

The power set of Ω is a σ-algebra.

From intervals (by means of the operations allowed in a σ-algebra), we can define a very important set, namely, the **Borel set** that we will discuss in Chapter 2. We will now try to define this set.

Definition 1.7a

Let X be a set. Consider a collection of open subsets of X, say $\mathcal{T}$, that satisfies the following conditions with open sets:

1. $\emptyset \in \mathcal{T}$.
2. $X \in \mathcal{T}$.
3. Let $A_1, A_2, \ldots, A_n$ be a finite number of sets in $\mathcal{T}$. Then, $A_1 \cap A_2 \ldots \cap A_n \in \mathcal{T}$.
4. Let $A_1, A_2, \ldots, A_n$ be an arbitrary number of sets in $\mathcal{T}$. Then, $A_1 \cup A_2 \ldots \in \mathcal{T}$.

Then, X together with $\mathcal{T}$ is called a **topological** or **abstract space**.

Instead of open sets as above, the definition also holds for closed sets. That is, the following definition:

Definition 1.7b

Let X be a set. Consider a collection of open subsets of X, say $\mathcal{T}$, that satisfies the following conditions with open sets:

1. $\emptyset \in \mathcal{T}$.
2. $X \in \mathcal{T}$.
3. Let $A_1, A_2, \ldots, A_n$ be an arbitrary number of sets in $\mathcal{T}$. Then, $A_1 \cap A_2 \ldots \in \mathcal{T}$.
4. Let $A_1, A_2, \ldots, A_n$ be a finite number of sets in $\mathcal{T}$. Then, $A_1 \cup A_2 \ldots A_n \in \mathcal{T}$.

Then, X together with $\mathcal{T}$ is called a **topological** or **abstract space**.

Definition 1.8

Let $(X, \mathcal{T})$ be a topological space. Then, the smallest σ-algebra containing the open sets in $\mathcal{T}$ is called the collection of **Borel sets** in X.

Example 1.17

To make sure the collection specified in the definition 1.8 exists, empty intersections should not be chosen.

Note 1.13

Closed sets are the example of Borel sets.

1.3 ZERMELO AND FRAENKEL (ZFC) AXIOMATIC SET THEORY

The Zermelo–Fraenkel set theory was based on symbolic mathematics, as we do nowadays. However, Bertrand Russel raised a question through an example in 1901 that later became known as **Russell's paradox.** He tried to show that some attempted formalizations of the set theory created by Cantor lead to a contradiction. **Russell's paradox** (which apparently had been discovered a year before by Ernst Zermelo, but never published) appeared in his celebrated paper *Principles of Mathematics* in 1903. The classic Russel's paradox is given through the following example.

Suppose there is a group of barbers who shave only those men who do not shave themselves. Now suppose there is a male barber in the group who does not shave himself. Then, according to the assumption, he must shave himself. But again no barber in the group can shave himself. So, we are arriving at a paradox, which is an intuitive contradiction.

In set notation, **Russell's paradox** may be stated as follows: Let S be the set of all sets, say A, that do not belong to themselves. If S does not belong to itself, then by its own definition, it must contain itself. However, if it does contain itself, then it will contradict its own definition. Symbolically, the description of **Russell's paradox** is as follows (Figure 1.7):

$$\text{If } S = \{A \mid A \notin A\}, \text{ then } S \in S \Leftrightarrow S \notin S. \tag{1.25}$$

Note 1.14

It seems that Russell's paradox, which is indeed the property of the set that was defined, could have been avoided should the formulation excluded "self-reference". This, in fact, is a **set-theoretic singularity** at the property of the set. In other words, Russel's paradox demonstrates that an entity defined through particular inconsistent conditions doesn't exist. The bottom line is to avoid "the set of all sets" that does not exist. Indeed, if such a "set" would exist, say V, then, applying Axiom A.3 (relative comprehension), within V, for the formula $A \in A$, we would obtain that S would exist as a set, a contradiction, as above.

FIGURE 1.7 (a) Ernst Zermelo, 1871–1953. (b) Abraham Halevi (Adolf) Fraenkel, 1891–1965. (c) Bertrand Russell, 1872–1970.

Note 1.15

An address of Russell's paradox was given by Russell, Zermelo, and others in 1908. Zermelo's axiomatic set theory (an **axiom** or **postulate** is a statement that is assumed to be true) tried to avoid the paradox, and Fraenkel modified it later in 1920.

Various forms of **axioms** have been developed for set theory. However, according to a previous "Note", the general axiom of comprehension does not specify which mathematical processes are free of set theoretic singularities.

These days, the set of axioms used are of Zermelo and Fraenkel combination referred to as **ZF axioms**, and if the **axiom of choice** (AC) is included, they are referred to as **ZFC**. Indeed, the most widely accepted form and strength of **axiomatic set theory** is ZFC (*Zermelo–Fraenkel choice*). In addition to many text references in English and other languages, Montazer-Haghighi (1973, in Farsi Language) has a complete discussion of the ZFC. We will state the axioms of sets, including the AC, later.

The following is a list of some notations from symbolic logic and set theory that will be used later.

We note that in listing ZFC axioms below, we are using sets whose elements are also sets, often referred to as the **pure** or **hereditary sets**.

Logic Symbol	Meaning	Description/Example
$\vee$	"or"	$A \vee B, A$ or B
$\wedge$	"and"	$A \wedge B, A$ and B
$\Rightarrow$ or $\models$	"implies"	$x^3 = 27 \Rightarrow x = 3$ $\alpha \models \beta$ means that β follows from α
$\rightarrow$	"approaches", "tends to"	$t \rightarrow +\infty$
$\leftarrow$	"gets"	$R \leftarrow S + T$
$\Leftrightarrow$	"if and only if", "iff"	$(P \Rightarrow Q$ and $Q \Rightarrow P) \Rightarrow (P \Leftrightarrow Q)$
$\forall$	"for all"	$\forall n, n \in \mathbb{N}$, for all natural numbers n
$\exists$	"there exists"	$\exists\, a$ Set with cardinality infinity
$\exists!$	"there exists exactly one"	$\exists! x \ni f(x)$ means that there is exactly one x such that $f(x)$ is true.
$\nexists$	"there is no"	There is no element in the empty set.
$\mid$ or $\ni$ or :	"such that"	$A = \left\{ x \in \mathbb{R} \middle| x \geq 1 \right\}, A = \{ x \in \mathbb{R} \ni x \geq 1 \}$ A is a set of real numbers greater than or equal to 1.
$\equiv$	"equivalent"	$A \equiv B$, sets A and B are the same sets
$\cup$	"union"	A set whose elements are in either of the two sets
$\cap$	"intersection"	A set containing elements that are in both sets
$\{\ \}$ or $\varnothing$	"the empty set"	A set with no element
$\subset$	"proper subset"	A part of a set
$\subseteq$	"subset"	A part of or the whole set
$\supset$	"supset"	"contained", "B is contained in A", $\{1,2,3\} \supset \{1,2\}$
$\langle$	"left angle bracket"	
$\rangle$	"right angle bracket"	
$\neg$	"not"	
$\bullet$	"bullet"	
$\propto$	"proportional to"	The rate of growth of the population, r, is proportional to the size of the population, N. $dN/dt \propto N$

A.1 AXIOM OF EXTENSIONALITY

If sets A and B have the same elements, then A and B are equal or identical; that is, $A = B$, and vice versa. Symbolically,

$$\forall A, B (A = B) \Leftrightarrow (\forall c, c \in A \Leftrightarrow c \in B). \tag{1.26}$$

The axiom of extensionality speaks about the type of things sets are. The elements of each set distinguish one set from another if they are to be different; otherwise, they are the same. So, when there is an element in one that is not in the other, the two sets are not equal.

A.2 AXIOM OF PAIRING (UNORDERED PAIR)

For sets A and B, there is a set $\{A, B\}$ that contains exactly A and B. In other words, for any sets A and B, there exists a set C such that an element D is in C if and only if D is either equal to A or equal to B. Symbolically,

$$(\forall A, B) \exists C \ni (\forall D [D \in C \Leftrightarrow (D = A \vee D = B)]. \tag{1.27}$$

We leave it as an exercise to prove that C is unique. Thus, (1.27) can be rewritten as:

$$\{A, B\} = \text{the unique } C \ni (\forall D [D \in C \Leftrightarrow (D = A \vee D = B)]. \tag{1.28}$$

We also leave it as an exercise to show that from (1.2.1) and (1.2.3), we have:

$$\{A, A\} = \{A\}. \tag{1.29}$$

A.3 AXIOM OF (SCHEMA OF) SEPARATION (OR RELATIVE COMPREHENSION, AUSSONDERUNG OR SPECIFICATION OR SUBSETS)

This axiom essentially states that a subcollection of a set is a set. Here is its symbolic representation: Let ϕ be a property with parameter p. Then, for a set A and the parameter p, there is a set B, $B = \{C \in A \ni \phi(C, p)\}$, that contains all elements in A that have the property ϕ.

$$\forall A \forall p \exists B \ni \forall C \{C \in B \equiv [C \in A \wedge \phi(C, p)]\}. \tag{1.30}$$

Note 1.16

It is the relative comprehension axiom that addresses "set-theoretic singularity", as mentioned earlier, and thus solves the paradox Russell raised.

A.4 AXIOM OF UNION (OR SUM SET)

This axiom essentially states that for any set A, there is a set $B = \cup A$, the **union** of A, such that B is the set of all elements of A, or equivalently, B is the **union** of all sets in A. Symbolically,

$$\forall A \exists B \ni \forall C[C \in B \Leftrightarrow \exists D(C \in D \wedge D \in A)]. \tag{1.31}$$

A.5 AXIOM OF POWER SET

This axiom states that for any set A, there is a set $B = P(A)$, the **power set** of A, such that B is a set that contains exactly all subsets of A. Symbolically,

$$\forall A \exists B \ni \forall C(C \in B \Leftrightarrow C \subseteq A). \tag{1.32}$$

A.6 AXIOM OF INFINITY (OR CLOSURE)

$$\textit{There exists an infinite set} \tag{1.33}$$

We leave it as an exercise to use the **axiom of infinity** and **axiom of separation** to prove the existence of the empty set. Thus, we really do not need the **axiom of empty set**. In other words, there exists a set that has no element. Symbolically,

$$\exists A \ni \forall B \ B \notin A, A \text{ is denoted by} \{ \ \} \text{ or } \varnothing. \tag{1.34}$$

We note that some of the axioms may vacuously be acceptable without the assumption of the existence of the empty set; however, some cannot be acceptable.

We also leave it as an exercise to prove that the empty set is unique.

A.7 AXIOM OF REPLACEMENT

This axiom states that if ϕ represents a function, then the image of a set in ϕ is a set. In other words, for any set A, there is a set B such that $\phi(A) = \{\phi(B) \ni B \in A\}$. Here is its symbolic representation:

$$\forall A, \forall B, \forall C[\phi(A, B, p) \wedge \phi(A, C, p) \Rightarrow B = C]$$

$$\Rightarrow \forall D \exists E \ni \forall B[B \in E \equiv (\exists A \in D)\phi(A, B, p)] \tag{1.35}$$

or if

$$\Rightarrow \forall A, \text{ there exists at most one } B \ni [\phi(A, B)],$$

then

$$\forall D \exists E \ni \forall C[C \in E \Leftrightarrow [(\exists G \in D)\phi(G, C)]]. \tag{1.36}$$

A.8 AXIOM OF FOUNDATION (OR REGULARITY)

The **axiom of foundation** states that every nonempty set belongs to the cumulative hierarchy. In other words, it turns out that this is equivalent to the statement: For every nonempty set, there is an $\in$-minimal member. Symbolically,

$$\forall A[A \neq 0 \Rightarrow [\exists B \in A \ni \cap B = \varnothing] \tag{1.37}$$

or

$$\forall A\{\exists B \ni B \in A \Rightarrow \exists B[B \in A \wedge \forall C(C \in B \Rightarrow C \notin A)]\}. \tag{1.38}$$

A.9 ZERMELO'S AXIOM OF CHOICE

The **axiom of choice**, formulated by Ernst Zermelo in 1904, abbreviated as AC, is sometimes known as the **choice principle of set theory**. It is, perhaps, the most discussed axiom of mathematics, second only to Euclid's **axiom of parallels**, which was introduced more than two thousand years ago, by Fraenkel, Bar-Hillel, and Levy (1973). On the other hand, AC perhaps is the most problematic axiom of set theory. Quite a lot have been written about AC as it is so controversial, a subject in pure mathematics.

To state the AC, we choose the relatively simple version used by Russell who referred to it as the **multiplicative axiom**. The **AC** states that, given any collection of mutually disjoint nonempty sets, it is possible to assemble a new set, referred to as a **transversal** or **choice set**. This set is to contain exactly one element from each member of the given collection. Symbolically, let C be a nonempty set. Then, there is a function, say f, defined as

$$f : C \Rightarrow \bigcup C \ni (\forall A \in C)[f(A) \in A]. \tag{1.39}$$

Such an f is then called a **choice function** for C, and its range is the corresponding **choice set, defined as above**. Traditionally, there are three major schools of thought regarding the use of AC:

1. Accept it as an axiom and use it without question.
2. Accept it as an axiom, but use it only when you have no other way in proving your theorem and, perhaps, also making a note of its usage.
3. AC is unacceptable.

Note 1.17

In recent times, there has been an increasing confidence in accepting AC as a valid principle, in general, within the ordinary practice of mathematics.

Example 1.18

As an example for AC, we choose the following from Bell, John L. (2008, revised 2015).

Let C be the collection of nonempty subsets of the set $\{0,1\}$. That is, C is:

$$C = \{\{0,1\},\{0\},\{1\}\}.$$

Hence, there are two distinct choice functions for C, say f_1 and f_2, defined as:

$$f_1(\{0\}) = 0,\, f_1(\{1\}) = 1,\, f_1(\{0,1\}) = 0,$$

$$f_2(\{0\}) = 0,\, f_2(\{1\}) = 1,\, f_2(\{0,1\}) = 1.$$

Example 1.19

Let C be the collection of all intervals $S_a = [a, a+1)$, partitioning the set of real numbers $\mathbb{R}$, where a is an integer (an element of $\mathbb{Z}$). Then, a choice function f on C can be defined by $f(S_a) = a$.

1.4   BASIC CONCEPTS OF MEASURE THEORY

Later in Chapter 2, dealing with "probability", we will need the idea of "**measure**", like measuring the size of an interval on the real axis. Conceptually, the **length** or a **measure**, of an interval, denoted by μ, is a function that assigns a nonnegative number to the interval.

Henri-Léon Lebesgue, a French Mathematician, rigorously considered the idea which we will discuss briefly below (Figure 1.8).

There are three types of intervals on the real axis:

1. **Open**, that is, an interval not including its boundaries;

FIGURE 1.8   Henri-Léon Lebesgue, 1875–1941.

2. **Half-open (left-open, right-open)**, that is, an interval that includes only one side;
3. **Closed (or bounded)**, that is, an interval that includes both boundaries.

Note 1.18

In real analysis (a part of pure mathematics), "∞" (read "infinity") is referred to as **a symbol unequal to any real number**.

Note 1.19

In addition to the three types of intervals mentioned above, there is another interval denoted by $[0, \infty]$, which consists of all nonnegative real numbers and ∞.

Note 1.20

For our purposes in this book, we use the following properties as conventional properties. Hence, we state them without proof.

$$c \cdot \infty = \infty \cdot c = 0, c = 0, \tag{1.40}$$

$$c \cdot \infty = \infty \cdot c = \infty, 0 < c \leq \infty, \tag{1.41}$$

and

$$c + \infty = \infty + c = \infty, 0 \leq c \leq \infty. \tag{1.42}$$

Example 1.20

The interval $(-1,5)$ is an open bounded interval, $[-1,5)$ is a half-open (right-open) bounded interval, $(-1,5]$ is a half-open (left-open) bounded interval, and $[-1,5]$ is a closed bounded interval. However, $(-\infty,4)$, $[4,\infty)$, and $(-\infty,+\infty)$ are the examples of unbounded (or non-bounded) intervals.

Definition 1.9

The **size of an interval** is its **length**. The length of a bounded interval of any kind with endpoints i and j is defined as the absolute value of the difference between the endpoints, that is, $|i - j|$. In case $i \leq j$, the absolute value may be dropped and it would simply be $j - i$. For an unbounded interval, the length is defined as ∞.

The idea of the length may be generalized to measure the size of a set, for instance, measuring the size of all irrational numbers within the interval $\left[1, \sqrt{5}\right]$.

Many, including Lebesgue, have made this generalization that is referred to as the **measure**. Essentially, "measure" is a generalization of the notions of length, area, and volume. The **measure theory** in mathematics studies the measures. Some particular measures are of special importance, such as Lebesgue measure, Jordan measure, Borel measure, Haar measure, complex measure, and probability measure. At this point, our interest is the Lebesgue measure. To define it, we will discuss some necessary ideas.

Definition 1.10

Measure of a set (or an interval), E, say μ, is a nonnegative real number such that $\mu(E) \in \mathbb{R}$. In other words, the measure of a set (or an interval) is a function that assigns to the set (or an interval) a nonnegative real number.

Note 1.21

It is provable that every countable set is of **measure 0**. The proof is left as an exercise.

Note 1.22

The measure of a set may be generalized to cover a collection of subsets of an interval $[a,b]$, denoted by $\mathcal{F}$, that is, $\mu : \mathcal{F} \to \mathbb{R}$. Thus, μ is a set function because its domain, $\mathcal{F}$, consists of subsets of $[a,b]$. If we impose conditions mentioned for a σ-algebra on μ, that is, if

  i. $\mathcal{F}$ is not empty; of course, this will be a fact when we include the empty set $\varnothing$ and the whole set $[a,b]$;
 ii. $\mathcal{F}$ is closed under complementation, that is,

$$A \in \mathcal{F} \Leftrightarrow A^C \in \mathcal{F};$$

iii. $\mathcal{F}$ is closed under countable unions, that is,

$$\left\{ A_i, i = 1, 2, \ldots, \infty \right\} \in \mathcal{F} \to \bigcup_{i=1}^{\infty} A_i \in \mathcal{F},$$

then μ becomes a σ-algebra.

Definition 1.11

Let the length of the interval $[a,b]$ be denoted by $l[a,b]$. The set function μ, $\mu : \mathcal{F} \to [0, \infty]$ (i.e., the range of μ is the set of nonnegative real numbers), is a **measure** if and only if it has the following properties:

i. The **extended length property**

$$\mu([a,b]) = l[a,b] = b - a; \tag{1.43}$$

ii. The **monotonicity property**

$$A \subseteq B \to \mu(A) \leq \mu(B); \tag{1.44}$$

iii. The **countable additivity property**

If $A_1, A_2, \ldots$ is a disjoint sequence of sets of $\mathcal{F}$, then the set function of union of A's is the sum of the set functions. In symbols:

$$\{A_i, i = 1, 2, \ldots, \infty\} \in \mathcal{F}, A_i \cap A_j = \varnothing, i \neq j, \to \mu\left(\bigcup_{i=1}^{\infty} A_i\right) = \sum_{i=1}^{\infty} \mu(A_i); \tag{1.45}$$

iv. The **nonnegativity property**

$$\mu(A) \geq 0, \forall A \in \mathcal{F}; \tag{1.46}$$

v.

$$\mu(A) \in [0, \infty], \text{ for any set } A \in \mathcal{F}; \tag{1.47}$$

vi.

$$\mu(\theta) = 0; \tag{1.48}$$

In this case, $\mu(A)$ is the **measure** of A, or A is **μ-measurable**.

Note 1.23

Using AC, it can be proved that if $\mathcal{F}$ is the set of all subsets of $[a,b]$, then there is no measure for $\mathcal{F}$. Hence, for each use, one needs to construct μ and $\mathcal{F}$.

Note 1.24

Based on the definition stated above, a measure ' is finite or infinite, that is, $\mu(A) < \infty$ or $\mu(A) = \infty$, respectively. $\mu(A) = 1$ is the case where it is called the **probability measure**. We will discuss other properties of probability measures in Chapter 2.

Definition 1.12

Consider a set A with finite measure. By a **partition** of A, it is meant a finite sequence of disjoint measurable sets $A_1, A_2, \ldots, A_n$ such that

$$A = \bigcup_{i=1}^{n} A_i. \tag{1.49}$$

Example 1.21

Consider the set of intervals

$$\left\{\left[1,\frac{6}{5}\right),\left[\frac{6}{5},\frac{7}{5}\right),\left[\frac{7}{5},\frac{8}{5}\right),\left[\frac{8}{5},\frac{9}{5}\right),\left[\frac{9}{5},2\right]\right\}.$$

Clearly, the sequence of the five intervals in the set are disjoint and their union is the interval [1,2]. Thus, the sequence of subintervals

$$\left[1,\frac{6}{5}\right),\left[\frac{6}{5},\frac{7}{5}\right),\left[\frac{7}{5},\frac{8}{5}\right),\left[\frac{8}{5},\frac{9}{5}\right),\left[\frac{9}{5},2\right]$$

with endpoints

$$\left\{1,\frac{6}{5},\frac{7}{5},\frac{8}{5},\frac{9}{5},2\right\}$$

is a partition of the interval $[1,2]$.

As a generalization of length to open set, let us consider an interval $[a,b]$. Every open set, say O, in $[a,b]$ is the disjoint union of intervals, that is, $O = \cup_i (b_i - a_i)$.

Definition 1.13

Let $A \subseteq [a,b]$ for some finite $a \neq b \in \mathbb{R}$. The **outer measure** of A, denoted by $\mu^*(A)$, is defined as

$$\mu^*(A) = \inf\left\{\mu^*(O) : A \subset O\right\}. \tag{1.50}$$

Definition 1.14

The **Lebesgue measure** of a set A of real numbers, denoted by $\mu(A)$, if it is defined, where μ is a set function such that the properties (i), (ii), (iii), and (vii) "translation invariant" hold, is defined as follows.

vii. The **translation invariant property**

$$\text{for } x_0 \in \mathbb{R},\, A + x_0 = \left\{a + x_0, a \in A\right\} \rightarrow \mu(A + x_0) = \mu(A), \tag{1.51}$$

holds.

Definition 1.15

The **Lebesgue outer measure** of a set A of real numbers, denoted by $\mu^*(A)$, is defined as follows:

$$\mu^*(A) = \inf\left\{\sum_{i=1}^{\infty} l(O_i) : A \subset \bigcup_{i=1}^{\infty} O_i\right\}, \tag{1.52}$$

where $\{O_i, i = 1, 2, \ldots, \infty\}$ is a sequence of open intervals.

Note 1.25

It is provable that if A is countable, then $\mu^*(A) = 0$. Also, $\mu^*(A)$ satisfies the properties (140), (141), (142), and (148). It further satisfies the following property:

viii. The **countably sub-additive property**

$$\mu^*\left(\bigcup_{i=1}^{\infty} A_i\right) \le \sum_{i=1}^{\infty} \mu^*(A_i), \quad \forall\{A_i\} \subset \mathbb{R}. \tag{1.53}$$

In other words, every subset of the set of real numbers has Lebesgue outer measure satisfying properties (i)–(iii), but does not fully satisfy property (iv).

Definition 1.16

A subset of the real numbers, say B, is referred to as **Lebesgue measurable** or simply **measurable**, if for every subset A of the set of real numbers we have the following:

$$\mu^*(A) = \mu^*(A \cup B) + \mu^*(A \cap B). \tag{1.54}$$

Note 1.26

If A is a Lebesgue measurable set, then its **Lebesgue measure** is defined as the outer measure of A, $\mu^*(A)$, and therefore written as $\mu(A)$.

Note 1.27

Let A and B be two measurable sets. Then, the set $A - B$ consisting of all elements of A not in B is measurable. The proof is left as an exercise.

Note 1.28

The set of all measurable sets of the n-dimensional real number set $\mathbb{R}^n$ is denoted by $\mathcal{M}\mathbb{R}^n$.

Note 1.29

Since every open set is Lebesgue measurable, the Borel sets are measurable.

Definition 1.17

For a set A with a σ-algebra on A, the pair (X, A) is referred to as the **measurable space**.

Example 1.22

Let $A = \{1, 3, 5\}$ and $S = \{A, \emptyset\}$ representing a σ-algebra on A. Then, the pair (A, S) is a measurable space.

Example 1.23

Let A be a singleton. Then, A is the Lebesgue measurable with measure 0.

Example 1.24

The set $A = \left(\dfrac{1}{2}, 1\right)$ is the Lebesgue measurable with measure $\dfrac{1}{2}$.

Example 1.25

The set

$$A = \left(\frac{2}{7}, \frac{3}{7}\right) \cup \left(\frac{5}{7}, \frac{6}{7}\right)$$

is the Lebesgue measurable with measure $\dfrac{2}{7}$.

Example 1.26

The set

$$A = \left(\frac{1}{15}, \frac{4}{15}\right) \cup \left(\frac{6}{15}; \frac{9}{15}\right) \cup \left(\frac{11}{15}; \frac{14}{15}\right)$$

is the Lebesgue measurable with measure $\dfrac{3}{5}$.

Example 1.27

A countable set like the set of rational numbers is measurable with measure 0.

Example 1.28

The n-dimensional Euclidean space is measurable with infinite measure.

The Lebesgue measure is often used in probability theory, physics, and other disciplines. Here, in this chapter, we confine it to probability. As we discussed earlier, the Lebesgue measure, say μ of the set of real numbers $\mathbb{R}$ (an unbounded open set), is infinity. However, **in probability theory, where the Lebesgue measure is a probability measure, it is required that** $\mu(\mathbb{R}) = 1$. The **Lebesgue measure** of most interest is that of assigning a measure to subsets of the n-dimensional Euclidean space. This general statement includes the cases of $n = 1$, 2, and 3, that is, length, area, and volume, respectively. In a one-dimensional case, the Lebesgue measure can be thought of as the usual concept of length, when we are dealing only with intervals. We will have a rigorous definition below.

Remark 1.2

The Lebesgue measure of a set A satisfies the properties (i), (ii), (iii), and (vii) on a collection $\mathcal{C}$ of measurable subsets of the set of real numbers. But not all subsets of the real number system are measurable.

Remark 1.3

Since all open and closed sets are measurable, and the collection $\mathcal{C}$ of measurable sets is closed under countable unions and intersections, properties (i), (ii), (iii), and (vii) imply that measurable sets do exist. The proof is left as an exercise.

Definition 1.18

Let X be a nonempty set, $\mathcal{F}$ a σ-algebra, and μ a measure on $(X, \mathcal{F})$. Then, the triple $(X, \mathcal{F}, \mu)$ is called a **measure space**.

Note 1.30

The **probability space** that we will discuss in Chapter 2 is an important example of a measure space.

Definition 1.19

Let $(X, \mathcal{T})$ be a topological space. Then, a **Borel measure** is a measure that is defined on all open sets. It is denoted by $\mathcal{B}(X)$. Equivalently, a **Borel measure** is a measure defined on the Borel set $\mathscr{B}$.

1.5 LEBESGUE INTEGRAL

In this section, we briefly discuss the Lebesgue integral for the real line, as we will need it later in other chapters. The area under nonnegative function with a continuously defined on a closed interval or a "smooth enough" graph above the x-axis is known as the **integral of the function**. Such an area can be approximated using various approximation techniques. The Riemann integral is one of the most popular methods that was proposed by German mathematician Bernhard Riemann (1826–1866) at the University of Göttingen in 1854 and published later in 1868.

The method includes partitioning the x-axis and finding the limit of sum of vertical rectangles rising above the x-axis. The integral of a given function $f(x)$ is then defined in terms of the limit of the Riemann sum, which is defined as follows.

Definition 1.20

Consider a given function $f(x)$ on an interval $[a,b]$. Partition the closed interval $[a,b]$ into n subintervals with boundary points $\{a, x_1, x_2, \ldots, x_{n-1}, b\}$ such that $a < x_1 < x_2 < \cdots x_{n-1} < b$. Let the length of these subintervals be denoted by $\Delta x_1, \Delta x_2, \ldots \Delta x_n$. That is, $\Delta_j = x_{j+1} - x_j$, $j = 1, 2, \ldots, n-1$. Let x_j^* also be an arbitrary point in the jth subinterval. Then,

$$\sum_{j=1}^{n} f\left(x_j^*\right) \Delta x_j \tag{1.55}$$

is referred to as a **Riemann sum** for $f(x)$ with the given partition. The quantity $\max \Delta x_j$ is called the **mesh size** of the partition. If the limit of the sum in (1.55) exists as $\max \Delta x_j \rightarrow 0$, then the limit is referred to as the **Riemann integral** of $f(x)$ over the interval $[a,b]$. In other words, the Riemann integral is defined as a set of bounded functions, say $g : [a,b] \rightarrow \mathbb{C}$, called the **Riemann integral functions**, that include all continuous functions. It may be extended to improper functions, as well.

Thus, we may define the Riemann integral as follows.

Definition 1.21

Let

$$\Omega = \left\{ (x,y) : \quad a \leq x \leq b, \quad 0 < y < f(x) \right\},$$

where $f(x)$ is a nonnegative real-valued function on a closed interval $[a,b]$. Then, the **Riemann integral** is defined as:

$$\int_a^b f(x)dx = \lim_{\max \Delta_j x \rightarrow 0} \sum_{j=1}^{n} f\left(x_j^*\right) \Delta x_j. \tag{1.56}$$

To overcome some shortcomings of the Riemann integral, Fourier series and Fourier transform were developed. However, integration of irregular functions, such as those particularly arose in probability theory, and of those on spaces more general than the real axis became needed. That is, how the Lebesgue integral was developed and presented in Lebesgue (1904) and was vastly used (in addition to the areas of real analysis and mathematical sciences) in probability theory. Essentially, in the Lebesgue integration, sums are over horizontal rectangles rather than vertical as in the Riemann case. In other words, in the Lebesgue case, the y-axis is partitioned versus the Riemann case where the x-axis is partitioned. Hence, there might be more than one interval on the x-axis for a y-value to be considered in this case. Then, a question may be asked: Would the value of the integral be the same in both cases? The answer is yes, most of the times. See the example below. However, Lebesgue will become useful when the Riemann does not exist.

Generally speaking, our purpose of discussion is to integrate unbounded functions and can take limits under the integral sign of an almost general nature. The Lebesgue integral is a Riemann integral of a function that exists over a set.

We formally define the Lebesgue integral, as follows.

Definition 1.22

Let $\mathcal{M}$ be a set of measurable sets on $[0,1]$. The function $f:[0,1]\to\mathbb{R}$ is called a **simple function** if and only if there are real numbers $a_1,a_2,\ldots,a_j$, not necessarily distinct, and measurable sets $A_1,A_2,\ldots,A_j\in\mathcal{M}$ such that $\mathcal{A}=\{A_1,A_2,\ldots,A_j\}$ is a partition of $[0,1]$ and

$$f(x)=\sum_j a_j I_{A_j}(x),\tag{1.57}$$

where I_A is an indicator function defined earlier by (1.11).

Example 1.29

Consider a function whose value is 1 at rational numbers and 0 elsewhere. Such a function is called the **Dirichlet function**, and it is a simple function. In other words, the function $g(x)$ defined on the interval $[0,1]$ by

$$g(x)=\begin{cases} 0 & x \text{ is rational} \\ 1 & x \text{ is irrational} \end{cases}\tag{1.58}$$

is called the **Dirichlet function**. Thus, the area should be 1. However, it can be shown (left as an exercise) that since the upper and lower Riemann integrals are not the same, the Riemann integral for $g(x)$ does not exist.

Note 1.31

Lebesgue showed that every Riemann integral is, indeed, a Lebesgue integral. But a Lebesgue integral is more general.

Definition 1.23

A simple function is called a **step function** if sets A_j are the intervals.

Example 1.30

Let $A_1 = \left[\frac{1}{5}, \frac{2}{5}\right]$, $A_2 = \left[\frac{1}{4}, 1\right]$, $a_1 = \frac{4}{5}$, and $a_2 = \frac{28}{25}$. Then, $\mu(A_1) = \frac{1}{5}$, $\mu(A_2) = \frac{3}{4}$, and the associated f is a step function.

Definition 1.24

Let $f:[0,1] \to \mathbb{R}$ be a simple function, as defined above, that is, $f(x) = \sum_j a_j I_{A_j}(x)$.

Then, the Lebesgue integral of f, denoted by $\int f$, is given by

$$\int_A f = \sum_j a_j \mu(A_j), \tag{1.59}$$

where $\mu(A)$ is the measure (size, a nonnegative number) of the set A.

Definition 1.25

Let $f(x)$ be a nonnegative measurable function. Then, $f(x)$ is called the **Lebesgue integrable** if and only if there is a sequence of nonnegative simple functions $\{f_j\}$ such that:

$$\sum_{j=1}^{\infty} \int f_j < \infty \tag{1.60}$$

and

$$f(x) = \sum_{j=1}^{\infty} \int f_j(x), \text{ almost everywhere.} \tag{1.61}$$

Example 1.31

Consider Example 1.26. Then,

$$\int f = \left(\frac{4}{5}\right)\left(\frac{1}{5}\right) + \left(\frac{28}{25}\right)\left(\frac{3}{4}\right) = \left(\frac{4}{25}\right) + \left(\frac{21}{25}\right) = 1.$$

Note 1.32

Sets of measure 0 really won't contribute to an integral. Thus, when we integrate two functions that are different only on a set of measure 0, we will find that their integrals will be the same.

As a generalization of the Lebesgue and Riemann integrals, we consider the **Lebesgue–Stieltjes measure**, which is a regular Borel measure. The **Lebesgue–Stieltjes integral** is the ordinary Lebesgue integral with respect to the Lebesgue–Stieltjes measure. To define the Lebesgue–Stieltjes integral, let $\mu : [a,b] \to \mathbb{R}$ be a Borel measure and bounded. Also, let $f : [a,b] \to \mathbb{R}$ be a bounded variation in [a, b]. By a function of **bounded variation**, it is meant a real-valued function whose total **variation** is **bounded (finite)**, for instance, functions of bounded variation of a single variable differentiable at almost every point of their domain of definition. Then, the **Lebesgue–Stieltjes integral** is defined as

$$\int_a^b \mu(x)\,df(x). \tag{1.62}$$

We now define the Lebesgue–Stieltjes integral in terms of step function that appears often in probability and stochastic processes.

Definition 1.26

Let $g(\cdot)$ be a right-continuous and nondecreasing step function with jumps at $x_1, x_2, \dots$ on the real line. Suppose $f(\cdot)$ is any function. Then, the **Lebesgue–Stieltjes integral**, denoted by

$$\int_a^b f(x)\,dg(x), \tag{1.63}$$

is defined as

$$\int_a^b f(x)\,dg(x) = \sum_{\substack{j \\ a \le x_j \le b}} f(x_j) \cdot \left[g(x_j) - g(x_j - 1)\right] = \sum_{\substack{j \\ a \le x_j \le b}} f(x_j) \cdot \Delta g(x_j), \tag{1.64}$$

where $\Delta g(x_j) = g(x_j) - g(x_j - 1)$.

Note 1.33

In case $g(\cdot)$ is a continuous differentiable function with its derivative denoted as $g'(\cdot)$, then the Lebesgue–Stieltjes integral is denoted by

$$\int_a^b f(x)\,dg(x) = \int_a^b f(x)g'(x)\,dx. \tag{1.65}$$

Hence, as it can be seen, the Lebesgue–Stieltjes integral has been defined for both step function and an absolutely continuous function.

Note 1.34

The existence of the Lebesgue–Stieltjes integral in (1.63) is guaranteed by the assumption that μ_A is Borel measurable.

1.6 COUNTING

Throughout probability and statistics, we need to know how to count. For example, in assigning probabilities to outcomes, we sometime need to count the number of outcomes of a chance experiment and/or the number of ways an outcome may occur. The main computational tool for the equiprobable measure is **counting**.

There are two basic rules and one principle for counting of tasks that we are to list here, namely, the sum rule, the product rule (principle of multiplication), and the inclusion–exclusion principle. These concepts are very closely related.

The Sum Rule

Suppose we have two tasks to conduct. The first task can be done in n_1 ways and the second in n_2 ways. The two tasks cannot be conducted at the same time. Then, either task can be conducted in $n_1 + n_2$ ways. This rule is referred to as the **sum rule**.

The sum rule can be extended as follows: Suppose there are n tasks $A_1, A_2, \ldots, A_n$, each can be conducted in $m_1, m_2, \ldots, m_n$, ways, respectively. It is assumed that no two tasks can be done at the same time. Then, the number of ways any one of the tasks can be done is $m_1 + m_2 + \cdots + m_n$.

Example 1.32

Let us assume that a department of mathematics currently has 25 faculty members and 50 majors. One of the members of the Mathematics Grade Challenge Committee has resigned and needs to be replaced by either a faculty or a major. Thus, there are $25 + 50 = 75$ choices for the mathematics department chairman to choose replacing the vacant position.

Example 1.33

In a statistics course, students are to conduct a project and submit it at end of semester that they are enrolled in. The professor of the class has a list of 5 sets of possible projects, each containing 5, 10, 15, 20, and 15 topics different from each other, respectively. How many choices are there for each student to conduct a project?

The answer is $5 + 10 + 15 + 20 + 25 = 75$.

The Product Rule or Multiplication Principle or The Fundamental Principal of Counting:

Suppose a task contains two subtasks. Processing of the task requires conducting the first subtask in n_1 ways and then the second one in n_2 ways. Then, completing processing of the task will have $n_1 \cdot n_2$ choices, independent of each other. This principle can be extended for a multiple tasks, that is, $n_1 \cdot n_2 \ldots$, as long as each task is performed independent of all preceding tasks. We can show this principle graphically as follows.

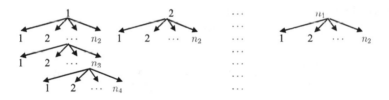

Example 1.34

Consider choosing a license plate in Texas, where it requires a sequence of three letters followed by three digits between 1 and 9. Let us ignore all possible restrictions of the state for plate number choice. Then, there are $26 \cdot 26 \cdot 26 \cdot 9 \cdot 9 \cdot 9 = 12,812,904$ ways to choose a license plate in Texas, under the assumption made.

Example 1.35

Let us now consider a finite set with n arbitrary elements. We can prove that this set has 2^n subsets.

Here is the proof by mathematical induction. Let the set be a singleton, say $S = \{a\}$. Then, the subsets of S are $\{a\}$ and $\varnothing$. That is, there are 2^1 subsets for a singleton. The set of all subsets of a set, including the empty set, is referred to as the **power set**. Thus, the power set of the singleton $\{a\}$ is $\{\{a\}, \varnothing\}$. Also, if $S = \{a, b\}$, then its power set is the set $\{\{a\}, \{b\}, \{a, b\}, \varnothing\}$, which has $4 = 2^2$ elements. We can easily prove, by mathematical induction, that for a finite set S of n elements, the power set of S has 2^n elements.

To show this fact, we have already seen that for a set with one element, the number of subsets was 2. Now, let us assume that the set S with $n - 1$ elements has 2^{n-1} subsets. Then, if S has n elements, the number of subsets will be 2^{n-1} for the first $n - 1$ elements and the same number in conjunction with the last one, making the total number double, that is, $2 \cdot 2^{n-1} = 2^n$.

The Inclusion–Exclusion Principle

In counting, there is a principle that is extremely important, called the **inclusion–exclusion principle**. This principle indicates the cardinality of a set. To explain this principle, remember that the sum rule is for doing one task or the other, not just for only one task and not either for both tasks at the same time. However, if both tasks were allowed to be done at the same time, then due to the double dipping of conduct, subtraction of the number of ways both tasks can be done at the same time is necessary.

With this new condition, we have the inclusion–exclusion principle. In set notation, the method for two finite sets A and B states the following:

$$|A \cup B| = |A| + |B| - |A \cap B|. \tag{1.66}$$

In case of three sets A, B, and C, we will have the following:

$$|A \cup B \cup C| = |A| + |B| + |C| - |A \cap B| - |A \cap C| - |B \cap C| + |A \cap B \cap C|. \tag{1.67}$$

This principle is generalized as follows:

Let $A_1, A_2, \ldots, A_n$ be n finite sets. Then,

$$\left| \bigcup_{i=1}^{n} A_i \right| = \sum_{i=1}^{n} |A_i| - \sum_{1 < i < j < n} |A_i - A_j| + \ldots + \sum_{1 \le i \le j \le k \le n} |A_i \cap A_j \cap A_k| - \cdots$$

$$+ (-1)^{n-1} |A_1 \cap A_2 \cap \cdots A_n|. \tag{1.68}$$

Example 1.36

Suppose we are to find the cardinality of sum of three subsets of the set $S = \{1,2,\ldots,10\}$. Let the three subsets be $A = \{1,3,5,8\}$, $B = \{3,5,7,8,10\}$, and $C = \{4,5,8,10\}$. Then, $A \cap B = \{3,5,8\}$, $A \cap C = \{5,8\}$, $B \cap C = \{5,8,10\}$ and $A \cap B \cap C = \{5,8\}$. Thus,

$$|A \cup B \cup C| = 4 + 5 + 4 - 3 - 2 - 3 + 2 = 7.$$

That is,

$$A \cup B \cup C = \{1,3,4,5,7,8,10\}.$$

Example 1.37

Suppose there are 20 students in a probability and statistics class. Of them, 10 students study mathematics, 12 students study engineering, and 4 students study both. How many of the class study neither mathematics nor engineering?

For this problem, let us denote C as the set of all students in the class. There are three subsets of C, say M, E, and N, where they stand for the sets of students majoring mathematics, engineering, and neither, respectively. The cardinality of C, M, E, and $M \cap E$ is 20, 10, 12, and 4, respectively. Therefore, according to

(1.68), we have $20 = 10 + 12 + |N| - 4$ from which the number of students who are neither majoring mathematics nor engineering is 2. We leave it as an exercise to show this result via a Venn diagram.

In counting, there are yet two other topics that we will be using later quite often in all three main subjects of this book, namely, probability, statistics, and stochastic processes. These are permutation and combination that are closely related. These two topics are normally studied in the field of **combinatorics** and in **sorting algorithm** area of computer science.

First, we define the "n factorial", denoted by $n!$, where n is a nonnegative integer, as

$$n! = 1 \cdot 2 \cdot 3 \cdots (n-1) \cdot n. \tag{1.69}$$

That is,

$$n! = n(n-1)! \tag{1.70}$$

Relation (1.70) may be expanded term by term rather than all at once, as in (1.69).

Note 1.35

By convention, it is agreed that $0! = 1! = 1$.

It should be noted that for large n, $n!$ can be approximated by **Stirling's formula** or **Stirling's approximation** as

$$n! \approx \sqrt{2\pi} n^{n+\frac{1}{2}} e^{-n}. \tag{1.71}$$

The symbol "$\approx$" means approximately equal to. In other words,

$$\lim_{n \to \infty} \frac{\sqrt{2\pi} n^{n+\frac{1}{2}} e^{-n}}{n!} = 1. \tag{1.72}$$

Historically, Stirling's formula originally came up from correspondences between James Stirling and Abraham de Moivre in the 1720s in regard to **de Moivre's** work on approximating a binomial distribution by a normal distribution. We will discuss these concepts later in subsequent chapters.

Using a calculator or a computer program, the actual value of 10! is 3,628,800. Using **Stirling's formula**, however, it is 3,598,695.61... with the difference of around 30,104, which is less than 1% (actually, 0.83%). For 100!, using the Stirling formula, we have $100! \approx 9 \times 10^{157}$ with an error of the same as for 10!, that is, 0.083%.

It was also shown that

$$n! \approx \sqrt{2n + \frac{1}{3}\pi} n^n e^{-n} \tag{1.73}$$

is a better approximation than the standard Stirling formula. An example to illustrate this preference is given as follows: Let $n = 0$, then from the standard Stirling formula (1.72), $n! = 0$. However, assuming that $\lim_{n \to 0} n^n = 1$ from the modified version

of the Stirling formula, Equation (1.73), we will have $n! = \sqrt{\dfrac{\pi}{3}} = 1.023333$, which is very close to the value of $0! = 1$. There is a problem though with this justification as the assumption for the Stirling formula is for large n and 0 is not a large number!!
    For large n, $n!$ may also be approximated by

$$n! = \int_0^\infty x^n e^{-x}\, dx. \tag{1.74}$$

The right-hand side of (1.74) is the **gamma function**, denoted by $\Gamma(n)$, and defined as follows.

Definition 1.27

Gamma function (one-parameter), $\Gamma(\alpha)$, is defined by

$$\Gamma(\alpha) = \int_0^\infty x^{\alpha-1} e^{-x}\, dx. \tag{1.75}$$

Note 1.36

From (1.75), using a change of variable, say $x = \lambda t$, for any positive integer n and $\alpha, \lambda > 0$, we have:

$$\Gamma(\alpha) = \lambda^\alpha \int_0^\infty t^{\alpha-1} e^{-\lambda t}\, dt, \tag{1.76}$$

from which integration by parts, yields

$$\Gamma(\alpha+1) = \alpha\Gamma(\alpha),\, \alpha > 0. \tag{1.77}$$

Thus, if $\alpha = n, n \in \mathbb{N}$, then from (1.77), we will have:

$$\Gamma(n+1) = n! \tag{1.78}$$

where $n!$ was defined in (1.69). As a reminder, by convention, $0! = 1$ and $1! = 1$. However, from (1.75), we see that

$$\Gamma(1) = \int_0^\infty e^{-x}\, dx = e^{-x}\Big|_0^\infty = 1. \tag{1.79}$$

Also, from (1.78), we have $\Gamma(n) = (n-1)!$, from which using $0! = 1$, we will have

$$\Gamma(1) = 0! = 1.$$

As a property of the gamma function, we leave it as an exercise to show that:

$$\Gamma\left(\frac{1}{2}\right) = \sqrt{\pi} \tag{1.80}$$

and generally,

$$\Gamma\left(\frac{1}{2}+n\right) = \left(\begin{array}{c} n-\frac{1}{2} \\ n \end{array}\right) n!\sqrt{\pi}, \text{ and } \Gamma\left(\frac{1}{2}-n\right) = \frac{\sqrt{\pi}}{\left(\begin{array}{c} -\frac{1}{2} \\ n \end{array}\right) n!}. \tag{1.81}$$

Example 1.38

Find $\Gamma\left(\dfrac{5}{2}\right)$.

Answer

From (1.77) for $\alpha = \dfrac{3}{2}$, we have:

$$\Gamma\left(\frac{5}{2}\right) = \Gamma\left(\frac{3}{2}+1\right) = \frac{3}{2}\left[\Gamma\left(\frac{3}{2}\right)\right]$$

$$= \frac{3}{2}\left[\Gamma\left(\frac{1}{2}+1\right)\right] = \left(\frac{3}{2}\right)\left(\frac{1}{2}\right)\Gamma\left(\frac{1}{2}\right) = \left(\frac{3}{4}\right)\sqrt{\pi}.$$

Example 1.39

Find the value of the following integral:

$$\int_0^\infty x^5 e^{-4x}\, dx.$$

Answer

Using (1.76), we let $\alpha = 6$ and $\lambda = 4$. Then, using (1.78), we have:

$$\int_0^\infty x^5 e^{-4x}\, dx = \frac{1}{4^6}\Gamma(6) = \frac{5!}{4^6} = \frac{120}{1{,}024} \approx 0.1172.$$

Arrangements of outcomes of an experiment may be desired to be in a certain order. The choice of arrangements may be done **with** or **without replacement**.

That is, an object may be selected and "put back" in place, and then another selection will be made. This selection is **with replacement**. However, choosing objects may be so as each is selected it will be remain out of the place until the next object is chosen, and so on. This type of process is referred to as the **selections without replacement**.

Definition 1.28

For a set of n distinct objects, a possible **ordered arrangement** of the objects is called a **permutation** and is denoted by $n!$.

An ordered arrangement of only some of the elements, say k, of a finite set, with say n elements, is called the **partial permutation** or **k-permutation**. It is denoted by $P(n,k)$, P_k^n, $_nP_k$, $(n)_k$, etc.

Note 1.37

The k-permutation is a weaker version of permutation in the sense that it chooses only those nonrepeated ordered elements of a set and not necessarily all elements of the set.

Theorem 1.1

The k-permutation:

$$P(n,k) = \begin{cases} \underbrace{n(n-1)(n-2)\cdots(n-k+1)}_{k \text{ factors}}, & 0 \le k \le n, \\ 0, & k > n. \end{cases} \tag{1.82}$$

Proof:
The first object out of n elements of the set can be chosen in n ways since there are n objects available. Putting the first object aside, there will be $n-1$ objects left to select the second objects. Continuing this method, the kth object will be chosen $(n-k+2)$ ways. Thus, by the product rule, (1.82) and (1.43) and with $0 \le k \le n$,

$$P(n,k) = n(n-1)(n-2)\cdots(n-k+1) = \frac{n!}{(n-k)!}. \tag{1.83}$$

From (1.83), when $k = n$, that is, permutation of the n objects, meaning the entire set, we see that:

$$P(n,n) = n!. \tag{1.84}$$

We note that, historically, the idea of permutation came from India around 1,150 in the twelfth century.

Example 1.40

Suppose there are four "best students" nominees (Andy, Barbara, Carolyn, and David) for the Best Students of the Year Award, and only three can be selected as winners. Thus, we can think of having three slots to fill from the available four students. For the first slot, we have all four students to choose from. As soon as the first winner is selected, only three remain for the second winner and finally only two to choose from for the third winner. Then, by the multiplication principle, the winners can be chosen in $4 \cdot 3 \cdot 2 = 24$ ways.

Symbolically, let $S = \{A, B, C, D\}$. The 3-permutation of the set S is:

$$P(4,3) = 4 \cdot 3 \cdot 2 = \frac{4!}{1!} = \frac{24}{1} = 24.$$

These elements are as follows:
    ABC, ABD, ACB, ADB, ACD, ADC,
    BAC, BAD, BCA, BDA, BCD, BDC,
    CAB, CAD, CBA, CDA, CBD, CDB,
    DAB, DAC, DBA, DCA, DBC, DCB.

Example 1.41

Suppose we want to choose two numbers randomly out of the natural numbers from 1 to 5 without replacement. Then, the different ways each of the two numbers can be chosen are 5 and 4, respectively. Thus, the total ways the two numbers can be chosen are $5 \times 4 = 20$. The possible arrangements are as follows:
    12, 13, 14, 15,
    21, 23, 24, 25,
    31, 32, 34, 35,
    41, 42, 43, 45,
    51, 52, 53, 54.

Example 1.42

Permutations of the letters a, b, and c (written as abc) are the possible ordered arrangements of a, b, and c. There will be $3! = 3 \cdot 2 \cdot 1 = 6$. They are as follows:
    $abc, acb, bac, bca, cab, cba.$

Definition 1.29

For a set of n distinct objects, an unordered selection of k objects from the set is called a **combination of k object selected from n objects or simply "n choose k",** denoted by $C(n; k)$ or $\begin{pmatrix} n \\ k \end{pmatrix}$ or C_k and others.

Also, an unordered selection of k objects from the set is called a **k-combination** of elements of the set. In other words, a k-combination of a set S is a subset of k distinct elements of the set. When S contains n elements, the number of k-combinations is as follows:

$$\left\{ \begin{array}{ll} \binom{n}{k} = \dfrac{n \cdot (n-1) \cdot (n-2) \cdots (n-k+1)}{k \cdot (k-1) \cdot (k-2) \cdots 2 \cdot 1} = \dfrac{n!}{k!(n-K)!} & k \leq n, \\[3mm] \qquad\qquad\qquad\qquad 0, & k > n. \end{array} \right. \qquad (1.85)$$

Note 1.38

Order is, indeed, what distinguishes between permutations and combinations. In other words, in permutation, we select r objects from a set of n objects with order, while for combination, we do the same except no ordering is required. This would mean that in counting the combination of n letters taking k letter at a time, we could look at the k-permutation and exclude those objects having the same letters. In fact, this idea is the basis of the proof of (1.83). Here is the proof: We take a k-combination of a set of n elements. This number includes all objects with no ordering. We, then, order the objects in each of the k-combination. This task can be done in $P(k,k)$ ways. Thus, $k < n$:

$$P(n,k) = C(n,k) \cdot P(k,k)$$

or

$$C(n,k) = \frac{P(n,k)}{P(k,k)} = \frac{\dfrac{n!}{(n-k)!}}{\dfrac{k!}{(k-k)!}},$$

which leads to (1.85).

Note that it is not too difficult to show that for $k \leq n$, the following is true:

$$C(n,k) = C(n,n-k). \qquad (1.86)$$

We also note that $\binom{n}{k}$ appears as the coefficient in the binomial formula.

The **binomial theorem** states:

$$\left(x+y\right)^n = \sum_{k=0}^{n} \binom{n}{k} x^k \cdot y^{n-k}. \qquad (1.87)$$

We further note that

$$\binom{n}{0} = \binom{n}{n} = 1. \qquad (1.88)$$

Example 1.43

Suppose we want to create a stock shares portfolio with four electronic items and three real estate pieces, when there are eight electronic items and five real estate buildings available. Here is how we calculate the answer:

1. Number of ways of choosing 4 electronic items from 8 is $C(8,4)$.
2. Number of ways of choosing 3 real estate buildings from 5 is $C(5,3)$.
3. Number of ways of choosing 4 electronic items from 8 and 3 real estate buildings from 5 is

$$C(8,4)\cdot C(5,3) = \left(\frac{8!}{4!(8-4)!}\right)\left(\frac{5!}{3!(5-3)!}\right) = 700.$$

Example 1.44

Suppose a company is to establish an Executive Committee consisting of two men and three women. There is a pool of ten candidates of whom there are six men and four women. In how many ways this committee can be established?

To answer the question, we argue as follows:

We need to choose three women out of four and two men out of six. Hence, we have:

$$C(4,3)\cdot C(6,2) = \frac{4!}{3!(4-3)!} \cdot \frac{6!}{2!(6-2)!}$$

$$= \frac{4\cdot 3!}{3!\cdot 1} \cdot \frac{6\cdot 5\cdot 4\cdot 3\cdot 2\cdot 1}{2\cdot 4\cdot 3\cdot 2\cdot 1} = \times \frac{4(6\cdot 5)}{2} = 60.$$

Thus, there are 60 ways of establishing the committee needed.

Now, we exemplify an application that involves both permutation and combination.

Definition 1.30

Suppose $r_1, r_2 \ldots r_k$ are k nonnegative integers such their sum is another nonnegative integer, say n. Then,

$$\binom{n}{r_1, r_2, \ldots, r_k} = \frac{n!}{r_1! r_2! \ldots r_k!} \tag{1.89}$$

defines the **multinomial coefficient**. As a special case, if all r's are zero, except one, say r, then (1.89) reduces to (1.52), that is, the **binomial coefficients**.

The following theorem proves what is known as the **permutation** or **distinguishable permutations of n objects** or **permutation with repetition** or **sampling repetition**.

Theorem 1.2

The number of permutations of n objects with r_1 alike, r_2 alike, ..., r_k alike, is given by the multinomial coefficient (1.86).

Proof:
Let S be a set of n objects such that r_1 of one kind and indistinguishable, r_2 of another kind and indistinguishable, ..., and yet r_k of another kind and alike with $r_1 + r_2 + ... r_k = n$. We need to find the number of distinct permutations of all n objects.

There are $\begin{pmatrix} n \\ r_1 \end{pmatrix}$ ways to choose the first r_1 objects. Having r_1 objects chosen,

there remains $n - r_1$ objects to choose the second set of r_2 objects with $\begin{pmatrix} n - r_1 \\ r_2 \end{pmatrix}$

ways to choose from. Following the same pattern, for the next selection, there will

be $n - r_1 - r_2$ objects left to choose r_3 objects in $\begin{pmatrix} n - r_1 - r_2 \\ r_3 \end{pmatrix}$ ways. Continuing the

process up to the case when there are $n - r_1 - r_2 - \cdots - r_{k-2}$ objects left to choose r_{k-1}

objects in $\begin{pmatrix} n - r_1 - r_2 - \cdots - r_{k-2} \\ r_{k-1} \end{pmatrix}$ ways. Thus, by the product rule, we have the num-

ber of distinct permutations of all n objects to be the:

$$\begin{pmatrix} n \\ r_1 \end{pmatrix} \begin{pmatrix} n - r_1 \\ r_2 \end{pmatrix} \begin{pmatrix} n - r_1 - r_2 \\ r_3 \end{pmatrix} \cdots \begin{pmatrix} n - r_1 - r_2 - \cdots - r_{k-2} \\ r_{k-1} \end{pmatrix}, \tag{1.90}$$

or

$$\frac{n!}{r_1!(n - r_1)!} \cdot \frac{(n - r_1)!}{r_2!(n - r_1 - r_2)!} \cdots \frac{(n - r_1 - \cdots - r_{k-2})!}{r_{n-1}!(n - r_1 - \cdots - r_{n-1})!}$$

$$= \frac{n!}{r_1! r_2! \cdots r_n!}, \tag{1.91}$$

which is the same as (1.56), completing the proof.

1.7 FUZZY SET THEORY, FUZZY LOGIC, AND FUZZY MEASURE

The classical set theory, as defined by Cantor and later axiomatized by Zermelo and Fraenkel, assumed that an element is either a member of a set or not. This standard assumption was severely rattled when Lotfi A. Zadeh (1965), in his paper entitled *Fuzzy Sets*, developed the **fuzzy set theory** by relaxing some of the assumptions in the standard set theory. That is, fuzzy set theory allows the "gradual assessment of the membership of elements in a set". For instance, the value of the membership in a

FIGURE 1.9   Lotfi Aliasghar Zadeh (1921–2017). Iranian engineer, computer scientist, and mathematician, professor at University of Berkeley, California, the United States. Born on February 4, 1921, and died on September 6, 2017, at age of 96, Photo taken in 2005. https://en.wikipedia.org/wiki/Lotfi_A._Zadeh.

set is a real number in the interval [0, 1]. Also, the weight of a person does not have to be in the light or heavy weight categorical set, but it could be a fraction of the highest possible weight set. The fuzzy set is defined as a pair with a set and grade of membership of its element (Figure 1.9).

In essence, what Zadeh expressed was the need for conceptual adjustments to the foundational mathematical entities, like sets, in order for them to acquire a closer approximation to the various actual circumstances in applications of mathematics, where factors such as lack of determinacy, uncertainty, and randomness come into consideration.

Definition 1.31 Zadeh (1965)

Zadeh (1965) in the abstract of his paper defined a **fuzzy set** as "a class of objects with a continuum of grades of membership. Such a set is characterized by a membership (characteristic) function which assigns to each object a grade of membership ranging between zero and one".

To explain "membership", he started the introduction as "More often than not, the classes of objects encountered in the real physical world do not have precisely defined criteria of membership. For example, the class of animals clearly includes dogs, horses, birds, etc. as its members, and clearly excludes such objects as rocks, fluids, plants, etc. However, such objects as starfish, bacteria, etc. have an ambiguous status with respect to the class of animals. The same kind of ambiguity arises in the case of a number like 10 in relation to the 'class' of all real numbers which are much greater than 1".

Note 1.39

In $\mathbb{R}^n$, a fuzzy set is defined by a characteristic function μ_A such that $\mu_A : \mathbb{R}^n \rightarrow [0,1]$, which associates with each x in $\mathbb{R}^n$ its "grade of membership", $\mu_A(x)$ in A.

Note 1.40

To distinguish between crisp and fussy sets, the word "membership function" is used for fuzzy set, while "characteristic function" is used for a crisp set.

Here is an example he presented for membership function.

Example 1.45 Zadeh (1968)

Consider the set of real numbers, $\mathbb{R}^1$. Then, $A = \{x | x \gg 0\}$ is a fuzzy set with its membership function, say

$$\mu_a(x) = \begin{cases} \left(1 + x^{-2}\right)^{-1}, & x \gg 0, \\ 0, & x < 0. \end{cases} \tag{1.92}$$

Zadeh redefined his definition of the fuzzy set symbolically as he stated.

Definition 1.32 Zadeh (1965)

"Let X be a space of points (objects), with a generic element of X denoted by x. Thus, $X = \{x\}$.

A *fuzzy set* (*class*) A in X is characterized by a *membership* (*characteristic*) *function* $f_A(x)$ which associates with each point (more generally, the domain of definition of $f_A(x)$ may be restricted to a subset of X) in X a real number in the interval [0, 1] (in a more general setting, the range of the membership function can be taken to be a suitable partially ordered set P. For our purposes, it is convenient and sufficient to restrict the range of f to the unit interval. If the values of $f_A(x)$ are interpreted as truth values, the latter case corresponds to a multivalued logic with a continuum of truth values in the interval [0, 1]), with the value of $f_A(x)$ at x representing the 'grade of membership' of x in A. Thus, the nearer the value of $f_A(x)$ to unity, the higher the grade of membership of x in A. When A is a set in the ordinary sense of the term, its membership function can take only two values 0 and 1, with $f_A(x) = 1$ or 0 whether x does or does not belong to A. Thus, in this case, $f_A(x)$ reduces to the familiar characteristic function of a set A".

Note 1.41

The contents in parentheses are from the footnotes of the paper.

Example 1.46 Zadeh (1965)

Let X be the real line $\mathbb{R}$. Let A also be a fuzzy set of numbers greater than 1. Then, $f_A(x)$ may be given as a function on $\mathbb{R}$ with representative values as $f_A(0) = 0$, $f_A(1) = 0$, $f_A(5) = 0.01$, $f_A(10) = 0.2$, $f_A(100) = 0.95$, and $f_A(500) = 1$.

Note 1.42

As Zadeh stated, "When there is a need to differentiate between such sets and fuzzy sets, the sets with two-valued characteristic functions will be referred to as **ordinary sets** or simply **sets**".

Note 1.43

Also, as Zadeh stated, "The notions of inclusion, union, intersection, complement, etc., are extended to such sets".

Note 1.44

The **distinction between a crisp set and a fuzzy set** is that the characteristic function of a crisp set, as we mentioned before, assigns only 1 or 0 to each element of the universal set, while for a fuzzy set, it assigns a range of numbers (partial membership) within the unit interval $[0,1]$. We may say that fuzzy sets allow vague concepts.

For example, for a crisp set, in the real-life situation, a grape is either sweet or not sweet; however, in a fuzzy set, a grape may be very sweet, approximately sweet, or not sweet at all.

As another example, we can consider the set of criminals versus innocents. It is the degree of evidence of guilt or innocence that puts the membership of the defendant in the crisp or fuzzy set.

Definition 1.33 Zadeh (1965)

1. A fuzzy set is **empty** if and only if its membership function is identically zero on X.
2. Two fuzzy sets A and B are **equal**, denoted by $A = B$, if and only if $f_A(x) = f_B(x)$, for all x in X.
3. The **complement** of a fuzzy set A, denoted by A', is defined by $f_{A'}(x) = 1 - f_A(x)$.
4. $A \subset B$, or **A, is smaller than or equal to B** if and only if $f_A(x) \le f_B(x)$.
5. The **union** of two fuzzy sets A and B with respective membership functions $f_A(x)$ and $f_B(x)$ is a fuzzy set C, denoted by $C = A \cup B$, whose membership function is related to those of A and B by $f_C(x) = \text{Max}\left[f_A(x), f_B(x)\right], x \in X$ or, symbolically, $f_C(x) = f_A(x) \vee f_B(x)$.

6. The **intersection** of two fuzzy sets A and B with respective membership functions $f_A(x)$ and $f_B(x)$ is a fuzzy set C, denoted by $C = A \cap B$, whose membership function is related to those of A and B by $f_C(x) = \text{Min}\left[f_A(x), f_B(x)\right]$, $x \in X$ or, symbolically, $f_C(x) = f_A(x) \wedge f_B(x)$.

7. The **algebraic sum** of fuzzy sets A and B, denoted by $A + B$, is defined in terms of the membership functions of A and B, as $f_{A+B} = f_A + f_B$ provided $f_A(x) + f_B(x) \leq 1$, $\forall x$.

8. The **algebraic product** of fuzzy sets A and B, denoted by AB, is defined in terms of the membership functions of A and B, as $f_{AB} = f_A f_B$. Note that the condition of the algebraic sum is not required in this case.

9. The **absolute difference** of fuzzy sets A and B, denoted by $|A - B|$, is defined by $f_{|A-B|} = |f_A - f_B|$.

10. A relation, as was originally defined by Halmos (1960), is defined as a set of ordered pairs. However, for fuzzy sets, a **fuzzy relation in X** is a fuzzy set in the product space $X \times X$.

For instance, in ordinary set theory, the set of all ordered pairs of real numbers x and y such that $x \geq y$ is a relation (a generalization of a function), while in fuzzy set theory, the relation, denoted by $x \gg y, x, y \in \mathbb{R}^1$, may be regarded as a fuzzy set $A \subset \mathbb{R}^2$ with the membership function of A, $f_A(x, y)$, having the following representative values as $f_A(10,5) = 0$; $f_A(100,10) = 0.7$; $f_A(100,1) = 1$; etc.

Note 1.45

As a description of the new direction, Zadeh pointed to the allowance of a "continuity" of the membership (characteristic) function of a set, beyond the standard binary form, as the principal, starting line of thought along these lines, which, indeed, seems to be a very reasonable and sound position. In fact, this is the line of thought through which **fuzzy set theory** can be described properly and accurately in a way that makes its nature and position in mathematics more immediately transparent.

Note 1.46

In earlier times, when Zadeh introduced the fuzziness idea in mathematics, some members of the mathematical community expressed statements, based on incomplete understanding of the theory and/or misconceptions. However, with appearances of applications of the theory, it found its place, and as we see these days, hundreds of papers are published as application of fuzzy set theory in mathematics. Having said that, perhaps a mathematically more transparent and precise definition of the theory may further help in avoiding such circumstances or misunderstandings in the future. Some even believe that the name should be reconsidered, for instance, "continuous set theory", as originally was considered by Zadeh. The word "continuous" here is used in its widest sense, namely, the existence of at least some intermediate values, beyond the standard binary ones. The main point in such a name suggestion for the theory, looking ahead, is for its name to already project, transparently, more exactly its exact mathematical nature and position in the mathematical spectrum, something that may be substantially beneficial.

Continuing his line of fuzzy development, in 1975, Zadeh introduced "fuzzy logic and approximate reasoning". The classic definition of logic is the study of principles of reasoning in any form. It deals with **propositions** that are required to be either true or false. For each proposition, there is its **negation**. Zadeh used the term "fuzzy logic" to describe an imprecise logical system, in which the true values are fuzzy subsets of the unit interval. In that context, terms such as "true", "false", "not true", "very true", "quite true", "not very true", and "not very false" have been used. Thus, the so-called truth tables in traditional logic with "and", "or", "implies", and "if and only if" in fuzzy logic become dependent upon the meaning of these terms.

What distinguishes fuzziness from traditional sets and logic is the "approximate" rather than "preciseness". Hence, we might say that fuzzy logic is "logic of approximate reasoning". Thus, the validity of rules of inference in fuzzy logic is approximate rather than exact. In other words, fuzzy set theory is based on intuitive reasoning. This reasoning involves the human subjectivity and imprecision, although the theory itself is a rigorous mathematical theory.

Example 1.47

As an example of a fuzzy set in conjunction with fuzzy logic, we could have the following implication:

> a is a *small positive number*; a and b are *almost equal*; therefore, b is *almost small*.

In this statement, both a and b are the fuzzy sets by virtue of the word "almost".

In the previous subsection, we discussed how the Lebesgue measure was defined. However, definition of the Lebesgue measure is based on the standard crisp sets. We now present two types of the fuzzy measure that are essentially the same.

Definition 1.34

Let U be a universal set and Ω a nonempty set of subsets of U. Then, a **fuzzy measure** on the space $\langle U, \Omega \rangle$ is a function, say μ, that satisfies the following properties:

1. $\mu(\varnothing) = 0$ and $\mu(U) = 1$ **boundary condition**.
2. $\forall A, B \in \Omega$, if $A \subseteq B$, then $\mu(A) \leq \mu(B)$ **monotonicity condition**.

3. For an increasing sequence $A_1 \subset A_2 \subset \cdots \in \Omega$,

$$\text{If } \bigcup_{i=1}^{\infty} A_i \in \Omega, \text{ then } \lim_{i \to \infty} \mu(A_i) = \mu\left(\bigcup_{i=1}^{\infty} A_i\right), \quad \textbf{continuity} \quad \textbf{from} \quad \textbf{below}$$

condition.

4. For a decreasing sequence $A_1 \supset A_2 \supset \cdots \in \Omega$,

$$\text{If } \bigcap_{i=1}^{\infty} A_i \in \Omega, \text{ then } \lim_{i \to \infty} \mu(A_i) = \mu\left(\bigcap_{i=1}^{\infty} A_i\right), \quad \textbf{continuity} \quad \textbf{from} \quad \textbf{above}$$

condition.

Definition 1.35 (Sugeno Fuzzy Measure)

Recall the definition of a σ-algebra or a σ-field on a set A that is a collection $\wp$ of subsets of A that includes the empty set, which is closed undercomplement, countable unions, and countable intersections. Now from Murofushi and Sugeno (1989), let $\wp$ be a σ-algebra on a universal set U. A **Sugeno fuzzy measure**, denoted by g, is defined by $g : \wp \to [0,1]$ such that:

1. $g(\varnothing) = 0$ and $g(U) = 1$ **boundary condition**.
2. If $A, B \in \wp$, and $A \subseteq B$, then $g(A) \leq g(B)$, **monotonicity condition**.
3. If $A_i \in \wp$, and $A_1 \subseteq A_2 \subseteq \cdots$, then $\lim_{i \to \infty} g(A_i) = g\left(\lim_{i \to \infty} A_i\right)$, **Sugeno's**

   convergence.

Note 1.47

We note that in **fuzzy measure**, some of the mentioned properties may not hold.

EXERCISES

1.1. Are $\{1,3,4\}$ and $\{4,3,1,1\}$ equal sets?
1.2. How many elements are in the set $\{1, a, 2, 1, a\}$?
1.3. A mathematics department has 40 full-time faculty members. For professional development and service to the profession, each may belong to one or more mathematics professional organizations such as
    1. AMS (American Mathematical Society),
    2. SIAM (Society for Industrial and Applied Mathematics),
    3. MAA (Mathematics Association of America),
    4. ASA (American Statistical Association),
    5. NTCM (National Council of Teachers of Mathematics).
    For this department, it is known that the membership of each of these associations is as follows:

AMS	SIAM	MAA	ASA	NCTM	(1) &	(1) &	(2) &	(3) &	(1), (2), &	All
(1)	(2)	(3)	(4)	(5)	(2)	(3)	(4)	(5)	(4)	
21	26	13	9	3	20	15	15	8	7	2

Draw a Venn diagram and find the number of those who do not belong to any one of these five organizations.

1.4. Let us denote a rainy day by R, a windy day by W, and a sunny day by S. Let E be the set of days in September. Draw a Venn diagram, if in Houston Texas in a September month, there are, on the average, 15 sunny days, 5 windy days, and 18 rainy days.

1.5. List the elements of a set A containing natural numbers between 20 and 200, divisible by 4.

1.6. Determine if the set is finite or infinite, and justify your answer:
i. The set of whole numbers between 1 and 10.
ii. The set of positive real numbers less than 9.
iii. The set of natural numbers greater than 9.
iv. The set $A = \{3,6,9,12,\ldots\}$.

1.7. For $A = \{2,4,5\}$ and $B = \{3,4,5\}$, answer the following questions:
i. Is A a subset of B?
ii. Is B a proper subset of A?
iii. Find the intersection of A and B.

1.8. Considering the set E of English alphabets, if V is the set of vowels, what is the complement of V called?

1.9. In a group of musicians, including singers, if S is the set of singers and I is the set of instrumentalists, in a set notation,
i. Write the set of singers who play an instrument.
ii. Write the set of musical performers.

1.10. Let $A = \{0,3,6,9\}$ and $B = \{1,2,4,7,8\}$. Draw the Venn diagram illustrating these sets with the union as the universal set.

1.11. Find the number of ways of choosing 2 consonants from 7 and 2 vowels from 4.

1.12. A game team consists of ten players, six male and four female players. In how many different ways can four players be selected such that at least one male player is included in the selected group?

1.13. In an election of the Student Government Association Executive Committee at a university, there are 13 student candidates, with 7 females and 6 males. Five persons are to be selected that include at least 3 females. In how many different ways can this committee be selected?

1.14. In how many different ways can the letters of the word "MATHEMATICS" be arranged so that the vowels always come together?

1.15. In how many ways can a group of five girls and two boys be made out of a total of seven boys and three girls?

1.16. Fifteen planes are commuting between two cities A and B. In how many ways can a person go from A to B and return by a different plane?

1.17. Three cars enter a parking lot with 11 empty parking spaces. Each car has an option of backing into its parking space or driving in forward. How many ways can the three drivers select
   i. Parking spaces?
   ii. Parking spaces and parking positions?

1.18. An executive lady has four clean skirts and five clean blouses to wear to work in a certain week. How many ways can she choose outfits for Monday, Wednesday, and Friday if
   i. She does not wish to wear the same skirt or blouse twice?
   ii. She is willing to repeat her attire?
   iii. She is willing to repeat her skirts but not her blouses?

1.19. Suppose stocks of eight electronic items and five real estate buildings are available in the stock market. We want to create a stock shares portfolio with four electronic items and three real estate pieces. How many different ways can we form this portfolio?

1.20. Suppose a company is to establish an Executive Committee consisting of two men and three women. There is a pool of ten candidates consisting of six men and four women. In how many ways can this committee be established?

1.21. Suppose we are to create 6-character codes (xxxxxx) for products of a manufacturer. The characters in each code are to be:
   A letter,
   Another letter different from the first choice,
   Another letter different from the first two choices,
   A nonzero digit
   Two digits.
   How many different codes could we have?

1.22. Using the Stirling formula, approximate the error of calculating 2! (2 factorial).

1.23. Consider the set {1,2,3}.
   i. Write the elements of the permutations by taking two digits from above three
   ii. Write the elements of the permutation without repetition of the 3 numbers taken all 3 at a time.
   iii. Do both cases (i) and (ii) have the same number of elements?

1.24. Suppose it is known that a box of 24 light bulbs contains five defective bulbs. A customer randomly chooses two bulbs from the box without replacement.
   i. What is the total number of ways the customer can choose the bulbs?
   ii. What is the number of ways to choose no defective bulb?

2 Basics of Probability

2.1 BASICS OF PROBABILITY

In this book, unless otherwise stated, we will assume that occurrences of events are not deterministic. In other words, the events cannot be entirely determined by the initial states and inputs. Instead, they are perceived to be random, probabilistic, or stochastic. Nature of these occurrences of events is random, probabilistic, or stochastic. We will start with basic definitions and develop, in the process, the vocabulary needed to discuss the subject matter with the clarity sufficient even to address the more advanced materials to be encountered later in this book.

Definition 2.1

When an experiment is performed, its results are referred to as the **outcomes**. If such results are not deterministic, the experiment is called a **chance experiment**, a **random experiment**, or a **trial**.

Thus, from now to the end of this book, we will be considering random experiments with uncertain outcomes.

Definition 2.2

A set of outcomes is called an **event**. An event with only one outcome, that is, a singleton, is referred to as an **elementary or simple event**. Hence, in general, an event is a collection of simple events. An event may be defined as an element of a σ-field, denoted by $\mathbb{S}$ (defined in Chapter 1) of subsets of the sample space Ω. **Occurrence** of an event depends on its member outcomes that take place. The collection of all possible events is called the **sample space**, usually denoted by Ω. Thus, an event is a subset of the sample space, whose elements are called the **sample points**.

Definition 2.3

A sample space that contains a finite or a countable collection of sample points is referred to as a **discrete sample space**. However, a sample space that contains an uncountable collection of sample points is referred to as a **continuous sample space**.

Example 2.1

The following are examples of finite and infinite discrete sample spaces:

i. Let Ω be the set of all natural numbers between 1 and 15. Then, the sample space has 15 sample points that are numbered from 1 through 15. And the set of all odd numbers between 1 and 15 is an event.
ii. As another example, consider an automatic switch for a traffic light with only four states "green", "yellow", "red", and "off". Let us, for instance, quantify the sample space Ω as $\{1,2,3,4\}$. There are $4! = 24$ possible outcomes for this random experiment: $\{1\}$, $\{2\}$, $\{3\}$, $\{4\}$, $\{1,2\}$, $\{1,3\}$, $\{1,4\}$, ... $\{4,3\}$, ... $\{1,2,3\}$, $\{1,2,4\}$, ... ,$\{1,2,3,4\}$, and $\varnothing$.
iii. As another example, consider the off and on cases of the traffic light mentioned above. In particular, consider the number of working days of the light. In this case, the sample space will be $\Omega = \{0,1,2,3,...\}$.

Example 2.2

In contrast to the discrete sample space discussed in Example 2.1, we may have continuous sample space. For example, consider a machine that may break while working. The length of breakdown of the machine is an interval of time on the real line, as the length of the working time of the machine. Suppose we are interested in the first time that the machine breaks down. Thus, the sample space for working time of the machine in this case is indicated by $[0,\infty)$, where the symbol ∞ denotes that the machine will never break.

Definition 2.4

Two events E_1 and E_2 from a sample space Ω are called **mutually exclusive events** if their intersection is the empty set; otherwise, they are referred to as **non-mutually exclusive events**. In the latter case, there is a chance that both events can occur simultaneously.

Definition 2.5

The set of n events $\{E_1, E_2, ..., E_n\}$ is called a **partition** of a sample space Ω if the events are **mutually exclusive**, that is, $E_1 \cup E_2 \cup, ..., \cup E_n = \Omega$.

Example 2.3

A random generator is referred to a process that can uniformly (with the same chance) generate numbers on a set. Using a random generator processor, we generate real numbers from an interval of $[1,2]$. We choose the point $\sqrt{2}$ from this interval. Then, the real numbers randomly selected from $[1,2]$ will be either in $\left[1,\sqrt{2}\right]$ or in $\left(\sqrt{2},2\right]$. Thus, the sample space in this case is the continuous infinite interval $\Omega = [1,2]$. The two events $\left[1,\sqrt{2}\right]$ and $\left(\sqrt{2},2\right]$, in this case, are mutually exclusive, that is, $\left[1,\sqrt{2}\right] \cap \left(\sqrt{2},2\right] = \varnothing$ and $\left[1,\sqrt{2}\right] \cup \left(\sqrt{2},2\right] = \Omega$. Thus, $\left\{\left[1,\sqrt{2}\right],\left(\sqrt{2},2\right]\right\}$ is a partition of the sample space Ω.

Example 2.4 Non-Mutually Exclusive Events

Suppose we roll a fair die. If the event shows an odd number and another event shows a number <5, then they are called as **non-mutually exclusive events**. Indeed, the numbers 1 and 3 are the common elements in these types of events.

Definition 2.6

Let E be an event. Let us also denote the probability of an event E by $P(E)$. We offer four definitions for $P(E)$ in the following.

Definition 2.6a First Definition

The **probability of E, $P(E)$**, is a **number between 0 and 1, inclusive**, that indicates the chance of occurrence of E. In other words, the probability of an event measures how likely it is for the event to occur.

There are two special types of events: one with probability 1 and another with probability 0. The former is called the **sure event**, and the latter is the **null** or **impossible event**. All other events are called the **proper events**.

Note 2.1

As mentioned, "null event" is typically called the **impossible event**. However, it may be argued that the **impossible event is the empty set** that it can never occur. On the other hand, a **null event is an event with probability 0** by the probability measure; it is not necessarily the empty set (event). For instance, as we will see in Chapter 3, for a **continuous random variable** X, the event $\{X = x\}$ is a null event because the probability of a point is 0. But the set representing an event is not empty.

Note 2.2

If a sample space is finite with n elements and all its outcomes have the same chance to occur, then the probability of any of the sample points occurring is assumed to be $1/n$, and the sample space is referred to as **equiprobable**. The concept of equiprobability in case of an infinite sample space is referred to as the **uniform measure**. Naturally, in this case, the aforementioned assumption of probability is not possible. We will later address this idea. The notion of equiprobable measure is an example of a discrete probability measure. In such a case, the choice of an event is referred to as the **selection at random**.

Example 2.5

We may want to choose a digit at random from 3 to 9. Thus, the outcomes may be the digits 3, 4, 5, 6, 7, 8, or 9 that have the same chance of 1/7 to be selected. In other words, all elementary events in {3},{4},....,{9} are equiprobable.

Definition 2.6b Second Definition

When an experiment is performed repeatedly, intuitively, the likelihood of occurrence of an outcome may be **approximated** by the proportion of the number of times that the outcome occurs. This proportion is called the **relative frequency**. Suppose a finite sample space has n equiprobable points with same k points. Then, the **probability of an event E, $P(E)$**, from such sample space is defined as **the relative frequency of the repetition**, that is, k/n.

Example 2.6

Suppose for the purpose of quality control of a product in a manufactory a sample size of 50 is chosen. Items in the sample are checked one at a time for matching the set criteria. Four items failed the match. Thus, the relative frequency of non-matched items is 4/50, or 8%.

   We should note the word "approximated" used in the definition above. That is, the probability in this case is "estimated by the **portion** or **proportion**". Thus, no matter how many repetitions of a trial are performed, the exact value of $P(E)$ is not known. Since the repetition of trials and stop at a point are equivalent to sampling in statistics (and simulation, in general), we will discuss finding this error when this subject comes up later.

Definition 2.6c Third Definition

Consider an equiprobable sample space. We now define the **probability of E, $P(E)$**, as follows:

$$P(E) = \frac{\text{Number of ways the event can occur}}{\text{Total number of ways outcomes can occur}}. \qquad (2.1)$$

Note 2.3

We note that Definition (2.1) is from the **combinatorics** viewpoint.

Definition 2.6d Fourth Definition (Geometric)

The number of elements of a finite set E is referred to as the **size of E**. Suppose we have an equiprobable set (space) Ω such that $E \subset \Omega$. Then, the **probability of E**, $P(E)$, is defined as follows:

$$P(E) = \frac{\text{Size of } E}{\text{Size of } \Omega}. \tag{2.2}$$

We note that Definition (2.2) can be used for infinite and continuous sample spaces.

Example 2.7

Let us consider I as an interval on the real line, A as an area of section in a two-dimensional space, and V as a volume in a three-dimensional space. We may want to select a few points from parts of each one of these entities; say J from I, α from A, and v from V. Then, based on Definition 2.6d, the probabilities of the subinterval J, subarea α, and subvolume v, assuming that probabilities are distributed uniformly over the entire interval, area, and volume, respectively, are as follows:

$$P(J) = \frac{\text{Length of } J}{\text{Length of } I}, P(\alpha) = \frac{\text{Area of } \alpha}{\text{Area of } A}, \text{ and } P(v) = \frac{\text{Volume of } v}{\text{Volume of } V}. \tag{2.3}$$

Example 2.8

Suppose we are to play an electronic game using a rectangular tablet of size 10 inches by 15 inches. A rectangular target of size 1 inch by 5 inches is marked on the screen. The idea is to land a random moving mouse in the marked area. What is the probability of this event?

Here is the solution to the problem: The areas of the screen and the marked area are $10 \cdot 15 = 150$ inches^2 and $1 \cdot 5 = 5$ inches^2, respectively. Thus, the probability of hitting the target, that is, be within the marked area, is $5/150 = 0.033$, or a very small chance of 3.3%.

As we discussed earlier, the probability was mainly used for gaming purposes. But that was the case until Laplace's book *Théorie Analytique des Probabilités* appeared in 1812. He applied probability to various scientific and practical issues such as actuarial mathematics, statistical mechanics, and theory of errors. Then, the probability theory started to develop more mathematically as its applications grew in various areas such as mathematical statistics and queueing theory, and later in engineering, economics, psychology, and the biological sciences.

The expansions and finally axiomatization of probability theory were developed by a number of famous mathematicians Chebyshev, Markov, von Mises, and Kolmogorov (Figure 2.1).

FIGURE 2.1   Pierre de Laplace 1749–1827.

The birth of modern probability theory occurred in the era of Andrey Nikolaevich Kolmogorov, a Russian mathematician. Kolmogorov finally ended all controversies in the various definitions of probability after nearly three centuries by presenting his axioms of probability in 1933. The monograph (available in English as *Foundations of Probability Theory*, Chelsea, New York, 1950) that we will highlight in Definition 2.7 is given below.

Again, we note that an axiom is a statement that cannot be proved or disproved; that is, it is an assumption that a person may or may not accept to use, such as the axiom of choice in axiomatic set theory. However, no probabilist has so far shown any objection to any and all the axioms of probability (Figure 2.2).

FIGURE 2.2   Andrey Kolmogorov 1903–1987.

Definition 2.7 Kolmogorov's Axioms of Probability

Let P be a function on a field $\mathcal{F}$. Then, P is called the **probability measure** (as defined in Chapter 1) if it satisfies the following properties:

(1) For any event, $E \in \mathcal{F}, 0 \le P(E) \le 1$, that is, the range of P is the interval $[0,1]$;
(2) $P(\varnothing) = 0$, that is, the empty set is an impossible event;
(3) $P(\Omega) = 1$, that is, the sample space is a sure event;
(4) If $\{E_n, n = 1,2,...\}$ is a countable collection of pairwise disjoint sets of the set function $\mathcal{F}$ and $\bigcup_{n=1}^{\infty} E_n \in \mathcal{F}$, then P is **countably additive**, that is,

$$P\left(\bigcup_{n=1}^{\infty} E_n\right) = \sum_{n=1}^{\infty} P(E_n). \tag{2.4}$$

We note that (1.61) holds for finitely countable additive, which is defined as follows:

$$P\left(\bigcup_{i=1}^{n} E_i\right) = \sum_{i=1}^{n} P(E_i). \tag{2.5}$$

Properties (1) through (3) and (2.5) are referred to as the **Kolmogorov axioms of probability**.

Example 2.9

Let $\Omega \times \{2,4,6\} \times$ be a set with probability measures of its elements as $P(\{2\}) = 1/5$,$P(\{4\}) = 2/5$, and $P(\{6\}) = 2/5$. Then, $P(\{2,4\}) = 1/5 + 2/5 = 3/5$.

Definition 2.8

Assuming $\mathcal{F}$ is a σ-field and let us denote it by, say, $\mathbb{S}$, in Ω and P is a probability measure on $\mathbb{S}$, the triple $(\Omega, \mathbb{S}, P)$ is referred to as a **probability measure space** or just **probability space**.

Let E_1 and E_2 be two events in $\mathbb{S}$. The following are some properties of the probability measure P. We leave the proofs of these properties as exercises.

If E_1 and E_2 are two events in $\mathbb{S}$ and P is a probability measure, then P is a monotonic function, that is,

$$P(E_1) \le P(E_2), \text{ if } E_1 \subset E_2. \tag{2.6}$$

$$P(E_1^c) = 1 - P(E_1). \tag{2.7}$$

$$P(E_1) + P(E_2) = P(E_1 \cup E_2) + P(E_1 \cap E_2). \tag{2.8}$$

$$P(E_1 \cup E_2) = P(E_1) + P(E_2) - P(E_1 \cap E_2). \tag{2.9}$$

General case of (2.9) is known as the **inclusion–exclusion** formula for a finite number n, which is as follows:

$$P\left(\bigcup_{i=1}^{n} E_i\right) = \sum_{i} P(E_i) - \sum_{i<j} P(E_i \cap E_j)$$

$$+ \sum_{i<j<k} P(E_i \cap E_j \cap E_k) + \cdots + (-1)^{n+1} P(E_1 \cap, \ldots, E_n). \quad (2.10)$$

We note that (2.10) can be rewritten as follows:

$$P\left(\bigcup_{i=1}^{n} E_i\right) = \sum_{i=1}^{n} (-1)^{i+1} S_i, \quad (2.11)$$

where

$$\begin{cases} S_1 = \sum_{i=1}^{n} P(E_i) \\ S_2 = \sum_{i<j} P(E_j \cap E_j), \ldots, \\ S_k = \sum_{1 \le i_1 < i_2 < \ldots < i_k \le n} P(E_{i_1} \cap E_{i_2}, \ldots, E_{i_k}). \end{cases} \quad (2.12)$$

Also, for non-necessarily disjoint sets $E_i, i = 1, 2, \ldots, n$,

$$P\left(\bigcup_{i=1}^{n} E_i\right) \le \sum_{i=1}^{n} P(E_i). \quad (2.13)$$

Note 2.4

We note that for the sake of simplicity, from now on, we refer to "p" as "probability", unless there is a need for clarification.

Example 2.10 Matching Problem

Another historic example in the foundation of probability is the **matching problem**, which is an application of the inclusion–exclusion principle. There are various versions of the classic matching problem, all of which involve the pairing of two orderings on a set of objects. There are many colorful descriptions of the matching problem.

An example is a well-known problem of n married couples going to a ball room dancing class. The instructor randomly assigns the women to the men as dancing partners. A match occurs if a wife is assigned as the dancing partner of her husband.

Another well-known example is that of a party whose guests arrive via their personal vehicles. At the arrival, each guest puts his or her car key in a uniform cover and drops it in a basket. As the party ends, each person blindly picks up a covered key from the basket. A picked key that belongs to the right person is a match.

We consider the following match problem. Suppose we have a set of ten cards with numbers one through ten written on them. There are also ten positions on a table showing one through ten. We randomly choose a card without replacement and place it on the table until all cards are chosen. What is the probability that each card is placed on its own numbered position on the table?

The problem may be generalized by replacing ten with a finite number n where $n \geq 2$. That is, there are n cards, numbered 1 through n, in a bucket. A game player is sitting at a table. The table has n positions numbered from 1 to n in an increasing order from left to right. Cards are drawn from the bucket randomly one at a time without replacement and are blindly placed on the positions. That is, the first card drawn will be placed at position 1, the next at position 2 and so on to the last one that will be placed at position n. Among various questions that may be asked, we choose the following 4, the first of which answers the question raised above. Hence, questions are as follows: What is the probability that

1. At least one of the n cards is a match?
2. Exactly k of the n cards are matches?
3. None of the n cards is a match?
4. Each of the n cards is a match?

We will answer the first question and will leave the other three as exercises. The answer is based on the inclusion–exclusion principle. Let C_i denote the event that the ith card drawn matches the space with the number written on the card. Then, the probability that there is at least one match is $P(C_1 \cup C_2 \cup ... \cup C_n)$. We note that when ith card is withdrawn, there are $n-1$ cards left. Then, using permutation of $(n-1)$, we can see that $P(C_i) = \dfrac{(n-1)!}{n!} = \dfrac{1}{n}$. With a similar argument, we will have $P(C_i \cap C_j) = \dfrac{(n-2)!}{n!} = \dfrac{1}{n(n-1)}$. Thus, for k cards withdrawn, we will have:

$$P(C_{i_1} \cap C_{i_2} \cap ... \cap C_{i_k}) = \frac{(n-k)!}{n!}. \tag{2.14}$$

Now, for $1 \leq k \leq n$, there are $\binom{n}{k}$ ways to draw k matches out of n cards. Thus, from (2.12),

$$S_k = \sum P(C_{i_1} \cap C_{i_2} \cap ... \cap C_{i_k}) = \binom{n}{k}\frac{(n-k)!}{n!} = \frac{1}{k!}. \tag{2.15}$$

Therefore, from (2.11) and (2.15), we have:

$$P(C_1 \cup C_2 \cup ... \cup C_n) = 1 - \frac{1}{2!} + \frac{1}{3!} - \cdots + (-1)^{n+1}\frac{1}{n!}, \tag{2.16}$$

from which

$$1 - P(C_1 \cup C_2 \cup \ldots \cup C_n) = 1 - 1 + \frac{1}{2!} - \frac{1}{3!} + \cdots + (-1)^n \frac{1}{n!}. \tag{2.17}$$

Note that Maclaurin expansion of e^x is:

$$e^x = \sum_{n=0}^{\infty} \frac{x^n}{n!} = 1 + x + \frac{x^2}{2!} + \frac{x^3}{3!} + \cdots, \quad \forall x. \tag{2.18}$$

Thus, the right-hand side of (2.17) is the first $(n+1)$ terms of the series expansion of e^{-1}. Therefore,

$$\lim_{n \to \infty} P(C_1 \cup C_2 \cup \ldots \cup C_n) = \lim_{n \to \infty} \left(1 - \frac{1}{2!} + \frac{1}{3!} - \cdots + (-1)^{n+1} \frac{1}{n!} \right) \tag{2.19}$$

$$= 1 - e^{-1} = 0.6321205\ldots$$

Note that the limit in (2.19) converges rapidly. For instance, for $n = 6$, 7, and 8, the values of the limit in (2.19) are 0.63194, 0.63214, and 0.63212, respectively.

It should also be noted that although the probability of having at least one match is a function of n (the number of cards), since the probability rapidly converges to $1 - e^{-1}$ for all practical purposes, we can say that for a very large n, the probability of having at least one match is 63% (about two-thirds).

For later use, we state the following theorem and leave the proof as an exercise.

Theorem 2.1 Distributive Laws of Probability

Let $E_1, E_2, \ldots$ be any events. Then, we have the following two properties:

$$\left(\bigcup_j E_j \right) \cap A = \bigcup_j (E_j \cap A) \tag{2.20}$$

and

$$\left(\bigcap_j E_j \right) \cup A = \bigcap_j (E_j \cup A). \tag{2.21}$$

For a finite number of events $E_j, j = 1, 2, \ldots, n$, (1.77) and (2.21) will be rewritten as follows:

$$\left(\bigcup_{j=1}^{n} E_j \right) \cap A = \bigcup_{j=1}^{n} (E_j \cap A) \tag{2.22}$$

and

$$\left(\bigcap_{j=1}^{n} E_j\right) \cup A = \bigcap_{j=1}^{n}(E_j \cup A). \tag{2.23}$$

For an infinite number of events $E_j, j = 1, 2, \ldots$, (2.22) and (2.23) will be rewritten as follows:

$$\left(\bigcup_{j=1}^{\infty} E_j\right) \cap A = \bigcup_{j=1}^{\infty}(E_j \cap A) \tag{2.24}$$

and

$$\left(\bigcap_{j=1}^{\infty} E_j\right) \cup A = \bigcap_{j=1}^{\infty}(E_j \cup A). \tag{2.25}$$

2.2   FUZZY PROBABILITY

To formally define fuzzy events, Zadeh (1968) assumed that the sample space is an n-dimensional Euclidean space, $\mathbb{R}^n$. Thus, the probability space becomes $(\mathbb{R}^n, \mathbb{S}, P)$, where, as we have defined before, $\mathbb{S}$ is the σ-field of Borel sets in $\mathbb{R}^n$ and P is the probability measure over $\mathbb{R}^n$.

Now let $A \in \mathbb{S}$ and μ_A be the characteristic function of A, as defined earlier. That is, for $x \in \mathbb{R}^n$,

$$\mu_A(x) = \begin{cases} 1 & \text{for } x \in A, \\ 0 & \text{for } x \notin A. \end{cases} \tag{2.26}$$

Then, $P(A)$ may be defined as

$$P(A) = \int_A dP \tag{2.27}$$

or equivalently as

$$P(A) = \int_{\mathbb{R}^n} \mu_A(x) dP, \quad x \in \mathbb{R}^n. \tag{2.28}$$

Note 2.5

As we will see later, relation (2.28) is defined as the expected value of μ_A. In other words, (2.28) equates the probability of A with the expected value of its characteristic function. And as Zadeh notes, it is this equation that can be generalized to fuzzy events through the concept of a fuzzy set.

Definition 2.9 (Zadeh (1968))

Let $(\mathbb{R}^n, \mathbb{S}, P)$ be a probability space, where $\mathbb{S}$ is the σ-field of Borel sets in $\mathbb{R}^n$ and P is the probability measure over $\mathbb{R}^n$. Then, a fuzzy event in $\mathbb{R}^n$ is a fuzzy set in $\mathbb{R}^n$ with membership function $\mu_A : \mathbb{R}^n \rightarrow [0,1]$ that is a Borel measurable. The probability of a fuzzy event is given by (2.28).

Zadeh (1965) noted that although the membership function of a fuzzy set has some similarities with a probability function when X is a countable set (or a probability density function when X is a continuous), there are essential differences between the two concepts. For instance, "the notion of a fuzzy set is completely nonstatistical in nature". It is perhaps because of this fact, in spite of its potential in dealing with uncertainties, very few works applying fuzzy logic concepts in biological system have been seen in the literature. For the first time, Zadeh in 1968 proposed the application of fuzzy set theory in life sciences, in a book titled *Biological Applications of the Theory of Fuzzy Sets and Systems*, which was edited by L. D. Proctor in 1969. It took almost two decades for the applications of fuzzy logic in biomedicine to develop.

However, Zadeh in his 1968 paper tried to show how fuzziness differs from probability. To do so, he showed the distinction between the events in traditional probability and the events in fuzzy probability. In the traditional probability, we deal with the chance of occurrence of a specific event, while in the fuzzy probability, we are talking about an almost event. For instance, rather than asking about the chance of rain, we ask the chance of "almost rain"!!

In other words, in the traditional probability, we define an event precisely as a class of points of a σ-field $\mathbb{S}$ in the sample space Ω. But in fuzzy probability, we define an event not precisely as a collection of points in the sample space, rather fuzzy. For instance, (1) it is mostly cloudy; (2) it is about to rain; (3) the student's test paper's score is almost 70 out of 100; (4) in repeating flipping a fair coin, the number of heads was a few more than the number of tails.

Note 2.6

According to (2.28), the probability of a fuzzy set A defined by (2.22) is a Lebesgue–Stieltjes integral.

We will return to fuzzy probability in Chapter 7.

2.3 CONDITIONAL PROBABILITY

We have already discussed various definitions of probability, from basic to measure theory with some basic properties, and finally mentioned Kolmogorov axioms of probability. In these discussions, we even mentioned the recent definition of fuzzy probability for later use. In Section 2.1, an event was defined and some properties were given. We went through all discussions of sets, logics, measures, axioms of sets, events, and probability on a probability space $(\Omega, \mathbb{S}, P)$, to build a strong background for the concept of probability. We now want to continue with events and probability without much sophistication and away from fuzziness to state some other properties of probability.

Let us start by asking the question if the probability of occurrence of an event may be affected by knowing occurrence of another event. In other words, so far we have been speaking about an unconditional probability $P(A)$ for an event A. We now want to impose a condition, say B on occurrence of A; denote the new probability by $P(A|B)$; and want an answer to the question posed. Here is an example.

Example 2.11

Let us consider a machine with three components C_1, C_2, and C_3 such that malfunctioning of any one component causes the machine to stop working. Here are two scenarios and questions to answer.

Case 1. Suppose C_2 and C_3 have the same chance of malfunction. Suppose also that the probability of malfunctioning of C_1 is 2/5 of the chance of malfunctioning of C_2 or C_3. A question is to find the probability of each part to be responsible for malfunctioning of the machine.

Answer

Let E_1, E_2, and E_3 denote the events that C_1, C_2, and C_3 malfunction and cause the machine to stop working, respectively. The problem assumes that $P(E_2) = P(E_3)$ and $P(E_1) = (2/5)P(E_2)$. Since the sum of probabilities is 1, we have:

$$P(E_1) + P(E_2) + P(E_3) = \frac{2}{5}P(E_2) + 2P(E_2) = \frac{12}{5}P(E_2) = 1.$$

Thus,

$$P(E_2) = \frac{5}{12}.$$

As a result,

$$P(E_3) = \frac{5}{12} \text{ and } P(E_1) = \frac{1}{6}.$$

Case 2. Suppose that the component C_3 is in a perfect condition and has no chance to break down. On the other hand, the chance of malfunctioning of C_1 is 1/5 of the chance of C_2 to malfunction. Again, the question is to find the probability of each part to be responsible for malfunctioning of the machine.

Answer

As we did with Case 1, in this case,

$$P(E_3) = 0 \text{ and } \frac{1}{5}P(E_2) + P(E_2) = \frac{6}{5}P(E_2) = 1.$$

Thus,

$$P(E_1) = \frac{5}{6} \text{ and } P(E_2) = \frac{1}{6}.$$

Note 2.7

It should be noted that as the conditions changed, the probabilities of the machine to stop working due to malfunctioning of components C_1 and C_2 changed. In other words, for the events under consideration, the occurrence of one event may have influence on the occurrence of the other events (the idea of dependence that we will discuss below), and consequently, a change in probabilities of those events is expected. This leads us to the following definition.

Definition 2.10

Consider a probability space $(\Omega, \mathbb{S}, P)$ and $E_1, E_2 \in \mathbb{S}$. Then, **the conditional probability of E_1 given E_2**, denoted by $P(E_2 \mid E_1)$, called a **probability measure** on Ω, is defined as follows:

$$P(E_2 \mid E_1) = \begin{cases} \dfrac{P(E_1 \cap E_2)}{P(E_1)}, & E_1, E_2 \in \mathbb{S}, P(E_1) > 0, \\ \text{Not defined}, & E_1, E_2 \in \mathbb{S}, P(E_1) = 0. \end{cases} \tag{2.29}$$

Note 2.8

It should be noted that the given condition reduces the size of the sample space Ω, since, in general, $P(E|E) = 1$. Hence, the sample space Ω will be replaced by E and the new probability space will be $(\Omega, \mathbb{S}, P(\cdot|E))$. It is referred to as **the conditional probability space induced on** $(\Omega, \mathbb{S}, P)$ **given E**. Consequently, the conditional probability has the same properties as ordinary probability, except that it is restricted to a smaller space, called the **conditional sample space**.

Example 2.12

Suppose we roll a fair die. Let E be the event of occurrence of even numbers. Note that this condition, immediately, reduces the original sample space, $\Omega = \{1, 2, 3, 4, 5, 6\}$, with six elements, to a new and smaller sample space, say Ω_c, which is $\Omega_c = \{2,4,6\}$ with three elements. Hence, for instance, $P(\{2,4\} \mid E) = \dfrac{2}{3}$, $P(\{2,4,6\} \mid E) = \dfrac{3}{3} = 1$, and $P(\{5\} \mid E) = 0$.

Example 2.13

Suppose we roll a fair die. By fairness, we mean all sides have the same chance to face up and, hence, the probability of each side to occur is 1/6. Now let us consider two events A and B such that A consists of occurrences of 1 or 6, that is, $A = \{1,6\}$, and B consists of occurrences of 3, 4, or 6, that is, $B = \{3,4,6\}$. Let us calculate the conditional probabilities $P(A|B)$ and $P(A^c \mid B)$.

To calculate these conditional probabilities, note that $A \cap B = \{6\}$, $A^c = \{2,3,4,5\}$, and $A^c \cap B = \{3,4\}$. Thus,

$$P(A \mid B) = \frac{P(A \cap B)}{P(B)} = \frac{\frac{1}{6}}{\frac{1}{2}} = \frac{1}{3},$$

and

$$P(A^c \mid B) = \frac{P(A^c \cap B)}{P(B)} = \frac{\frac{4}{6}}{\frac{1}{2}} = \frac{2}{3}.$$

Observe that since $A = \{1, 6\}$ and $A^c = \{2,3,4,5\}$, we have $P(A) = \frac{2}{6} = \frac{1}{3}$ and $P(A^c) = \frac{4}{6} = \frac{2}{3}$. Thus, the condition B has no effect on the conditional probabilities of A and A^o given B.

Note 2.9

We leave it as an exercise to prove that the conditional probability satisfies all three axioms of Kolmogorov stated earlier. In other words, let us consider an event E with probability $P(E) > 0$. Then,

i. For any event A, $0 \leq P(A|E) \leq 1$.
ii. For a sample space Ω, $P(\Omega \mid E) = 1$.
iii. For any disjoint events $A_1, A_2, \ldots$, $P\left(U_{i=1}^{\infty} A_i \mid E\right) = \sum_{i=1}^{\infty} P(A_i \mid E)$.

Note 2.10

Definition of conditional probability given by (2.29) can equally be defined for $P(E_1) > 0$ as

$$P(E_1 \cap E_2) = P(E_1)P(E_2 \mid E_1), E_2 \in \mathbb{S}. \tag{2.30}$$

The relation (2.30) states that the probability that two events E_1 and E_2 occur at the same time needs the event E_1 to occur. When that happens, the chance of occurrence of E_2 is $P(E_2 \mid E_1)$.

Note 2.11

For two events E_1 and E_2, the conditional probabilities $P(E_2 \mid E_1)$ and $P(E_1 \mid E_2)$ are not the same.

2.4 INDEPENDENCE

The idea we presented in Example 2.15 is one of many cases where the occurrence of one event **does not influence** the occurrence of another event. This is what is referred to as the **independence** of two events.

Definition 2.11

Events A and B are called **independent** if and only if

$$P(\cap B) = P(A)P(B). \tag{2.31}$$

If (2.31) is not true, that is, if

$$P(A \cap B) \neq P(A)P(B), \tag{2.32}$$

then A and B are referred to as **not independent** or **dependent**.

Example 2.14

Let us go back to Example 2.13. Recall that $A = \{1,6\}$, $B = \{3,4,6\}$, and $A \cap B = \{6\}$. Now, based on fairness assumption of the die, we can see that

$$P(A) = \frac{2}{6} = \frac{1}{3}, P(B) = \frac{3}{6} = \frac{1}{2}, \text{ and } A \cap B = \frac{1}{6}.$$

Thus,

$$P(A \cap B) = \frac{1}{6} = P(A)P(B) = \frac{1}{3} \cdot \frac{1}{2}.$$

Therefore, A and B are independent.

Note 2.12

Definition 2.11 may be extended to more than two events, finitely many or infinitely many events. In that case, we will be talking about **mutual independence** or **pairwise independence**. For instance, independence of three events A, B, and C is defined as

$$P(A \cap B \cap C) = P(A)P(B)P(C). \tag{2.33}$$

More generally,

i. A **finite collection of events** $A_1, A_2, \ldots, A_n$ is called **mutually independent** if for each subcollection of $A_1, A_2, \ldots, A_n$, the following is true:

$$P(A_1 \cap A_2 ... \cap A_n) = P(A_1)P(A_2)...P(A_n); \qquad (2.34)$$

ii. An infinite collection of events $A_1, A_2,...$ is called **mutually independent** if every finite collection $A_1, A_2,..., A_n$, of $A_1, A_2,...$ is independent.

Example 2.15

Let us return to Case 2 in Example 2.11. Recall that we created three events E_1, E_2, and E_3 and assumed $P(E_3) = 0$. There, we found the probability of the other two events as $P(E_1) = \dfrac{5}{6}$ and $P(E_2) = \dfrac{1}{6}$. A consequence of $P(E_3) = 0$ is that malfunctioning of only one of the components C_1 or C_2 will cause the machine to stop functioning. Hence, let us denote this last event by E_4. In other words, $E_4 = E_1 \cup E_2$. Based on the assumption stated in Example 2.11, $P(E_1 \cap E_2) = 0$. Hence, from the inclusion–exclusion formula, $P(E_3 \mid E_4) = 0$ and $P(E_4) = P(E_1) + P(E_2)$. So, let us calculate the probability of stopping the machine by component C_1 or C_2 in Case 2, that is, when $P(E_3) = 0$.

Answer

To calculate the probabilities in question, we need to find $P(E_1 \mid E_4)$ and $P(E_2 \mid E_4)$. Note that $P(E_1 \cap E_2) = 0$ implies that E_1 and E_2 are independent. Note also that $E_1 \cap E_4 = E_1$ and $E_2 \cap E_4 = E_2$. That is, $E_4 \subset E_1$ and $E_4 \subset E_2$. Since $P(E_1) = \dfrac{5}{6}$ and $P(E_2) = \dfrac{1}{6}$, $P(E_4) = 1 > 0$. Thus, from (2.29), we have:

$$P(E_1 \mid E_4) = \frac{P(E_1 \cap E_4)}{P(E_4)} = \frac{P(E_1)}{P(E_4)} = \frac{5}{6}$$

and

$$P(E_2 \mid E_4) = \frac{P(E_2 \cap E_4)}{P(E_4)} = \frac{P(E_2)}{P(E_4)} = \frac{1}{6}.$$

Example 2.16

Let us consider the set of all families of two children in a particular location. Let us also consider the order of the two children in each family as the first born and the second born. Finally, we assume that chosen boys and girls have the same chance to be born. Thus, denoting a girl by G and a boy by B, the sample space for the choice of children of a family will be $\Omega = \{GG, GB, BG, BB\}$.

We now randomly select a family. We define three events E_1, E_2, and E_3 as follows:

$E_1 \equiv$ The first child is a boy.
$E_2 \equiv$ The first child is a girl.
$E_3 \equiv$ A child is a girl and the other is a boy.

Hence, from these definitions, we can see that

$$(E_1 \cap E_2) = \emptyset, E_1 \cap E_3 = \{BG\} \text{ and } E_2 \cap E_3 = \{GB\}$$

and from Ω, we have:

$$P(E_1) = \frac{1}{2}, P(E_2) = \frac{1}{2}, P(E_1 \cap E_2) = P(\emptyset) = 0$$

$$P(E_1 \cap E_3) = P(\{BG\}) = \frac{1}{4} \text{ and } P(E_1 \cap E_2) E_2 \cap E_3 = \{GB\} = \frac{1}{4}.$$

Thus, $P(E_1 \cap E_3) = P(E_1)P(E_3) = \frac{1}{4}$ implies that E_1 and E_3 are independent. Similarly, E_2 and E_3 are independent. However, since $P(E_1 \cap E_2) = 0 \neq P(E_1)P(E_2) = \frac{1}{4}$, E_1 and E_2 are not independent.

Example 2.17

Suppose a computer program selects digits randomly from 1 through 7, that is, from the set $\{1,2,3,4,5,6,7\}$. That is, each digit will be selected with equal chance and independently. We consider two events: (1) selection of two digits 3 and 5, denoted by E_1, and (2) selection of three digits 3, 4, and 9, denoted by E_2. Since digits are selected randomly, E_1 can occur in two different ways: (a) 3 is selected first and 5 after and (b) 5 is selected first and 3 after. Thus, $E_1 = \{(3,5)\}$ or $E_1 = \{(5,3)\}$, each with the same probability $\frac{1}{7} \cdot \frac{1}{7}$, that is, $P(E_1) = 2\left(\frac{1}{7}\right)\left(\frac{1}{7}\right)$. For the second case, E_2 can occur in six different ways: $\{3,4,7\}$, $\{3,7,4\}$, $\{4,3,7\}$, $\{4,7,3\}$, $\{7,3,4\}$, and $\{7,4,3\}$, each with the same probability $\frac{1}{7} \cdot \frac{1}{7} \cdot \frac{1}{7}$, that is, $P(E_2) = 6\left(\frac{1}{7}\right)\left(\frac{1}{7}\right)\left(\frac{1}{7}\right)$.

Remark 2.1

Two events each with positive probability cannot be both disjoint and independent.

Proof:
Recall that events A and B are disjoint if $A \cap B = \emptyset$. Thus, $P(A \cap B) = P(\emptyset) = 0$. If A and B were independent, then $P(A \cap B) = P(A)P(B) = 0$, which means either $P(A) = 0$ or $P(B) = 0$. Therefore, the two events can be considered as disjoint and independent if the probability of one of the events will be zero.

Example 2.18

Consider rolling a fair die and two events A and B consist of odd and even outcomes, respectively. That is, $A = \{1,3,5\}$ with $P(A) = \frac{1}{2}$ and $B = \{2,4,6\}$ with $P(B) = \frac{1}{2}$.

But, obviously, A and B are disjoint, that is, $A \cap B = \emptyset$ with $P(A \cap B) = P(\emptyset) = 0$ while $P(A)P(B) = \dfrac{1}{2} \cdot \dfrac{1}{2} = \dfrac{1}{4}$. Thus, $P(A \cap B) \neq P(A)P(B)$, and therefore, while A and B are disjoint, they are dependent.

Remark 2.2

If an event has a probability of zero, then it is independent of any other events because in that case the product of probabilities, that is, the probability of intersection of the two sets, will be zero.

Remark 2.3

If the probability of an event is restrictively between 0 and 1, then the event and its complement are dependent.

Proof:

Since $P(A) \neq 0, 1, P\left(A^c\right) \neq 0$ and $P(A)P\left(A^c\right) \neq 0$. However, $P\left(A \cap A^c\right) = P(\emptyset) = 0$.

Thus, $P\left(A \cap A^c\right) \neq P(A)P\left(A^c\right)$.

Example 2.19

Let A and B represent the falling of rain on a certain day in a certain location and not falling of rain on the same day in the same location, respectively. That is, $B = A^c$. Let us also suppose that on the same day in the same location, the chance of falling of rain is 60% and that of falling of no rain is 40%. Then, $0 < P(A) = 0.60 < 1$ and $P(B) = P\left(A^c\right) = 0.40 \neq 0$. Therefore, the falling of rain and the not falling of rain are the dependent events.

As a consequence of the definition of independence, we have the following.

Theorem 2.2

Events E_1 and E_2 with $P(E_1) > 0$ are independent if and only if

$$P(E_2) = P(E_2 \mid E_1). \tag{2.35}$$

Proof:

If E_1 and E_2 are independent, then according to (2.31), $P(E_1 \cap E_2) = P(E_1)P(E_2)$. Hence,

$$P(E_2) = \frac{P(E_1 \cap E_2)}{P(E_1)}. \tag{2.36}$$

But the right-hand side of (2.36) is based on (2.31), which is the conditional probability of E_2, given E_1, that is, $P(E_2 \mid E_1)$. Hence, (2.35). On the other hand, if (2.35) is the case, from the right-hand side of it, we will have:

$$P(E_2) = P(E_2 \mid E_1) = \frac{P(E_1 \cap E_2)}{P(E_1)}.$$

Thus, $P(E_1)P(E_2) = P(E_1 \cap E_2)$. This completes the proof of the theorem.

Example 2.20

Suppose the data show that the first call arrives at an office within 60 minutes (12.5-minute unit times) from the start of the office hours. Let A denote the waiting time of the first call within 30 minutes (six unit times) and B the waiting time within 10 minutes (two unit times). Finally, suppose we were told that the first call arrived within 30 minutes (six unit times) from the start of office hours.
Find the following probabilities:

(1) The call had arrived within 10 minutes (two unit times), the grace period for staff to arrive and, thus, missing the call;
(2) The call had arrived between 10 minutes (two unit times) and 30 minutes (six unit times) from the start of the office hours.

Answer

To answer the questions, note that the questions should be subjected to the conditional probabilities (1) $P(B \mid A)$ and (2) $P(B^c \mid A)$, where B^c is the complement of B with respect to A that had occurred. In terms of time units, $A = \{5, 10, 15, 20, 25, 30\}$ and $B = \{5, 10\}$. Once we know that A had occurred, we can ignore the waiting times larger than 30 minutes. Hence, to answer question (1), we have $P(B) = P(B \mid A) = \frac{2}{6} = \frac{1}{3}$.

Therefore, to answer question (2), we have $P(B^c) = P(B^c \mid A) = \frac{4}{6} = \frac{2}{3}$ or simply,

$$P(B^c) = 1 - P(B \mid A) = 1 - \frac{1}{3} = \frac{2}{3}.$$

2.5 THE LAW OF TOTAL PROBABILITY AND BAYES' THEOREM

Recall that a stipulation on conditional probability defined by (2.29) was that $P(E_1) > 0$. With this condition holding, we obtain further properties for conditional probability that are discussed below.

Remark 2.4 The Multiplicative Law

Let E_1 and E_2 be two events with conditional probabilities $P(E_2 \mid E_1)$ and $P(E_1 \mid E_2)$. Then,

$$P(E_1 \cap E_2) = P(E_2 \mid E_1) \cdot P(E_1) = P(E_1 \mid E_2) \cdot P(E_2). \tag{2.37}$$

Proof:
The proof follows from the definition of conditional probability given earlier.

Note 2.13

It is interesting to note that for this theorem, satisfying the stipulation of conditional probability is not necessary. Since if $P(E_1)$ or $P(E_2)$ or both are zero, (2.37) still holds and it simply says that E_1 and E_2 are disjoint, that is, $P(E_1 \cap E_2) = 0 = P(\emptyset)$, implying that $(E_1 \cap E_2) = \emptyset$.

Example 2.21

Suppose a package containing 35 tablets from a distribution center is delivered to a computer store. It has been reported to the store manager that four of the tablets may be virus-infected. To test the validity of the rumor, the manager of the store randomly picks two tablets from the package (one after the other without putting it back, once picked). The first pick results in an infected one. The question is, what is the probability that the second one is also infected?

To answer the question, let E_1 and E_2 represent the two picks, the first and the second, respectively. Now the probability of the first pick to be infected, individually, is $\frac{4}{35}$, and for the second, since one item has already been taken out (for the condition), three possible infected is left in the remaining 34 tablets. Hence, the probability of the second pick to be infected, that is, $P(E_2 \mid E_1)$, is $\frac{3}{34}$. Thus, the probability that the first is an infected pick and the second pick is also infected is as follows:

$$P(E_1 \cap E_2) = \frac{4}{35} \cdot \frac{3}{34} = (0.114)(0.0088) = 0.010 = 1\%.$$

A generalization of Remark 2.4 for a finite number of events is the following theorem.

Remark 2.5 A Generalization of the Multiplicative Law

For n events $E_1, E_2, \ldots, E_n$ with nonempty intersections, we have the following:

$$P(E_1 \cap E_2 \ldots \cap E_n) = P(E_1)P(E_2 \mid E_1)P(E_3 \mid E_1 \cap E_2)\ldots P(E_n \mid E_1 \cap E_2 \ldots \cap E_{n-1}). \tag{2.38}$$

This relation is a multiplication with conditional probability.

Proof:
The proof follows the mathematical induction using the definition of conditional probability given earlier. For instance, for $n = 2$, we have Remark 2.4. For $n = 3$, let

the events be E_1, E_2, and E_3. Hence, for all three events to occur, if E_1 has occurred, then E_2 must occur, as well. If both E_1 and E_2 have occurred, then E_3 must occur, too. We leave it as an exercise to complete the mathematical induction and consequently the proof.

Example 2.22

Let us return to Example 2.22. Consider the same package with the same number of tablets and the same number of virus infections reported. However, this time the manager decides to test three tablets from the package (one after the other without replacement, once picked). The question is, what is the probability that all three draws in a row are infected?

To answer the question, let E_1, E_2, and E_3 represent the three draws, in a row. Now we have the following probabilities:

a. Probability of the first draw $P(E_1) = \dfrac{4}{35}$.

b. Probability of the second draw after the first draw without replacement $P(E_2 \mid E_1) = \dfrac{3}{34}$.

c. Probability of the third draw after the first and second draws happened without replacements $= P(E_3 \mid E_1 \cap E_2) = \dfrac{2}{33}$.

Thus, from (2.38), the probability that all three tablets are infected is as follows:

$$P(E_1 \cap E_2 \cap E_3) = \frac{4}{35} \cdot \frac{3}{34} \cdot \frac{2}{33} = (0.114)(0.088)(0.061) = 0.0006 = 0.06\%.$$

We now state and prove an important theorem with one of the most important applications of the conditional probability. This theorem is particularly useful when direct computation of probability of an event is challenging. It will help us use conditional probability to break down the probability into simpler components.

Theorem 2.3 The Law of Total Probability

Let $\{A_j, j = 1, 2, \ldots\}$ be a finite or countably infinite partition of a sample space Ω of the probability space $(\Omega, \mathbb{S}, P)$ with $P(A_j) > 0, j = 1, 2, \ldots$ That is, as defined in an earlier section, $\{A_j, j = 1, 2, \ldots\}$ is a set of pairwise disjoint events (in other words, A_j and A_k are disjoint) whenever $(j \neq k)$ such that their union is the sample space Ω, that is, $U_{j=1}^{\infty} A_j = \Omega$. It is also assumed that each $A_j, j = 1, 2, \ldots$, is a measurable event. Then, for any set B from the same probability space $(\Omega, \mathbb{S}, P)$, for a finite number of events, say n (similar result for a countably infinite number of events), we have:

$$\text{For finite case,} \quad P(B) = \sum_{j=1}^{n} P(B|A_j) P(A_j) \qquad (2.39)$$

and

$$\text{For infinite case,} \quad P(B) = \sum_{j=1}^{\infty} P(B|A_j)P(A_j). \tag{2.40}$$

Proof:
Consider the infinite case. From a distributive law mentioned in an earlier section, we will have:

$$B = B \cap \Omega = \bigcup_{j=1}^{\infty} (B \cap A_j). \tag{2.41}$$

Thus, the events $B \cap A_1, B \cap A_2, \ldots$ being disjoint imply that

$$P(B) = \sum_{j=1}^{\infty} P(B \cap A_j) = \sum_{j=1}^{\infty} P(B \mid A_j)P(A_j) \tag{2.42}$$

completing the proof for this case. The proof for the finite case is similar.

Remark 2.6

Consider a particular case when $n = 2$. Let the two events be E and its complement E^c. Clearly, these events are a partition of the sample space Ω. Thus, for an event B in the probability space $(\Omega, \mathbb{S}, P)$, we have the following special case of the law of total probability:

$$P(B) = P(B|E)P(E) + P(B|E^c)P(E^c). \tag{2.43}$$

Note 2.14

Using the definition of conditional probability, (2.43) can be rewritten as follows:

$$P(B) = P(B \cap E) + P(B \cap E^c). \tag{2.44}$$

Example 2.23

Let us consider a course of probability in a department at a university in the United States. The course has three sections S_1, S_2, and S_3 with the number of enrollments in each section as 25, 30, and 35, respectively. From students' transcripts, we know that among students enrolled in each section, the number of students with GPA (grade point average) of 3.5 or higher out of 4.0 is 3, 5, and 8, respectively. A section is randomly selected, and a student is randomly chosen from that section.

We want to compute the probability that the chosen student has a GPA between 3.5 and 4.0, inclusive.

The summary of the given information is as follows:

Section S_1 has 25 students, of which 3 are with a GPA of 3.5 or higher.
Section S_2 has 30 students, of which 5 are with a GPA of 3.5 or higher.
Section S_3 has 35 students, of which 8 are with a GPA of 3.5 or higher.

To compute the probability in question, let E represent the event that the chosen student has a GPA between 3.5 and 4.0, inclusive. Thus, we have the following conditional probabilities:

$$P(E|S_1) = \frac{3}{25}, P(E|S_2) = \frac{5}{30}, \text{and } P(E|S_3) = \frac{8}{35}.$$

Assuming that the likelihood of each section to be selected is the same, we have:

$$P(S_1) = P(S_2) = P(S_3) = \frac{1}{3}.$$

Hence, from the law of total probability (2.39), we have:

$$P(E) = P(E|S_1)P(S_1) + P(E|S_2)P(S_2) + P(E|S_3)P(S_3)$$

$$= \frac{3}{25} \cdot \frac{1}{3} + \frac{5}{30} \cdot \frac{1}{3} + \frac{8}{35} \cdot \frac{1}{3} = \frac{1}{25} + \frac{1}{18} + \frac{8}{105} = 0.040 + 0.056 + 0.076$$

$$= 0.172 \approx 1.7\%.$$

The law of total probability leads to a result that allows us to flip conditional probabilities. This result is known as **Bayes' theorem** (named after Reverend Thomas Bayes, 1701 – 1761) that is stated and proved below. His work was published as a book in (1763) titled *An Essay towards Solving a Problem in the Doctrine of Chances*. The theorem allows us to compute the probability of the "cause" A, when we are aware of the observed "effect" B. In other words, Bayes' rule describes the probability of an event when there is prior information or conditions that could be related to the event. It is also useful to revise probabilities as new information reveals. For instance, as it is known, cancer is related to smoking. Hence, applying Bayes' rule, a smoker may be used for more accurate assessment of a smoker to have cancer compared to not having such information. It is interesting that Pierre Simon Laplace extended Bayes' theorem in his paper published in 1812 entitled *Thérie ananlytique des probabilités*.

Theorem 2.4 Bayes' Theorem or Formula or Law

As in the law of total probability, let $\{A_j, j = 1, 2, \ldots\}$ be a finite (when j stops at a point, say n) or a countably infinite partition of a sample space Ω of the probability space $(\Omega, \mathbb{S}, P)$. Let B be an event from the probability space with $P(B) > 0$. Then, for any A_j, we have the following:

$$\text{For finite case,} \quad P(A_j|B) = \frac{P(B|A_j)P(A_j)}{\sum_{k=1}^{n} P(B|A_k)P(A_k)}, j = 1,2,...,n \qquad (2.45)$$

and

$$\text{For infinite case,} \quad P(A_j|B) = \frac{P(B|A_j)P(A_j)}{\sum_{k=1}^{\infty} P(B|A_k)P(A_k)}, j = 1,2,... \qquad (2.46)$$

Proof:
From the definition of conditional probability for two events B and any of the n events $A_j, j = 1,...,n$, we have:

$$P(A_j|B) = \frac{P(B|A_j)P(A_j)}{P(B)}, j = 1,2,...,n.$$

The proof will not be completed using the law of total probability. Similar proof can be performed for the infinite case $j = 1,2,...$

Note 2.15

The $P(A_j)$ is referred to as the **a priori** or **prior probability of the cause**, and $P(A_j|B)$ is called the **a posteriori** or **posterior probability of the cause**.

Note 2.16

Bayes' theorem may be interpreted as the relationship between the probability of the hypothesis, say an event A, before obtaining an evidence, E, that is, $P(A)$, the **prior probability**, and the probability of the hypothesis after obtaining the evidence, that is, the conditional probability of A, given E, $P(A|E)$, the **posterior probability**, which is as follows:

$$P(A|E) = \frac{P(E|A)}{P(E)} \cdot P(A). \qquad (2.47)$$

The fraction $\frac{P(E|A)}{P(E)}$ is called the **likelihood ratio** and is the factor that relates the two events A and E. Relation (2.47) is Bayes' theorem for $n = 2$.

Example 2.24

Let us return to Example 2.23. Recall that there are three sections of the class denoted by S_1, S_2, and S_3. Let A represents the events that the student chosen had 3.5 or higher GPA. We want to answer the following questions:

Q1. What is the probability that the student chosen had 3.5 or higher GPA?
Q2. Suppose that the student chosen had 3.5 or higher GPA. What is the probability that she/he was selected from the
   Q2a. First section, S_1?
   Q2b. Second section, S_2?
   Q2c. Third section, S_3?

Answer to Question 1

The question is to find $P(B)$. To answer this question, we use the law of total probability, relation (2.39), as follows:

$$P(B) = P(B|S_1)P(S_1) + P(B|S_2)P(S_2) + P(B|S_3)P(S_3)$$

$$= \frac{3}{25} \cdot \frac{1}{3} + \frac{5}{30} \cdot \frac{1}{3} + \frac{8}{35} \cdot \frac{1}{3} = 0.172 = 17.2\%.$$

Answer to Question 2

Note that this is an example for finite case of Bayes' theorem when $n = 2$. Also, note that assuming that all three sections have the same chance to be selected (i.e., 1/3) is what we referred to as **a priori** probability, and probabilities in questions are the examples of "a posteriori" probabilities.

Thus, applying Bayes' theorem, relation (2.46), with each A and B, $n = 3$, and $j = 1, 2, 3$, we have the following conditional probabilities questioned in Q2a, Q2b, and Q2c, as follows:

$$\text{Q2a } P(S_1|A) = \frac{P(A|S_1)P(S_1)}{P(A|S_1)P(S_1) + P(A|S_2)P(S_2) + P(A|S_3)P(S_3)}$$

$$= \frac{\dfrac{3}{25} \cdot \dfrac{1}{3}}{\dfrac{3}{25} \cdot \dfrac{1}{3} + \dfrac{5}{30} \cdot \dfrac{1}{3} + \dfrac{8}{35} \cdot \dfrac{1}{3}} = \frac{126}{541} = 23\%.$$

$$\text{Q2b } P(S_2|A) = \frac{P(A|S_2)P(S_2)}{P(A|S_1)P(S_1) + P(A|S_2)P(S_2) + P(A|S_3)P(S_3)}$$

$$= \frac{\dfrac{5}{30} \cdot \dfrac{1}{3}}{\dfrac{3}{25} \cdot \dfrac{1}{3} + \dfrac{5}{30} \cdot \dfrac{1}{3} + \dfrac{8}{35} \cdot \dfrac{1}{3}} = \frac{175}{541} = 32\%.$$

$$\text{Q2c } P(S_3|A) = \frac{P(A|S_3)P(S_3)}{P(A|S_1)P(S_1)+P(A|S_2)P(S_2)+P(A|S_3)P(S_3)}$$

$$= \frac{\dfrac{8}{35}\cdot\dfrac{1}{3}}{\dfrac{3}{25}\cdot\dfrac{1}{3}+\dfrac{5}{30}\cdot\dfrac{1}{3}+\dfrac{8}{35}\cdot\dfrac{1}{3}} = \frac{336}{541} = 45\%.$$

Example 2.25

Let us return to Example 2.16. Consider a couple who has two children. Let us also consider order of the two children in each family as the first born and the second born. Finally, we assume that chosen boys and girls have the same chance to be born. Thus, denoting a girl by G and a boy by B, the sample space for the choice of a children of a family will be $\Omega = \{GG, GB, BG, BB\}$. Now, let us consider three events E_1, E_2, and E_3, which are defined as follows:

$E_1 =$ one of the children is a girl,
$E_2 =$ both children are girls,
$E_3 =$ the younger child is a girl.

We want to calculate the following probabilities:

1. Given that one of the children is a girl, what is the probability that both children are girls? That is, to find $P(E_2|E_1)$.
2. Given that the younger child is a girl, what is the probability that both children are girls? That is, to find $P(E_2|E_3)$.

Answer to 1

To calculate $P(E_2|E_1)$, note that $P(E_1)$ is the prior probability that the couple has at least one girl. Hence, E_1 is the complement of the event "both children are boys". Hence,

$$P(E_1) = 1 - P(\text{both children are boys}) = 1 - \frac{1}{4} = \frac{3}{4}.$$

Thus,

$$P(E_2|E_1) = \frac{P(E_1|E_2)\cdot P(E_2)}{P(E_1)} = \frac{1\cdot\dfrac{1}{4}}{\dfrac{3}{4}} = \frac{1}{3}.$$

Answer to 2

To calculate $P(E_2|E_1)$, we use Bayes' theorem. Hence,

$$P(E_2|E_3) = \frac{P(E_3|E_2) \cdot P(E_2)}{P(E_3)} = \frac{1 \cdot \dfrac{1}{4}}{\dfrac{1}{2}} = \frac{1}{2}.$$

Example 2.26

It is common practice in medicine to be concerned about the percent of accuracy of a medical diagnostic test, that is, the probability that the test result is correct. For instance, the process of using low-energy X-rays is to test human's breast for early detection of breast cancer, and it is referred to as **mammography** or **mastography.** In recent studies, for women in Canada, the United States, and Europe, this test is recommended every 2–3 years for ages between 40 and 74. Sometimes the person taking the test whose test results shows a positive, though the person actually does not have the disease. This is called **false positive.** Percent of this error in the United States is about 7%. The opposite is also possible; that is, the test may show that a woman does not have the disease while she have the disease. This error of the test is referred to as **false negatives.** Percent of this type of error is not easily found and is not available.

So, suppose an advertisement states that the result of mammogram test is 99% accurate. In other words, the test will result in 99% positive for those who have the disease and 99% negative who do not have the disease.

Now suppose that based on a research organization's survey, it is known that 7% of the population under consideration have the disease. Suppose a woman randomly selected (according to the rule of statistics that we will study in Chapter 4) from the population under consideration undergone the test with a positive result. What is the probability that she actually has the disease?

Let us denote the woman having the disease by D and the woman not having the disease by $\overline{D}$. We also denote the positive test result by T^+ and the negative test result by T^-. Then, using Bayes' theorem, we will have the following:

$$P(D|T^+) = \frac{P(T^+|D) \cdot P(D)}{P(T^+)} = \frac{P(T^+|D) \cdot P(D)}{P(T^+|D) \cdot P(D) + P(T^+|\overline{D}) \cdot P(\overline{D})}$$

$$= \frac{0.99 \cdot 0.07}{0.99 \cdot 0.07 + 0.01 \cdot 0.93} = 0.882 = 88.2\%.$$

The result states that if the test result is positive, there is an 88.2% chance that she has the disease. Thus, for example, if 1,000 women take the mammogram test, based on the research, it is expected to see 930 healthy women and 70 women having the disease. But out of 930 healthy women, there is a chance that (930). (0.01) = about 9 women's test results positive. On the other hand, from 70 women with the disease, there is a chance that (70).(0.99) = about 69 women's test results positive. Thus, from 9+69 = 78 test positives, only (78).(0.882) = about 69 women actually have the disease.

Example 2.27

Consider a mathematics department at a university that three sections of probability and statistics course, say A, B, and C, are taught by the same professor. By the end of a semester, 160 students complete this professor's sections, and it is known how they can be accessed after leaving this course. Of the 160 students that completed each section by the final examination, there were 45, 55, and 60 from sections A, B, and C, respectively.

After the final grades are posted, it is noted that the percent of the number of failures from each section is 10, 12, and 15, respectively. For the purpose of assessing the cause of failures, from 160 students that completed the course, a student is selected randomly. We want to calculate the following two probabilities: (1) the probability that the selected student is one of the failures and (2) the probability that the selected student is from section A, B, or C, knowing that he/she was one of the failures.

To calculate the inquired probabilities, let us denote the event of failing the course by F. Then, for the first question, we have to calculate $P(F)$. To do this, we apply the law of total probability, $P(F)$, as follows:

$$P(F) = P(F|A) \cdot P(A) + P(F|B) \cdot P(B) + P(F|C) \cdot P(C)$$

$$= (0.10)\left(\frac{45}{160}\right) + (0.12)\left(\frac{55}{160}\right) + (0.15)\left(\frac{60}{160}\right)$$

$$= 0.028125 + 0.04125 + 0.05625 = 0.126 = 12.6\%.$$

For the second question, we have to calculate the conditional probabilities $P(A|F)$, $P(B|F)$, and $P(C|F)$. To calculate $P(A|F)$, we apply Bayes' theorem and use the probability found for the first question. Hence, we have:

$$P(A|F) = \frac{P(F|A) \cdot P(A)}{P(F)}$$

$$= \frac{(0.10)\left(\frac{45}{160}\right)}{0.126} = \frac{0.02825}{0.126} = 0.223 = 22.3\%$$

$$P(B|F) = \frac{P(F|B) \cdot P(B)}{P(F)}$$

$$= \frac{(0.12)\left(\frac{55}{160}\right)}{0.126} = \frac{0.04125}{0.126} = 0.327 = 32.7\%$$

$$P(C|F) = \frac{P(F|C) \cdot P(C)}{P(F)}$$

$$= \frac{(0.15)\left(\frac{60}{160}\right)}{0.126} = \frac{0.05625}{0.126} = 0.45 = 45\%.$$

Of course, as expected, the sum of probabilities is 1.

EXERCISES

2.1. Consider an experiment of tossing two fair coins. Suppose you observe two outcomes of both coins. State the sample space.

2.2. Consider an experiment of observing the gender of a child of a family of three children.
   i. State the sample space.
   ii. Define an event as a family with only one boy. State the sample points for this event.

2.3. State the sample space when a coin and a die are tossed together.

2.4. A pair of two fair dice is rolled and the outcomes are observed.
   i. State the sample space for the experiment.
   ii. Define an event that represents the sum of two numbers is 5. Find the probability of this event.
   iii. Define an event that represents the absolute value of the difference is 1. Find the probability of this event.

2.5. To assess the quality of a certain product of a company, the controller of the company uses a sample of size 3. His choices for rating are as follows:
   i. Above the average,
   ii. Average,
   iii. Below the average.
      Answer the following questions:
   a. State the sample space of the rating.
   b. Define an event that he rates only one item above the average. Calculate the probability of this event.
   c. Define an event that he rates at least two items above the average. Calculate the probability of this event.
   d. Define an event that he rates at most one item above the average. Calculate the probability of this event.

2.6. What is the sample space when measuring the lifetime of a light bulb?

2.7. Consider a standard deck of 52 playing cards. Randomly select a card. What is the probability that the
   i. Card is a queen?
   ii. Card is a queen and a spade?
   iii. Card is a queen or a spade?
   iv. Card is a queen but not a spade?

2.8. Suppose there is an urn with 20 equal-sized balls. Of them, 12 are red, 5 are blue, and the rest are of different colors. If you randomly select a ball, what is the probability that the ball picked
   i. Is red?
   ii. Is neither red nor blue?

2.9. There are 100 mathematics majors and 200 engineering majors in a group of 500 students. If a student is randomly selected, answer the following questions:
   i. What is the probability that the student is not an engineering major?
   ii. What is the probability that the student is a mathematics major?

iii. What is the probability that the student is neither a mathematics major nor an engineering major?

2.10. A professor in a college posted the following grade distribution for one of his courses he teaches,

Grade	Probability
A	0.15
B	0.25
C	0.50
D	0.10

with "D" to be considered as failure grade. If a student is selected randomly, find the probability of the following events:
i. The student received a grade higher than a "C".
ii. The student does not fail the course.
iii. The student received a "C" or a "B".
iv. The student received at most a "B".

2.11. A bookstore carries some academic books, including mathematics and biology. The kind of books purchased are noted. Let us focus on the next customer who is at the register and is buying books. A denotes the event that the customer purchases a mathematics book, and B denotes the event that the person purchases a biology book. Based on the bookstore's history of selling subjects books, we have the following probabilities:

$$P(A) = 0.20, P(B) = 0.40, \text{ and } P(A \cap B) = 0.05.$$

i. Explain why $P(A) + P(B)$ is not equal to 1.
ii. What is the probability that the person will not buy a mathematics book?
iii. What is the probability that the person will buy at least a mathematics or a biology book?
iv. What is the probability that the person will buy a different type of book?

2.12. Consider a shipment of computers to an electronic store. Historical data indicates that a shipment of this type contains defects of different types. Here are the numbers: CPU only: 5%, RAM: 7%, and both CPU and RAM: 3%. To assess the defectiveness, a computer is selected randomly. Find the following probabilities regarding the selected computer:
i. Only one type of defect,
ii. Both types of defects,
iii. Either CPU defect or RAM defect,
iv. Neither type of defects.

2.13. Consider the results of a survey conducted by a researcher regarding the owning of the latest smart phones. This study was conducted using a group of high school students.

	Gender		
Owning the Latest Smart Phone	Male	Female	Total
Yes	52	30	82
No	28	48	76
Total	80	78	158

If a student is randomly selected, calculate the probabilities of the following events:
   i.   The student is a male.
   ii. The student does not own the latest smart phone.
   iii. The student is a female and owns the latest smart phone.
   iv. The student is a female or owns the latest smart phone.

2.14. A mechanic shop has three mechanics, identified as #1, #2, and #3. The percentage of all jobs assigned are 40, 35, and 25, to mechanics #1, #2, and #3, respectively. From the history of experience, the owner of the shop knows the probability of errors for each of the mechanics as 0.05, 0.03, and 0.03, for mechanics #1, #2, and #3, respectively. Answer the following questions:
   i. What is the probability that the service stations make a mistake with a given job?
   ii. If a customer complains about a mistake in a prior completed job, who might have done the above mistake?

2.15. A quality controller finds three types of defects with a particular product in a company. Based on her experiences, she expects percentages of these defects as 7, 8, and 9, for type 1, 2, and 3 defects, respectively. Also, the probability of defects for both types 1 and 2 is 0.03, for types 2 and 3 is 0.04, and for types 1 and 3 is 0.02. In addition, the quality controller finds that all the three types of defects occur 1% of the time. Using this information, answer the following questions:
   i.   What is the probability that the quality controller experiences any type of defects?
   ii. What is the probability that the quality controller experiences only one type of defect?
   iii. What is the probability that the quality controller experiences only two types of defects?
   iv. What is the probability that the quality controller does not experience any type of defects?
   v.   What is the probability that the quality controller observes type 1 and 2 defects but not type 3 defect?

2.16. In an experiment of rolling two fair dice simultaneously, what is the probability that both show 5?

2.17. At a small college, 10% of the students are mathematics majors. Answer the following questions:
  i. If a student is selected at random from this college, what is the probability that the student is not a mathematics major?
  ii. If three students are randomly selected, what is the probability that all three are not mathematics majors?

2.18. According to the estimation of a particular insurance company, in a particular year, 1 out of each 300 houses can experience fire at some point of time. If there are eight houses with insurance protection for fire, what is the probability that the insurance company needs to pay for all the eight houses for claims regarding the fire?

2.19. Suppose Steve and his sister have motorcycles. The probability that Steve's cycle has a mechanical problem is found to be 0.3, whereas the probability that his sister's cycle has a mechanical problem is 0.4. Answer the following questions:
  i. What is the probability that both of motorcycles have mechanical problems?
  ii. What is the probability that either Steve's or his sister's motorcycle has a mechanical problem?
  iii. What is the probability that only one of the motorcycles has a mechanical problem?
  iv. What is the probability that both of their motorcycles do not have mechanical problems?

2.20. Consider two events A and B with $P(A|B) = 0.6$ and $P(B) = 0.5$. Calculate $P(A \cap B)$.

2.21. It is known that 10% of new computers arriving at a computer store are defective of some type. Answer the following questions:
  i. Two computers of some new arrivals have been selected consecutively to check for defectiveness without replacement. What is the probability that the first computer is non-defective and the second is defective?
  ii. If three computers are selected in the same way as in (i), what is the probability that the first and second are defective and third is non-defective?

2.22. A survey was conducted about the three hobbies, namely, reading books, watching movies, and listening to music. The following statistics were calculated using a collection of 1,000 college students. Of them, 13%, 22%, and 36% read books, watch movies, and listen to music, respectively. Also, 7% of the students read books and watch movies; 8% read books and listen to music, and 12% watch movies and listen to music. Finally, 4% of the students do all the three hobbies. Using these pieces of information, answer the following questions:
  i. If a student is selected and he watches movies, what is the probability that he reads books, also?

     ii.   If a student is selected and she watches movies and listens to music, what is the probability that she also reads books?

2.23. Let A be the event that a person drives an automatic-geered car, and B be the event that the person drives a manual-geered car. Suppose $P(A) = 0.5$, $P(B) = 0.4$, and $P(A \cap B) = 0.25$.

     i.   Interpret the event $B|A$ and calculate its probability.

     ii.   Interpret the event $A|B$ and calculate its probability.

2.24. A group of students comprise 30% biology majors, 25% math majors, and 45% engineering majors. Suppose that GPA of 30% of mathematics and biology majors is above 3.0 out of 4.0 points. For mathematics and engineering students, the numbers are 40% and 30%, respectively. Answer the following questions:

     i.   If a student is randomly selected, what is the probability that the GPA is over 3.0?

     ii.   Given that the student has a GPA over 3.0, what is the probability that the student is a mathematics major?

2.25. There are three brands of soda, say A, B, and C, available in a store. Assume that 40%, 35%, and 25% of the soda drinkers prefer brands A, B, and C, respectively. When a brand A soda drinker is considered, only 30% of them use ice cubes to drink the soda. For brands B and C, these percentages are 60% and 50%, respectively. Using this information, answer the following questions:

     i.   What is the probability that a randomly selected soda drinker is a brand A drinker who uses ice cubes to drink soda?

     ii.   What is the probability that a soda drinker drinks soda with ice cubes?

     iii.   If a soda drinker drinks with ice, what is the probability that the drinker is of brand A?

     iv.   If a soda drinker drinks with ice, what is the probability that the drinker is of brand C?

2.26. Let E_1 and E_2 be two events in the space $\mathbb{S}$. The following are some properties of the probability measure P. Prove the following:

     If E_1 and E_2 are two events in $\mathbb{S}$ and P is a probability measure, then P is a monotonic function, that is,

     i. $P(E_1) \leq P(E_2)$, if $E_1 \subset E_2$

     ii. $P(E_1^c) = 1 - P(E_1)$

     iii. $P(E_1) + P(E_2) = P(E_1 \cup E_2) + P(E_1 \cap E_2)$

     iv. $P(E_1 \cup E_2) = P(E_1) + P(E_2) - P(E_1 \cap E_2)$

2.27. Prove the distributive laws of probability.

     i.   Let $E_1, E_2, \ldots$ be any events. Then, we have the following two properties:

        a. $\left(\bigcup_j E_j \right) \cap A = \bigcup_j (E_j \cap A)$

        b. $\left(\bigcap_j E_j \right) \cup A = \bigcap_j (E_j \cup A)$

ii. For a finite number of events $E_j, j = 1, 2, \ldots, n$, prove the following:

  a. $\left(\bigcup\limits_{j=1}^{n} E_j \right) \cap A = \bigcup\limits_{j=1}^{n} \left(E_j \cap A \right).$

  b. $\left(\bigcap\limits_{j=1}^{n} E_j \right) \cup A = \bigcap\limits_{j=1}^{n} \left(E_j \cup A \right).$

iii. For an infinite number of events $E_j, j = 1, 2, \ldots, n$, prove the following:

  a. $\left(\bigcup\limits_{j=1}^{\infty} E_j \right) \cap A = \bigcup\limits_{j=1}^{\infty} \left(E_j \cap A \right)$

  b. $\left(\bigcap\limits_{j=1}^{\infty} E_j \right) \cup A = \bigcap\limits_{j=1}^{\infty} \left(E_j \cup A \right)$

2.28. Prove the generalization of the multiplicative law: For n events $E_1, E_2, \ldots, E_n$ with nonempty intersections, prove the multiplication law with conditional probability:

$$P(E_1 \cap E_2 \ldots \cap E_n) = P(E_1) P(E_2 | E_1) P(E_3 | E_1 \cap E_2) \ldots P(E_n | E_1 \cap E_2 \ldots \cap E_{n-1}).$$

3 Random Variables and Probability Distribution Functions

3.1 INTRODUCTION

Since the sample space is a part of the probability space, we now seek to quantify the sample space whose elements may be qualitative or quantitative or mixed. To do must define a function, called a **random variable**, that maps a sample space to a **set of numbers** or **sets**. However, we should be reminded that there are two types of variables:

i. **Quantitative** or **numerical** such as height, weight, student population at a university, or number of children in a family,
ii. **Qualitative** or **categorical** such as colors, names, and human characteristics and behaviors. A categorical data is a grouping of data into discrete groups, such as age group and dress sizes.

A quantitative data set often appears in the theory of probability, while a qualitative data set appears more in statistics when data gathering and data analysis are under consideration. Thus, a sample space may contain a purely qualitative or quantitative outcome or of course a mixture of both. For instance, the outcomes of an experiment of tossing a coin are head and tail that are qualitative, while the height that the coin reaches in air when it is flipped is a quantitative measure. Thus, a sample space containing outcomes of either of the two will require a different type of a random variable.

Definition 3.1

A **random variable** is a measurable function from a set of possible outcomes (quantitative or qualitative), that is, the sample space, to a measurable space like the sets of real numbers.

In other words, a random variable is the value of a measurement associated with a random experiment, such as the time a customer has to wait to check out in a busy grocery store or the number of cars stopped at a traffic red light. That is, random variables may be used to define events.

Symbolically, let X and Ω denote a random variable and the sample space, respectively. If $\omega \in \Omega$, then $X(\omega) = x$, where x is a real number (finite or countably infinite). The **domain** of X is the sample space, and its **range** is the set of its possible values.

The values of a random variable could be **discrete** such as integers (finite or countably infinite), **continuous** such as intervals on the real line, or **mixed** (discrete and continuous). For instance, values of the random variable could be an interval as images of some events and natural numbers for others. The sets could be crisp or fuzzy. We will discuss these cases in separate subsections. There are properties of random variables that are common to all kinds. In such cases, we avoid the adjective and just say "random variable".

Note 3.1

The range of a random variable is determined before the random experiment is performed. But the value of the random variable is known after performance of the experiment.

Definition 3.2

By a **discrete random variable**, it is meant a **function** (or a **mapping**), say X, from a sample space Ω, into the set of real numbers. Symbolically, if $\omega \in \Omega$, then $X(\omega) = x$, where x is a real number.

Note 3.2

If a random variable can take only a finite number of discrete values, then it is discrete.

Example 3.1

In tossing a coin, the sample space is $\Omega = \{H, T\}$, where H and T, the outcomes of the random experiment of tossing, stand for "heads" and "tails", respectively. Then, a discrete random variable X may be defined as 1 and 0 (or other digits) for H and T, respectively. Symbolically, we have:

$$X : \begin{cases} H \rightarrow 1, \\ T \rightarrow 0. \end{cases}$$

Example 3.2

Similar to Example 3.1, let us consider tossing two coins. Then, the set of possible outcomes, that is, the sample space $\Omega = \{(H,H),(H,T),(T,H),(T,T)\}$. Hence, the elements of the sample space this time is a set of ordered pairs. Thus, we define a discrete random variable X such that it assigns a number to each of the ordered pair in Ω as follows:

$$X : \begin{cases} (H,H) \to 1, \\ (H,T) \to 2, \\ (T,H) \to 3, \\ (T,T) \to 0. \end{cases}$$

Example 3.3

A fair die is a small cube with a natural number from 1 to 6 engraved on each side equally spaced without repetition. The fairness means that a die is made so that its weight is equally spread and, thus, all six faces are equally likely to face when rolled. So, if rolled, the set of numbers $\{1,2,3,4,5,6\}$ is the sample space of this experiment.

Now let's consider the experiment of rolling a pair of fair dice. Then, the set of possible outcomes, that is, the sample space Ω, contains 36 pairs,

$$\Omega = \begin{cases} (1,1),(1,2),(1,3),(1,4),(1,5),(1,6),(2,1),(2,2),(2,3),(2,4),(2,5),(2,6), \\ (3,1),(3,2),(3,3),(3,4),(3,5),(3,6),(4,1),(4,2),(4,3),(4,4),(4,5),(4,6), \\ (5,1),(5,2),(5,3),(5,4),(5,5),(5,6),(6,1),(6,2),(6,3),(6,4),(6,5),(6,6) \end{cases}.$$

In each pair, the first element represents the number appearing on one die and the second appearing on the other. We can define a discrete random variable X such that it assigns numbers 1 through 36 to the ordered pairs in Ω from the beginning to the end, respectively, as follows:

$(1,1) \to 1,$	$(1,2) \to 2,$	$(1,3) \to 3,$	$(1,4) \to 4,$	$(1,5) \to 5,$	$(1,6) \to 6,$
$(2,1) \to 7,$	$(2,2) \to 8,$	$(2,3) \to 9,$	$(2,4) \to 10,$	$(2,5) \to 11,$	$(2,6) \to 12,$
$(3,1) \to 13,$	$(3,2) \to 14,$	$(3,3) \to 15,$	$(3,4) \to 16,$	$(3,5) \to 17,$	$(3,6) \to 18,$
$(4,1) \to 19,$	$(4,2) \to 20,$	$(4,3) \to 21,$	$(4,4) \to 22,$	$(4,5) \to 23,$	$(4,6) \to 24,$
$(5,1) \to 25,$	$(5,2) \to 26,$	$(5,3) \to 27,$	$(5,4) \to 28,$	$(5,5) \to 29,$	$(5,6) \to 30,$
$(6,1) \to 31,$	$(6,2) \to 32,$	$(6,3) \to 33,$	$(6,4) \to 34,$	$(6,5) \to 35,$	$(6,6) \to 36,$

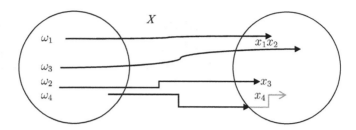

FIGURE 3.1   Discrete random variable.

Example 3.4

Consider hair colors of people living in Houston, Texas, the United States, as black, blond, brown, and red. Thus, the sample space, Ω, consists of four elements that are the aforementioned four colors. Let us denote these elements by $\omega_1, \omega_2, \omega_3$, and ω_4, for black, blonde, brown, and red, respectively. We now define the function X to map $\omega_1, \omega_2, \omega_3$, and ω_4 to x_1, x_3, x_2, and x_4, respectively. That is, $\omega_1 \rightarrow x_1, \omega_2 \rightarrow x_3, \omega_3 \rightarrow x_2$, and $\omega_4 \rightarrow x_4$. Hence, x_1, x_3, x_2, and x_4 constitute the range of X. X is a finite discrete random variable. Figure 3.1 shows what X does.

Example 3.5

Consider the following sample space of pairs:

$$\Omega = \{(-1,4),(0,4),(1,2),(1,9),(5,-1),(6,3),(7,-2)\}.$$

We may now define a random variable X as the absolute value of the difference of two numbers, that is, $X = \{|i - j|; (i, j) \in \Omega\}$. Under this definition of X, the mapping will be as follows:

$$X : \begin{cases} (-1,4) & \rightarrow & 5, \\ (0,4) & \rightarrow & 4, \\ (1,2) & \rightarrow & 1, \\ (1,9) & \rightarrow & 8, \\ (5,-1) & \rightarrow & 6, \\ (6,3) & \rightarrow & 3, \\ (7,-2) & \rightarrow & 9. \end{cases}$$

Figure 3.2 shows the values of the random variable X. The numbers 1 through 7 on the horizontal axis show the corresponding elements (the ordered pairs) of the sample space Ω. The numbers 1 through 9 on the vertical axis constitute a set containing the images of the ordered pairs, which are the values of X. The values of X are connected by lines to better illustrate the plot. The vertical axes could be extended, but it is not necessary since the maximum number needed is 9.

It should be recalled that in Chapter 1, we defined a characteristic or indicator function of a set A, denoted by I_A as a function that identifies an element

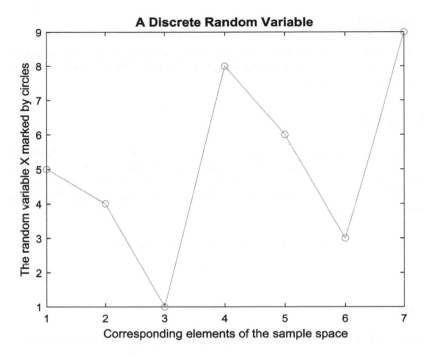

FIGURE 3.2 Discrete random variable, Example 3.5.

of I, say i, as either in A or not in A. In other words, an indicator random variable is one associated with the occurrence of an event. Symbolically, we present it in the following definition.

Definition 3.3

Let the set A be an event from a sample space Ω. Then, for each element ω of Ω, the random variable I_A is called an **indicator function** or **indicator random variable** of A if for $\omega \in \Omega$, we have:

$$I_A(\omega) = \begin{cases} 1, \text{ for } \omega \in A, \\ 0, \text{ for } \omega \notin A. \end{cases} \tag{3.1}$$

That is, the values of the random variable I_A are 1 or 0, depending upon whether the event A occurred or not occurred, respectively.

Note 3.3

$$\forall \omega \in \Omega, \begin{cases} I_\Omega(\omega) = 1, \\ I_\varnothing(\omega) = 0. \end{cases} \tag{3.2}$$

Example 3.6

A student is taking a test with "true" or "false" questions. Suppose she answered a question as "true". The event that her answer is, in fact, true can be represented by an indicator function.

Example 3.7

Let the set A represent the population of a particular state in the United States, say $A = \{a_1, a_2, \ldots, a_n\}$, where $a_i, i = 1, 2, \ldots, n$ are the people living in the state and n is the total population of the state. We define I_A as the indicator random variable representing the event that all persons in A have the same birthday. Then, I_A, in this case, is defined as:

$$I_A = \begin{cases} 1, & \text{if all } a_1, a_2, \ldots, a_n \text{ have the same birthday,} \\ 0, & \text{otherwise.} \end{cases}$$

Definition 3.4

A random variable X is called **continuous** if it takes all possible values in its range.

Note 3.4

A continuous random variable is, as we will see later, usually, defined over an interval on the number line or a union of disjoint intervals. The probability of a continuous random variable will be defined by the area under the curve. The probability distribution of a continuous random variable can be represented by a smooth curve over an interval.

Note 3.5

When the outcomes of a chance experiment can assume real values (not necessarily integers or rational), the sample space, Ω, is a **continuous sample space**; that is, Ω is the entire real number set, $\mathbb{R}$, or a subset of it (an interval).

Note 3.6

Since the set consisting of all subsets of $\mathbb{R}$ is extremely large, it will be impossible to assign probabilities to all. It has been shown in the theory of probability that a smaller set, say B, may be chosen that contains all events of our interest. In this case, B is referred to as the **Borel field**, similar to what was defined for discrete random variables. The triplet (Ω, B, P) is called the **probability space**, where P is the probability (measure) of events.

Example 3.8

Suppose the temperatures in Houston in the month of July in the past many years have always been between 95°F and 105°F. This means that the temperature can take any value between the ranges 94.5°F and 105.5°F. When we say that the temperature is 100°F, it means that the temperature lies somewhere between 99.5°F and 100.5°F. This is an example of a continuous random variable.

3.2 DISCRETE PROBABILITY DISTRIBUTION (MASS) FUNCTIONS (PMF)

A discrete random variable assigns discrete values to the sample points. As a random variable represents an event, we can assign probabilities to its values. A function describing how probabilities are distributed across the possible values of the random variable is referred to as the **probability distribution of the random variable**. In other words, in a random experiment, we might want to look at situations from which we can observe the predictable patterns and can use them as models. The patterns of such models are called the **probability distributions**. As a random variable may be discrete or continuous, so is its distribution. In this section, we discuss the discrete case.

Note 3.7

A pmf of a random variable can be represented as a mathematical formula, table, or a graph.

Definition 3.5

Let X be a discrete random variable (finite or countably infinite) defined on the sample space Ω with range $R = \{x_1, x_2, ..., x_n\}$, where n is a nonnegative integer. Then, p_X defined by

$$p_X = P\{X = x_k\}, \quad k = 1, 2, ..., n, \tag{3.3}$$

is called the **probability mass function** (*pmf*) or **discrete probability distribution** of X if

i. For each value of X, say x,

$$0 \leq p_x \leq 1 \tag{3.4}$$

and
ii.

$$\sum_{x \in R} p_x = 1. \tag{3.5}$$

Note 3.8

Although the values of a discrete random variable are integers, the values of pmf are real numbers.

Note 3.9

In general, to specify the probability distribution (mass) function of a random variable X, discrete or continuous, denoted by $\rho_X(x)$ and $f_X(x)$, respectively, X could be denoted by

$$X \sim \rho_X(x) \text{ and } X \sim f_X(x), \tag{3.6}$$

respectively.

Example 3.9

Let us consider a manufacturer whose products are classified as excellent, acceptable, and defective. We define the random variable X representing these three cases and map them to 5, 3, and 1, respectively. Data for the past three years of the production shows the chance for each one of the three cases as 60%, 30%, and 10%, respectively. Thus,

$$\Omega = \left\{ \text{defective, acceptable, excellent} \right\} \xrightarrow{\ x\ } \{1,3,5\},$$

that is,

$$X : \begin{cases} \text{defective} \rightarrow 1, \\ \text{acceptable} \rightarrow 3, \\ \text{excellent} \rightarrow 5. \end{cases}$$

In other words, the image, R, of X is {1, 3, 5}. The pmf, in this case, in a table format, is as follows:

x	1	3	5
p_x	0.1	0.3	0.6

Of course, the sum of the probabilities on the second row is $0.1 + 0.3 + 0.6 = 1$.

Example 3.10

Consider an experiment consisting of rolling two fair dice. Let the random variable X represent the sum of the number of points that appear on both dice after each roll. What is the pmf for X?

Answer

In Table 3.1, we list the possible outcomes of the experiment as ordered pair (x_1, x_2), where x_1 is the number of points on one die and x_2 on the other die.

Entries of Table 3.2 show the sum of components of elements of Table 3.1.

Probability of each entry of Table 3.1 and, hence, of Table 3.2 is 1/36. In Table 3.3, we list the sums, their frequencies, and the pmf.

TABLE 3.1

Possible Outcomes of Rolling Two Dice

			x_1			
x_2	1	2	3	4	5	6
1	(1,1)	(1,2)	(1,3)	(1,4)	(1,5)	(1,6)
2	(2,1)	(2,2)	(2,3)	(2,4)	(2,5)	(2,6)
3	(3,1)	(3,2)	(3,3)	(3,4)	(3,5)	(3,6)
4	(4,1)	(4,2)	(4,3)	(4,4)	(4,5)	(4,6)
5	(5,1)	(5,2)	(5,3)	(5,4)	(5,5)	(5,6)
6	(6,1)	(6,2)	(6,3)	(6,4)	(6,5)	(6,6)

TABLE 3.2

Sum of Numbers Appear on Rolling Two Dice

			x_1			
x_2	1	2	3	4	5	6
1	2	3	4	5	6	7
2	3	4	5	6	7	8
3	4	5	6	7	8	9
4	5	6	7	8	9	10
5	6	7	8	9	10	11
6	7	8	9	10	11	12

TABLE 3.3

Summary of Information from Table 3.2

Sum	2	3	4	5	6	7	8	9	10	11	12
Frequency	1	2	3	4	5	6	5	4	3	2	1
Probability	1/36	2/36	3/36	4/36	5/36	6/36	5/36	4/36	3/36	2/36	1/36

TABLE 3.4

Tabular Representation of pmf and cdf

x	x_1	x_2	$\cdots$	x_n	$\cdots$
p_x	p_1	p_2	$\cdots$	p_n	$\cdots$
$F_X(x)$	p_1	$p_1 + p_2$	$\cdots$	$p_1 + p_2 + \cdots + p_n$	$\cdots$

Definition 3.6 (Cumulative Probability Mass Function, cmf)

Let X be a discrete random variable, x a real number from the infinite interval $(-\infty, x]$, and p_X the pmf of X. The **cumulative mass (or distribution) function, cmf (or cdf) of X**, denoted by $F_X(x)$, is defined by

$$F_X(x) = P(X \le x) = \sum_{k=-\infty}^{x} p_k. \tag{3.7}$$

Note 3.10

Based on the axioms of probability, for all x, $p_X > 0$ and $\sum_x p_x = 1$.

Note 3.11

Representing the pmf and cdf combined using a table with values of the random variable X, as $x_1, x_2, \ldots, x_n, \ldots$, and corresponding pmf values as $p_1, p_2, \ldots, p_n, \ldots$, we will have the following.

3.3   MOMENTS OF A DISCRETE RANDOM VARIABLE

3.3.1   ARITHMETIC AVERAGE

Let us start this section with a very basic relation that we all use it in our daily lives regardless of our education levels, like arithmetic mean.

Definition 3.7

The **arithmetic mean** of n numbers $x_1, x_2, \ldots, x_n$ denoted by $\bar{x}$ is defined as:

$$\bar{x} = \frac{x_1 + x_2 + \cdots x_n}{n} \tag{3.8}$$

Relation (3.8) can also be written as:

$$\bar{x} = \frac{1}{n}x_1 + \frac{1}{n}x_2 + \cdots + \frac{1}{n}x_n \tag{3.9}$$

or

$$\bar{x} = \frac{1}{n} \sum_{k=1}^{n} x_k. \tag{3.10}$$

The coefficient $\frac{1}{n}$ in (3.8)–(3.10), for each $x_k, k = 1,2,...,n$, is referred to as the **weight**, which in this case is the same for each $x_k, k = 1,2,...,n$. If the weight varies for each $x_k, k = 1,2,...,n$, then (3.8) is called the **weighted average** of n numbers $x_1, x_2,...,x_n$. In this case, (3.8) and (3.9), respectively, become

$$\bar{x} = \frac{1}{k_1} x_1 + \frac{1}{k_2} x_2 + \cdots + \frac{1}{k_n} x_n \tag{3.11}$$

and

$$\bar{x} = \sum_{i-1}^{n} \frac{1}{k_i} x_i. \tag{3.12}$$

Example 3.11

The syllabus for a probability class indicates the following:

1. There will be five sets of homework (HW) assignments with equal weight of 15 percent.
2. Four tests with weights as follows:

Test No.	Weight (%)
1 (before Midterm Exam)	15
2 (Midterm Exam)	20
3 (after Midterm Exam)	15
4 (Final Exam)	25

3. Attendance has a weight of 10%.

The final score is calculated based on the weighted average of HW, test scores, and attendance. The final grade, denoted by G, is given letter grade according to the following:

$$\begin{cases} A, & G \geq 90, \\ B, & 80 \geq G < 90, \\ C, & 70 \geq G < 80, \\ D, & 60 \geq G < 70, \\ F, & G < 60. \end{cases}$$

A student receives her scores on the three categories as follows:

1. HW No.	Score (%)	2. Test No.	Score (%)	3. Attendance (%)
1	91	1 (before Midterm Exam)	80	95
2	85	2 (Midterm Exam)	86	
3	46	3 (after Midterm Exam)	58	
4	100	4 (Final Exam)	89	

Calculate the student's

i. Arithmetic HW average,
ii. Weighted test scores average,
iii. Final letter grade.

Answers

i. There are four HW assignments with equal weight. Thus, the arithmetic average, denoted by $\overline{HW}$, from (3.8) is:

$$\overline{HW} = \frac{91+85+46+100}{4} = 80.5.$$

ii. From the student's test scores, the weight for each test, and (3.12), we have:

Test No.	Test Score	Weight (%)	Weighted Test Score (%)
1 (before Midterm Exam)	80	15	12.00
2 (Midterm Exam)	86	20	17.20
3 (after Midterm Exam)	58	15	08.70
4 (Final Exam)	89	25	22.25
Weighted average of test score (%)			60.15

iii. From the information above, we have the following:
    HW having a weight of 15%, the weighted HW average, denoted by $W\overline{HW}$, is:

$$W\overline{HW} = (80.5)(0.15) = 12.075.$$

Also, for attendance, we have 10% of the student's score, that is, $(95)(0.10) = 9.5$. Thus, we can now calculate the final letter grade as the sum of the three weighted scores as follows:

1. (%) HW Score Weighted Average	2. (%) Test Score Weighted Average	3. (%) Attendance Weighted Score	(%) Total Final Weighted Score	Final Letter Grade
12.075	60.15	9.5	81.725	B

3.3.2 MOMENTS OF A DISCRETE RANDOM VARIABLE

We now extend the idea of weighted average of numbers to random variables. Recall that a random variable, essentially, is the numerical measure of the outcomes of a chance experiment. Thus, it should be expected to have the weighted average for its values. However, the weight for each value is the probability of that value. Here is the formal definition.

Definition 3.8

Let X be a discrete random variable defined on a sample space Ω with its pmf, denoted by $p_X(x)$. The **expected value** or **mean** or **expectation** of X, denoted by $E(X)$, is defined as:

$$E(X) = \sum_{x \in \Omega} p_X(x) \cdot x, \tag{3.13}$$

provided the sum converges absolutely; otherwise, X does not have an expected value. In other words, **the mean of a random variable X is a weighted average** and its weight of the possible values of X, say x, is its probability $p_X(x)$.

In case Ω is finite, say n elements, or infinite, (3.13) can be represented as:

$$E(X) = \sum_{i=1}^{n} x_i p_X(x_i) \text{ and } E(X) = \sum_{i=1}^{\infty} x_i p_X(x), \tag{3.14}$$

respectively.

Example 3.12

Consider an experiment consisting of rolling two fair dice. Let the random variable X represent the sum of the number of points that appear on both dice after each rolling. What is the expected value of X?

Answer

Recall Example 3.9 and Tables 3.1–3.4, where we listed the possible outcomes of the experiment. In Table 3.4, we listed the values of the random variable and its pmf. Thus, the expected value of X, $E(X)$, is:

$$E(X) = \left(\frac{1}{36}\right)(2) + \left(\frac{2}{36}\right)(3) + \left(\frac{3}{36}\right)(4) + \left(\frac{4}{36}\right)(5) + \left(\frac{5}{36}\right)(6) + \left(\frac{6}{36}\right)(7)$$

$$+ \left(\frac{5}{36}\right)(8) + \left(\frac{4}{36}\right)(9) + \left(\frac{3}{36}\right)(10) + \left(\frac{2}{36}\right)(11) + \left(\frac{1}{36}\right)(12)$$

$$= \frac{2}{36} + \frac{6}{36} + \frac{12}{36} + \frac{20}{36} + \frac{30}{36} + \frac{42}{36} + \frac{40}{36} + \frac{36}{36} + \frac{30}{36} + \frac{22}{36} + \frac{12}{36} = \frac{252}{36} = 7.$$

So, we can summarize these values in a new row of Table 3.5 as follows.

TABLE 3.5

Expected Value of *X* in Example 3.12

Sum	2	3	4	5	6	7	8	9	10	11	12
Frequency	1	2	3	4	5	6	5	4	3	2	1
pmf	1/36	2/36	3/36	4/36	5/36	6/36	5/36	4/36	3/36	2/36	1/36

$E(X) = 7 = 252/36 = $ 2/36 + 6/36 + 12/36 + 20/36 + 30/36 + 42/36+ 40/36+36/36 + 30/36 + 22/36 + 12/36

Example 3.13

An investor in the stock market is trying to choose a company between two available ones. His anticipations are as follows:

Company No. 1:
    i. 25% chance of profit of $80,000.
    ii. 40% chance of breakeven.
    iii. 35% chance of loss of $25,000.

Company No. 2:
    i. 20% chance of profit of $90,000.
    ii. 65% chance of breakeven.
    iii. 15% chance of loss of $65,000.

Question: Which company, 1 or 2, is better to invest?

Answer

Let us define X as a random variable representing its values for the amount of money earned as x_1 for profit (positive earning), x_2 for breakeven (0 earning), and x_3 for loss (negative earning). Thus, using these notations, we summarize for the cases in Table 3.6.

TABLE 3.6

Example 3.13

	Company				
		No. 1		No. 2	
X	Amount, $	Probability, $p_X(x)$, %	Amount, $	Probability, $p_X(x)$, %
x_1	+80,000.00	25	+90,000.00	20
x_2	0	40	0	65
x_3	−25,000.00	35	−65,000.00	15

Thus, calculating the expected values for each company, we will have:

Company No. 1

$$E(X) = (0.25)(80,000) + (0.40)(0) + (0.35)(-25,000)$$
$$= 20,000 + 0 - 8,750 = 11,250.$$

Company No. 2

$$E(X) = (0.20)(90,000) + (0.65)(0) + (0.15)(-65,000)$$
$$= 18,000 + 0 - 9,750 = 8,250.$$

Since the expected value for the first company is $11,250 and for the second company is $8,250, that is, $3,000 differences, it is more beneficial for the investor to choose Company No. 1.

Example 3.14

A school district superintendent is investigating the teacher attrition and retention rate in her district for new hires. She gathers data from the past 5 years and finds the following:

X: No. of years stayed	1	2	3	4	5
p_X: % Rate of stay in years	35	30	18	10	7

What is the

  i. Probability that a new hire stays at work for at least 2 years with the first 5 years of hire,
  ii. Mean number of years a new hire will stay within the first 5 years of hire?

Answer

  i. From the given data, table above, we have:

$$p_X(2 \le x \le 5) = 0.30 + 0.18 + 0.10 + 0.07 = 0.65 \approx 65\%$$

We could also answer this question using the complement rule as follows:

$$P(X = \text{at least } 2) = P(X \ge 2) = 1 - P(X < 2) = 1 - P(X = 1) = 1 - 0.35 = 0.65.$$

  ii. Applying (3.14), we use the table above to answer as follows:
       In other words, new teachers hired on the average, within the first 5-year period, will stay 28 months (Table 3.7).
       If X is a random variable, whose values are nonnegative integers, then

$$P(X > x) = P(X = x+1) + P(X = x+2) + \cdots = 1 - P(X \le x)$$
$$= 1 - P(X = 0) - P(X = 1) - P(X = 2) - \cdots - P(X = x). \quad (3.15)$$

TABLE 3.7
Example 3.14

X: No. of	x_1	x_2	x_3	x_4	x_5	Expected number of
years stayed	1	2	3	4	5	years $= \sum\limits_{x=1}^{4} xp_X(x)$
P_X: % Rate of	$p_X(x_1)$	$p_X(x_2)$	$p_X(x_3)$	$p_X(x_4)$	$p_X(x_5)$	
stay in years	35	30	18	10	7	
$xp_X(x)$	0.35	0.60	0.54	0.40	0.35	2.24

Theorem 3.1

If X is a random variable, whose values are nonnegative integers, then

$$E(X) = \sum_{x=0}^{\infty} P(X > x). \tag{3.16}$$

Proof:
Applying (3.16), we will have:

$$\sum_{x=0}^{\infty} P(X > x) = P(X > 0) + P(X > 1) + P(X > 2) + \cdots$$

$$= P(X = 1) + P(X = 2) + P(X = 3) + \cdots$$

$$+ P(X = 2) + P(X = 3) + P(X = 4) + \cdots$$

$$+ P(X = 3) + P(X = 4) + P(X = 5) + \cdots + \vdots \cdots$$

$$= 1 \cdot P(X = 1) + 2 \cdot P(X = 2) + 3P(X = 3) + \cdots$$

$$\sum_{x=0}^{\infty} x \cdot P(X = x) = E(X).$$

Note 3.12

We leave it as an exercise to show that expected value of an indicator random variable of an event (defined in (3.1)) from a sample space is the probability of the event. In other words, for

$$I_A(\omega) = \begin{cases} 1, \text{ for } \omega \in A, \\ 0, \text{ for } \omega \notin A, \end{cases}$$

$$E(I_A) = P(A). \tag{3.17}$$

Definition 3.9

Let X be a discrete random variable defined on a sample space Ω with pmf denoted by $P_X(x)$. Let Z also be a new random variable as a **function of the random variable X**, denoted by $Z = z(X)$. Then, the expected value of Z is defined by

$$E(Z) = E[Z(X)] = \sum_{x \in \Omega} Z(x) p_X(x). \qquad (3.18)$$

This definition can be extended for n random variables $X_1, X_2, \ldots, X_n$. The random variable Z could be the sum or product of two or more random variables, for instance. In this case, (3.18) for a sum is referred to as the **linearity property of the expected value**, which is as follows: For constants a and b, we have:

$$E(aX + b) = aE(X) + b. \qquad (3.19)$$

Generally, for constant numbers, a and b, we have:

$$E[aZ(X) + b] = aE[z(X)] + b. \qquad (3.20)$$

Also, for a sequence of random variables, $X_1, X_2, \ldots$, we have:

$$E\left(\sum_{i=1}^{\infty} X_i\right) = \sum_{i=1}^{\infty} E(X_i), \qquad (3.21)$$

provided that $\sum_{i=1}^{\infty} E(X_i)$ converges. In a finite case, for constants $a_1, a_2, \ldots, a_n$, and b, we have:

$$E(a_1 X_1 + a_2 X_2 + \cdots + a_n X_n + b) = a_1 E(X_1) + a_2 E(X_2) + \cdots + a_n E(X_n) + b. \qquad (3.22)$$

In case $X_1, X_2, \ldots, X_n$ are n independent random variables, then

$$E(X_1 X_2 \ldots X_n) = E(X_1) E(X_2) \ldots E(X_n). \qquad (3.23)$$

Example 3.15

Let us toss a fair coin twice. Let X be a random variable representing the number of heads that appear in this experiment (Table 3.8). Find

    i. $E(X)$
    ii. $E(X^2)$, expected value of a function of the random variable X.

TABLE 3.8
Example 3.15

X	0	1	2	$E(X)$	x^2	0	1	4	$E(X^2)$
$p_X(x)$	1/4	1/2	1/4		$p_X(x)$	1/4	1/2	1/4	
$x p_X(x)$	0	1/2	1/2	1	$x^2 p_X(x)$	0	1/2	1	3/2

Answer

The sample space in this case is:

$$\Omega = \{HH, HT, TH, TT\},$$

where H and T stand for head and tail, respectively. Thus, $X = 0,1,2$. Since the coin is fair, we will have:

We leave the derivation of these properties as exercises.

Note 3.13

$$E\left(\frac{1}{X}\right) \neq \frac{1}{E(X)}. \tag{3.24}$$

Example 3.16

Consider the random variable X with values 1, 2, and 3 and probability 1/3 for each. We list the values of $1/X$, pmfs, and expected values in Table 3.9. The last row confirms (3.19).

In this subsection so far, we have computed the expected value of a random variable X and a function of a random variable, specifically X^2. We have also seen some examples. For instance, in Example 3.21, we found $E(X^2)$. We now want to generalize this concept.

TABLE 3.9
Example 3.16

X	1	2	3		$1/X$	1	1/2	1/3
$p_X(x)$	1/3	1/3	1/3		$p_X(x)$	1/3	1/3	1/3
$x p_X(x)$	1/3	2/3	1		$x p_X(x)$	1/3	1/6	1/9

$$\frac{1}{E(X)} = \frac{1}{\frac{1}{3}+\frac{2}{3}+1} = 1/2 \qquad\qquad \neq \qquad\qquad E\left(\frac{1}{X}\right) = 5/9$$

Definition 3.10

Consider a discrete random variable X with its pmf $p_X(x) = P(X = x)$ defined on a sample space Ω and a positive integer r; then, $E(X')$ is called the **rth moment of X** or **moment of order r of X**. Symbolically:

$$E(X^r) = \sum_{x \in \Omega} x^r p_X(x). \tag{3.25}$$

Specifically,

$$E(X^r) = \sum_{i=1}^{n} x_i^r p_X(xi), \ E(X^r) = \sum_{i=1}^{\infty} x_i^r p_X(xi). \tag{3.26}$$

In (3.26), if $r = 1$, the expected value of X is called the **first moment** of X, and if $r = 2$, $E(X^2)$ is referred to as the **second moment** of X.

Note 3.14

When X is a random variable and μ (pronounces as mu) is a constant, the first moment of X ($E(X)$) is denoted by μ, based on properties of the expected values, and $X - \mu$ is also a random variable and $E(X - \mu) = E(X) - \mu = 0$. This shows that we can center X by shifting X and choosing $X - \mu$.

Definition 3.11

Consider the random variable $X - \mu$. Then, $E\left[(X - \mu)^r\right]$ is the rth moment of $X - \mu$ and is called the **central rth moment of X**. The random variable $X - \mu$ measures the deviation of X from its expected value or the mean of X. Since this deviation may be positive or negative depending upon the values of $X - \mu$, the absolute value of $X - \mu$, $|X - \mu|$, is the **absolute measure of deviation** of X from its mean, μ. For the sake of convenience in calculations, the **mean square deviation**, $E\left[(X - \mu)^2\right]$, which is the **second central moment** of X, is used rather than absolute value, $|X - \mu|$.

The second moment leads to another very important concept as defined in the following definition.

Definition 3.12

Consider a random variable X with a finite mean $E(X)$. The **variance of X**, denoted by $Var(X)$ or $\sigma^2(X)$ or if there is no danger of confusion, just σ^2, is defined as follows:

$$\sigma^2(X) = E\left[(X - \mu)^2\right]. \tag{3.27}$$

Relation (3.27) states that the variance measures the average **deviation** or **dispersion of the random variable X from its mean**.

Note 3.15

For a random variable X with mean μ, the variance of X can be calculated as follows:

$$\sigma_X^2 = E\left[(X-\mu)^2\right] = E\left[X^2 - 2\mu X + \mu^2\right]$$

$$= E(X^2) - 2\mu E(X) + \mu^2$$

$$= E(X^2) - 2\mu^2 + \mu^2$$

$$= E(X^2) - \mu^2. \tag{3.28}$$

Note 3.16

In case the value of variance is small, it would mean that the values of the random variable are clustered about its mean. This could also happen when the sample size is too large.

We list other **properties of variance** of a random variable X and leave the proofs as exercises. For a constant a,

$$1. \quad Var(X+a) = Var(X). \tag{3.29}$$

$$2. \quad Var(aX) = a^2 Var(X). \tag{3.30}$$

Note 3.17

It is interesting to note that mean and variance of two random variables may be equal while their pmfs are different. Here is an example.

Example 3.17

Let X and Y be two discrete random variables representing two biased dice with the following different pmfs (Tables 3.10 and 3.11).

This raises a question that we will later answer: For a random variable X, if the mean and variance are known, can we find the pmf?

From (3.36), it can be seen that the variance of a random variable X measures the deviation from the mean by squares. Thus, correcting the squaring process, it is needed to take square root. This leads us to the following definition.

TABLE 3.10
Example 3.17

X	1	2	3	4	5	6	Sum
p_X	$\dfrac{1}{2}$	0	$\dfrac{1}{6}$	0	0	$\dfrac{1}{3}$	p_X
xp_X	$\dfrac{1}{2}$	0	$\dfrac{1}{2}$	0	0	2	xp_X
X^2	1	4	9	16	25	36	X^2
$x^2 p_X$	$\dfrac{1}{2}$	0	$\dfrac{3}{2}$	0	0	12	$x^2 p_X$

$\mu_X = E(X) = 3$
$E(X^2) = 14$
$\sigma^2(X) = E(X^2) - [E(X)]^2 = 14 - 9 = 5$

TABLE 3.11
Example 3.17

Y	1	2	3	4	5	6	Sum
p_Y	$\dfrac{8}{15}$	0	0	$\dfrac{1}{6}$	0	$\dfrac{3}{10}$	1
yp_y	$\dfrac{8}{15}$	0	0	$\dfrac{2}{3}$	0	$\dfrac{9}{5}$	$\mu_Y = 3$
Y^2	1	4	9	16	25	36	
$y^2 p_y$	$\dfrac{8}{15}$	0	0	$\dfrac{8}{3}$	0	$\dfrac{54}{3}$	$E(Y^2) = 14$

$\mu_y = E(Y) = 3$
$E(Y^2) = 14$
$\sigma^2(Y) = E(Y^2) - [E(Y)]^2 = 14 - 9 = 5$

Definition 3.13

For a random variable X, the positive square root of its variance is referred to as the **standard deviation of X** and is denoted by $\sigma(X)$.

Example 3.18

As in Example 3.17, suppose an investor is trying to earn with the following percents of profit, breakeven, and loss:

i. 25% chance of profit of $80,000.

TABLE 3.12
Example 3.18

	x_1	x_2	x_3	
x	+80,000	0	−25,000	
x^2	64×10^8	0	6.25×10^8	Sum
$p_X(x)$	0.25	0.40	0.35	$
$xp_X(x)$	+20,000	0	−8,750	$\mu \equiv E(X) = 11,250$
$x^2 p_X(x)$	16×10^8	0	2.1875×10^8	$E(X^2) = 18.1875 \times 10^8$

    ii. 40% chance of breakeven.
    iii. 35% chance of loss of $25,000.

Find the mean, variance, and standard deviation of his gain.

Answer

As before, let us define X as a random variable representing its values for money earned as x_1 for profit (positive earning), x_2 for breakeven (0 earning), and x_3 for loss (negative earning). In Table 3.12, we insert the values of X and its pmf. We also include the values to calculate the second moment and, hence, the variance.
    Thus, from (3.34), we have:

$$\sigma^2(X) = E(X^2) - \mu^2 = 234.75 \times 10^6 - (11,250)^2$$

$$= 1,818,750,000 - 126,562,500 = 1,692,177,500$$

and

$$\sigma(X) = 41,136.09.$$

The variance and, hence, the standard deviation are largely due to a small range of possible values of X. This type of problem may appear with very small variance. Here is an example.

Example 3.19

Let us consider a game with four possible outcomes as −2, −1, 3, and 5 with respective probabilities of 0.1, 0.3, 0.4, and 0.2. We want to calculate the mean and standard deviation for this distribution.

Answer

As in Example 3.18, we let X be a random variable taking the values of outcomes and $p_X(x)$ as its pmf. This time we use (3.46), $\sigma^2(X) = E\left[(X - \mu)^2\right]$. Thus, we will have Table 3.13.
    Thus, mean = 1.7, variance = 6.41, and $\sigma \equiv$ standard deviation = $\sqrt{\sigma^2} = 2.53$.

TABLE 3.13
Example 3.19

	x_1	x_2	x_3	x_4	
x	-2	-1	3	5	
$p_X(x)$	0.1	0.3	0.4	0.2	Sum
					\$
$xp_X(x)$	-0.2	-0.3	1.2	1	$\mu \equiv E(X) = 1.7$
$X - \mu$	-3.7	-2.7	1.3	3.3	
$(X - \mu)^2$	13.69	7.29	1.69	10.89	
$(X - \mu)^2 p_X(x)$	1.369	2.187	0.676	2.178	$\sigma^2 \equiv E[X - \mu]^2 = 6.41$

Despite the expected value that had linearity property $E(aX + b) = aE(X) + b$, variance does not have such property. That is,

$$\sigma^2(uX + b) = u^2\sigma^2(X). \tag{3.31}$$

We leave the proof of (3.31) as an exercise.

Example 3.20

Consider Example 3.19. Suppose the reward of the game is changed to increase the payout in half, then add a dollar. The pmf remains as before. Again, we want to calculate the mean and standard deviation for this distribution.

Answer

Once again, let Y be a random variable taking the values of the new outcomes and $p_Y(y)$ as its pmf. Thus, we have Table 3.14.

TABLE 3.14
Example 3.20

	x_1	x_2	x_3	x_4	
X	-2	-1	3	5	
$X/2$	-1	-0.5	1.5	2.5	
	y_1	y_2	y_3	y_4	
$Y = X/2 + 1$	0	0.5	2.5	3.5	
$p_Y(y) = p_X(x)$	0.1	0.3	0.4	0.2	
$yp_Y(y)$	0	0.15	1	0.7	$\mu_Y = E(Y) = 1.85$
$Y - \mu_Y$	-1.85	-1.35	0.65	1.65	
$(Y - \mu_Y)^2$	3.4225	1.8225	0.4225	2.7225	
$(Y - \mu_Y)^2 p_Y(y)$	0.34225	0.54675	0.169	0.5445	$\sigma^2 \equiv E\left[(Y - \mu_Y)^2\right] = 1.6025$

Thus, mean = 1.35, variance = 3.6025, and $\sigma \equiv$ standard deviation = $\sqrt{\sigma^2} = 1.2659$. As it can be seen, in this case, the variance is one-fourth of the original value of the variance, in Example 3.19. This is because from (3.49), $\sigma^2(X/2+1) = \sigma^2(X)/4$.

Even the following definitions (up to Definition 3.15.) should be under discrete random variables. In addition to (3.37), it can be proved (left as an exercise) that variance has the following properties: If X and Y are two **independent** random variables, then

$$\sigma^2(X \pm Y) = \sigma^2(X) + \sigma^2(Y). \qquad (3.32)$$

In other words, the variance of sum and difference of two independent random variables X and Y is the sum of the variances of X and Y.

We may wonder if two random variables are not independent of how we measure the dependencies. The answer is in the definition below.

Definition 3.14

For two random variables X and Y, let $\mu_X = E(X)$ and $\mu_Y = E(Y)$ denote the means of X and Y, respectively. Then, covariance of X and Y, denoted by $Cov(X,Y)$, is defined as

$$Cov(X,Y) = E\big[(X - \mu_X)(Y - \mu_Y)\big]. \qquad (3.33)$$

In other words, covariance measures the linear agreement between two random variables. This value can be any real value. Positive values indicate the positive agreement between two random variables (i.e., when one value increases, the other value also increases), negative values indicate a negative agreement between two random variables (i.e., when one value increases, the other value decreases), and zero indicates the nonexistence of a linear relationship.

Example 3.21

Suppose an electronic repair store has 20 computers for sale, 8 of which are repaired, 5 are new, and 7 are non-repairable (sold as parts). Three computers are randomly chosen to be tested, one at a time without replacement. We consider two random variables X and Y with joint pmf and marginal pmfs as in Table 3.15.

We want to calculate the covariance of X and Y.

Solution:

We now calculate $X - \mu_X$ and $Y - \mu_Y$ (Table 3.16).
Thus,

$$Cov(X,Y) = \sum_x \sum_y (X - \mu_X)(Y - \mu_Y) p_{XY}(x,y) = -0.0585.$$

TABLE 3.15

Example 3.21

		X			
Y	**0**	**1**	**2**	**3**	$p_Y(y)$
0	0.03	0.09	0.06	0.01	0.19
1	0.15	0.25	0.07	0	0.47
2	0.17	0.12	0	0	0.29
3	0.05	0	0	0	0.05
$p_X(x)$	0.40	0.46	0.13	0.01	1 (total)

TABLE 3.16

Example 3.21

X	**0**	**1**	**2**	**3**	
$p_X(x)$	0.40	0.46	0.13	0.01	
$x p_X(x)$	0	0.46	0.26	0.03	$\mu_X = 0.75$
$X - \mu_X$	−0.75	0.25	1.25	2.25	
Y	**0**	**1**	**2**	**3**	
$p_Y(y)$	0.19	0.47	0.29	0.05	
$y p_Y(y)$	0	0.47	0.58	0.15	$\mu_Y = 1.2$
$Y - \mu_Y$	−1.2	−0.2	0.8	1.8	

Let X, Y, U, and V be random variables and a, b, c, and d constants. Then, the following are some properties of covariance, which we leave the proof as exercises:

$$Cov(X,a) - 0. \tag{3.34}$$

$$Cov(X,X) = Var(X). \tag{3.35}$$

$$Cov(X,Y) = Cov(Y,X). \tag{3.36}$$

$$Cov(X,Y) = E(XY) - \mu_X \mu_Y. \tag{3.37}$$

$$Cov(X + a, Y + b) = Cov(X,Y). \tag{3.38}$$

$$Cov(aX,bY) = ab\,Cov(Y,X). \tag{3.39}$$

$$Cov(aX,Y) = a\,Cov(X,Y). \tag{3.40}$$

$$Cov(aX + bY, cU + dV) = ac\,Cov(X,U) + ad\,Cov(X,V)$$
$$+ bc\,Cov(Y,U) + bd\,Cov(Y,V). \tag{3.41}$$

$$Var(X + Y) = Var(X) + Var(Y) + 2Cov(X,Y). \tag{3.42}$$

One of the main issues with the covariance between two random variables is the difficulty of interpreting the value of covariance. The magnitude of the value of covariance does not represent the strength of the association between the two random variables, as the value of the covariance depends on the units of both random variables. Therefore, we need a better measurement to represent the linear relationship between two random variables. This is the subject of the following definition.

Definition 3.15

Consider two random variables X and Y with their respective standard deviations as σ_X and σ_Y. The **coefficient of correlation** of X and Y, denoted by $\rho_{X,Y}$, is given by:

$$\rho_{X,Y} = \frac{Cov(X,Y)}{\sigma_X \sigma_Y}, \tag{3.43}$$

provided that $\sigma_X \sigma_Y \neq 0$.
   It is left as an exercise to prove that

$$-1 \leq \rho_{X,Y} \leq 1. \tag{3.44}$$

Note 3.18

   $\rho > 0$ represents a positive linear correlation, means that both random variables increase or decrease together.
   $\rho < 0$ represents a negative linear correlation, means that the random variables are dependent in opposite direction, so that if the value of one variable increases, the value of the other one will decrease.
   $\rho = 0$ means that the random variables are **linearly uncorrelated** (or there may be a nonlinear correlation).

Unlike the value of covariance, correlation coefficient represents the strength of the relationship between two random variables. As a convention, we adhere to the following classification (see Table 3.17); though, the values may change slightly depending on the discipline and the application (Table 3.18).
   Here are some other **properties of the coefficient of correlation** ρ:

$$\rho_{X,Y} = \rho_{Y,X}. \tag{3.45}$$

$$\rho_{X,X} = 1. \tag{3.46}$$

$$\rho_{X,-X} = -1. \tag{3.47}$$

For real numbers a, b, and c, and $a, c \neq 0$,

$$\rho_{aX+b,cY+d} = \rho_{Y,X}. \tag{3.48}$$

TABLE 3.17
Example 3.21

	Y		0	1	2	3	Sum
	$Y - \mu_Y$		-1.2	-0.2	0.8	1.8	
X	$X - \mu_X$	$(X - \mu_X)(Y - \mu_Y)p_{XY}(x,y)$					
0	-0.75		0.027	0.15675	0,036	0.0135	0.23325
1	0.25		-0.045	-0.0125	0.014	0	-0.0435
2	1.25		-0.11475	-0.03	0	0	-0.14475
3	2.25		-0.135	0	0	0	-0.0135
	Sum		-0.26775	0.11425	0.05	0.0135	0.0315
	Cov(X,Y)						-0.0585

TABLE 3.18
Correlation Coefficient Convention

Value of ρ	Interpretation
$-1 \leq \rho < 0.7$	Strong-negative relationship
$-0.7 \leq \rho < -0.5$	Moderate-negative relationship
$-0.5 \leq \rho < -0.3$	Weak-negative relationship
$-0.3 \leq \rho \leq 0$	None or very weak-negative relationship
$0 < \rho \leq 0.3$	None or very weak-positive relationship
$0.3 < \rho \leq 0.5$	Weak-positive relationship
$0.5 < \rho \leq 0.7$	Moderate-positive relationship
$0.7 < \rho \leq 1$	Strong-positive relationship

Example 3.22

In a quality control process at a factory, a trouble-shooting instrument is used to test two similar products. Let the random variables X and Y represent the status of the products 1 and 2, respectively. For both products, being good in quality is described by above average or 4, and being poor in quality is described by below average, or 2. Based on the factory's data collected in past years, the joint pmf of X and Y is given as follows (Table 3.19).

TABLE 3.19
Example 3.22

		Y		
X	$p_{X,Y}(x,y)$	2	4	p_X (Marginal of X)
	2	0.01	0.10	0.11
	4	0.09	0.80	0.89
p_Y (Marginal of Y)		0.1	0.9	1 (Sum)

Hence, we calculate the mean, variance, covariance, and the correlation coefficient of X and Y, as follows (Table 3.20–3.22).

Now,

$$Cov(X,Y) = \sum_{x=2,4} \sum_{y=2,4} p_{X,Y}(x,y)(x-\mu_X)(y-\mu_Y):$$

$$\rho_{X,Y} = \frac{Cov(X,Y)}{\sigma_X \sigma_Y} = \frac{-0.017}{(0.62)(0.60)} = -0.046.$$

Here, we state two inequalities for the application of both mean and variance that we will use in the later chapters.

TABLE 3.20
Example 3.22

X	2	4			Y	2	4	
X^2	4	16			Y^2	4	16	
$p_X(x)$	0.11	0.89			$p_Y(y)$	0.10	0.90	
$xp_X(x)$	0.22	3.56	$\mu_X = 3.78$		$yp_Y(y)$	0.20	3.60	$\mu_Y = 3.80$
$x^2 p_X(x)$	0.44	14.24	$\mu_{X^2} = 14.68$		$y^2 p_Y(y)$	0.40	14.40	$\mu_{Y^2} = 14.80$

$\sigma_X^2 Var(X) = E(X^2) - [E(X)]^2$ $\sigma_Y^2 Var(Y) = E(Y^2) - [E(Y)]^2$

$\qquad\qquad = 14.68 - 14.29 = 0.39$ $\qquad\qquad = 14.80 - 14.44 = 0.36$

$\sigma_X = \sqrt{0.39} = 0.62$ $\sigma_Y = \sqrt{0.36} = 0.60$

TABLE 3.21
Example 3.22

X		2	4	Y	2	4
$X - \mu_X$		$2 - 3.78 = -1.78$	$4 - 3.78 = 1.22$	$Y - \mu_Y$	$2 - 3.80 = -1.80$	$4 - 3.80 = 1.20$

(X, Y)	$(2, 2)$	$(2, 4)$	$(4, 2)$	$(4, 4)$
$(X - \mu_X)(Y - \mu_Y)$	$(-1.78)(-1.80)$	$(-1.78)(1.20)$	$(1.22)(-1.80)$	$(1.22)(1.20)$
	$= 3.20$	$= -2.16$	$= -2.20$	$= 1.46$

TABLE 3.22
Example 3.22

(X, Y)	$(2, 2)$	$(2, 4)$	$(4, 2)$	$(4, 4)$	Sum $= Cov(X,Y)$
$p_{X,Y}(x,y)$	$(0.01)(3.20)$	$(0.1)(-2.16)$	$(0.09)(-2.20)$	$(0.8)(1.46)$	-0.017
$(X - \mu_X)(Y - \mu_Y)$	$= 0.032$	$= -0.216$	$= -0.198$	$= 0.117$	

Theorem 3.2 Markov's Inequality

If the expected value of a nonnegative random variable is small, then the random variable must itself be small with high probability. The following is the formulation of this statement that is referred to as **Markov's inequality**:

Let X be a nonnegative continuous random variable with finite mean, $E(X)$. Then,

$$P\left(X \geq \epsilon\right) \leq \frac{E(X)}{\epsilon}, \quad \epsilon > 0. \tag{3.49}$$

Proof:

Using the fact that the integer and defining a continuous expected value is nonnegative, and the values of X are bounded from below by ϵ, since X is non-negative by assumption and $P(x)$ is also non-negative by definition, $xP(x)$ is nonnegative. Thus, X being a continuous random variable, by assumption, we have:

$$E(x) = \int_0^\infty x \, dP(x) \geq \int_\epsilon^\infty x \, dP(x) \geq \int_\epsilon^\infty \epsilon \, dP(x) = \epsilon \, P(X \geq \epsilon).$$

Note 3.19

Markov's inequality(3.49) can be defined and proved for discrete and mixed random variables, as well.

The Markov inequality can be stated for a nonnegative increasing function of X, say $g(X)$. In that case, (3.49) becomes what is referred to as a **derivative inequality**, defined by:

$$P(X \geq \epsilon) \leq \frac{E[g(X)]}{g(\epsilon)}, \quad \epsilon > 0. \tag{3.50}$$

Example 3.23

Let X be a binomial random variable with parameters n and p. Using Markov's inequality, we want to

   i. Find An upper bound on $P(X \geq \epsilon n)$, where $p < \epsilon < 1$,

   ii. Evaluate the bounds for the values of $p = \dfrac{1}{4}$ and $\epsilon = \dfrac{3}{10}$.

Answer

   i. Since X is a binomial random variable, it is nonnegative and its mean is np. Hence,

$$P(X \geq \epsilon n) \leq \frac{E(X)}{\epsilon n} = \frac{pn}{\epsilon n} = \frac{p}{\epsilon}.$$

Thus, an upper bound in question is $\dfrac{p}{\epsilon}$.

ii. $P\left(X \geq \dfrac{3}{10}n\right) \leq \dfrac{5}{6}$.

As a special case of (3.50), we prove the following theorem.

Theorem 3.3 Chebyshev's Inequality

Let X be a random variable with finite mean, μ, and finite and nonzero variance, σ^2. Let $\in$ also be a positive real number. Then,

$$P\left(|X - \mu| \geq \in\right) \leq \frac{\sigma^2}{\in^2}, \quad \forall_\in \in \mathbb{R}^+ \tag{3.51}$$

or

$$P\left(|X - \mu| \geq \in \sigma\right) \leq \frac{1}{\in^2}, \quad \forall_\in \in \mathbb{R}^+. \tag{3.52}$$

In other words, the probability that the value X lies at least $\in$ from its mean is at least $\dfrac{\sigma^2}{\in^2}$, or the probability that the value X lies at least $\in$ standard deviation from its mean is at least $\dfrac{1}{\in^2}$.

Proof:
Let $g(X) = [x - E(X)]^2$, which is a nonnegative function regardless of values of X being nonnegative or not, be a random variable with mean μ and variance σ^2. Then,

$$P\left[(X - \mu)^2 \geq \in^2\right] \leq \frac{E\left[(X - \mu)^2\right]}{\in^2} = \frac{\sigma^2}{\in^2}, \quad \text{for any real } \in > 0.$$

Thus,

$$P\left(|X - \mu| \geq \in\right) \leq \frac{\sigma^2}{\in^2}, \quad \text{for any real } \in > 0.$$

Example 3.24

Let X be a binomial random variable with parameters n and p. Using Chebyshev's inequality, we want to

    i. Find an upper bound on $P(X \geq \in n)$, where $p < \in < 1$,
    ii. Evaluate the bounds for values of $n = 20$, $p = \dfrac{1}{4}$, and $\in = \dfrac{3}{10}$.

Answer

i. Since X is a binomial random variable, it is nonnegative and its mean is np. Hence,

$$P(X \geq \in n) = P(X - np \geq \in n - np) \leq P(|X - np| \geq \in n - np) \leq \frac{\sigma^2}{(\in n - np)^2}$$

$$= \frac{p(1-p)}{n(\in -p)^2}.$$

Thus, an upper bound in question is $\dfrac{p(1-p)}{n(\in -p)^2}$.

ii. $P\left(X \geq \dfrac{3}{10} 20\right) \leq \dfrac{\left(\dfrac{1}{4}\right)\left(\dfrac{3}{4}\right)}{20\left(\dfrac{3}{10} - \dfrac{1}{4}\right)^2} = \dfrac{13}{4} = 3.75.$

Note 3.20

The name "Chebyshev" has been written in a variety of ways in the literature. These include Tchebicheff, Čebyšev, Tschebyscheff, and Chebishev (Clenshaw 1962), also Cheney and Kincaid (1994, Ex. 14, p. 16). We use Chebyshev in this book.

Note 3.21

In (3.52), only the value of $\in = 1$ can be useful since for their values the inequality will be trivial. For instance, if $\in = 2$, then the inequality says that the probability that the value of X falls outside the interval $(\mu - 2\sigma, \mu + 2\sigma)$ does not exceed $\dfrac{1}{4}$.

Example 3.25

Consider the population of a city whose household income is $50,000.00 with the standard deviation of $25,000.00. An individual is selected from this city randomly. What is the probability that the average income for this individual is either less than $15,000.00 or greater than $75,000.00?

Answer

Since the distribution of income for households in this city is not known, we cannot compute the exact value of the probability in question. But using Chebyshev's inequality, we can compute an upper bound to it.

Thus, let X be the random variable representing the average household income of this city. Let $\mu = 45,000$ and $\in = 30,000$. Hence, $|X - \mu| \geq \in$. Therefore, we have:

$$P(|X - \mu| \geq \in) \leq \frac{\sigma^2}{\in^2} = \frac{(25000)^2}{(30000)^2} = \frac{625}{900} = \frac{25}{36}.$$

In other words, the probability of the selected individual's household income to be either less than \$15,000 or greater than \$75,000 is $\dfrac{25}{36}$.

Example 3.26

Suppose the distribution of the grades of an examination is skewed with mean and standard deviations of 75 and 5, respectively. What percent of students have received grade between 50 and 100?

Answer

Using Chebyshev's theorem, we know $1 - \dfrac{1}{k^2}$ proportion of data is in the interval of $(\mu - ks, \mu + ks)$. Therefore, $\mu + ks = 100$ or $75 + k(5) = 100$. Hence, $k = 5$. Therefore, the proportion of grades between 50 and 100 is $1 - \dfrac{1}{k^2} = 1 - \dfrac{1}{25} = 0.96$.

The following term is a special case of moment of a random variable X.

Definition 3.16

Let X be a discrete random variable with *pmf* p_X. Let t also be a real number or a complex number with nonnegative real part. Suppose that $E(e^{tX})$ exists. Then, the **moment generating function of X**, denoted by $M(t)$, is defined as follows:

$$M(t) = E(e^{tX}) = \sum_x e^{tx} px. \tag{3.53}$$

Note 3.22

From Definition 3.16, we see that

$$M(0) = \sum_x p_x = 1 \tag{3.54}$$

and for a finite natural number n,

$$\lim_{t \to 0^-} \frac{d^n M(t)}{dt^n} = \sum_x e^{tx} p_x = E(X^n). \tag{3.55}$$

From (3.55), we see that the moment generating function generates all the moments of X by differentiating $M(t)$ defined in (3.54) and devaluing at $t = 0$. The first derivative will give the expected value, that is, $E(X) = M'(0)$. Relation (3.55) leads us to the following generalization of this special case:

$$M^{(n)}(0) = E(X^n) = \sum_{n=0}^{\infty} x^n p_x. \tag{3.56}$$

Example 3.27

Let X be a random variable with n values $1,2,...,n$ and p_X defined as:

$$p_X = P(X = x) = \begin{cases} \dfrac{1}{n}, & x = 1,2,3,...,n, \\ 0, & \text{otherwise.} \end{cases} \tag{3.57}$$

Later, we will see that (3.57) is referred to as the **discrete uniform pmf**. Then, the moment generating function of X is:

$$M(t) = \sum_{k=1}^{n} e^{tk} \frac{1}{n}$$

$$= \frac{e^t + e^{2t} + \cdots + e^{nt}}{n} \tag{3.58}$$

$$= \frac{e^t}{n}\left(1 + e^t + \cdots + e^{(n-1)t}\right). \tag{3.59}$$

Note 3.23

The random variable defined in Example 3.25 is called the **discrete uniform random variable** defined between 1 and n.

Now, we want to rewrite (3.58) for easier use. Hence, we define the geometric progression.

The sequence $ar, ar^2,...,ar^n$ $a, ar, ar^2,...,ar^n$ is called a finite **geometric progression with ratio r**. Its first term is a and its nth term, denoted by a_n, is $a_n = ar^{n-1}$ as n approaches infinity, and the geometric progression takes the form $a, ar, ar^2,...,ar^n,....$ We leave it as exercises to show that the sums of geometric progressions, finite and infinite, denoted by S_n and S_∞, respectively, are as follows:

$$S_n = \sum_{k=1}^{n} ar^{k-1} = \frac{a(1-r^n)}{1-r}, \quad r \neq 1, \tag{3.60}$$

and

$$S_\infty = \sum_{n=1}^{\infty} ar^{n-1} = \frac{a}{1-r}, \quad -1 < r < 1. \tag{3.61}$$

Thus, the sum $1 + e^t + \cdots + e^{(n-1)t}$ in (3.57) is a finite geometric progression with ratio $r = e^t$. Hence, from (3.59), we have:

$$M(t) = \left(\frac{e^t}{n}\right)\left(\frac{1-e^{nt}}{1-e^t}\right). \tag{3.62}$$

We can now calculate the mean and variance of X. To do that, we need to use the sum and the sum of squares of the first n natural numbers. They are as follows:

$$1+2+\cdots+n = \frac{n(n+1)}{2} \text{ and } 1+2^2+\cdots+n^2 = \frac{n(n+1)(2n+1)}{6}. \tag{3.63}$$

Now from (3.58), using (3.63), we have:

$$E(X) = M'(0) = \left. \frac{e^t + 2e^{2t} + \cdots + ne^{nt}}{n} \right|_{t=0}$$

$$= \frac{1+2+\cdots+n}{n} = \frac{n(n+1)}{2n} = \frac{(n+1)}{2}. \tag{3.64}$$

$$E(X^2) = M''(0) = \left. \frac{e^t + 2^2 e^{2t} + \cdots + n2 e^{nt}}{n} \right|_{t=0}$$

$$= \frac{1+2^2+\cdots+n^2}{n} = \frac{n(n+1)(2n+1)}{6n}$$

$$= \frac{(n+1)(2n+1)}{6}.$$

$$Var(X) = M''(0) - M'[(0)]^2 = \frac{(n+1)(2n+1)}{6} - \left(\frac{n+1}{2}\right)^2$$

$$= \frac{(n^2-1)}{12}. \tag{3.65}$$

To encapsulate all the information about a random variable, (3.65) is expressed differently as follows.

Definition 3.17

Let X be a discrete random variable with *pmf* $p_n = P(X = n)$, where n is a non-negative integer. Consider the random variable z^X, which is a function of random variable X and z is a complex variable. Then, the **probability generating function** (or if there is no confusion, **generating function**) (pgf) of X, denoted by $G(z)$, is defined as:

$$G(z) = E(z^X) = \sum_{n=0}^{\infty} z^n p_n, \quad |z| < 1. \tag{3.66}$$

From (3.66), it can be seen that for $z = 1$ and p_n being a pmf, we have:

$$G(1) = \sum_{n=0}^{\infty} p_n = 1 \tag{3.67}$$

In case p_n is not a pmf, (3.67) will not necessarily hold; it will be of the form referred to as the **generating function**, but not a **probability generating function**. Here is an example.

Example 3.28

Consider the sequence of 1s, $\{a_n\} = 1,1,\ldots = \{1\}, \forall_n = 1,2,\ldots$. The generating function of this sequence is the infinite power series,

$$G(z) = \sum_{n=0}^{\infty} 1.z^n = 1 + z + z^2 + \cdots,$$

whose sum is $\dfrac{1}{1-z}$. But this series is not a pgf since $G(1) \neq 1$.

We now discuss some **properties of pgf**:

1. The pmf of the random variable can be obtained from the nth derivative of its pgf with respect to z, evaluated at $z = 0$. That is,

$$p_n = P(X = n) = \frac{G^{(n)}(0)}{n!}. \tag{3.68}$$

2. Obtaining the nth **factorial moment** of X, that is, $E[X(X-1)\ldots(X-n+1)]$, is another **property of pgf**. This can be easily obtained by the nth derivative of the pgf, evaluated at 1. Here is an illustration: Using the fact that $X! = X(X-1)\ldots 2 \cdot 1$, we can complete the expression $X(X-1)\ldots(X-n+1)$ and divide by terms introduced as follows:

$$E[X(X-1)\ldots(X-n+1)] = E\left[\frac{X!}{(X-n)!}\right]$$

$$= \frac{d^n G(z)}{dz^n}\Bigg|_{z=1} = \sum_{n=k}^{\infty} n! \, p_n \frac{z^{n-k}}{(n-k)!}\Bigg|_{z=1}$$

$$= \sum_{n=k}^{\infty} \frac{n! \, p_n}{(n-k)!}. \tag{3.69}$$

As a consequence of (3.69), when $n = 1$, we obtain the expected value of X as the first factorial moment. That is,

$$E(X) = G'(1). \tag{3.70}$$

Also,

$$Var(x) = E\left(X^2\right) - (E(X))^2$$

$$= E[X(X-1)] + E(X) - [E(X)]^2$$

$$= G''(1) + G'(1) - [G'(1)]^2. \tag{3.71}$$

3. Yet, another **property of the pgf** is the generating function of sum of n independent random variables, as well as the product. We prove the following two theorems.

Theorem 3.2

The **pgf of the sum** of n independent random variables is the product of the pgf of each.

Proof:
Let $X_1, X_2, \ldots, X_n$ be n independent random variables with corresponding pgfs, respectively, as

$$G_1(z) = E\left(z^{X_1}\right), G_2(z) = E\left(z^{X_2}\right), \ldots, G_n(z) = E\left(z^{X_n}\right).$$

Also, let $X = X_1 + X_2 + \ldots + X_n$. Then,

$$E\left(z^X\right) = E\left(z^{X_1 + X_2 + \cdots + X_n}\right) = E\left(z^{X_1} z^{X_2} \ldots z^{X_n}\right)$$

$$= E\left(z^{X_1}\right) E\left(z^{X_1}\right) \ldots E\left(z^{X_n}\right)$$

$$= G_1(z) G_2(z) \ldots G_n(Z).$$

Theorem 3.3

The **product of two pgfs** is another generating function.

Proof:
Let $G(z)$ and $H(z)$ be two generating functions, respectively, defined as

$$G(z) = \sum_{i=0}^{\infty} a_i Z^i \text{ and } H(z) = \sum_{j=0}^{\infty} b_j z^j \text{ with } |z| < 1.$$

Then, their product is

$$G(z) \cdot H(z) = \left(\sum_{i=0}^{\infty} a_i z^i\right)\left(\sum_{j=0}^{\infty} b_j z^j\right) = \sum_{i=0}^{\infty} \sum_{j=0}^{\infty} a_i b_j z^{i+j}. \tag{3.72}$$

Now, let $k = i + j$. Then, the right-hand side of (3.72) can be rewritten as a single power series as follows:

$$G(z) \cdot H(z) = \sum_{k=0}^{\infty} c_k z^k, \tag{3.73}$$

where

$$c_k = \sum_{i=0}^{k} a_i b_{k-i}, \quad k = 1, 2, \dots, \tag{3.74}$$

Definition 3.18

From (3.74), the sequence $\{c_k\}$ is called the **convolution** of the two sequences $\{a_i\}$ and $\{b_j\}$. In other words, let X_1 and X_2 be two nonnegative integer-valued independent random variables with pmfs $p_{X_1} = P(X_1 = i)$, $i = 0, 1, 2, \dots$ and $q_{X_2} = P(X_2 = j)$, $j = 0, 1, 2, \dots$. Then, $X = X_1 + X_2$ is another nonnegative integer-valued random variable with pmf defined by:

$$r_k = P(X_1 + X_2 = k) = \sum_{i+j=k} p_i q_j = \sum_{i=0}^{k} p_i q_{k-i}, \quad k = 0, 1, 2, \dots \tag{3.75}$$

Hence, we write $\{r_k\} = \{p_i\} * \{q_j\}$ and say $\{r_k\}$ is the **convolution** of $\{p_i\}$ and $\{q_j\}$.

Note 3.24

Essentially, the convolution of two functions, say f and g, is a mathematical operation to produce a new function.

3.4 BASIC STANDARD DISCRETE PROBABILITY MASS FUNCTIONS

We now present several basic standard pmfs that are widely used. They are as follows:

1. Discrete uniform,
2. Bernoulli,
3. Binomial,
4. Geometric,
5. Negative binomial,
6. Hypergeometric,
7. Poisson.

3.4.1 DISCRETE UNIFORM PMF

The simplest random variable that has both discrete and continuous distribution functions is uniform. We saw an example of the discrete case earlier, (3.54). We now discuss this pmf in detail.

Definition 3.19 Discrete Uniform pmf

Let X be a random variable. We divide the $[a,b]$ into k subintervals, each of size $\dfrac{b-a}{k}$, where k is a natural number such that $a \le k \le b$. That is, the endpoints of the subintervals are $a, a+k, a+2k, \ldots, b$. Then, the probability, p_X, of X, is defined as:

$$p_X = P(X = x) = \begin{cases} \dfrac{k}{b-a+k}, & a \le k \le b, x = a, a+k, a+2k, \ldots, b, \\ 0, & \text{otherwise.} \end{cases} \tag{3.76}$$

It is called the **discrete uniform pmf**. X is referred to as the **uniformly distributed random variable**.

Note 3.25

We can clearly see that (3.76) is positive. It can easily be shown (left as an exercise) that $\displaystyle\sum_{k=a}^{b} \dfrac{k}{b-a+k} = 1$. Thus, (3.76) is a pmf.

Let us consider a special case of (3.76) when $a = 1$, $k = 1$, and $b = n$. Then, (3.76) becomes:

$$p_X = P(X = x) = \begin{cases} \dfrac{1}{n}, & x = 1, 2, 3, \ldots, n, \\ 0, & \text{otherwise.} \end{cases} \tag{3.77}$$

It is clear that since $\dfrac{1}{n} > 0$, for each $x = 1, 2, 3, \ldots, n$, and $\displaystyle\sum_{k=1}^{n} \dfrac{1}{n} = 1$, (3.77) is a pmf.

For this special case, we leave it as an exercise to show that the mean, variance, and moment generating function of X, respectively, are as follows:

$$E(X) = \frac{(n+1)}{2}, \tag{3.78}$$

$$Var(X) = \frac{(n^2 - 1)}{12}, \tag{3.79}$$

$$M(t) = \left(\frac{e^t}{n} \right)\left(\frac{1 - e^{nt}}{1 - e^t} \right). \tag{3.80}$$

Example 3.29

Let the random variable X represent the outcome of the experiment of rolling a fair die. Then, X is a discrete uniform random variable. Figure 3.3 shows the pmf of X.

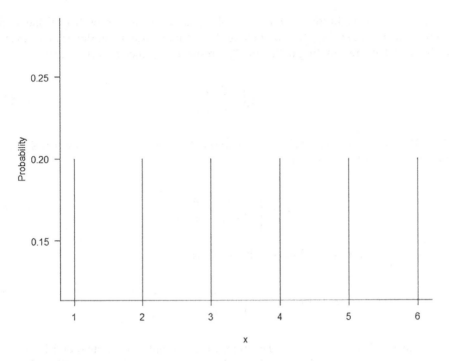

FIGURE 3.3 pmf of uniform random variable

Example 3.30

Let X be a finite random variable representing the outcomes of rolling a fair die. That is, the values of X are $1,2,3,4,5,6$. Hence, $p_X = P(x = i) = \frac{1}{6}, i = 1,2,\dots,6$. Thus, from (3.77) through (3.78), we have:

$$E(X) = \frac{(6+1)}{2} = \frac{7}{2}, Var(X) = \frac{(36-1)}{12} = \frac{35}{12}, \text{ and } M(t) = \left(\frac{e^t}{6}\right)\left(\frac{1-e^{6t}}{1-e^t}\right).$$

The cmf of discrete uniform random variable X, denoted by $F_X(x)$, is

$$F_X(x) = P(X \le x) = \begin{cases} \dfrac{x-a+k}{b-a+k}, & a < k < b, x = x_1, x_2, \dots, x_n \\ 0, & \text{otherwise.} \end{cases} \qquad (3.81)$$

3.4.2 BERNOULLI PMF

Definition 3.20 Bernoulli pmf

If a trial has exactly two possible outcomes, it is called a **Bernoulli trial**. The possible outcomes of a Bernoulli trial are referred to as **success** and **failure**. The random variable representing such a trial is called a **Bernoulli random variable**. Because the random variable is discrete, its pmf is referred to as the **Bernoulli pmf**.

According to Definition 3.9, the sample space Ω for a Bernoulli trial has two sample points as $\Omega = \{s, f\}$. If we denote the random variable on this sample space with 1 and 0, corresponding to "s" and "f", respectively, then $\Omega = \{0,1\}$ and

$$X : \begin{cases} S \to 1, \\ f \to 0. \end{cases} \tag{3.82}$$

For a **Bernoulli random variable** X, defining the probability of "1" by p, $0 \le p \le 1$, then the Bernoulli pmf will be:

$$p_x = \begin{cases} p, & x = 1, \\ 1 - p, & x = 0. \end{cases} \tag{3.83}$$

Letting $q = 1 - p$, formula (3.83) can be rewritten as:

$$p_x = P(X = x) = \begin{cases} p^x q^{1-x}, & x = 0,1, \\ 0, & \text{otherwise.} \end{cases} \tag{3.84}$$

To see that (3.82) and (3.83) are the same, we can simply substitute 0 and 1 for x in (3.83), and we obtain (3.83). Formula (3.83) is referred to as the **Bernoulli pmf** since

a. $p^x q^{1-x} > 0$, for both $x = 0$ and $x = 1$,

b. $\sum_{x=0}^{1} p^x q^{1-x} = p + q = 1$.

From (3.84), the cmf for Bernoulli random variable is as follows:

$$cmf(X) = \begin{cases} 0, & x < 0, \\ q, & 0 \le x < 1, \\ 1, & x \ge 1. \end{cases} \tag{3.85}$$

From (3.83) and (3.84), to find the mean, variance, and moment generating function of X, we go through Table (3.23) as follows.
Thus,

$$E(X) = p. \tag{3.86}$$

$$Var(X) = pq. \tag{3.87}$$

$$M(t) = q + pe^t. \tag{3.88}$$

$$G(z) = q + pz. \tag{3.89}$$

TABLE 3.23
Moments of Bernoulli pmf

X	1	0	Sum
$P_X(x)$	p	q	1
$xP_X(x)$	p	0	p
X^2	1	0	
$x^2 P_X(x)$	p	0	p
$e^{tx} p_X(x)$	pe^t	q	$q + pe^t$
$z^x p_X(x)$	zp	q	$q + pz$

Example 3.31

Suppose we toss a biased coin such that the probability of a head shows up is $p = 2/3$ and a tail shows up is $q = 1/3$, $p + q = 2/3 + 1/3 = 1$. Let X be a random variable representing such a random experiment. Letting 0 and 1 representing tail and head for showing up a tail and a head, respectively, X has values as 0 and 1. Thus, $P(X = 0) = \dfrac{1}{3}$ and $P(X = 1) = \dfrac{2}{3}$. In other words,

$$p_x = P(X = x) = \begin{cases} \dfrac{1}{3}, & x = 0, \\ \dfrac{2}{3}, & x = 1, \end{cases}$$

or

$$p_x = P(X = x) = \left(\frac{2}{3}\right)^x \left(\frac{1}{3}\right)^{1-x}, x = 0, 1.$$

We now calculate the mean, variance, moment generating function, and generating function using (3.85)–(3.88) as follows (Table 3.24).

TABLE 3.24
Example 3.31

X	0	1	Sum	
$p_X(x)$	1/3	2/3	1	
$xp_X(x)$	0	2/3	2/3	$E(X) = 2/3 = p$
X^2	0	1		
$x^2 p_X(x)$	0	2/3	2/3	$Var(X) = 2/3 - (2/3)^2 = 2/9 = pq$
$e^{tx} p_X(x)$	$e^t/3$	$2e^t/3$	e^t	$M(t) = 1/3 + 2e^t/3 = q + pe^t$
$z^x p_X(x)$	$z/3$	$2z/3$	z	$G(z) = 1/3 + 2z/3 = q + pz$

If a Bernoulli trial is repeated, independently, n times with the same probabilities of success and failure in each trial, then the sample space will contain 2^n sample points.

Example 3.32

If we toss a coin, the outcomes of each trial will be heads (H) or tails (T). Hence, for one toss, we will have $\Omega = \{H, T\}$ that contains 2^1 elements. For 2 coin tosses, we will have $\Omega = \{(H,H),(H,T),(T,H),(T,T)\}$ that contains $4 = 2^2$ elements. That is, the first and the second tosses turn out heads, the first a head and second tail, the first tail and second head, and both tails, respectively; that is, each appears as an ordered pair elements of Ω.

If we let the head and tail appear on the coin after each toss with probabilities p and q, respectively, then the probabilities of elements of the sample space Ω are listed in Table (3.25).

Similarly, for three tosses of the coin, the sample space will have $8 = 2^3$ elements as:

$$\Omega = \{(H,H,H),(H,H,T),(H,T,H),(T,H,H),(H,T,T),(T,H,T),(T,T,H),(T,T,T)\}$$

Then, probabilities of elements of Ω are triplets as listed in Table 3.26 (the parentheses are dropped for the sake of ease).

We leave it as an exercise to show that the sum of the probabilities in Table 3.26 is 1; that is, $p^3 + 3p^2q + 3pq^2 + q^3 = 1$.

Now let us suppose that the coin is biased with the probability of a head to occur as 5/8, and it is flipped three times independently; then, the probability of a head

appearing only on the second trial would be $P(THT) = pq^2 = \left(\frac{5}{8}\right)\left(\frac{3}{8}\right)^2 = \frac{45}{125} \approx 36\%$.

Similarly, the probability of three heads in a row is $\left(\frac{5}{8}\right)^3 \approx 24\%$, while that of having three tails in a row is $\left(\frac{3}{8}\right)^3 \approx 5\%$.

TABLE 3.25
Example 3.32

1	2	3	4	Sum
(H, H)	(H, T)	(T, H)	(T, T)	$(p+q)^2 = 1$
p^2	pq	pq	q^2	

TABLE 3.26
Example 3.32

1	2	3	4	5	6	7	8	Sum
HHH	HHT	HTH	THH	HTT	THT	TTH	TTT	1
p^3	p^2q	p^2q	p^2q	pq^2	pq^2	pq^2	q^3	

Example 3.33

In this example, we will simulate a Bernoulli counting process to count the number of success in N trials, say n, $n = 0,1,2,\ldots N$. For this purpose, we choose $N = 50$ and the probability of success in each trial as 0.6. Figure 3.4 shows the number of success for select numbers of trials.

To obtain Figure 3.4, we have used MATLAB® version 2017b. Table 3.27 describes the graph.

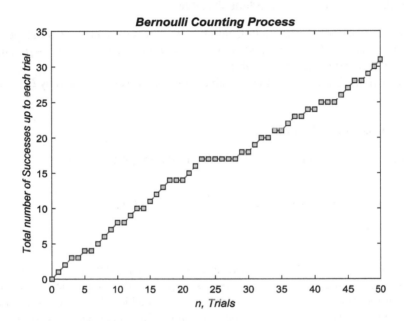

FIGURE 3.4   Bernoulli counting process simulation.

TABLE 3.27
Example 3.33

Trial No.	No. of Successes
1	1
5	4
10	8
15	11
20	14
25	17
30	18
35	21
40	24
45	37
50	31

3.4.3 Binomial pmf

Instead of looking at the order of occurrences of outcomes in repeated independent trials, we might ask about the number of successes or failures in n independent trials. The variable representing this concept is referred to as the **counting variable**. For starters, we define the following random variable.

Definition 3.21 (Binomial Random Variable)

Let random variable X represent the number of successes in n repeated independent Bernoulli trials with the probability of success p and of failure $q = 1 - p$. This is termed a **binomial random variable**.

The pmf of a binomial random variable X for j successes is denoted by $\mathcal{B}_j(j; n, p)$.

Theorem 3.4 Binomial pmf

Let X be a binomial random variable. Then, **binomial pmf** is given by:

$$P(X = j) \equiv \mathcal{B}_j(j; n, p) = \binom{n}{j} p^j q^{n-j}, \quad j = 0, 1, 2, \ldots, n, q = 1 - p. \quad (3.90)$$

Proof:
It was noted that if a Bernoulli trial is repeated independently n times, then the sample space will contain 2^n sample points. Repeating the independent trials yields a row of success and failures, as we saw in the two tables above for tossing a coin repeatedly. We also illustrated that the probability of a sequence in a row is the product of outcomes in that row. Since we are counting only successes in the n number of repeated trials, the order of occurrence of outcomes does not matter. Thus, we have to choose j successes out of the n trials, that is, $\binom{n}{j}$, $j = 0, 1, 2, \ldots, n$. Hence, for j successes, it remains $n - j$ failures, and using the product of probabilities, p^j and q^{n-j}, respectively, we have (3.90). It remain to show that (3.90) constitutes a discrete distribution function. For each term of $\binom{n}{j}$, p^j and q^{n-j} are nonnegative and, thus, the right-hand side of (3.90) is nonnegative.

Finally, we have to show that the sum is 1 for (3.90) to be a pmf.

A well-known algebraic expression with two terms, say x and y, can be raised to power k, $k = 0, 1, 2, \ldots, n$, where n is a positive integer, inductively, and we obtain:

$$(x + y)^n = \sum_{k=0}^{n} \binom{n}{k} x^k y^{n-k}, \ n \text{ is a positive integer} \quad (3.91)$$

$$= x^n + nx^{n-1}y + \frac{n(n-1)}{2!}x^{n-2}y^2 + \frac{n(n-1)(n-2)}{3!}x^{n-3}y^3 + \cdots + y^n.$$

The latter is termed the **binomial theorem** or **binomial expansion**.

Note 3.26

Formula (3.90) was somewhat generalized for $y = 1$ and n as a nonnegative real number by Arflken (1985). Later, Graham et al. (1994) generalized it further for n as a nonnegative real number, say v, as follows:

$$(x+y)^v = \sum_{k=0}^{\infty} \binom{v}{k} x^k y^{v-k}, \tag{3.92}$$

assuming the series converges for nonnegative integer v or $|x/y| < 1$.

By (3.91),

$$\sum_{j=0}^{n} \binom{n}{j} p^j q^{n-j} = (p+q)^n = 1. \tag{3.93}$$

In summary, the binomial pmf $B_j(j; n, p)$ describes the behavior of a random variable (or **count variable**) X assuming the following:

  i. The number of trials or observations, n, is fixed.
 ii. Each trial is independent of others.
iii. Each trial has two outcomes.
 iv. The probability of success in each trial is p.

From (3.90), the cmf for binomial random variable X is defined as:

$$P(X \le j) = \sum_{k=0}^{[j]} \binom{n}{k} p^k q^{n-k}, \tag{3.94}$$

where $q = 1 - p$ and $[j]$ is the greatest integer less than or equal to j.

Note 3.27

Historically, a special case of the binomial theorem was known by the Greek mathematician Euclid, about fourth century B.C. The coefficients of the binomial theorem form a combination of "n choose k" that is known as **Pascal's triangle**. It is related to Hindu lyricis Pingala, Ramakrishna Rao (2007), about 200 B.C.

$$
\begin{array}{ccccccccccccc}
 & & & & & & 1 & & & & & & \\
 & & & & & 1 & & 1 & & & & & \\
 & & & & 1 & & 2 & & 1 & & & & \\
 & & & 1 & & 3 & & 3 & & 1 & & & \\
 & & 1 & & 4 & & 6 & & 4 & & 1 & & \\
 & 1 & & 5 & & 10 & & 10 & & 5 & & 1 & \\
1 & & 6 & & 15 & & 20 & & 15 & & 6 & & 1
\end{array}
$$

FIGURE 3.5   Pascal–Khayyam triangle.

It has also been evidenced that the theorem for $n = 3$ was known by Indians in the sixth century. The binomial theorem as we know it today is in the work of Persian mathematician Al-Kharaji. The coefficients of the binomial theorem is a combination of "n choose k", that is, $\begin{pmatrix} n \\ k \end{pmatrix}$, and in the recent texts is known as **Pascal's triangle**. It was included in Al-Kharaji's work describing the binomial theorem. In fact, Al-Kharaji proved both the binomial theorem and the so-called Pascal triangle. Additionally, the well-known Persian mathematician, astronomer, and poet Omar Khayyam was familiar with the binomial theorem and its expansion with some higher order of n (see Figure 3.5).

Example 3.34

Suppose the grade point average (GPA) of 20% of students in a certain college is more than 3.5 on 4-point scale. Ten students are randomly chosen from all the students at this college.

Let the random variable X represent the total number of students out of 10, whose GPA is more than 3.5. Then, X is a binomial random variable, and graph of its pmf is shown in Figure 3.6.

Example 3.35

Let us consider a probability and statistics class with 30 students enrolled. A student registered for such a course has a 70% chance of passing the final examination. The professor is interested to know the chance of 20 students passing his final examination. In this case, using a binomial distribution, the random variable represents the number of students (out of 30) who pass the examination. To find the desired probability, we use (3.90) with $n = 30$, $j = 20$, and $p = 0.7$. Thus, we have:

$$
B_{20}(20; 30, 0.7) = \begin{pmatrix} 30 \\ 20 \end{pmatrix} (0.7)^{20} (0.3)^{10} = 0.1416 \approx 14\%.
$$

Example 3.36

Let us consider a large family of five children, which includes girls and boys. We want to find the pmf of boys and girls in the family, assuming equal probabilities for both.

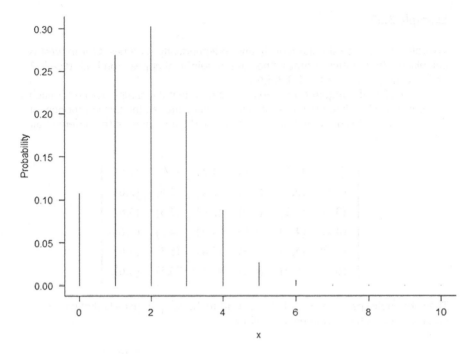

FIGURE 3.6 pmf of binomial random variable.

Answer

Let X represent the number of girls in the family. We consider each of the children a trial with a probability of success (a girl) p and failure (a boy) $1-p$. Assume that successive births are independent with equal probability for girls and boys as 1/2. Thus, from (3.90), we have:

$$P\left(\text{exactly } x \text{ girls}\right) = P(X = x) = \left(\begin{array}{c} 5 \\ x \end{array}\right)\left(\frac{1}{2}\right)^x \left(\frac{1}{2}\right)^{5-x}$$

$$= \left(\begin{array}{c} 5 \\ x \end{array}\right)\left(\frac{1}{2}\right)^5.$$

Choosing different value for x, we will have Table 3.28.

TABLE 3.28
Example 3.36

X	0	1	2	3	4	5
pmf, $p_X(x)$	1/32	5/32	10/32	10/32	5/32	1/32

Example 3.37

We roll two fair dice simultaneously and independently six times. Our interest is calculating the number of times "the sum of point that appears on both dice is 7 or 12 occurs j times, $j = 0,1,2,3,4,5,6$".

To calculate the inquired probability, we note that the sample space for each trial contains 36 ordered pairs as listed in the following set, the first component of the ordered pair from the first die and the second component of the ordered pair from the second die:

$$
\Omega = \left\{ \begin{array}{cccccc}
(1,1) & (1,2) & (1,3) & (1,4) & (1,5) & (1,6) \\
(2,1) & (2,2) & (2,3) & (2,4) & (2,5) & (2,6) \\
(3,1) & (3,2) & (3,3) & (3,4) & (3,5) & (3,6) \\
(4,1) & (4,2) & (4,3) & (4,4) & (4,5) & (4,6) \\
(5,1) & (5,2) & (5,3) & (5,4) & (5,5) & (5,6) \\
(6,1) & (6,2) & (6,3) & (6,4) & (6,5) & (6,6)
\end{array} \right\}.
$$

The subsample space, say Ω_1, of elements from the sample space Ω that have sum equal to 7 or 12 has 7 elements as follows:

$$
\Omega = \left\{ \begin{array}{cccccc}
 & & & & & (1,6) \\
 & & & & (2,5) & \\
 & & & (3,4) & & \\
 & & (4,3) & & & \\
 & (5,2) & & & & \\
(6,1) & & & & & (6,6)
\end{array} \right\}.
$$

Thus, the probability of getting a sum of 7 or 12 (i.e., a success) in each trial is $p = \dfrac{7}{36}$. Now, let us denote the random variable X representing the number of times "the sum of point that appears on both dice is 7 or 12 occurs j times". Thus, X is a binomial random variable taking its values as j. In this case, $j = 6$.

$$
B_j\left(j; 6, \frac{7}{36} \right) = \binom{6}{j}\left(\frac{7}{36} \right)^j \left(\frac{29}{36} \right)^{6-j}, \quad j = 0,1,2,3,4,5,6.
$$

For instance, for $j = 2$, that is, appearance of 7 or 12 in 2 rolls of the dice, we will have:

$$
B_2\left(2; 6, \frac{7}{36} \right) = \binom{6}{2}\left(\frac{7}{36} \right)^2 \left(\frac{29}{36} \right)^4 \approx 24\%.
$$

Now, what is the mean of the number of successes in n independent repetitions of a Bernoulli trial with the probability of a success as p?

The following theorem will aid in calculations involving moments and moment generating functions.

Theorem 3.5

For a binomial random variable X with $\mathcal{B}_j(j;n,p)$, $n = 0,1,\ldots$, letting $q = 1-p$, we have the following properties:

$$E(X) = np. \tag{3.95}$$

$$Var(X) = np(1-p) = npq. \tag{3.96}$$

$$M(t) = \left[pe^t + (1-p)\right]^n = \left(pe^t + q\right)^n. \tag{3.97}$$

Proof:
We first prove the moment generating function, from which we derive the first and second moments and, consequently, the variance.

$$M(t) = E\left(e^{tX}\right) = \sum_{j=0}^{n} e^{tj} p_j = \sum_{j=0}^{n} e^{tj} \binom{n}{j} p^j q^{n-j}$$

$$= \sum_{j=0}^{n} \binom{n}{j} \left(pe^t\right)^j (1-p)^{n-j} = \left[pe^t + (1-p)\right]^n.$$

$$E(X) = \sum_{j=0}^{n} j \binom{n}{j} p^j q^{n-j} = M'(0) = n(p+1-p)p = np.$$

$$E(X^2) = \sum_{j=0}^{n} j^2 \binom{n}{j} p^j q^{n-j} = M''(0)$$

$$= n(n-1)\left[pe^t + (1-p)\right]^{n-2} \left(pe^t\right)^2 + n\left[pe^t + (1-p)\right]^{n-1} \left(pe^t\right)\Big|_{t=0}$$

$$= n(n-1)p^2 + np.$$

$$Var(X) = n(n-1)p^2 + np - (np)^2 = np(1-p).$$

Note 3.28

The proof of the expected value of the binomial random variable is as follows:
From (3.90), for $n = 0,1,\ldots$, we have:

$$E(X) = \sum_{j=0}^{n} j \binom{n}{j} p^j (1-p)^{n-j}$$

$$= np \sum_{j=0}^{n} j \frac{(n-1)!}{(n-j)j!} p^{j-1} (1-p)^{(n-1)-(j-1)}$$

$$= np \sum_{j=1}^{n} \frac{(n-1)!}{\left[(n-1)-(j-1)\right]!(j-1)!} p^{j-1} (1-p)^{(n-1)-(j-1)}$$

$$= np \sum_{j=1}^{n} \binom{n-1}{j-1} p^{j-1} (1-p)^{(n-1)-(j-1)}$$

$$= np \sum_{k=0}^{n-1} \binom{n-1}{k} p^{k} (1-p)^{(n-1)-k}, \quad k = j-1$$

$$= np \sum_{k=0}^{l} \binom{l}{k} p^{k} (1-p)^{l-k}, \quad l = n-1$$

$$= np \left[p + (1-p) \right]^{l}$$

$$= np.$$

Example 3.38

We return to Example 3.28 and consider tossing a fair coin three times. That is, the probability of each outcome in each independent trial is 1/2. The sample space has $2^3 = 8$ elements. Suppose we are interested to know the probabilities of the number of heads to appear in the end of tossing process, that is, the pmf of number of heads.

Answer

We revise Table 3.29 for this example and probabilities inquired.

Let X denote a random variable representing the number of heads in the three tosses of a fair coin. Hence, we summarize parts of the information in Tables 3.29a and 3.29b.

We now want to use binomial pmf, relation (3.90), to answer the question.

$$P(X = 0) \equiv B_0(0; 3, 0.5) = \binom{3}{0}(0.5)^0 (0.5)^{3-0} = \frac{1}{8} = 0.125.$$

TABLE 3.29A
Example 3.38

No. of Elements of Ω	1	2	3	4	5	6	7	8
Outcomes	*HHH*	*HHT*	*HTH*	*HTT*	*THH*	*THT*	*TTH*	*TTT*
pmf with $P(H) = P$	p^3	p^2q	p^2q	pq^2	p^2q	pq^2	pq^2	q^3
pmf with $P(H) = \dfrac{1}{2}$	$\dfrac{1}{8}$	$\dfrac{1}{8}$	$\dfrac{1}{8}$	$\dfrac{1}{8}$	$\dfrac{1}{8}$	$\dfrac{1}{8}$	$\dfrac{1}{8}$	$\dfrac{1}{8}$
Number of heads	3	2	2	1	2	1	1	0

TABLE 3.29B
Example 3.38

X	0	1	2	3	Sum
No. of heads outcomes	1	3	3	1	8
pmf of No. of heads	$\dfrac{1}{8}$	$\dfrac{3}{8}$	$\dfrac{3}{8}$	$\dfrac{1}{8}$	1

$$P(X=1) \equiv B_1(1;3,0.5) = \begin{pmatrix} 3 \\ 1 \end{pmatrix}(0.5)^1(0.5)^{3-1} = \frac{3}{8} = 0.375.$$

$$P(X=2) = B_2(2;3,0.5) = \begin{pmatrix} 3 \\ 2 \end{pmatrix}(0.5)^2(0.5)^{3-2} = \frac{3}{8} = 0.375.$$

$$P(X=3) \equiv B_3(3;3,0.5) = \begin{pmatrix} 3 \\ 3 \end{pmatrix}(0.5)^3(0.5)^{3-3} = \frac{1}{8} = 0.125.$$

These are the same as the last row in Table 3.29b.
    As before, the mean and variance are given in Table 3.29c.

$$E(X) = 1.5 = np = (3)(0.5).$$

$$E(X^2) = 3.$$

$$Var(X) = 3 - (1.5)^2 = 0.75$$

$$= np(1-p) = 3 \cdot 0.5 \cdot 0.5 = 0.75.$$

TABLE 3.29C
Example 3.38

X	0	1	2	3	Sum
p_x	$\dfrac{1}{8}$	$\dfrac{3}{8}$	$\dfrac{3}{8}$	$\dfrac{1}{8}$	1
xp_x	0	$\dfrac{3}{8}$	$\dfrac{6}{8}$	$\dfrac{3}{8}$	1.5
X^2	0	1	4	9	
$x^2 p_x$	0	$\dfrac{3}{8}$	$\dfrac{12}{8}$	$\dfrac{9}{8}$	3

Example 3.39

Suppose it is known that 19% of individuals with Crohn's disease (a chronic inflammatory disease of the large intestine) who are using a certain medicine experience side effects. To analyze this case, a sample of 100 Crohn's disease patients using this drug are studied. Let X be a random variable representing the number of patients who will experience some known side effects. If we are interested in a specific number j of X, then the pmf of X will be $B_j(j;100,0.19)$. For instance, for $j = 10$, we will have:

$$B_{10}(10;100,0.19) = \begin{pmatrix} 100 \\ 10 \end{pmatrix}(0.19)^{10}(0.81)^{100-10} = 0.0062 \approx 0.62\%.$$

Of course, the mean, variance, moment generating function, and generating function of X, from (3.102) to (3.105), are, respectively, as:

$$E(X) = np = 100(0.19) = 19.$$

$$Var(X) = np(1-p) = 100(0.19)(0.81) = 15.39.$$

$$M(t) = \left[pe^t + (1-p)\right]^n = \left[0.19e^t + 0.81\right]^{100}.$$

$$G(z) = 0.81 + 0.19z.$$

Example 3.40

Suppose we toss a biased coin as in the previous example. However, in this example, we let the probability of a head to show up as $p = 0.4$ and of a tail to show up as $1 - p = 0.6$. We repeat tossing, where j is the number of successes in 10 trials. In this case, pmf will be

$$B_j(j;10,0.4) = \binom{10}{j}(0.40)^j(0.6)^{10-j}, \quad j = 0,1,2,\dots 10.$$

Table 3.30 and Figure 3.7 illustrate this example.

For this example, the mean, variance, moment generating function, generating function, and cmf are, respectively, as follows:

$$E(X) = np = 10(0.4) = 4,$$

$$Var(X) = np(1-p) = 10(0.4)(0.6) = 2.4,$$

$$M(t) = \left[pe^t + (1-p)\right]^n = \left[0.4e^t + 0.6\right]^{10},$$

$$G(z) = 0.6 + 0.4z,$$

TABLE 3.30
Binomial pmf of Example 3.40

n	0	1	2	3	4	5	6	7	8	9	10
1	0.6000	0.4000	0	0	0	0	0	0	0	0	0
2	0.3600	0.4800	0.1600	0	0	0	0	0	0	0	0
3	0.2160	0.4320	0.2880	0.0640	0	0	0	0	0	0	0
4	0.1296	0.3456	0.3456	0.1536	0.0256		0	0	0	0	0
5	0.0778	0.2592	0.3456	0.2304	0.0768	0.0102	0	0	0	0	0
6	0.0467	0.1866	0.3110	0.2765	0.1382	0.0369	0.0041	0	0	0	0
7	0.0280	0.1306	0.2613	0.2903	0.1935	0.0774	0.0172	0.0016	0	0	0
8	0.0168	0.0896	0.2090	0.2787	0.2322	0.1239	0.0413	0.0079	0.0007	0	0
9	0.0101	0.0605	0.1612	0.2508	0.2508	0.1672	0.0743	0.0212	0.0035	0.0003	0
10	0.0060	0.0403	0.1209	0.2150	0.2508	0.2007	0.1115	0.0425	0.0106	0.0016	0.0001

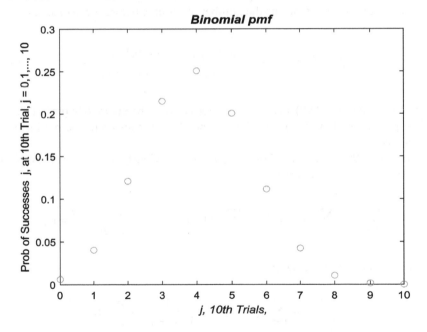

FIGURE 3.7 Graph of a binomial pmf, Example 3.40.

$$cmf = \sum_{j=0}^{10} \binom{10}{j} (0.40)^j (0.6)^{10-j}.$$

In other words, the probability of having no girl among 5 children is 1/32, the same as having all 5 to be girls, etc.

3.4.4 GEOMETRIC PMF

Definition 3.22 Geometric Random Variable

Consider independently repeating a Bernoulli trial with probability p of success in each trial. Let X be a discrete random variable that represents the first success after a number of failures or the number of failures before occurrence of the first success. In this case, X is called a **geometric random variable** with either one of the two cases:

Case 1. The first success is included in the number of trials; that is, the x number of trials consists of $x-1$ failure followed by a success and pmf $p_x, x = 1,2,\ldots,$ given by:

$$p_x = P(X = x) = \begin{cases} p(1-p)^{x-1}, & x = 1,2,\ldots, \\ 0, & \text{otherwise.} \end{cases} \tag{3.98}$$

Case 2. The number of trials is up to and not including the first success and pmf p_x, given by here X means the number of failures before the fist success:

$$p_x = P(X = x) = \begin{cases} p(1-p)^{x}, & x = 0,1,2,\ldots, \\ 0, & \text{otherwise.} \end{cases} \tag{3.99}$$

Each of the relations (3.98) and (3.99) is referred to as the **geometric pmf** of X.

The pmf defined in (3.98) describes the number of trials up to and not including the first success.

Relation (3.98) is a pmf. For starters, it's obviously nonnegative. Applying geometric series sum, we have:

$$\sum_{x=1}^{\infty} p_x = \sum_{x=1}^{\infty} p(1-p)^{x-1} = \sum_{x=0}^{\infty} p(1-p)^{x} = p\frac{1}{1-(1-p)} = \frac{p}{p} = 1.$$

Similarly, from (3.99), we have:

$$\sum_{x=0}^{\infty} p_x = \sum_{x=0}^{\infty} p(1-p)^{x} = p\frac{1}{1-(1-p)} = \frac{p}{p} = 1.$$

Example 3.41 Geometric Distribution

Suppose 20% of the students in a certain college have a GPA more than 3.5. You select students randomly until you meet the first student with at least a GPA of 3.5. Let the random variable X represent the total number of students you select until you find the first student with GPA 3.5 or higher. Then, the random variable X follows a geometric distribution with pmf as in Figure 3.8.

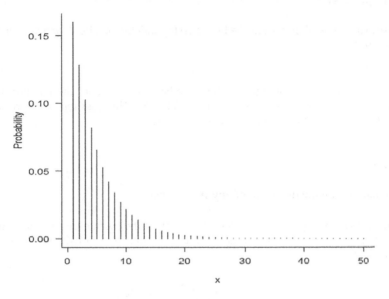

FIGURE 3.8   The geometric pmf, Example 3.41.

The **cmfs for the geometric random variable** are:

$$\sum_{x=1}^{n} px = \sum_{x=1}^{n} p(1-p)^{x-1} = 1-(1-p)^n, \quad n=1,2,\dots \tag{3.100}$$

and

$$\sum_{x=0}^{n} p_x = \sum_{x=0}^{n} p(1-p)^x = 1-(1-p)^{n+1}, \quad n=0,1,2,\dots \tag{3.101}$$

Relation (3.100) is because:

$$\sum_{x=1}^{n} px = \sum_{x=1}^{n} p(1-p)^{x-1}$$

$$= p\sum_{x=0}^{n-1}(1-p)^x$$

Now, using the sum of a finite geometric progression with ratio $r \neq 1$, we have:

$$p\sum_{x=0}^{n-1}(1-p)^x = p\frac{1-(1-p)^n}{1-(1-p)} \tag{3.102}$$

$$= 1-(1-p)^n, \quad n=1,2,\dots$$

We leave the proof of (3.102) as an exercise.

Example 3.42

How many times should a fair die be rolled for until the number 3 appears for the first time?

Answer

We denote the event of "the number of other numbers until the number 3 to appear for the first time by a random variable X". The question is finding the number of failures before the first success occurs.

Note 3.29

The number on the die is not of any importance.

Now, the die being fair, the appearance of any of the six numbers has the same probability, 1/6. Thus, X has a geometric pmf, and from (3.100), we have the following:

$$p_n = P(X = n) = \left(\frac{1}{6}\right)\left(\frac{5}{6}\right)^n, \quad n = 0,1,2,....$$

For each of the numbers 1 through 6, the probability of obtaining that number for the first time after n rolls is:

$$p_0 = P(X = 0) = \left(\frac{1}{6}\right)\left(\frac{5}{6}\right)^0 = \left(\frac{1}{6}\right) \approx 17\%.$$

$$p_1 = P(X = 1) = \left(\frac{1}{6}\right)\left(\frac{5}{6}\right)^1 = \left(\frac{5}{36}\right) \approx 14\%.$$

$$p_2 = P(X = 2) = \left(\frac{1}{6}\right)\left(\frac{5}{6}\right)^2 = \left(\frac{25}{216}\right) \approx 12\%.$$

$$p_3 = P(X = 3) = \left(\frac{1}{6}\right)\left(\frac{5}{6}\right)^3 = \left(\frac{125}{1,296}\right) \approx 10\%.$$

$$p_4 = P(X = 4) = \left(\frac{1}{6}\right)\left(\frac{5}{6}\right)^4 = \left(\frac{625}{7,776}\right) \approx 8\%.$$

$$p_5 = P(X = 5) = \left(\frac{1}{6}\right)\left(\frac{5}{6}\right)^5 = \left(\frac{3125}{46,656}\right) \approx 7\%.$$

$$p_6 = P(X = 6) = \left(\frac{1}{6}\right)\left(\frac{5}{6}\right)^6 = \left(\frac{15625}{279,936}\right) \approx 6\%.$$

$$p_7 = P(X = 7) = \left(\frac{1}{6}\right)\left(\frac{5}{6}\right)^7 = \left(\frac{78125}{1,679,616}\right) \approx 5\%.$$

As n approaches infinity, $\left(\dfrac{5}{6}\right)^n$ approaches 0, and hence, $p_\infty = 0$. Of course,

$$\sum_{n=0}^{\infty}\left(\frac{1}{6}\right)\left(\frac{5}{6}\right)^n = \left(\frac{1}{6}\right)\sum_{n=0}^{\infty}\left(\frac{5}{6}\right)^n = \left(\frac{1}{6}\right)\left(\frac{1}{1-\dfrac{5}{6}}\right) = 1.$$

Example 3.43

Consider a biased coin with probabilities of a head and a tail as 3/4 and 1/4, respectively. What is the probability of a head to appear after five tosses of this coin for the first time?

Answer

Let the random variable X denote the event "the number of tails until a 'head' appears for the first time". Then,

$$p_n = P(X = n) = \left(\frac{3}{4}\right)\left(\frac{1}{4}\right)^n, \quad n = 0, 1, 2, \ldots.$$

The probability of obtaining a head for the first time after n tosses is:

$$p_0 = P(X = 0) = \left(\frac{3}{4}\right)\left(\frac{1}{4}\right)^0 = \left(\frac{3}{4}\right) = 75\%.$$

$$p_1 = P(X = 1) = \left(\frac{3}{4}\right)\left(\frac{1}{4}\right)^1 = \left(\frac{3}{16}\right) \approx 19\%.$$

$$p_2 = P(X = 2) = \left(\frac{3}{4}\right)\left(\frac{1}{4}\right)^2 = \left(\frac{3}{64}\right) \approx 5\%.$$

$$p_3 = P(X = 3) = \left(\frac{3}{4}\right)\left(\frac{1}{4}\right)^3 = \left(\frac{3}{256}\right) \approx 1\%.$$

$$p_4 = P(X = 4) = \left(\frac{3}{4}\right)\left(\frac{1}{4}\right)^4 = \left(\frac{3}{1,024}\right) \approx 0\%.$$

Of course,

$$0.75 + 0.19 + 0.05 + 0.01 + 0 + \cdots + 0 = 1$$

or

$$\sum_{n=0}^{\infty}\left(\frac{3}{4}\right)\left(\frac{1}{4}\right)^n = \left(\frac{3}{4}\right)\sum_{n=0}^{\infty}\left(\frac{1}{4}\right)^n = \left(\frac{3}{4}\right)\left(\frac{1}{1-\dfrac{1}{4}}\right) = 1.$$

For the geometric random variable X, expected value means the expected number of trials before the first success.

Theorem 3.6

The expected value of a geometric random variable X, in cases of (3.98) and (3.99), is, respectively, as follows:

$$E(X) = \frac{1}{p}, \tag{3.103}$$

and

$$E(X) = \frac{1-p}{p}, \quad x = 0,1,2,\ldots, \tag{3.104}$$

Proof:
Case 1

For the case of (3.103), we have the following:

$$E(X) = \sum_{x=1}^{\infty} x p_x = \sum_{x=1}^{\infty} x p (1-p)^{x-1}$$

$$= p \left[\sum_{x=1}^{\infty} (1-p)^{x-1} + \sum_{x=2}^{\infty} (1-p)^{x-1} + \sum_{x=3}^{\infty} (1-p)^{x-1} + \cdots \right]$$

$$= p \left[\frac{1}{p} + \frac{1-p}{p} + \left(\frac{1-p}{p} \right)^2 + \cdots \right]$$

$$= 1 + (1-p) + (1-p)^2 + \cdots$$

$$= \frac{1}{p}.$$

Case 2

For the case of (3.104), we have the following:

$$E(X) = \sum_{x=0}^{\infty} x p_x = \sum_{x=0}^{\infty} x p (1-p)^x$$

$$= p \sum_{x=0}^{\infty} x (1-p)^x$$

$$= p(1-p) \sum_{x=0}^{\infty} x (1-p)^{x-1}.$$

But

$$\sum_{x=0}^{\infty} x(1-p)^{x-1} = \frac{d}{dp}\left(\sum_{x=0}^{\infty}(1-p)^{x}\right).$$

The series on the right is a geometric series that converges for $1-p < 1$. Assuming the convergence of the series, the summation and differentiation can be interchanged.

Now using the infinite geometric progression, we have:

$$\sum_{x=0}^{\infty} x(1-p)^{x-1} = -\frac{d}{dp}\left(\frac{1}{1-(1-p)}\right)$$

$$= \frac{1}{p^2}.$$

Thus,

$$p(1-p)\sum_{x=0}^{\infty} x(1-p)^{x-1} = \frac{p(1-p)}{p^2}$$

$$= \frac{1-p}{p}.$$

Example 3.44

Let us consider tossing a biased coin, with probability of a head as 2/3 and a tail as 1/3.

1. What is the probability of occurring the first head at the fifth trial?
2. What is the expected number of tosses?

Answer

Let X be a random variable representing the number of tosses until a head appears. Thus, we define the pmf of X by

$$p_X(x) = \left(\frac{1}{3}\right)\left(\frac{2}{3}\right)^{x-1}, \quad x = 1,2,\ldots$$

1. Hence, probability that we obtain a head, with the probability of 2/3 in each trial, after four failure trials, that is, on the fifth trail, is

$$p_X(5) = P(X = 5) = \left(\frac{2}{3}\right)^4\left(\frac{1}{3}\right) = \left(\frac{16}{81}\right)\left(\frac{1}{3}\right) = \frac{16}{243} = 7\%.$$

2. Now, expected value of the number of trials from (3.103) is:

$$E(X) = \frac{1}{p} = \frac{3}{2}.$$

Example 3.45

A relatively inexperienced basketball player is to enter a game. His past exercises show that his chance of success at a throw through the basket is one out of 4. How many times would the player fail on average before his first success occurs?

Answer

Let the random variable X describe the number of times player fails before a success. Hence, under the assumptions of the problem, X is a geometric pmf. The question is finding the mean of X.

Assuming having the ball in hand as the first failure, we should consider Case 2. Thus, we have:

$$E(X) = \frac{1 - 0.25}{0.25} = 3.$$

Example 3.46

Suppose we roll a fair die until a 2 appears for the first time. What is the mean number of number of trials?

Answer

From (3.104), we have:

$$p_X(x) = \left(\frac{1}{6}\right)\left(\frac{5}{6}\right)^{x-1}, \quad x = 1, 2, 3, \ldots$$

$$E(X) = \frac{1 - p}{p} = \frac{\frac{5}{6}}{\frac{1}{6}} = 5.$$

Theorem 3.7

Let X be a geometric random variable. Then,
For Case 1:

$$M(t) = \frac{pe^t}{1 - (1-p)e^t}, \quad 0 < p < 1. \tag{3.105}$$

$$Var(X) = \frac{1-p}{p^2}. \tag{3.106}$$

$$G(z)\frac{pz}{1 - (1-p)z}, \quad 0 < p < 1. \tag{3.107}$$

For Case 2:

$$M(t) = \frac{p}{1-(1-p)e^t}, \quad 0 < p \le 1. \tag{3.108}$$

$$Var(X) = \frac{1-p}{p^2}. \tag{3.109}$$

$$G(z) = \frac{p}{1-(1-p)z}, \quad 0 < p \le 1. \tag{3.110}$$

Proof:
We prove (3.108) and leave the proof of others as exercises.
   Based on definition of pgf, we have:

$$G(z) = \sum_{x=1}^{\infty} z^x p_X(x)$$

$$= \sum_{x=1}^{\infty} z^x p(1-p)^{x-1}$$

$$= pz \sum_{x=1}^{\infty} z^{x-1}(1-p)^{x-1}$$

$$= pz \sum_{x=1}^{\infty} \left[z(1-p)\right]^{x-1}$$

$$= \frac{pz}{1-(1-p)z}, \quad 0 < p < 1.$$

Example 3.47

Referring to Example 3.41, we have:

$$Var(X) = \frac{1-\dfrac{1}{6}}{\left(\dfrac{1}{6}\right)^2} = 30.$$

Another very important property of the geometric pmf is its memoryless.

Theorem 3.8

Let X be a geometric random variable and r and s two integers. Then,

$$P(X \ge r+x \mid X \ge r) = P(X \ge x). \tag{3.111}$$

(We will see later, in Chapter 7, that (3.111) is referred to as the **memoryless property** for X; i.e., the system does not remember its history.)

Proof:

From the general properties of a pmf, and for $x > 1$ and Case 1 of geometric pmf (and similarly for Case 2), we have:

$$P(X \geq x) = 1 - P(X < x) = P(X \leq x - 1).$$

$$= 1 - \left(1 - (1-p)^{(x-1)+1}\right) = (1-p)^x.$$

Thus, we need to prove the following:

$$P(X \geq r + x \mid X \geq r) = P(X \geq x) = (1-p)^x.$$

Now,

$$P(X \geq r + x \mid X \geq r) = \frac{P(\{X \geq r + x\} \cap \{X \geq r\})}{P(X \geq r)}$$

$$= \frac{P(X \geq r + x)}{P(X \geq r)}$$

$$= \frac{(1-p)^{(r+x)}}{(1-p)^r}$$

$$= (1-p)^x.$$

3.4.5 NEGATIVE BINOMIAL PMF

The following definition is a generalization of geometric pmf. Previously, we were interested in the probability of the first success in a sequence of repeated independent Bernoulli trials. Here, however, we are interested in the number of successes before a specific number (nonrandom) of failures to occur, for instance, tossing a coin repeatedly before the fifth tail occurs.

Definition 3.23 (Negative Binomial pmf)

A discrete random variable X representing the number of failures x before a specific, nonrandom, number of successes r in a sequence of independent Bernoulli trials with the probability of success p is called a **negative binomial** with pmf, denoted by $\mathcal{NB}(x; r, p)$, as:

$$\mathcal{NB}(x; r, p) = \binom{x + r - 1}{x} p^r (1-p)^x, \quad r > 0, \quad x = 0, 1, \dots \quad (3.112)$$

Note 3.30

Relation (3.112) is referred to as the **Pascal negative binomial pmf** with real r referred to as the **Pólya negative binomial pmf**.

Note 3.31

We leave it as an exercise to show that (3.112) can be rewritten as:

$$\mathcal{NB}(x;r,p)=(-1)^x\begin{pmatrix} -r \\ x \end{pmatrix}p^r(1-p)^x, \quad r>0 \quad x=0,1,.... \qquad (3.113)$$

To show that $\mathcal{NB}(x,r,p)$ defined by (3.112), indeed, is a pmf, we note that the right-hand side of (3.113) is positive and

$$\sum_{x=0}^{\infty}\mathcal{NB}(x;r,p)=\sum_{x=0}^{\infty}\begin{pmatrix} x+r-1 \\ x \end{pmatrix}p^r(1-p)^x,$$

$$=\sum_{x=0}^{\infty}(-1)^x\begin{pmatrix} -r \\ x \end{pmatrix}p^r(1-p)^x$$

$$=(1-p)^{-r}(1-p)^r=1.$$

Example 3.48

A company conducts a geological study in an agriculture search for underground water. It is known that exploring the water well has 25% chance of striking water.

Question 1. What is the probability that the first strike occurs on the fourth well drilled?

Question 2. What is the probability that the fourth strike occurs on the tenth well drilled?

Answers

Answer to Question 1
We could use a geometric pmf here since we are looking for the first success after three failures. Hence, having $p=0.25$, $1-p=0.75$, $n=4$, and $x=1$, as $r=n-x=4-1=3$ we have:

$$P(X=3)=\begin{pmatrix} 3+1-1 \\ 3 \end{pmatrix}p^1(1-p)^{4-1}=\begin{pmatrix} 3 \\ 3 \end{pmatrix}(0.25)(0.75)^3=0.1055\approx11\%.$$

Using geometric pmf and letting X representing the number of trials before the first success (strikes), we will have:

$$P(X=4)=(0.25)(0.75)^3=0.1055\approx11\%$$

Answer to Question 2

For this case, from (3.112), we have:

$$P(X=4)=\binom{6+4-1}{6}p^4(1-p)^{10-4}=\binom{9}{6}(0.25)^4(0.75)^6=0.0584\approx 6\%.$$

Let X be a discrete random variable with negative binomial pmf, representing the number of success x before a specific, nonrandom, number of failures r, $r > 0$, in a sequence of independent Bernoulli trials with the probability of success p, $0 < p < 1$. We leave proofs of the following as exercises:

$$E(X)=\frac{pr}{(1-p)},\qquad\qquad(3.114)$$

$$V(X)=\frac{pr}{(1-p)^2},\qquad\qquad(3.115)$$

$$M(t)=\left(\frac{1-p}{1-pe^t}\right)^r,\quad t<\ln p,\text{ and}\qquad\qquad(3.116)$$

$$G(z)=\left(\frac{1-p}{1-pz}\right)^r,\quad |z|<\frac{1}{p}.\qquad\qquad(3.117)$$

Example 3.49 Negative Binomial Distribution

Suppose 20% of the students in a certain college have a GPA more than 3.5. Let the random variable X represent the total number of students you select until you find 5 with GPA 3.5 or higher. Then, the random variable X follows a negative binomial distribution with pmf of X as in Figure 3.9.

3.4.6 HYPERGEOMETRIC PMF

The random variable we are to describe is very much similar to the binomial random variable we discussed earlier. As we will see in Chapter 5, a statistical **population** is a collection or a set of all individuals, objects, or measurements of interest. A **sample** is a portion, subset, or a part of the population of interest (finite or infinite number of them). A sample selected such that each element or unit in the population has the same chance to be selected is called a **random sample**. Sampling from a population is sometimes referred to as **random draw**.

Now, suppose we have a finite population of size N, $N = 0,1,2,\ldots$. We draw a sample of size n from this population without replacement. There are two types of outcomes from each draw in this experiment that we call them **success** and **failure**, in their general sense. Thus, each sample consists of two subsamples, one the success observations and the other failure observations, with total number of outcomes as n, which is the total number of draws, where $n = 0,1,2,\ldots,N$.

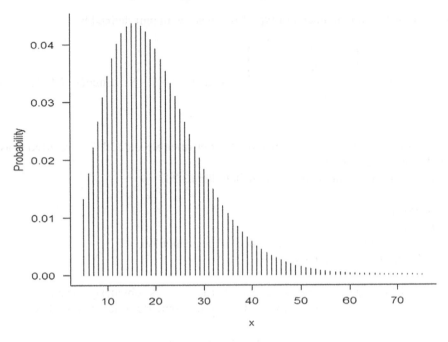

FIGURE 3.9   Negative binomial pmf, Example 3.49.

Note 3.32

Based on the description of the scenario, the probability of success and failure changes on each draw due to the change in the remaining population as a draw is made. This feature is the difference between the binomial pmf and hypergeometric pmf.

Now, we let S denote the total number of successes in the population, that is, $S = 0,1,2,\ldots,N$.

Of interest is calculating the probability of having x, $\max(0, n+S-N) \leq x \leq \min(n,S)$, successes in the sample.

To respond, we first argue as follows: Let us denote the random variable X to represent the number of successes. Looking for x successes would mean $n - x$ failures. The number of ways x successes can occur from the total successes in the population, S, is $\binom{S}{x}$ and of failures is $\binom{N-S}{n-x}$. Occurrence of the successes and failures is the product of the two, that is, $\binom{S}{x}\binom{N-S}{n-x}$. It is assumed that choosing n object from a total of N objects is equiprobable. Thus, we can now state the following definition.

Definition 3.24 Hypergeometric pmf

Let X represent the number x of type 1 outcomes with pmf defined by:

$$p_x = P(X = x) = \frac{\binom{S}{x}\binom{N-S}{n-x}}{\binom{N}{n}}, x = \max\left[0, n+S-N, \ldots, \min(n, S)\right]. \quad (3.118)$$

Then, X is called the **hypergeometric random variable** and (3.118) is referred to as the **hypergeometric pmf**.

Let us list the symbols involved in Definition 3.24.

N: Population,
n: Total number of draws,
S: The number of successes included in the population,
x: Number of observed successes.

$p_X(x)$ defined in (3.118) is a pmf because it is obviously positive when $\max(0, n+S-N) \le x \le \min(n, S)$, and applying geometric series sum, we have:

$$\sum_{0 \le x \le n} \frac{\binom{S}{x}\binom{N-S}{n-x}}{\binom{N}{n}} = 1.$$

Example 3.50

Consider a junior statistics class with 30 students enrolled that consists of 20 mathematics majors and 10 other science majors. To assess some measures regarding students' learning, we draw 10 students without replacement. What is the probability that there are exactly six mathematics majors in the drawing?

Answer

In this drawing, we consider the inclusion of a mathematics major as a success. We now summarize the problem's information in the following table:

$N = 30, x = 6, S = 20, n = 10$

	Drawn	Not Drawn	Total
Mathematics majors	$x = 6$	$S - x = 20 - 6 = 14$	$S = 20$
None mathematics majors	$n - x = 4$	$N + x - n - S = 6$	$N - S = 10$
Total	$n = 10$	$N - n = 30$	$N = 30$

Thus, from (3.118), we have:

$$p_X(6) = P(X = 6) = \frac{\binom{20}{6}\binom{30-20}{10-6}}{\binom{30}{10}}$$

$$= \frac{\binom{20}{6}\binom{10}{4}}{\binom{30}{10}} = \frac{\frac{20!}{6!14!}\frac{10!}{4!6!}}{\frac{30!}{10!20!}}$$

$$= \frac{8,139,600}{30,045,015} = 0.2709134943018001$$

$$\approx 27\%.$$

Example 3.51

Playing cards were found in thirteenth century AD, according to Wikipedia, (https://en.wikipedia.org/wiki/Playing_card). Their foundation goes back to the ninth century AD with 30 cards in Asia. About the eleventh century, it spreads from Asia, through Egypt and Europe, to the United States by the eighteenth century.

The currently used deck of playing card was found in the fifteenth century AD in Rouen, France, and comprises 52 cards with four suits: hearts, diamonds, cubs, and spades. English version of this deck appeared in the sixteenth century AD.

Now, suppose we draw five cards without replacement. What is the probability of getting three hearts?

Answer

To answer the question, we note that $N = 52$, the number of failures $= s_2 = 39$, the number of successes $= s_1 = 13$, $n = 5$, and $x = 3$. Hence,

$$p_3 = P(3 \text{ hearts}) = \frac{\binom{13}{3}\binom{52-13}{5-3}}{\binom{52}{5}} = 0.0815 \approx 8\%.$$

Example 3.52

Consider a College of Arts and Sciences at a small university with 300 faculty members, 30 of whom are international. A random sample of size 25 is taken from the faculty of this college. What is the probability of observing four international faculty members in the sample?

Answer

Once again, to answer the question in this example, we note that $N = 300$. The number of failures (non-international faculty) = $s_2 = 270$, the number of successes (international faculty) = $s_1 = 30$, $n = 25$, random variable X takes values between 0 and 25, and $x = 4$. Hence,

$$p_3 = P\left(4\ \text{international}\right) = \frac{\dbinom{30}{4}\dbinom{300-30}{25-4}}{\dbinom{300}{25}} = 0.1415 \approx 14\%.$$

Example 3.53 Hypergeometric Distribution

Consider a box with 30 light bulbs and assume that there are 20 non-defective and 10 defective bulbs. Suppose a quality controller randomly selects five bulbs without replacements. Let X be the random variable that represents the number of defective bulbs he finds in the selected five. Then, the random variable X follows a hypergeometric distribution, and the pmf of X can be graphed as in Figure 3.10.

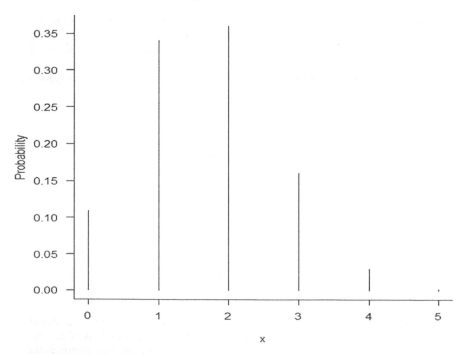

FIGURE 3.10   pmf of hypergeometric, Example 3.53

3.4.7 POISSON PMF

The final standard discrete random variable we will discuss is the Poisson. This is a very important discrete random variable in the theory of probability and stochastic processes, as an application of the theory of probability like arrival process of tasks.

A Poisson pmf assists in describing the chances of occurrences of a number of events in some given time interval or in a given space when the value of average number of occurrences of the events is known. Consider, for instance, a random experiment where we observe the number of occurrences of independent events during a given time interval such that two or more events do not occur at the same time, and the events are occurring at the same rate.

In many standard queues (waiting lines), the arrival to a waiting room (buffer) is assumed to follow Poisson pmf. Here are other examples of number of events occurring in an interval that Poisson pmf can be applied: telephone calls ringing in an office per hour, the number of cars arriving at a traffic light every 3 minutes, the number of patients arriving to an emergency room every hour, the number of mutations on a given strand of DNA over a unit time, the number of particles emitted by a radioactive source in a given time, and the number of network breakdowns over a 24-hour period.

The Poisson random variable and pmf is named after the French mathematician Siméon Denis Poisson (1781–1840) (Figure 3.11).

We now define the Poisson pmf.

Definition 3.25 Poisson pmf

A nonnegative random variable X, with a one parameter λ and pmf

$$p_j = P(X = j) = \frac{e^{-\lambda}\lambda^j}{j!}, \quad j = 0,1,\ldots, \tag{3.119}$$

FIGURE 3.11 Siméon Denis Poisson (1781–1840).

is called the **Poisson random variable**. The parameter λ is the shape parameter, which indicates the average number of events in the given time interval. The base e is **Euler's constant**, which is approximately 2.71828.

Note 3.33

It is left as an exercise to prove that for a Poisson random variable X, $E(X) = \lambda$ and $Var(X) = \lambda$.

It can easily be seen from (3.119) that $p_j \geq 0$. We also see that

$$\sum_{j=0}^{\infty} p_j = \sum_{j=0}^{\infty} \frac{e^{-\lambda} \lambda^j}{j!} = e^{-\lambda} \sum_{j=0}^{\infty} \frac{\lambda^j}{j!} = e^{-\lambda} e^{\lambda} = 1.$$

Thus, (3.119), indeed, is a pmf. The cdf for Poisson pmf is

$$P(X \leq j) = \sum_{j=0}^{j} \frac{e^{-\lambda} \lambda^j}{j!}, \quad j = 0, 1, \dots. \tag{3.120}$$

Example 3.54

Suppose that computers in a computer laboratory at a university break on an average of 1.5 computers per day. What is the probability of breakdown of

a. Three computers in a particular day?
b. No more than four breakdowns in a day?
c. At least five breakdowns on a particular day?

Answer

Let X be a random variable representing the number of breakdown of computers in the laboratory per day. We assume that breakdown of computers follows a Poisson pmf with the average rate of $\lambda = 1.5$.

Now for part (a), from (3.119), we have:

$$p_3 = P(X = 3) = \frac{e^{-1.5}(1.5)^3}{3!} = 0.1255 \approx 13\%.$$

For part (b), we are to find

$$P(0 \leq X \leq 4) = P(X = 0) + P(X = 1) + P(X = 2) + P(X = 3) = P(X = 4)$$

$$p_0 = P(X = 0) = \frac{e^{-1.5}(1.5)^0}{0!} = 0.2231 \approx 22\%.$$

$$p_1 = P(X = 1) = \frac{e^{-1.5}(1.5)^1}{1!} = 0.3347 \approx 33\%.$$

$$p_2 = P(X = 2) = \frac{e^{-1.5}(1.5)^2}{2!} = 0.2510 \approx 25\%.$$

$$p_3 = P(X = 3) = \frac{e^{-1.5}(1.5)^3}{3!} = 0.1255 \approx 13\%.$$

$$p_4 = P(X = 4) = \frac{e^{-1.5}(1.5)^4}{4!} = 0.0471 \approx 5\%.$$

Thus,

$$P(0 \le X \le 4) = 0.2231 + 0.3347 + 0.2510 + 0.1255 + 0.0471 = 0.9814 \approx 98\%.$$

For part (c), we are to find

$$P(X \ge 5) = P(X = 5) + P(X = 6) + \ldots$$

$$p_5 = P(X = 5) = \frac{e^{-1.5}(1.5)^5}{5!} = 0.0141 \approx 1\%.$$

$$p_6 = P(X = 6) = \frac{e^{-1.5}(1.5)^6}{6!} = 0.0035 \approx 0\%.$$

$$\vdots \qquad \vdots \quad \vdots$$

Therefore,

$$P(X \ge 5) = 0.0141 + 0.0035 + \cdots = 0.0176 + \cdots.$$

However, to obtain a precise value, we can use:

$$P(X \ge 5) = 1 - P(X < 5) = 1 - P(X \le 4)$$

$$= 1 - (P(X = 0) + P(X = 1) + P(X = 2) + P(X = 3) + P(X = 4))$$

$$= 1 - 0.9814 = 0.0186 \approx 2\%.$$

Of course, as expected, from parts (b) and (c), the sum of probabilities is 1.

Note 3.34

The idea in Example 3.53 is that the Poisson pmf determines the number (events occurring in a certain time), for instance, the number of telephone calls to a business office per hour or per day. Other examples are the number of cars arriving at a red traffic light and the number of computers break down in a computer laboratory per day.

Example 3.55

Suppose that production of an item from a manufacturer follows a Poisson pmf with average number of items produced each day as 100. The history of the production in this factory shows that 5% of items produced are defective. Suppose the number of production in a particular day is n items. What is the distribution of the total number of defective items for that day?

Answer

To answer the question, note that there are two random variables: (1) the daily production and (2) the total number of defective items per day. Thus, let X and Y be random variables representing (1) and (2), respectively. Hence, the question is to find $P(Y = k)$, $k = 0,1,2,....$ These items imply that:

$$P(Y = k | X = n) = B_k (k; n, 0.05)$$

$$= \binom{n}{k}(0.05)^k (0.95)^{n-k}, \quad k = 0,1,...,n.$$

Now applying the law of total probability, we will have:

$$P(Y = k) = \sum_{n=k}^{\infty} P(Y = k | X = n) \cdot P(X - n)$$

$$= \sum_{n=k}^{\infty} \binom{n}{k}(0.05)^k (0.95)^{n-k} \left(\frac{e^{-100} 100^n}{n!} \right)$$

$$= (0.05)^k e^{-100} \sum_{n=k}^{\infty} \binom{n}{k}(0.95)^{n-k} \left(\frac{100^n}{n!} \right)$$

$$= 0.05^k e^{-100} \sum_{n=k}^{\infty} \frac{n!}{k!(n-k)!}(0.95)^{n-k} \left(\frac{100^n}{n!} \right).$$

$$= \frac{0.05^k e^{-100}}{k!} \sum_{n=k}^{\infty} \frac{(0.95)^{n-k}}{(n-k)!} 100^{n-k} 100^k$$

$$= \frac{0.05^k 100^k e^{-100}}{k!} \sum_{n=k}^{\infty} \frac{0.95^{n-k}}{(n-k)!} 100^{n-k}$$

$$= \frac{5^k e^{-100}}{k!} \sum_{n-k=0}^{\infty} \frac{(95)^{n-k}}{(n-k)!}$$

$$= \frac{5^k e^{-100}}{k!} \sum_{m=0}^{\infty} \frac{95^m}{m!}.$$

Using the McLaurin expansion of exponential function, we will have:

$$P(Y=k)=\frac{5^{-5}5^{k}}{k!}.$$

As it can be seen, pmf of Y is a Poisson with parameter $\lambda = 5$. For instance, probability that there will be 10 defective items found among the production of a particular day is:

$$p_{10}=P(Y=10)=\frac{5^{-5}5^{10}}{10!}=\frac{5^{5}}{10!}=8.6117e^{-04}\approx 0.007.$$

Example 3.56

Consider a bank at a location that closes at 6:00 pm on weekdays. Teller windows are set in parallel, and each has its own queue and its own overhead survival camera with a counter. An arriving client may join a line at his/her choice. It is the bank's policy for a teller to leave at the closing; he/she should close the line at 5:45 so that no new arrival joins the queue. However, the teller has to complete his/her services of all clients in the line before leaving. It is 5:30 and a teller who wants to leave closing is to know if he/she can complete services to all clients in line by 6:00. Data from the survival cameras show that on average eight clients enter each individual line between 5:30 and 5:45 pm on weekdays. Assuming that the arrival of clients follows a Poisson pmf, we want to answer the following questions:

1. What is the probability that exactly five clients arrive the line of a teller who wants to leave on time between 5:30 and 5:45 pm?
2. What is the probability that the teller will not be able to leave on time?

Answer

Based on the assumption that arrival distribution is Poisson, the parameter $\lambda = 8$. Let X be a random variable representing the number of arrivals between 5:30 and 5:45. Then, we have:

1. $p_5 = P(X=5)=\dfrac{e^{-8}8^{5}}{5!}=0.0916\approx 9\%.$
2. The question essentially means that the probability of more than 8 clients arrives. Hence,

$$P(X>8)=1-P(X\le 8)$$

$$= P(X=0)+P(X=1)+\cdots+P(X=8)$$

$$= \sum_{k=0}^{8}\frac{e^{-8}8^{k}}{k!}$$

$$= 1-0.5925=0.4075\approx 41\%.$$

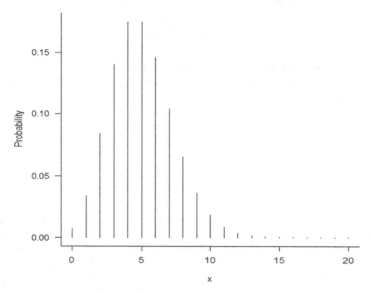

PMF of the Poisson Random Variable

FIGURE 3.12   pmf of Poisson pmf, Example 3.57.

Example 3.57 Poisson Probability Mass Function

Let us consider an office, which experiences approximately 5 telephone calls in an hour on a working day. We assume that the office does not receive more than one telephone call at a time and the rate of receiving calls is a constant. In this case, the random variable X represents the number of telephone calls received by the office and X is a Poisson random variable. Hence, based on the assumptions, the probability mass function (pmf) of X (see Figure 3.12) with arrival rate (parameter) $\lambda = 5$ is

$$P(X = x) = \sum_{j=0}^{x} \frac{e^5 5^j}{j!}, \ j = 0,1,2,\mathrm{L}.$$

Note 3.35

An important relationship between binomial and Poisson distributions is that when the product np is kept fixed, the binomial pmf approaches the Poisson pmf.

Theorem 3.9

If X is a binomial random variable with pmf $\mathcal{B}_j(j;n,p)$, $n \to \infty$, $p \to 0$ (i.e., n becomes large and p becomes small) and $\lambda = np$ is fixed, then the binomial pmf with parameters n and p approaches Poisson pmf with parameter $\lambda = np$. In other words,

$$\lim_{\substack{n \to \infty \\ p \to 0}} \mathcal{B}_j(j;n,p) = \frac{e^{-\lambda} \lambda^j}{j!}, \quad j = 0,1,2,\dots. \tag{3.121}$$

Proof:

From the definition of binomial pmf and the fact that from $\lambda = np$, that is, $p = \dfrac{\lambda}{n}$, we have:

$$B_j(j;n,p) = P(Y = j | X = n)$$

$$= \binom{n}{j} p^j (1-p)^{n-j} = \frac{n!}{j!(n-j)!} p^j (1-p)^{n-j}, \quad j = 0,1,\ldots,n.$$

$$= \frac{n!}{j!(n-j)!} \left(\frac{\lambda}{n}\right)^j \left(1-\frac{\lambda}{n}\right)^{n-j}, \quad j = 0,1,\ldots,n.$$

$$= \frac{n(n-1)\ldots(n-j+1)}{nj} \frac{\lambda^j}{j!} \frac{\left(1-\dfrac{\lambda}{n}\right)^n}{\left(1-\dfrac{\lambda}{n}\right)^j}, \quad j = 0,1,\ldots,n.$$

Now as n and p approach infinity and 0, respectively, the last relation will be undefined. Using L'Hôspital's rule, we have:

$$\lim_{n \to \infty} \left(1-\frac{\lambda}{n}\right)^n \approx e^{-\lambda}, \quad \lim_{n \to \infty} \frac{n(n-1)\ldots(n-j+1)}{n^j} \approx 1. \tag{3.122}$$

Inserting these values in the last statement will complete the proof.

Note 3.36

The relation $\lim\limits_{x \to \infty} \left(1+\dfrac{1}{x}\right)^x$, where x is a real number, is sometimes taken as the definition of the exponential number e. That is,

$$\lim_{x \to \infty} \left(1+\frac{1}{x}\right)^x = e \approx 2.7183. \tag{3.123}$$

Note 3.37

Let us define Y such that:

$$Y = 1 - \frac{1}{x} = \frac{x-1}{x} = \frac{1}{\dfrac{x}{x-1}}. \tag{3.124}$$

Letting $u = x - 1$, (3.124) becomes:

$$Y = \frac{1}{\left(1 + \dfrac{1}{u}\right)}. \tag{3.125}$$

Thus, from (3.124) and (3.125), we will have:

$$\lim_{x \to \infty}\left(1 - \frac{1}{x}\right)^x = \lim_{u \to \infty}\left(1 + \frac{1}{u}\right)^{-(1+u)}. \tag{3.126}$$

$$= \lim_{u \to \infty}\left(1 + \frac{1}{u}\right)^{-1}\left(1 + \frac{1}{u}\right)^{-u} = 1 \cdot e^{-1} = e^{-1}.$$

Hence, from (3.122), we have:

$$e^{-\lambda} = \left(e^{-1}\right)^{\lambda} = \lim_{x \to \infty}\left(1 - \frac{1}{x}\right)^{\lambda x}. \tag{3.127}$$

Letting $n = \lambda x$, from (3.127), we will have:

$$e^{-\lambda} = \left(e^{-1}\right)^{\lambda} = \lim_{n \to \infty}\left(1 - \frac{\lambda}{x}\right)^n. \tag{3.128}$$

Example 3.58

Suppose that in a factory of producing light bulbs, it is known that 1 in 1,000 light bulbs is defective. A group of 8,000 light bulbs from the storage room is taken randomly for sale delivery. What is the probability of the group containing at least five defectives?

Answer

To answer the question, we assume that all light bulbs have the same chance to be selected. Hence, the probability of a light bulb to be defective is $p = 0.001$, and the sample size $n = 8,000$, that is, $np = \lambda = 8$. Letting the random variable X representing the number of defective light bulbs in the sample, we will have:

$$p_5 = P(X = 5) = B_5(5; 8,000, 0.001) = \frac{e^{-8}8^5}{5!} = 0.0916 \approx 9\%.$$

Theorem 3.10

$$E(X) = Var(X) = \lambda. \tag{3.129}$$

$$M(t) = e^{\lambda\left(e^t - 1\right)}. \tag{3.130}$$

$$G(z) = e^{\lambda(z-1)}.$$ (3.131)

Proof:
We prove that the mean and variance of a Poisson random variable are λ and leave the proof of the other two as exercises.

Now, from Poisson pmf, $p_j = P(X = j) = \dfrac{e^{-\lambda}\lambda^j}{j!}$, $j = 0,1,\ldots$, the **rth moment** of the Poisson random variable X is

$$E(X^r) = \sum_{x=0}^{\infty} x^r \frac{e^{-\lambda}\lambda^j}{x!}.$$ (3.132)

Hence, for $r = 1$, we have:

$$\mu = E(X) = \sum_{x=0}^{\infty} x \frac{e^{-\lambda}\lambda^x}{x!} = \lambda e^{-\lambda} \sum_{x=0}^{\infty} x \frac{\lambda^{x-1}}{x(x-1)!}$$

$$= \lambda e^{-\lambda}\left[1 + \lambda + \frac{\lambda^2}{2!} + \frac{\lambda^3}{3!}\right] = \lambda e^{-\lambda} e^{\lambda} = \lambda.$$

Now to find the second moment, we let $r = 2$ and obtain:

$$E(X)^2 = \sum_{x=0}^{\infty} x^2 \frac{e^{-\lambda}\lambda^x}{x!} = \sum_{x=0}^{\infty}\left([x(x-1)+x]\frac{\lambda^x e^{-\lambda}}{x!}\right)$$

$$= \sum_{x=0}^{\infty}\left(x(x-1)\frac{\lambda^x e^{-\lambda}}{x!}\right) + \sum_{x=0}^{\infty}\left(x\frac{\lambda^x e^{-\lambda}}{x!}\right)$$

$$= \sum_{x=0}^{\infty}\left(x(x-1)\frac{\lambda^x e^{-\lambda}}{x(x-1)(x-2)!}\right) + \sum_{x=0}^{\infty}\frac{\lambda^x e^{-\lambda}}{(x-1)!}$$

$$= \sum_{x=0}^{\infty}\frac{\lambda^x e^{-\lambda}}{(x-2)!} + \sum_{x=0}^{\infty}\frac{\lambda^x e^{-\lambda}}{(x-1)!}$$

$$= \lambda^2 e^{-\lambda}\sum_{x=0}^{\infty}\frac{\lambda^{x-2}}{(x-2)!} + E(X)$$

$$= \lambda^2 e^{-\lambda}\left[1 + \lambda + \frac{\lambda^2}{2!} + \frac{\lambda^3}{3!} + \cdots\right] + \lambda$$

$$= \lambda^2 e^{-\lambda} e^{\lambda} + \lambda = \lambda^2 + \lambda.$$

Hence,

$$Var(X) = E(X)^2 - \left[E(X)^2\right] = \lambda^2 + \lambda - \lambda^2 = \lambda.$$

Example 3.59

Let us suppose that sale of particular brand of television by a salesperson at an electronic store follows a Poisson pmf with an average of 6 per week. For the salesperson,

i. What is the probability that in a given week she will sell at least one television?
ii. What is the probability that in a given week she will sell between 1 and 5 televisions?
iii. Assuming that the salesperson works 5 days per week, what is the probability that in a given day of a given week, she will sell one television?

Answer

Using MATLAB, we can use the code "poisspdf(x,λ)" to find various values we need to answer the three questions. Of course, in this problem, $\lambda = 6$.

i. Let X be the random variable representing the sales of the televisions. Then, from the properties of pmf of a random variable, the probability of at least one means that:

$$P(X > 0) = 1 - P(X = 0).$$

Now from Poisson pmf with $\lambda = 6$, we have:

$$P(X = 0) = \frac{e^{-6}6^0}{0!} = 0.0025.$$

Hence,

$$P(X > 0) = 1 - 0.0025 = 0.9975.$$

ii. The probability of selling between 1 and 5 televisions is:

$$P(1 < X < 5) = P(X = 2) + P(X = 3) + P(X = 4)$$

$$= \frac{e^{-6}6^2}{2!} + \frac{e^{-6}6^3}{3!} + \frac{e^{-6}6^4}{4!}$$

$$= 0.0446 + 0.0892 + 0.1339 = 0.2677.$$

iii. The average number of televisions sold per day will be $\frac{6}{5} = 1.2$. Hence, the average number of televisions sold per day is:

$$P(X = 1) = \frac{e^{-1.2}(1.2)^1}{1!} = 0.3614.$$

Note 3.38

One should not expect that the probability of selling one television per day would be the same as five per week. In fact, the probability of selling one television per day would be 0.3614, as found for case (iii). However, the probability of selling five televisions in a week for 5 days per week will be 0.1606.

3.5 PROBABILITY DISTRIBUTION FUNCTION (CDF) FOR A CONTINUOUS RANDOM VARIABLE

In the previous sections, we discussed discrete random variables and their distributions. Numerous examples were presented to illustrate these random variables, among which coin tossing was frequently used. The reason for this choice was that there are only two outcomes for this experiment, simply heads and simply tails. However, what if we were to consider the height the coin reaches in air when it is tossed? Since the measure of the height is a real number, not necessarily a nonnegative integer, it is not discrete, and indeed, it is continuous. In other words, continuous random variables are random quantities that are measured on a continuous scale. They usually can take on any value over some interval, which distinguishes them from discrete random variables, which can take on only a sequence of values, usually integers. Examples of continuous random variables other than the height are waiting times to receive service in a waiting line, length of a telephone call, time between two calls arriving at an office, length of time it will take for a signal to go to a satellite and return to a receiver, height, weight, the amount of sugar in an orange, the time required to run a mile, and so on.

Earlier in this chapter, we discussed the pmf for a discrete random variable. In this section, we will discuss the probability distribution function for a continuous random variable. Hence, we let X be a random variable defined on a continuous sample space Ω. Then, we define the probabilities associated with the points of Ω for which the values of X fall on the interval $[a,b]$ by $P(a \leq X \leq b)$.

Definition 3.26

Let S be the set of possible outcomes of a random experiment whose measures are real numbers. That is, S is a single interval or a union of intervals on the number line. Let X be a random variable defined on S with its values as x. Such an X is referred to as a **continuous random variable**. The **probability density function, pdf**, of X is an integrable function $f_X(x)$ satisfying the following:

i. $f_X(x)$ has no negative values on S, that is,

$$f_X(x) \geq 0, \forall x \in S. \tag{3.133}$$

ii. The total area under the curve $f_X(x)$ equals 1, that is,

$$\int_s f_X(x)\,dx = 1. \tag{3.134}$$

iii. If $f_X(x)$ is a pdf of X and A is an interval, then the probability that x belongs to A is the integral of $f_X(x)$ over A, that is,

$$P(X \in A) = \int_A f_X(x)dx. \qquad (3.135)$$

In other words, the probability of X being in A is the area under pdf $f_X(x)$ over A.

Note 3.39

If A is the interval $[a,b]$, then from the last property of pdf, we have:

$$P(a \le X \le b) = \int_a^b f_X(x)dx. \qquad (3.136)$$

That is, the probability of X between a and b is the area under $f_X(x)$ over the interval $[a,b]$. If the interval A is unbounded, that is, $(-\infty, \infty)$, then from (3.134) and (3.136), we will have:

$$P(-\infty < X < \infty) = \int_{-\infty}^{\infty} f_X(x)dx = 1. \qquad (3.137)$$

Although (3.137) suggests that the area under the pdf on the entire real axis is 2, it also suggests that a continuous random variable may be unbounded, even though rarely.

Note 3.40

The probability of observing any single value of the continuous random variable is 0 since the number of possible outcomes of a continuous random variable is uncountable and infinite. That is, for a continuous random variable, we must calculate a probability over an interval rather than at a particular point. This is why the probability for a continuous random variable can be interpreted as an area under the curve on an interval. In other words, we cannot describe the probability distribution of a continuous random variable by giving probability of single values of the random variable as we did for a discrete random variable. This property can also be seen from the fact that

$$P(X = c) = P(c \le X \le c) = \int_c^c f_X(x)dx = 0, \qquad (3.138)$$

for any real number c.

Definition 3.27 Cumulative Probability Distribution Function, cdf

Let X be a continuous random variable, t a real number from the infinite interval $(-\infty, x]$, and $f_X(x)$ the pdf of X. The **cumulative distribution function (cdf) of** X, denoted by $F_X(x)$, is defined by

$$F_X(x) = P(X \le x) = \int_{-\infty}^{x} f_X(t)dt. \tag{3.139}$$

As seen from the definition of the pdf, $f_X(x)$, we can obtain it from the cumulative probability distribution function (cdf) simply by taking the derivative of $F_X(x)$ with respect to x. Conversely, we can obtain $F_X(x)$ by integrating $f_X(x)$ with respect to x. That is,

$$f_X(x) = F_X'(x). \tag{3.140}$$

Other properties of a continuous random variable X, based on properties of definite integrals, are:

$$P(a \le X \le b) = P(a \le X < b) = P(a < X \le b) = P(a < X < b). \tag{3.141}$$

$$P(a \le X \le b) = P(X \le b) - P(X \le a) = F_X(b) - F_X(a). \tag{3.142}$$

$$P(X > b) = 1 - P(X \le b) = 1 - F_X(b). \tag{3.143}$$

Also, if $F_X(x)$ is the differential at every point of an interval, then it is continuous and at a point, say x_0, we have:

$$\left. \frac{dF_X(x)}{dx} \right|_{x=x_0} = F_X'(X_0) = f_X(x_0). \tag{3.144}$$

Note 3.41

An obvious difference between pdf and pmf should be noted. That is, for a discrete random variable, $P(X = x)$ is indeed pmf, say $F_X(x)$. However, for a continuous random variable X, $P(X = x)$ is not the pdf, say $f_X(x)$. Here is an example.

Example 3.60

Let X be a continuous random variable whose pdf is given by

$$f_X(x) = 2x, \quad 0 \le x \le 1.$$

Hence, if we were to find the probability of $X = 0.7$ by simply substituting 0.7 in $f_X(x)$, as we would do in the discrete case, we would have gotten $f_X(0.7) = 2 \times 0.7 = 1.4$, which is not a probability value since the value is more than 1.

Of course, as we mentioned above, for a continuous random variable probability at a point, say $x = 0.70$, the sample space is 0. Thus, in an example like this, we should ask for the probability of one interval, say 0.65–0.75. In such a case, we would have:

$$P(0.65 < x < 0.75) = \int_{0.65}^{0.75} (2x - 1)dx$$

$$= 0.75^2 - 0.65^2 = 0.14.$$

Note 3.42

Generally, if $F_X(x)$ and $f_X(x)$ are stepwise continuous functions, which are not differentiable at some points, then $F_X'(x_0)$ does not necessarily equal $f_X(x_0)$. This is a property known for distribution functions that are not necessarily pdfs. The idea comes from a well-known fact in continued fractions theory. For reference, for instance, see Prats'ovytyi (1996), *Singularity of Distributions of Random Variables Given by Distributions of Elements of the Corresponding Continued Fraction*

More specifically, by a **singular continuous probability distribution function** is meant a cdf on a set of Lebesgue measure zero, in which the probability of each point is zero. Obviously, such a distribution is not absolute continuous with respect to the Lebesgue measure. The **Cantor distribution function** is the famous well-known example that has neither a pdf nor a pmf.

Example 3.61

Let X be a continuous random variable defined on $[0,10]$. Define $f_X(x)$ as below:

$$f_X(x) = \begin{cases} c(x+2), & 0 \le x \le 10, \\ 0, & \text{otherwise.} \end{cases} \qquad (3.145)$$

Find:

  i. A constant c such that $f_X(x)$ defined in (3.143) is a pdf,
  ii. $P(X \ge 7)$,
  iii. $P(2 < X \le 7)$.

Answers

i. For $f_X(x)$ to be a pdf, one condition is that the integral of $f_X(x)$ defined by (3.145) over $[0,10]$ be 1. Thus,

$$\int_{-\infty}^{\infty} f_X(x)dx = 1 = \int_0^{10} c(x+2)dx,$$

$$= c\left(\frac{x^2}{2}+2x\right)_0^{10} = 70c = 1. \tag{3.146}$$

From (3.139), it can be seen that

$$c = \frac{1}{70}. \tag{3.147}$$

Now, from (3.147), substituting c in (3.145), we obtain:

$$f_X(x) = \begin{cases} \dfrac{1}{70}(x+2), & 0 \le x \le 10, \\ 0, & \text{otherwise,} \end{cases} \tag{3.148}$$

which is surely nonnegative. Hence, $f_X(x)$ defined in (3.145) defines a pdf.

ii. From (3.148), we have:

$$P(X \ge 7) = \int_7^{10} \frac{1}{70}(x+2)dx = \frac{1}{70}\left(\frac{x^2}{2}+2x\right)_7^{10} = \frac{63}{140} = 0.45.$$

iii. Also, from (3.148), we have:

$$P(2 \le X \le 7) = \int_2^7 \frac{1}{70}(x+2)dx = \frac{1}{70}\left(\frac{x^2}{2}+2x\right)_2^7 = \frac{65}{140} = 0.46.$$

Example 3.62

We consider the total assets dollar amount (in hundreds of thousands of dollars) of residents of a nursing home community in a city. We assume that the dollar amount in this case can be approximated by a continuous random variable with pdf defined by

$$f_X(x) = \begin{cases} \dfrac{2}{x^3}, & x \ge 1 \\ 0, & x < 1. \end{cases}$$

i. Show that $f_X(x)$ is a pdf.
ii. Find the probability that a randomly selected resident has a total asset between \$300,000 and \$500, 000.
iii. Find the probability that a randomly selected resident has a total asset of at least \$200,000.
iv. Find the probability that a randomly selected resident has a total asset of at most \$400,000.

Answers

i. From the definition of $f_X(x)$, it is positive. Also,

$$\int_{-\infty}^{\infty} f_X(x)dx = \int_1^{\infty} \frac{2}{x^3} dx = -\frac{1}{x^2}\bigg]_1^{\infty} = 1.$$

Hence, $f_X(x)$ is a pdf.

ii. $P(3 \le X \le 5) = \int_3^5 \frac{2}{x^3} dx = -\frac{1}{x^2}\bigg]_3^5 = -\left(\frac{1}{25} - \frac{1}{9}\right) = 0.07 = 7\%.$

iii. $P(X \ge 2) = \int_2^{\infty} \frac{2}{x^3} dx = -\frac{1}{x^2}\bigg]_2^{\infty} = -\left(0 - \frac{1}{4}\right) = 0.25 = 25\%.$

iv. $P(X \le 4) = \int_{-\infty}^4 \frac{2}{x^3} dx = 0 + \int_1^4 \frac{2}{x^3} dx = -\frac{1}{x^2}\bigg]_1^4 = -\left(\frac{1}{16} - \frac{1}{1}\right) = \frac{15}{16} = 93.75\%.$

Example 3.63

Suppose the random variable X represent the lifetime, in hours, of a light bulb. Let the cdf of X be given as:

$$F_X(x) = \begin{cases} 0, & x < 100, \\ \dfrac{x^2 + x}{990,900}, & 100 \le x \le 1,000, \\ 1, & x > 1000. \end{cases}$$

i. Find the pdf.
ii. Find $P(X \ge 800)$.

Answers

i. Taking the derivative of (3.147) with respect to x, we will have:

$$f_X(x) = \begin{cases} \dfrac{2x + 1}{990,900}, & 100 \le x \le 1,000, \\ 0, & \text{otherwise.} \end{cases}$$

ii. $P(X \ge 800) = 1 - P(X < 800) = 1 - P(X \le 800) = 1 - F_X(800)$

$$= 1 - \frac{(800)^2 + 800}{990900} = 1 - \frac{640800}{990900}$$

$$\approx 0.3533 = 35.33\%.$$

3.6 MOMENTS OF A CONTINUOUS RANDOM VARIABLE

The expected value of a continuous random variable is very much similar to that one of a discrete random variable except that the integral sign is used in place of sigma (the summation). The properties are also very similar. As we had to discuss the convergence of series in the case of discrete random variables, we should do something similar for the continuous case. We now start, by formally, defining the terms.

Definition 3.28

Let X be a continuous random variable with pdf as $f_X(x)$. The **expected value** (expectation or mean) of X is defined as the following Lebesgue integral:

$$E(X) = \int_{\mathbb{R}} x\, f_X(x)\, dx. \tag{3.149}$$

As it can be seen from (3.149), the procedure for finding expected values of continuous random variables is similar to the one for finding expected values of discrete random variables. The difference is that (1) the sum in the formula for a discrete random variable is replaced by integral and (2) the pmf in the formula for discrete random variable is replaced by the pdf. Similar differences are there for the variance and standard deviation of the two kinds of random variables.

More specifically, if X is a random variable with pdf as $f_X(x)$, then the expected value of X, denoted by μ, is given by

$$\mu = E(X) \int_{-\infty}^{\infty} x\, f_X(x)\, dx, \tag{3.150}$$

only if the integral in (3.150) exists.

Generalization of Definition 3.28 for a function of a continuous random variable is as follows.

Definition 3.29

Let X be a continuous random variable with pdf as $f_X(x)$. Similar to the discrete case, the **expected value** (**expectation** or **mean**) of the random variable $g(X)$, which is a **function of random variable** X, is defined by the following Lebesgue integral:

$$E(g(X)) = \int_{\mathbb{R}} g(x)\, f_X(x)\, dx, \tag{3.151}$$

only if the integral in (3.151) exists.

An important property of the moment of a continuous random variable is that of **linearity**. Applying this property on the expected value, we have:

$$E(c_1 g_1(x) + c_2 g_2(x) + c_3) = c_1 E(g_1(x)) + c_2 E(g_2(x)) + c_3. \tag{3.152}$$

Note 3.43

Not all continuous random variables have moments. Here is an example.

Example 3.64

This example shows a random variable with a pdf and a nonexistent expected value. Let X be a random variable with pdf defined as

$$f_X(x) = \frac{1}{\pi} \frac{1}{1+x^2}, \quad -\infty < x < \infty. \tag{3.153}$$

The function given in (3.153) is clearly positive, and it can easily be seen that its integral over the range of X is 1. We leave the proof as an exercise. Hence, it is a pdf. The random variable with pdf given in (3.153) is referred to as the **Cauchy random variable**. What is $E(X)$?

Answer

$$E(X) = \int_{-\infty}^{\infty} x f_X(x) dx = \int_{-\infty}^{\infty} \frac{1}{\pi} \frac{x}{1+x^2} dx$$

$$= \frac{1}{2\pi} \ln|1 + x|\Big|_{-\infty}^{\infty},$$

which is undefined.

Definition 3.30

The **variance** of a continuous random variable X, denoted by σ_X^2, is given by

$$\sigma_X^2 = E\left[(X - \mu)^2\right] = \int_{-\infty}^{\infty} (x - \mu)^2 f_X(x) dx. \tag{3.154}$$

From Chapter 2, $\sigma_X^2 = E(X^2) - \mu^2$. Hence, for a continuous random variable, in its general form:

$$\sigma_X^2 = \int_{-\infty}^{\infty} x^2 f_X(x) dx - \mu^2. \tag{3.155}$$

Example 3.65

A school bus arrives shortly (about 5 minutes) after 7:00 am each non-holiday weekdays during an academic year. The number of minutes after 7:00 that the bus arrives can be modeled as a continuous random variable X with pdf as

$$f(x) = \begin{cases} \dfrac{2}{25}(5-x), & 0 \le x \le 5, \\ 0, & \text{otherwise.} \end{cases}$$

Answer the following:

   i. Graph the pdf,
  ii. Find the cdf,
 iii. Graph the cdf,
 iv. Find the mean,
  v. Find the standard deviation of the number of minutes after 7:00 that the bus arrives.

Answers

   i. See Figure 3.13

  ii. $F_X(x) = P(X \le x) = \displaystyle\int_0^x f_X(t)\,dt$

$$= \int_0^x \frac{2}{25}(5-t)\,dt = \frac{2}{25}\int_0^x (5-t)\,dt$$

$$= \frac{2}{25}\left[5t - \frac{t^2}{2}\right]_0^x = \frac{2}{25}\left[5x - \frac{x^2}{2}\right]$$

$$= \frac{1}{25}\left(10x - x^2\right), 0 \le x \le 5.$$

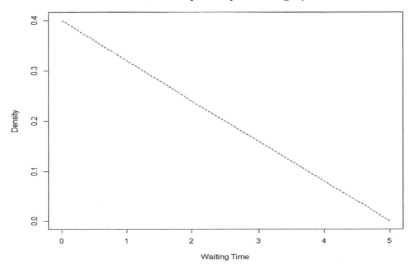

Probability Density Function (pdf)

FIGURE 3.13   Graph of pdf, Example 3.65

iii. See Figure 3.14
iv. From (3.148), we have:

$$\mu = E(X) = \int_0^5 x f_X(x)\,dx$$

$$= \frac{2}{25}\int_0^5 x(5-x)\,dx$$

$$= \frac{2}{25}\left[\frac{5x^2}{2} - \frac{x^3}{3}\right]_0^5$$

$$= \frac{5}{3} \approx 1.7\,\text{min.}$$ (3.156)

v. From (3.154) and (3.156), we have:

$$\sigma_x^2 = \int_0^5 x^2 f_X(x)\,dx - \mu^2$$

$$= \frac{2}{25}\int_0^5 x^5 (5-x)\,dx - \left(\frac{5}{3}\right)^2$$

$$= \frac{2}{25}\int_0^5 x^2 (5-x)\,dx - \left(\frac{5}{3}\right)^2$$

$$= \frac{2}{25}\left[\frac{5x^3}{3} - \frac{x^4}{4}\right]_0^5 - \left(\frac{5}{3}\right)^2 = \frac{25}{18}\,\text{min}^2.$$

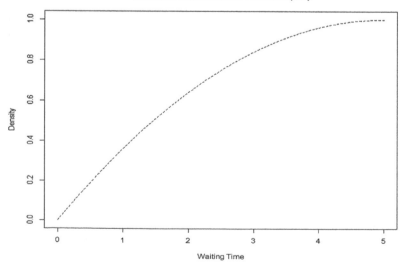

Cumulative Distribution Function (cdf)

FIGURE 3.14   Graph of cdf, Example 3.65

Hence, the standard deviation of X is:

$$\sigma_X = \sqrt{\frac{25}{18}} = \frac{5\sqrt{2}}{6} \approx 1.18 \text{ min}.$$

3.7 CONTINUOUS MOMENT GENERATING FUNCTION

We have seen the moment generating function for a discrete random variable. We will now discuss this function for a continuous random variable.

Definition 3.31

Let X be a continuous random variable. Then, the **moment generating function (mgf) of** X denoted by $M_X(t)$ (as for a discrete random variable) is defined as $M_X(t) = E\left(e^{tX}\right)$. That is, if the pdf of X is denoted by $f(x)$, then

$$M_X(t) = \int_{-\infty}^{\infty} e^{tx} f(x) dx. \tag{3.157}$$

Some properties of mgf

1. Differentiating the mgf r times, we have:

$$M_X^{(r)}(t) = \frac{d^r}{dt^r} \int_{-\infty}^{\infty} e^{tx} f(x) dx = \int_{-\infty}^{\infty} \left(\frac{d^r}{dt^r} e^{tx} f(x) \right) dx$$

$$= \int_{-\infty}^{\infty} x^r e^{tx} f(x) dx. \tag{3.158}$$

Letting $t = 0$ in (3.158) yields:

$$M^{(r)}(0) = \int_{-\infty}^{\infty} x^r e^{tx} f(x) dx = E\left(X^r\right). \tag{3.159}$$

2. Let X and Y be two independent random variables with $M_X(t)$ and $M_Y(t)$ as their respective mgfs. Let, also, $Z = X + Y$. Then, the moment generating function of the product of the sum is the product of mgfs. That is,

$$M_Z(t) = E\left(e^{tZ}\right) = E\left(e^{tX+tY}\right)$$

$$= E\left(e^{tX} e^{tY}\right) = E\left(e^{tX}\right) E\left(e^{tY}\right) = M_X(t) M_Y(t). \tag{3.160}$$

This property can be extended for n independent random variables. For example, let $X_1, X_2, \ldots, X_n$ be n independent random variables. Let also $X = X_1 + X_2 + \ldots + X_n$. Then,

$$M_X(t) = \prod_{k=1}^{n} M_{X_k}(t). \tag{3.161}$$

The proof of this property incorporates the previous one as a special case. Here, it is:

$$M_X(t) = E\left(e^{tX}\right) E\left[e^{t(X_1 + X_2 + \cdots + X_n)}\right]$$

$$= E\left[e^{tX_1 + tX_2 + \cdots + tX_n}\right]$$

$$= E\left(e^{tX_1}\right) E\left(e^{tX_2}\right) \cdots E\left(e^{tX_n}\right)$$

$$= M_{X_1}(t) + M_{X_1}(t) \cdots M_{X_n}(t).$$

3. Let X be a random variable with $M_X(t)$. Also, let Y be a random variable such that $Y = a + bX$, where a and b are the constants. Then,

$$M_Y(t) = e^{at} M_X(bt). \tag{3.162}$$

This is because

$$M_Y(t) = E\left(e^{tY}\right) = E\left[e^{t(aX+b)}\right]$$

$$= e^{tb} E\left(e^{taX}\right) = e^{tb} M_X\left(e^{ta}\right).$$

4. Let F_n be a sequence of cdfs with corresponding mgfs $M_n(t)$. Let F also be a cdf with mgf M. Further, $M_n(t) \rightarrow M(t)$, for all t within an open interval containing 0. Then, $F_n(x) \rightarrow F(x)$, for all x at which F is continuous.

3.8   FUNCTIONS OF RANDOM VARIABLES

In this section, we will discuss the joint distribution for both discrete and continuous random variables due to their similarities. Hence, we consider a sample space Ω with a general element ω. Also, let X be a random variable. Recall that X is a function that associates the real number $X(\omega)$ with an outcome ω with pdf or pmf $f_X(x)$ and cdf or cmf $F_X(x)$.

Now suppose $y = g(x)$ is a real-valued function of the real variable x. Then, $Y = g(X)$ is a transformation of the random variable X into the random variable Y. Hence, as $X(\omega)$ is a random variable, so is $g(X(\omega))$ called the **function of the random variable X**. Then, the question is how to find pdf or pmf $f_Y(y)$ and cdf or cmf $F_Y(y)$ of $g(X(\omega))$. Of course, the domain of function g should contain the range of X.

Definition 3.32

If $\mathbf{X} = (X_1, X_2, \ldots X_n)$ is a random vector of n continuous random variables that associate the sample space Ω with the space $\mathbb{R}^n$ of real n-tuples, then the joint distribution function of X is defined as:

$$F_{X_1, X_2, \ldots, X_n}(x_1, x_2, \ldots x_n) = P\{X_1 \le x_1, X_2 \le x_2, \ldots, X_n \le x_n\}. \tag{3.163}$$

The pdf of X is defined as:

$$F_{X_1, X_2, \ldots, X_n}(\mathbf{X}) \frac{\partial^n}{\partial x_1, \ldots \partial x_n} F_{X_1, \ldots, X_n(x_1, \ldots x_n)}. \tag{3.164}$$

In the case of one variable, the marginal pdf for each random variable is obtained as follows:

$$F_{X_1}(x_1) = \int \cdots \int f_{X_1, \ldots X_n}(x_1, \ldots, x_n) dx_2 \ldots dx_n. \tag{3.165}$$

In particular, we consider only two continuous variables X and Y with $f_X(x)$ and $f_Y(y)$. Then, the **joint bivariate pdf of X and Y** is denoted by $f_{XY}(x,y)$. From pmf, we will have:

$$P(X \le x \text{ and } Y \le y) \approx f_{X,Y}(x, y) dx dy. \tag{3.166}$$

Definition 3.33

Similar to (3.163), the joint pmf for two discrete random variables X and Y is defined by

$$P_{XY} = P(X = x \text{ and } Y = y),$$

or

$$P_{XY} = P(X = x, Y = y). \tag{3.167}$$

Definition 3.34

Let X and Y be two continuous random variables. Then, the **joint pdf of X and Y** is an integrable function, say $f_{XY}(x,y)$ or just $f(x,y)$, such that:

i. $f_{X,Y}(x, y) \ge 0$, $\tag{3.168}$

ii. $\displaystyle\int_{-\infty}^{\infty} \int_{-\infty}^{\infty} f_{X,Y}(x, y) dx\, dy = 1$, $\tag{3.169}$

   and for an event $\{(X,Y) \in S\}$ defined in the xy-plane,

iii. $\displaystyle P\{(X,Y) \in S\} = \int\int_S f_{X,Y}(x, y)\, dx\, dy. \tag{3.170}$

The **conditional pdf of X given Y** can be used to find the joint pdf. Here is how it is defined.

Definition 3.35

The **conditional pmf** or pdf of a random variable X **given a random variable Y** is defined as:

$$f_{X|Y}(x,y) = \frac{f_{X,Y}(x,y)}{f_Y(y)} \quad \text{if } f_Y(y) > 0. \tag{3.171}$$

Thus, from (3.171), the joint pdf or pmf of X and Y is:

$$f_{XY}(x,y) = f_{X|Y}(x,y) \cdot f_Y(y). \tag{3.172}$$

The **marginal** pdf of X (or Y) can be obtained from the joint pdf $f_{XY}(x,y)$, (3.172), denoted by:

$$\begin{cases} f_X(x) = \displaystyle\int_{-\infty}^{\infty} f_{X,Y}(x,y)\,dy, \\[2mm] f_Y(y) = \displaystyle\int_{-\infty}^{\infty} f_{X,Y}(x,y)\,dx. \end{cases} \tag{3.173}$$

Similarly, the **marginal** pmf of X (or Y) can be obtained from the joint pdf $f_{XY}(x,y)$, (3.171), denoted by:

$$\begin{cases} f_X(x) = \displaystyle\sum_{-\infty}^{\infty} f_{X,Y}(x,y), \\[2mm] f_Y(y) = \displaystyle\sum_{-\infty}^{\infty} f_{X,Y}(x,y). \end{cases} \tag{3.174}$$

Example 3.66

Let X and Y have a joint pdf defined by

$$f_{XY}(x,y) = e^{-x-y}, \quad 0 < x, y < \infty$$

with

$$S = \left\{ (x,y) \ni 0 < y < \frac{x}{7} \right\}.$$

Find the probability that (X, Y) falls in S.

Answer

$$P\{(XY) \in S\} = \int_0^\infty \int_0^{\frac{x}{7}} e^{-x-y} dx dy$$

$$= \int_0^\infty e^{-x} \left[-\left(e^{\frac{-x}{7}} - 1 \right) \right] dx = \int_0^\infty e^{-x} \left(1 - e^{\frac{-x}{7}} \right) dx$$

$$= \int_0^\infty \left(e^{-x} - e^{\frac{-8x}{7}} \right) dx = \left[-e^{-x} + \frac{7}{8} e^{\frac{-8x}{7}} \right]_0^\infty = \frac{1}{8}.$$

Example 3.67

Let X and Y have a joint pdf defined as

$$f_{XY}(x,y) = \frac{3}{2}, \quad x^2 \le y \le 1, \quad 0 < x < 1. \tag{3.175}$$

We want to

i. Find the conditional pdf of Y given X,
ii. Find the conditional pdf of Y, $\frac{1}{9} \le y \le 1$, given $x = \frac{1}{3}$,
iii. Check to make sure the answer in (ii) is a pdf.

Answer

i. From (3.173), we have:

$$f_X(x) = \int_{-\infty}^\infty f_{X,Y}(x,y) dy$$

$$= \int_{x^2}^1 \frac{3}{2} dy = \frac{3}{2} y \Big|_{x^2}^1 = \frac{3}{2}(1-x^2), \quad 0 < x < 1.$$

Thus, from (3.171) and (3.175), we also have:

$$f_{Y|X}(x,y) = \frac{f_{X,Y}(x,y)}{f_X(x)}, \quad \text{if } f_X(x) > 0$$

$$= \frac{\frac{3}{2}}{\frac{3}{2}(1-x^2)} = \frac{1}{(1-x^2)}, \quad 0 < x < 1, x^2 \le y \le 1. \tag{3.176}$$

ii. For $x = \frac{1}{3}$, (3.176) yields

$$f_{Y|X}\left(\frac{1}{3}, y\right) = \frac{1}{\left(1 - \frac{1}{9}\right)} = \frac{9}{8}, \quad x^2 \le y \le 1. \tag{3.177}$$

iii. To see that the function found in (3.177) is a pdf, we note that $\frac{9}{8} > 0$.
Then, integrating with respect to y and evaluating for $x^2 \le y \le 1$, we have:

$$\int_{\frac{1}{9}}^{1} \frac{9}{8} dy = \frac{9}{8}\left(1 - \frac{1}{9}\right) = 1.$$

Definition 3.36

Let X and Y be two discrete random variables. Let the sample space of X, Y and the joint X and Y be denoted by S_X, S_Y, and $S_{X,Y}$, respectively. Then, the **joint pmf** of X and Y is given by the conditional probability as follows:

$$P_{XY} = P(X = x, Y = y)$$

$$= P(Y = y | X = x) \cdot P(X = x) \tag{3.178}$$

$$= P(X = x | Y = y) \cdot P(Y = y)$$

$$0 \le P_{XY}(x, y) \le 1, \text{ and } \sum_{x \in S_X} \sum_{y \in S_Y} P_{XY}(x, y) = 1. \tag{3.179}$$

From the law of total probability, the pmf of either X or Y, that is, **marginal pmf**, can be found as follows:

$$P_X(x) = P(X = x) = \sum_{y \in S_Y} P(X = x, Y = y) = \sum_{y \in S_Y} P_{XY}(x, y). \tag{3.180}$$

Similarly,

$$P_Y(y) = P(Y = y) = \sum_{x \in S_X} P(X = x, Y = y) = \sum_{x \in S_X} P_{XY}(x, y). \tag{3.181}$$

Example 3.68

Suppose the joint pmf of two discrete random variables X and Y is given in Table 3.31.
Find:

   i. The marginal pmf of X and Y,
   ii. $P(X = 0, Y > 1)$,
   iii. $P(X = 1, Y \le 2)$,
   iv. $P(Y = 1 | X = 1)$.

TABLE 3.31

Joint pmf of Two Random Variables X and Y

	$Y = 1$	$Y = 2$	$Y = 3$
$X = 1$	$\dfrac{1}{5}$	$\dfrac{1}{7}$	$\dfrac{1}{9}$
$X = 0$	$\dfrac{1}{10}$	$\dfrac{1}{5}$	$\dfrac{31}{126}$

Answer

From Table 3.31, we have: $S_X = \{0,1\}$ and $S_Y = \{1,2,3\}$.

i. $P_X(0) = P_{XY}(0,1) + P_{XY}(0,2) + P_{XY}(0,3)$

$$= \frac{1}{10} + \frac{1}{5} + \frac{31}{126} = \frac{143}{315}.$$

$P_X(1) = P_{XY}(1,1) + P_{XY}(1,2) + P_{XY}(1,3)$

$$= \frac{1}{5} + \frac{1}{7} + \frac{1}{9} = \frac{172}{315}.$$

Hence,

$$P_X(x) = \begin{cases} \dfrac{143}{315} & x = 0 \\[2mm] \dfrac{172}{315} & x = 1 \\[2mm] 0 & \text{otherwise.} \end{cases}$$

$$P_Y(1) = P_{XY}(0,1) + P_{XY}(1,1) = \frac{1}{10} + \frac{1}{5} = \frac{3}{10}.$$

$$P_Y(2) = P_{XY}(0,2) + P_{XY}(1,2) = \frac{1}{5} + \frac{1}{7} = \frac{12}{35}.$$

$$P_Y(3) = P_{XY}(0,3) + P_{XY}(1,3) = \frac{31}{126} + \frac{1}{9} = \frac{5}{14}.$$

Hence,

$$P_Y(y) = \begin{cases} \dfrac{3}{10}, & y = 1, \\[2mm] \dfrac{12}{35}, & y = 2, \\[2mm] \dfrac{5}{14}, & y = 3, \\[2mm] 0, & \text{otherwise.} \end{cases}$$

ii. $P(X = 0, Y > 1) = P_{XY}(0,2) + P_{XY}(0,3) = \dfrac{1}{5} + \dfrac{31}{126} = \dfrac{157}{630}$.

iii. $P(X = 0, Y \le 2) = P_{XY}(0,1) + P_{XY}(0,2) = \dfrac{1}{10} + \dfrac{1}{5} = \dfrac{3}{10}$.

iv. $P(Y = 1 | X = 1) = \dfrac{P(X = 1, Y = 1)}{P(X = 1)} = \dfrac{\dfrac{1}{5}}{\dfrac{172}{315}} = \dfrac{63}{172}$.

Example 3.69

Consider tossing two fair coins separately. Let the random variable X represent the occurrence of tails (failure) of the first coin and the random variable Y represent the occurrence of heads of the second coin. Find the joint pmf of X and Y.

Answer

The sample space in this case is $\{(H,H),(H,T),(T,H),(T,T)\}$. Let us denote the values of X and Y as follows:

$$X = \begin{cases} 1, & \text{if } T \text{ occurs,} \\ 0, & \text{if } H \text{ occurs,} \end{cases} \quad \text{and} \quad Y = \begin{cases} 1, & \text{if } H \text{ occurs,} \\ 0, & \text{if } T \text{ occurs.} \end{cases}$$

Since X and Y are independent, as the problem states, we have:

$$P(X = x, Y = y) = \begin{cases} \dfrac{1}{4}, & x = 0, y = 0, \\ \dfrac{1}{4}, & x = 0, y = 1, \\ \dfrac{1}{4}, & x = 1, y = 0, \\ \dfrac{1}{4}, & x = 1, y = 1. \end{cases}$$

We can extend (3.167) to a finite discrete random vector, say with n elements, as follows.

Definition 3.37

For a discrete random vector $\{X_1, X_2, \ldots, X_n\}$ with n elements, the **joint probability mass function** $P_{X_1, X_2, X_3}(x_1, x_2, \ldots, x_n)$ is a function that completely characterizes the distribution of a discrete random vector and is defined as:

$$P_{X_1, X_2, \ldots, X_n} = P(X_1 = x_1, X_2 = x_2, \ldots, X_n = x_n). \tag{3.182}$$

As in bivariate cases, discrete or continuous, it is important to note the relationship among components of the vector. One particular important case is the one with no relationship at all.

Definition 3.38

Let $x_1, x_2, ..., x_n$ be n **independent random variables** with pdfs or pmfs as $f_{X_1}(x_1), f_{X_2}(x_2), ..., f_{X_n}(x_n)$, respectively. Then,

$$f_{X_1, X_2, ... X_n}(x_1, x_2, ... x_n) = f_{X_1}(x_1) \cdot f_{X_2}(x_2) \cdots f_{X_n}(x_n) = \prod_{i=1}^{n} f_{X_i}(x_i). \quad (3.183)$$

Note 3.44

In case the pdf or pmf is parametric, say with one parameter μ, denoted by $f(x|0)$, then with the same parameter value μ in each marginal pdf or pmf, the joint pdf or pmf will be

$$f(x_1, x_2, ... x_n | \theta) = f_{X_1}(x_1|\theta) \cdot f_{X_2}(x_2|\theta) \cdots f_{X_n}(x_n|\theta) = \prod_{i=1}^{n} f_{X_i}(x_i|\theta). \quad (3.184)$$

Note 3.45

We leave it as an exercise to prove, by a counterexample, that pairwise independence does not imply mutual independence.

Note 3.46

If each one of the random variables X_i, $i = 1, 2, ..., n$, is a random vector, then we obtain a mutual random vector.

Example 3.70

Let the random variables X and Y have a joint pdf defined as

$$f_{X,Y}(x, y) = 2, \quad 0 \le x \le y < 1.$$

Find $f_X(x)$, $f_Y(y)$, $E(X)$, $E(Y)$, $Var(X)$, and $Var(Y)$

Answer

From the given joint pdf, the marginal pdfs are:

$$f_X(x) = \int_x^1 2\,dy = 2y\Big]_x^1 = 2(1-x), \quad 0 \le x \le 1,$$

and

$$f_Y(y) = \int_0^y 2\,dx = 2x\Big]_0^y = 2y, \quad 0 \le y \le 1.$$

For the moments, we can use either the joint pdf or the marginal. Hence, for the first moments, we have:

$$E(X) = \int_0^1 2x(1-x)\,dx = 2\left[\frac{x^2}{2} - \frac{x^3}{3}\right]_0^1$$

$$= 2\left(\frac{1}{2} - \frac{1}{3}\right) = \frac{1}{3}$$

and

$$E(Y) = \int_0^1 2y \cdot y\,dy = 2\left[\frac{y^3}{3}\right]_0^1 = \frac{2}{3}.$$

Now, for the second moments, we have:

$$E(X^2) = \int_0^1 2x^2(1-x)\,dx = 2\left[\frac{x^3}{3} - \frac{x^4}{4}\right]_0^1$$

$$= 2\left(\frac{1}{3} - \frac{1}{4}\right) = \frac{1}{6}$$

and

$$E(Y^2) = \int_0^1 2y^2 \cdot y\,dy = 2\left[\frac{y^4}{4}\right]_0^1 = \frac{1}{2}.$$

Thus, for the variances, we have:

$$Var(X) = E(X^2) - [E(X)]^2 = \frac{1}{6} - \left(\frac{1}{3}\right)^2 = \frac{1}{18}$$

and

$$Var(Y) = E(Y^2) - [E(Y)]^2 = \frac{1}{2} - \left(\frac{2}{3}\right)^2 = \frac{1}{18}.$$

Below, we now present some properties of functions of random variables in both the discrete and continuous cases.

Definition 3.39

Let X and Y be two independent discrete random variables. Denote the sum by $Z = X + Y$. Then, the **probability of the sum**, Z, is given as follows:

$$P(Z = z) = \sum_{j=-\infty}^{\infty} P(X = j)P(Y = z - j). \tag{3.185}$$

Definition 3.40

Let X and Y be two discrete random variables defined on a sample space Ω with joint pmf denoted by $p_{XY}(x,y)$. Let Z also be a new bivariate as a function of X and Y, denoted by $Z = z(X,Y)$. Then, the **expected value of sum**, Z, is defined by

$$E(Z) = E[z(X,Y)] = \sum_{(x,y)\in\Omega} z(x,y)p_{XY}(x,y). \tag{3.186}$$

Properties of expected value of sum:

Definition 3.41

The **convolution** of two independent continuous random variables with pdfs $f_X(x)$ and $f_Y(y)$, denoted by $h(z) = (f * g)(z)$, is defined as:

$$h(z) = (f * g)(z) = \int_{-\infty}^{\infty} f(z - u)g(u)\,du. \tag{3.187}$$

We leave it as an exercise to prove that

$$h(z) = (g * f)(z) = \int_{-\infty}^{\infty} f(u)g(z - u)\,du. \tag{3.188}$$

The idea of the convolution being connected to the probability of the sum of two independent random variables came about as follows. Let Z be the random variable representing the sum of two random variables X and Y, that is, $Z = X + Y$. Then, pdf of Z, denoted by $h_Z(z)$, is:

$$h_Z(z) = \int_{-\infty}^{\infty} f_X(z - y)g_Y(y)\,dy, \tag{3.189}$$

or

$$h(z) = \int_{-\infty}^{\infty} g_Y(z-x) f_X(x)\, dx. \tag{3.190}$$

In case the joint pdf of X and Y, denoted by $H_Z(z)$, is given, then

$$P\{Z \le z\} = P\{X + Y \le z\}, \tag{3.191}$$

and

$$H_Z(z) = \int_{-\infty}^{\infty} \int_{-\infty}^{z-y} h_{X,Y}(x,y)\, dx\, dy. \tag{3.192}$$

Therefore,

$$h_Z(z) = \frac{dH_Z(z)}{dz} = \int_{-\infty}^{\infty} h_{X,Y}(x, z-x)\, dx. \tag{3.193}$$

Example 3.71

Let X and Y be two uniformly distributed random variables distributed with cdf as $U(0,5)$ and $U(0,10)$, respectively. We want to find the distribution of $Z = X + Y$.

Answer

For the pdf of $Z = X + Y$, applying the convolution, we have the following three cases:

$$h_Z(z) = \int_0^z g_Y(z-x) f_X(x)\, dx$$

$$= \int_0^z \frac{1}{10} \frac{1}{5}\, dx = \frac{z}{50}.$$

$$h_Z(z) = \int_0^5 g_Y(z-x) f_X(x)\, dx$$

$$= \int_0^5 \frac{1}{5} \frac{1}{10}\, dx = \frac{1}{10}.$$

$$h_Z(z) = \int_{z-10}^5 g_Y(z-x) f_X(x)\, dx$$

$$= \int_{z-10}^5 \frac{1}{5} \frac{1}{10}\, dx = \frac{15-z}{50}.$$

Hence,

$$
h_Z(z) = \begin{cases} \dfrac{z}{50}, & 0 \le z < 5, \\[2mm] \dfrac{1}{10}, & 5 \le z < 10, \\[2mm] \dfrac{15-z}{50}, & 10 \le z < 15. \end{cases}
$$

Example 3.72

Let X and Y be two independent exponential random variables with common pdf as $f_X(x) = \lambda e^{-\lambda x}$ with $\lambda > 0$. Let Z also be the random variable representing the sum of X and Y; that is, $Z = X + Y$. Then, the pdf of Z, denoted by $h_Z(z)$, that is, the pdf of the sum of two iid (independent, identically distributed random variables) exponentially distributed random variables, is a gamma pdf with parameters 2 and λ, as follows:

$$
h(z) = \int_0^z f_Y(z-x) f_X(x)\, dx
$$

$$
= \int_0^z \lambda e^{-\lambda(z-x)} \lambda e^{-\lambda x}\, dx
$$

$$
= \lambda^2 z e^{-\lambda z}.
$$

3.9   SOME POPULAR CONTINUOUS PROBABILITY DISTRIBUTION FUNCTIONS

As was the case with discrete random variables, when we considered some special cases, specifically, the geometric, binomial, and Poisson distributions, we do the same for continuous distributions, too. Probably the three most important continuous distributions are uniform, exponential, and normal distribution. However, we will discuss more types of continuous pdf, some of which are vastly used in the industry and applications, and some that will be used in Chapter 5 of this book.

1. Uniform,
2. Gamma,
3. Exponential,
4. Beta,
5. Erlang,
6. Normal,
7. Chi-squared,
8. F,
9. Student's t,
10. Lognormal,

11. Logit,
12. Extreme value,
13. Weibull,
14. Phase type.

3.9.1 CONTINUOUS UNIFORM DISTRIBUTION

Definition 3.42

A random variable X has a **uniform** or **rectangular** distribution on $[a, b]$ if X has the following density function (Figures 3.15 and 3.16):

$$f_X(x) = \begin{cases} \dfrac{1}{b-a}, & a \leq x \leq b, \\ 0, & \text{otherwise.} \end{cases} \tag{3.194}$$

From (3.194), we see that $f_X(x) > 0$ and

$$\int_a^b \frac{1}{b-a} dx = \frac{b-a}{b-a} = 1.$$

From the pdf, the **cumulative probability distribution function for uniform random variable**, cdf, denoted by $F_X(x)$, is

$$F_X(x) = \begin{cases} \dfrac{x-a}{b-a}, & a \leq x \leq b, \\ 0, & \text{otherwise.} \end{cases} \tag{3.195}$$

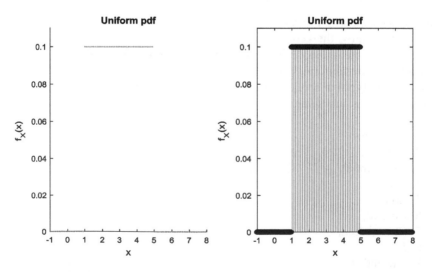

FIGURE 3.15 Uniform pdf.

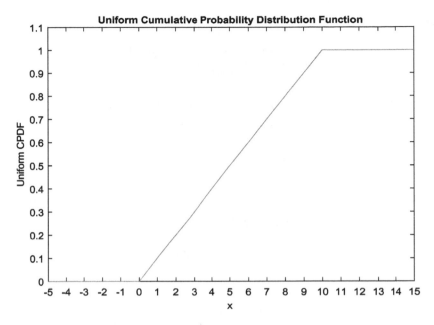

FIGURE 3.16 Uniform cumulative probability distribution function.

Note 3.47

The uniform distribution may be interpreted as representing a number chosen uniformly at random from the interval $[a, b]$.

Theorem 3.11

The expected value and variance of a uniform random variable X are given as follows:

$$\mu_X = E(X) = \frac{b+a}{2} \text{ and } \sigma_X^2 = Var(X) = \frac{(b-a)^2}{12}. \qquad (3.196)$$

Proof:
From (3.196) and Definition 3.36, the mean and variance of the uniform random variance X can be found as follows:

$$E(X) = \int_a^b x\left(\frac{1}{b-a}\right)dx = \frac{x^2}{2(b-a)}\Big|_a^b$$

$$= \frac{b^2-a^2}{2(b-a)} = \frac{b+a}{2}.$$

Now,

$$E(X^2) = \int_a^b x^2 \left(\frac{1}{b-a}\right) dx = \frac{x^3}{3(b-a)}\Bigg|_a^b$$

$$= \frac{b^3 - a^3}{3(b-a)} = \frac{(b-a)(b^2 + ab + a^2)}{3(b-a)}$$

$$= \frac{b^2 + ab + a^2}{3}.$$

Hence,

$$\sigma_X^2 = E(X^2) - \mu^2$$

$$= \frac{b^2 + ab + a^2}{3} - \left(\frac{b+a}{2}\right)^2$$

$$= \frac{(b-a)^2}{12}.$$

Example 3.73

Let X be a uniform random variable on [1, 6]. Find $P(X > 3)$, μ_X, σ_X^2, and standard deviation of X.

Answer

Given the range of X as [1, 6], $a = 1$ and $b = 6$. The pdf of this uniform random variable is given by

$$f(x) = \begin{cases} \dfrac{1}{5}, & 1 \le x \le 6, \\ 0, & \text{otherwise.} \end{cases}$$

Hence,

$$P(X > 3) = \int_3^\infty \frac{1}{5} dx = \int_3^6 \frac{1}{5} dx = \frac{3}{5} = 60\%.$$

The expected value of this random variable is

$$\mu_X = \frac{6+1}{2} = \frac{7}{2} = 3.5.$$

The second moment of X is

$$\mu_{X^2} = \frac{b^2 + ab + a^2}{3} = \frac{36 + 6 + 1}{3} = \frac{43}{3}.$$

Hence,

$$\sigma_X^2 = \frac{43}{3} - \left(\frac{7}{2}\right)^2 = \frac{25}{12} \approx 2.08$$

and

$$\sigma_X = \sqrt{2.09} = 1.44.$$

3.9.2   Gamma Distribution

There are distributions with one-, two-, three-, and even four-parameter family with each having its own applications. We now consider a gamma distribution with three parameters α (the **shape** parameter), λ (the **rate** parameter), and l (the **location** parameter), followed by its special cases of two and one parameters and then three special cases reduced from the one-parameter case, namely, exponential, Erlang, and chi-squared.

Note 3.48

We note that the parameter λ in some publications is used as $1/\lambda$ so that it would show the mean rather than the rate, as we will see later.

Definition 3.43

Let α and λ be the positive real numbers. A random variable X with pdf defined by

$$f_X(x;\lambda;\alpha,l) = \begin{cases} \dfrac{\lambda^\alpha}{\Gamma(\alpha)}(x-l)^{\alpha-1}e^{-\lambda(x-l)}, & x \geq l, \alpha, \lambda > 0, \\ 0, & x < 0, \end{cases} \tag{3.197}$$

is referred to as a **three-parameter gamma random variable**, where $\Gamma\alpha$ is the gamma function given in Chapter 1 as

$$\Gamma(\alpha) = \int_0^\infty x^{\alpha-1}e^{-x}\,dx. \tag{3.198}$$

The distribution function of X is referred to as a **three-parameter gamma pdf** with parameters α, λ, and l.

If $l = 0$, then (3.189) reduces to a **two-parameter gamma pdf** given as below:

$$f_X(x;\lambda;\alpha) = \begin{cases} \dfrac{\lambda^\alpha}{\Gamma(\alpha)}x^{\alpha-1}e^{-\lambda x}, & \alpha, \lambda > 0, x > 0, \\ 0, & x < 0, \end{cases} \tag{3.199}$$

where $\Gamma(\alpha)$ is given by (3.198). Now, if we let $t = \lambda x$ in (3.192) and write $x = \dfrac{t}{\lambda}$, we will have:

$$\int_0^\infty \frac{\lambda}{\Gamma(\alpha)} t^{\alpha-1} e^{-t} dt = 1 \tag{3.200}$$

and the integrand is nonnegative. Thus, (3.198) is a pdf.

If we let $\alpha = 1$ in (3.198), then we will have the **one-parameter gamma pdf**, given by

$$f_X(x; \lambda) = \begin{cases} \lambda e^{-\lambda x}, & \lambda > 0, x \geq 0, \\ 0, & x < 0. \end{cases} \tag{3.201}$$

Theorem 3.12

For a two-parameter gamma random variable X, the expected value and variance are given, respectively, as follows:

$$\mu_X = E(X) = \frac{\alpha}{\lambda} \text{ and } \sigma_X^2 = Var(X) = \frac{\alpha}{\lambda^2}. \tag{3.202}$$

Proof:
From (3.201), we have:

$$E(X) = \int_{-\infty}^\infty x \frac{\lambda^\alpha}{\Gamma(\alpha)} x^{\alpha-1} e^{-\lambda x} dx = \int_0^\infty x \frac{\lambda^\alpha}{\Gamma(\alpha)} x^{\alpha-1} e^{-\lambda x} dx.$$

Using a substitution $t = \lambda x$ and write $x = \dfrac{t}{\lambda}$, hence, $dx = \dfrac{dt}{\lambda}$, we will have:

$$E(X) = \int_0^\infty \frac{t}{\lambda} \frac{\lambda^\alpha}{\Gamma(\alpha)} \left(\frac{t}{\lambda}\right)^{\alpha-1} e^{-t} \frac{dt}{\lambda} = \frac{1}{\lambda \Gamma(\alpha)} \int_0^\infty t^\alpha e^{-t} dt. \tag{3.203}$$

From (3.202) and the property of the gamma function mentioned in Chapter 2, the integral in the right-hand side of (3.195) is $\Gamma(\alpha+1) = \alpha\Gamma(\alpha)$. Hence, we will have:

$$E(X) = \frac{1}{\lambda \Gamma(\alpha)} \alpha\Gamma(\alpha) = \frac{\alpha}{\lambda}.$$

Now, using again the definition of the second moment, substituting $t = \lambda x$, and writing $x = \dfrac{t}{\lambda}$, so that $dx = \dfrac{dt}{\lambda}$, we have:

$$E(X^2) = \int_0^\infty x^2 \frac{\lambda}{\Gamma(\alpha)} (\lambda x)^{\alpha-1} e^{-\lambda x} dx = \int_0^\infty \frac{t^2}{\lambda^2} \frac{\lambda}{\Gamma(\alpha)} t^{\alpha-1} e^{-t} \frac{dt}{\lambda}$$

$$= \frac{1}{\lambda^2 \Gamma(\alpha)} \int_0^\infty t^{\alpha+1} e^{-t} dt = \frac{1}{\lambda^2 \Gamma(\alpha)} \Gamma(\alpha+2) = \frac{\alpha(\alpha+1)}{\lambda^2}.$$

Hence,

$$\sigma_X^2 = Var(X) = E(X^2) - [E(X)]^2 = \frac{\alpha(\alpha+1)}{\lambda^2} - \left(\frac{\alpha}{\lambda}\right)^2 = \frac{\alpha}{\lambda^2}.$$

This proves the theorem.

Note 3.49

The gamma pdf is, sometimes, written differently. This can be done by changing the variable, $t = \lambda x$, and writing $x = \frac{t}{\lambda}$. Here is what the gamma pdf with 2 parameters k and θ, where α is replaced by k and β is replaced by θ, looks like:

$$f_X(x;\theta;k) = \begin{cases} \dfrac{1}{\theta^k \Gamma(k)} x^{k-1} e^{-\frac{x}{\theta}}, & k, \theta > 0, x > 0, \\ 0, & x \le 0, \end{cases} \tag{3.204}$$

where $\Gamma(\alpha)$ is the gamma function, defined earlier. From (3.204), we will have $E(X) = k\theta$ and $Var(X) = k\theta^2$. We leave the derivations as exercises.

3.9.3 EXPONENTIAL DISTRIBUTION

Definition 3.44

A random variable X with a one-parameter gamma pdf as given in (3.201) is referred to as the **exponential** (and sometimes **negative exponential**) **random variable**. From (3.201), its cpdf, denoted by $F_X(x;\lambda)$ (or if there is no confusion just simply $F_X(x)$), is given by

$$F_X(x;\lambda) = \int_0^x \lambda e^{-\lambda u} du = \begin{cases} 1 - e^{-\lambda x}, & x \ge 0, \\ 0, & x < 0. \end{cases} \tag{3.205}$$

It is referred to as the **exponential probability distribution function with parameter** (or **rate**) λ.

Note 3.50

Recall from the Chapter 2 about Poisson pmf that it is an integer-valued (discrete) pmf. It was noted there that such a pmf arises from **counting the number of occurrences of events in a period of time**.

Recall also that the geometric random variable measures the **waiting time for the first success** after a sequence of failures in a Bernoulli trial.

The exponential random variable with a continuous distribution, however, represents the **waiting time between occurrence of a sequence of events that happen** at the rate λ per unit time, such as the time between arriving customers in a store, between the calls received to an office, and between the scoring of a "goal" in a soccer game.

Example 3.74

Let us consider a set of iid random variables $X_1, X_2, \ldots, X_n$, each distributed exponentially with the same parameter λ. The question is finding the distribution of the smallest random variable among the set, that is, $X = \min(X_1, X_2, \ldots, X_n)$.

Answer

Let t be a real number. Since $X = \min(X_1, X_2, \ldots, X_n)$, for X to be greater than t, all elements of the set must also be greater than t, that is,

$$P(X > t) = P\big((X_1 > t \text{ and } X_2 > t) \text{ and} \cdots \text{and } X_n > t\big). \qquad (3.206)$$

But $X_1, X_2, \ldots, X_n$ are assumed to be independent. Thus,

$$P(X > t) = P(X_1 > t) P(X_2 > t) \cdots P(X_n > t). \qquad (3.207)$$

Let us denote the cdf of X by $F_X(t)$. Then,

$$P(X_k > t) = 1 - F(t), \quad k = 1, 2, \ldots, n$$

$$= 1 - \left(1 - e^{\lambda t}\right)$$

$$= e^{\lambda t}.$$

Thus,

$$P(X > t) = P(X_1 > t) P(X_2 > t) \cdots P(X_n > t)$$

$$= \left(e^{-\lambda t}\right)^n = e^{-n\lambda t}. \qquad (3.208)$$

Recall from the Chapter 2 that the most important property of the geometric pmf was memoryless. For continuous distribution, the exponential distribution is known as the only probability distribution function that has the **memoryless property**. This property says that if the time to wait for an event to occur has the exponential distribution, then the probability that we have to wait an additional s unit of time is the same no matter how long we have already waited. In other words, the history of the process is of no importance for this distribution. Formally, as for geometric pmf, we have:

Note 3.51

As mentioned earlier, geometric pmf has **memoryless** property, and it is the only discrete pmf having such property. That is, it does not remember the history. For the continuous case, on the other hand, the exponential distribution function has the same property. It is the only continuous distribution having such property. The proof of this property is given in the following theorem.

Theorem 3.13

Let X be an exponential random variable and t and s two nonnegative real numbers. Then, the **memoryless property** for X holds. That is,

$$P(X > t + s | X > t) = P(X > s), \qquad (3.209)$$

where

$$F_X(X > s) = 1 - \left(1 - e^{-\lambda s}\right) = e^{-\lambda s}.$$

In other words, the conditional probability in (3.177) does not depend on t.

Proof:
As we noted in the Chapter 2, from conditional property, for $P(X > t) > 0$, we have:

$$P(X > t + s | X > t) = \frac{P(\{X > t + s\} \cap \{X > t\})}{P(X > t)}. \qquad (3.210)$$

It is clear that $(X > s + t) \rightarrow (X > t)$. Hence, from (3.209), we have:

$$P(X > t + s | X > t) = \frac{P(X > t + s)}{P(X > t)}. \qquad (3.211)$$

From (3.207), we have $F_X(X > t) = 1 - \left(1 - e^{-\lambda t}\right)$. Hence, from (3.211), we will have:

$$P(X > t + s | X > t) = \frac{P(X > t + s)}{P(X > t)}$$

$$= \frac{e^{-\lambda(s+t)}}{e^{-\lambda t}} = e^{-\lambda s}.$$

In real-life situations, (3.211) says that the probability of waiting for an additional s unit of time, given that we have already waited for t units of time, is the same as the probability at the start that we would have had to wait for t units of time. In other words, every epoch is like the beginning of a new random period, which has the same distribution regardless of how much time has already past.

Example 3.75

In a real-life setting, suppose you are in a waiting line at a checkout register at a grocery store. A customer is checking out and 5 minutes has past. What is the probability of her checking out taking another 3 minutes?

Answer

The answer is that the 5 minutes passed does not count. The probability of the question is the probability that it would take 3 minutes, that is, $e^{-3\lambda}$.

Note 3.52

If there is no confusion in denoting the exponential pdf, we may drop the index X and the parameter λ and write $f(x)$.

Theorem 3.14

The nth moment, expected value, variance, and moment generating function of an exponential random variable X are given, respectively, as follows:

$$E\left(X^{n}\right) = \frac{n!}{\lambda^{n}}. \tag{3.212}$$

$$\mu_{X} = E(X) = \frac{1}{\lambda}. \tag{3.213}$$

$$\sigma_{X}^{2} = \frac{1}{\lambda^{2}}. \tag{3.214}$$

$$M_{X}(t) = \frac{\lambda}{\lambda - t}, \quad t < \lambda. \tag{3.215}$$

Proof:
We first prove the nth moment

$$E\left(X^{n}\right) = \int_{0}^{\infty} x^{n} \lambda e^{-\lambda x} dx.$$

In order to use the gamma function, as the left-hand side looks like, we let $t = \lambda x$. That is, $x = \dfrac{t}{\lambda}$ and $dt = \lambda dx$. Then,

$$E\left(X^n\right) = \int_0^\infty x^n \lambda e^{-\lambda x} dx = \frac{1}{\lambda^n} \int_0^\infty t^n e^{-t} dt$$

$$= \frac{1}{\lambda^n} \Gamma(n+1) = \frac{n!}{\lambda^n},$$

where Γ stands for the gamma function defined in Chapter 1. This proves (3.212). From (3.214), when $n = 1$, we obtain (3.215). Also, from (3.213), we have:

$$E\left(X^2\right) = \frac{2}{\lambda^2}. \qquad (3.216)$$

From (3.215) and (3.216), we have:

$$\sigma^2 = \frac{2}{\lambda^2} - \left(\frac{1}{\lambda}\right)^2 = \frac{1}{\lambda^2},$$

as stated in (3.207).

Now, the moment generating function, $M_X(t)$.

$$M_X(t) = E\left(e^{tX}\right) = \int_{-\infty}^\infty e^{tx} f_X(x) dx$$

$$= \int_0^\infty e^{tx} \lambda e^{-\lambda x} dx = \int_0^\infty \lambda e^{(t-\lambda)x} dx$$

$$= \frac{\lambda}{t-\lambda} \left[e^{(t-\lambda)x} \right]_0^\infty = \frac{\lambda}{\lambda - t}, \quad t < \lambda.$$

This completes the proof of the theorem.

Note 3.53

As we have seen, the mean value of the exponential random variable is the inverse of its parameter λ. Hence, in observing a stream of arrival of events, the expected time until the next event is always $1/\lambda$. In other words, the mean time to wait for an event to occur is inversely proportional to the rate at which the event occurs.

Example 3.76

Suppose the customers at the grocery store in the previous example arrive at the register at the rate of 3 per every 10 minutes. What is the probability that the cashier will have to wait between 3 and 5 minutes for the next customer to arrive?

Answer

Let X be the random variable representing the time the cashier has to wait for the next customer to arrive. Then, X is exponentially distributed with parameter $\lambda = 3$ per 10 minutes, equivalent to 18 per hour. Hence, the pdf of X, denoted by $f_X(x)$, is

$$f_X(x;\lambda) = \begin{cases} 18e^{-18\lambda}, & x \geq 0, \\ 0, & x < 0. \end{cases}$$

Note 3.54

$\lambda = 3$ per 10 minutes implies that the average number of customers to arrive at the register is 18 per hour. Also that 3–5 minutes of waiting time for a customer to arrive at the register is equivalent to 1/20 to 1/12 of an hour.

Hence, we will have:

$$P\left(\frac{1}{20} \leq X \leq \frac{1}{12}\right) = \int_{\frac{1}{20}}^{\frac{1}{12}} 18e^{-18x}\, dx$$

$$= -e^{-18x}\Big|_{\frac{1}{20}}^{\frac{1}{12}} = -\left[e^{-\frac{3}{2}} - e^{-\frac{9}{10}}\right]$$

$$\approx 0.4066 - 0.2231 \approx 0.1835 \approx 18.35\%.$$

That is, assuming that on the average, three customers arrive at the register every 10 minutes, the probability of the cashier waiting 3–5 minutes for the next customer is about 0.18, or 18%.

Example 3.77

Similar to the previous example, suppose telephone calls in a business office arrive at the rate of 6 per hour. Taking waiting time for the call as the random variable, find

    i. The probability of waiting between 5 and 15 minutes for the next call,
    ii. The mean,
    iii. The variance,
    iv. The standard deviation,
    v. The moment generating function.

Answer

Let X be a random variable representing the time it takes for the next call to arrive. Hence, X has the exponential pdf with parameter $\lambda = 6$, that is,

$$f_X(x;\lambda) = \begin{cases} 6e^{-6\lambda}, & x \geq 0, \\ 0, & x < 0. \end{cases}$$

i. The probability to wait between 5 and 15 minutes for the next call is:

$$P\left(\frac{1}{12} \leq X \leq \frac{1}{4}\right) = \int_{\frac{1}{12}}^{\frac{1}{4}} 6e^{-6x}\,dx$$

$$= -e^{-6x}\Big|_{\frac{1}{12}}^{\frac{1}{4}} = -\left[e^{-\frac{3}{2}} - e^{-\frac{1}{2}}\right]$$

$$\approx 0.6065 - 0.2231 \approx 0.3834 \approx 38.34\%.$$

ii. $\mu_X = E(X) = \dfrac{1}{\lambda} = \dfrac{1}{6} \approx 0.17$ calls per hour.

iii. $\sigma_X^2 = Var(X) = \dfrac{1}{\lambda^2} = \dfrac{1}{36} \approx 0.0278.$

iv. $\sigma_X = STD(X) = \dfrac{1}{\lambda} = \dfrac{1}{6} \approx 0.17.$

v. $M_X(t) = \dfrac{6}{6-t}, \quad t < 6.$

Theorem 3.15

Consider a set of n iid random variables $X_1, X_2, \ldots, X_n$, each distributed exponentially with the same parameter λ, denoted by $f(x)$, given as in (3.195). Then, the pdf of the sum of these iid random variables, denoted by $f_n(x)$, is the convolution of $f(x)$ by itself n times, that is, the n-fold convolution, given as:

$$f_n(x) \equiv \left(\underbrace{f * f * \cdots * f}_{n \text{ times}}\right)(x) = (f * f_{n-1})(x)$$

$$= \begin{cases} \dfrac{\lambda^n x^{n-1}}{(n-1)!} e^{-\lambda x}, & \lambda > 0, x \geq 0, \\ 0, & x < 0. \end{cases} \qquad (3.217)$$

Proof:
We prove the theorem by **mathematical induction**. So, we denote the pdf for each of $X_1, X_2, \ldots, X_n$, from (3.1775), by $f_1(x)$, as:

$$f_1(x) \equiv \begin{cases} \lambda e^{-\lambda x}, & \lambda > 0, x \geq 0, \\ 0, & x < 0. \end{cases} \qquad (3.218)$$

Then, the pdf of the sum $X_1 + X_2$ would be the convolutions of $f(x)$ with itself, denoted by $f_2(x)$, that is,

$$f_2(x) \equiv (f*f)(x) = \int_0^x f(x-t)f(t)\,dt$$

$$= \begin{cases} \lambda^2 e^{-\lambda x}, & \lambda > 0, x \geq 0, \\ 0, & x < 0. \end{cases} \qquad (3.219)$$

To see the pattern, let us find the pdf of the sum $X_1 + X_2 + X_3$, which would be the convolutions of $f(x)$ with itself three times, denoted by $f_3(x)$, that is,

$$f_3(x) \equiv (f*f*f)(x) = (f_1 * f_2)(x)$$

$$= \begin{cases} \dfrac{\lambda^3 x^2}{2} e^{-\lambda x}, & \lambda > 0, x \geq 0, \\ 0, & x < 0. \end{cases} \qquad (3.220)$$

Now using (3.210)–(3.212) and applying mathematical induction on n, we will have (3.189) completing the proof of the theorem. We leave the details of the proof as an exercise.

3.9.4 BETA DISTRIBUTION

Very interestingly, when there is a variety of guessing about the occurrence of an event, the beta distribution walks in and presents them as a probability distribution. In other words, the beta distribution is a presentation *of probabilities*! For instance, in the 2016 presidential election in the United States, almost every single media was guessing about the so-called nontraditional candidate that eventually surprised them all by being elected as the president. There were chances given by pollster as 51%, 52%, 53%, etc. for the more traditional politician who eventually lost the election. Additionally, there were quite a bit of guessing among the voters in the country that varied a lot about the winner, from almost zero chance to 40% chance. Yet as another example, to market a product, a representative of a food company asks people in a shopping mall passing her desk to try a sample and then asks the tester what the chance of purchasing the product would be. In response, the representative receives answers such as 10%, 5%, 35%, 20%, 0%, 50%, 10%, 15%, 0%, and 5%. These probabilities would lead to a beta distribution and the company can make a decision on that basis. Beta distribution is one that represents such a diversified guesswork or subjective beliefs (in a Bayesian sense). In other words, the beta distribution function is a way to represent outcomes like proportions or probabilities. It is clear as the responses that beta distribution is defined on the unit interval $[0,1]$. In summary, the **beta distribution** is most commonly used to model one's uncertainty about the probability of success of a random experiment. We now define this distribution.

Definition 3.45

A continuous random variable X, defined on $[0,1]$, has **beta pdf with two parameters p and q** given by:

$$f_X(x; p,q) = \begin{cases} \dfrac{x^{p-1}(1-x)^{q-1}}{B(p,q)}, & 0 \le x \le 1, \\ 0, & \text{otherwise,} \end{cases} \qquad (3.221)$$

where $B(p,q)$ is called the **beta function**, which is defined as

$$B(p,q) = \int_0^1 u^{p-1}(1-u)^{q-1}\, du = \frac{\Gamma(p)\Gamma(q)}{\Gamma(p+q)}, \qquad (3.222)$$

where $\Gamma(t)$ is the gamma function defined earlier and the positive parameters p and q are referred to as the **shape parameters** and $p-q$ is referred to as the **scale parameter**.

Note 3.55

In case of $0 < p, q < 1$, the unit interval should be an open one, that is, $(0,1)$, for the beta pdf to be defined.

Note 3.56

The beta function appearing in the definition of the beta pdf is, indeed, a normalization constant to make (3.220) a pdf.

Note 3.57

We caution the reader to distinguish between the beta function and the beta distribution function.

Note 3.58

The number of parameters of the beta distribution may be more than 2 that we used in the definition above. For instance, see McDonald and Xub (1995), wherein the authors introduce a five-parameter beta distribution (GB) (referred to as the **generalized beta distribution**) related to the gamma distributions. The GB includes more than 30 distributions as limiting or special cases. They consider applications of the models to investigating the distribution of income, stock returns, and regression analysis.

Note 3.59

The uniform distribution on $(0,1)$ is a degenerate case of the beta pdf, where $p = 1$ and $q = 1$.

Note 3.60

The parameters of a beta random variable could be vectors, matrices, or multidimensional arrays.

Example 3.78

Let the parameters p and q be matrices, $p = \begin{pmatrix} 1 & 2 \\ 3 & 4 \end{pmatrix}$ and $q = \begin{pmatrix} 0.4 & 0.8 \\ 2 & 3 \end{pmatrix}$.

Then, for $x = 0.75$, $B(x; p, q)$, using MATLAB code beta pdf (0.75, p, q) would be:

$$B(0.75; p, q) = \begin{pmatrix} 0.9190 & 1.4251 \\ 1.6875 & 1.5820 \end{pmatrix}.$$

Example 3.79

Let us consider a beta random variable with four pairs of parameter values as follows:

$$p_1 = 0.3, q_1 = 1.5, p_1 = 0.5, q_1 = 2.0,$$

$$p_1 = 0.7, q_1 = 1.0, p_1 = 1.0, q_1 = 3.0.$$

We also choose the values for X as $X = 0: 0.01: 1.5$. Then, we use the following MATLAB (version 2018b) codes to graph the pdf for beta with four different choices of parameters. See Figure 3.17.

```
p_1 = 0.3; q_1 = 1.5;
p_2 = 0.5; q_2 = 2.0;
p_3 = 0.7; q_3 = 1.0;
p_4 = 1.0; q_4 = 3.0;

X = 0:.01:1.5;
y_1 = betapdf(X,p_1,q_1);
y_2 = betapdf(X,p_2,q_2);
y_3 = betapdf(X,p_3,q_3);
y_4 = betapdf(X,p_4,q_4);

plot(X,y_1,'Color','r')
hold on
plot(X,y_2,'Color','b')
hold on
plot(X,y_3,'Color','k')
hold on
```

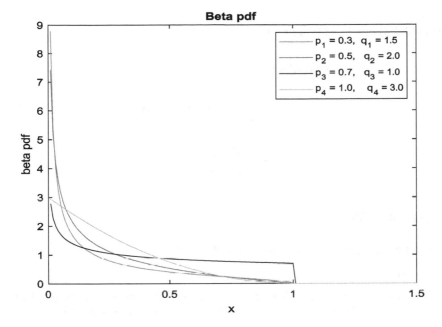

FIGURE 3.17   Beta pdf, Example 3.79.

```
plot(X,y_4,'Color','m')

legend({'p_1 = 0.3, q_1 = 1.5', 'p_2 = 0.5, q_2 = 2.0', 'p_3 =
0.7, q_3 = 1.0', 'p_4 = 1.0, q_4 = 3.0'},'Location','NorthEast');
hold off
```

Theorem 3.16

The cdf of the beta random variable, denoted by $F_X(x; p, q)$, is given by:

$$F_X(x; p, q) = \begin{cases} 0, & x < 0, \\ \dfrac{B(x; p, q)}{B(p, q)}, & 0 \le x \le 1, \\ 0, & x > 1, \end{cases} \qquad (3.223)$$

where

$$B(x; p, q) = \int_0^x u^{p-1}(1-u)^{q-1}\, du \qquad (3.224)$$

is the so-called incomplete beta function.

Proof:

From (3.221) and (3.224), we have:

$$F_X(x;p,q) = \int_{-\infty}^{x} f_x(u,p,q)\,du$$

$$= \frac{1}{B(p,q)} \int_{0}^{x} u^{p-1}(1-u)^{q-1}\,du$$

$$= \frac{B(x;p,q)}{B(p,q)}.$$

Example 3.80

Refer to Example 3.79, and now graph cdf.

Answer

The MATLAB (version 2018b) program for this example would be as the previous one except that the pdf will change to cdf. Thus, we will have Figure 3.18.

Theorem 3.17

First, the nth moment, moment generating function, expected value, and variance of a beta random variable X are, respectively, follows:

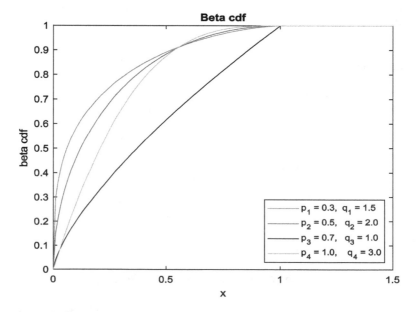

FIGURE 3.18   Beta cdf, Example 3.80

$$E\left(X^{n}\right)=\prod_{k=0}^{n-1}\frac{p+k}{p+q+k}. \tag{3.225}$$

$$M\left(t\right)=E\left(e^{tX}\right)=1+\sum_{n=1}^{\infty}\frac{t^{n}}{n!}\prod_{k=0}^{n-1}\frac{p+k}{p+q+k}. \tag{3.226}$$

$$\mu_{X}=E\left(X\right)=\frac{p}{p+q}. \tag{3.227}$$

$$\sigma_{X}^{2}=Var\left(X\right)=\frac{pq}{\left(p+q+1\right)\left(p+q\right)^{2}}. \tag{3.228}$$

Proof:
First the *n*th moment.

$$E\left(X^{n}\right)=\int_{-\infty}^{\infty}x^{n}f_{X}\left(x;p,q\right)dx$$

$$=\int_{0}^{1}x^{n}f_{X}\left(x;p,q\right)dx=\int_{0}^{1}x^{n}\frac{x^{p-1}\left(1-x\right)^{q-1}}{B\left(p,q\right)}dx$$

$$=\frac{1}{B\left(p,q\right)}\int_{0}^{1}x^{p+n-1}\left(1-x\right)^{q-1}dx$$

$$=\frac{B\left(p+n,q\right)}{B\left(p,q\right)}=\frac{\Gamma\left(p+q\right)\Gamma\left(p+n\right)\Gamma\left(q\right)}{\Gamma\left(p\right)\Gamma\left(q\right)\Gamma\left(p+q+n\right)}$$

$$=\frac{\Gamma\left(p+q\right)}{\Gamma\left(p\right)\Gamma\left(q\right)}\cdot\frac{\Gamma\left(p\right)\cdot p\cdot\left(p+1\right)\cdots\left(p+n-1\right)\cdot\Gamma\left(q\right)}{\Gamma\left(p+q\right)\cdot\left(p+q\right)\cdot\left(p+q+1\right)\cdots\left(p+q+n-1\right)\cdot\Gamma\left(q\right)}$$

$$=\frac{p\cdot\left(p+1\right)\cdots\left(p+n-1\right)}{\left(p+q\right)\cdot\left(p+q+1\right)\cdots\left(p+q+n-1\right)}$$

$$=\prod_{k=0}^{n-1}\frac{p+k}{p+q+k}.$$

Now the moment generating function.

$$M\left(t\right)=E\left(e^{tX}\right)=\int_{-\infty}^{\infty}e^{tx}f_{X}\left(x,p,q\right)dx$$

$$=\int_{0}^{1}e^{tx}\frac{x^{p-1}\left(1-x\right)^{q-1}}{B\left(p,q\right)}dx=\frac{1}{B\left(p,q\right)}\int_{0}^{1}e^{tX}x^{p-1}\left(1-x\right)^{q-1}dx.$$

Using the Taylor expansion of the exponential function, we continue as follows:

$$= \frac{1}{B(p,q)} \int_0^1 \left(\sum_{n=0}^{\infty} \frac{(tx)^n}{n!} \right) x^{p-1} (1-x)^{q-1} \, dx$$

$$= \frac{1}{B(p,q)} \sum_{n=0}^{\infty} \int_0^1 \frac{(tx)^n}{n!} x^{p-1} (1-x)^{q-1} \, dx$$

$$= \frac{1}{B(p,q)} \sum_{n=0}^{\infty} \frac{t^n}{n!} \int_0^1 x^{p+n-1} (1-x)^{q-1} \, dx$$

$$= \frac{1}{B(p,q)} \sum_{n=0}^{\infty} \frac{t^n}{n!} B(p+n,q)$$

$$= \sum_{n=0}^{\infty} \frac{B(p+n,q)}{B(p,q)} \frac{t^n}{n!} = 1 + \sum_{n=1}^{\infty} \frac{B(p+n,q)}{B(p,q)} \frac{t^n}{n!}$$

$$= 1 + \sum_{n=1}^{\infty} E(x^n) \frac{t^n}{n!}$$

$$= 1 + \sum_{n=1}^{\infty} \frac{t^n}{n!} \prod_{k=0}^{n-1} \frac{p+k}{p+q+k}.$$

To prove the mean, we can substitute $n = 1$ in (3.183). However, it can be found directly from the definition of expected value. We leave this proof as an exercise.

Finally, the variance which can also be proven by finding the second moment using $n = 2$ in (3.184) and then using (3.186) and the fact that $\sigma_X^2 = E(X^2) - (E(X))^2$.

Note 3.61

Relation (3.219) shows that the variance of the beta random variable X depends on the mean of X. Using (3.218) and rewriting (3.219), we have:

$$Var(X) = \frac{q}{(p+q+1)(p+q)} \mu_X, \tag{3.229}$$

which shows that for a fixed pair of parameters (p,q), the variance is proportional to the mean.

We now illustrate the relationship between the beta distribution and three other distributions, discrete and continuous.

Theorem 3.18

The beta distribution function approaches the Bernoulli pmf as both shape parameters p and q approach zero.

Proof:
See Johnson and Kotz (1995).

Theorem 3.19

The beta distribution function reduces to the continuous uniform distribution function on the unit interval $[0,1]$ when $p = q = 1$.

Proof:
From (3.183), (3.184) and properties of the gamma function described in Chapter 1, we have:

$$\frac{x^{p-1}(1-x)^{q-1}}{B(p,q)} = \frac{\Gamma(p+q)}{\Gamma(p)\Gamma(q)} x^{1-p}(1-x)^{1-q}$$

$$= \frac{\Gamma(1+1)}{\Gamma(1)\Gamma(1)} x^0 (1-x)^0 = \frac{\Gamma(2)}{\Gamma(1)\Gamma(1)} = \frac{\Gamma(2)}{0!0!} = 1.$$

Hence,

$$f_X(x;1,1) = \begin{cases} 1, & x \in [0,1], \\ 0, & \text{otherwise,} \end{cases}$$

which is the uniform pdf on $[0,1]$.

Theorem 3.20

Let X be a beta random variable whose distribution is with two parameters p and q. We also suppose that Y is a random variable such that its conditional distribution given X is a binomial distribution with parameters n and s. Then, the conditional distribution of X given $Y = y$ is a beta distribution with parameters $1 + y$ and $1 + n - y$.

Proof:
Left as an exercise.

Example 3.81

Suppose it is found that the defectiveness of items shipped by a carrier is distributed according to a beta distribution with two parameters $p = 2$ and $q = 6$. Answer the following questions:

   i. What is the probability that a particular shipment contains 10%–20% defective items?
   ii. What is the probability that a particular shipment contains 20%–30% defective items?
   iii. Find the mean and variance of the distribution.

Answer

Let the random variable X represent the percent defective items in the shipment. Then, using the given information, from (3.222) and (3.196), we will have:

   i. $P(0.1 \leq X \leq 0.2) = \sum\limits_{x=0.1}^{0.2} \dfrac{x^{2-1}(1-x)^{6-1}}{B(2,6)} = 0.248 \approx 25\%.$

   ii. $P(0.2 \leq X \leq 0.3) = \sum\limits_{x=0.2}^{0.2} \dfrac{x^{2-1}(1-x)^{6-1}}{B(2,6)} = 0.5505 \approx 55\%.$

   iii. From (3.227) and (3.228), we have:

$$\mu_X = E(X) = \frac{2}{2+6} = 0.25 = 25\%.$$

$$\sigma_X^2 = Var(X) = \frac{2 \cdot 6}{(2+6+1)(2+6)^2} = \frac{12}{(9)(64)} = \frac{1}{48} = 0.02083 \approx 0.21\%.$$

Note 3.62

Using (3.195) and a calculator, we can answer both questions. However, using MATLAB, it can be easily done in a matter of a second. Below is the MATLAB (version 2018b) quote for case (ii), similarly for case (i):

```
B = beta(2,6);
X = 0.2:.01:0.3; $ Beta function
for x = 1:size(X)
 N(x)=(X(x))^2*((1-X(x))^5); $ Numerator
 f(x)=N(x)/B; $ The fraction inside the sigma.
end
P=sum(f) $ The sum
```

B in this case (using calculator) is as follows

$$\frac{\Gamma(p)\Gamma(q)}{\Gamma(p+q)} = \frac{\Gamma(2)\Gamma(6)}{\Gamma(8)} = \frac{(1)(5!)}{7!} = \frac{1}{42} \approx 0.02380952381,$$

which is about the same using MATLAB, that is, 0.0238.

3.9.5 ERLANG DISTRIBUTION

During the early part of the twentieth century, A. K. Erlang was to examine the number of telephone calls arriving to the telephone switchboard station (possibly at the same time). This consideration was the beginning of development of the queueing processes, which is almost well developed by now. We refer the reader to some of the publications, including books, of the first author on this subject such as:

> *Queuing Models in Industry and Business,* Second Edition 2014. Nova Science Publishers, Inc., a New York. https://www.novapublishers.com/catalog/product_info.php?products_id=42055&osCsid=7ed271dac267e38ea22b939 87f939075
> and
> *Delayed and Network Queues,* John Wiley & Sons Inc., New Jersey, 2016. http://www.wiley.com/WileyCDA/WileyTitle/productCd-1119022134,subjectCd-BA11.html

Definition 3.46

Let X be a random variable with a two-parameter pdf, shape parameter k (k is a positive integer) and scale parameter λ, defined as

$$f_X(x; k, \lambda) = \begin{cases} \dfrac{\lambda^k x^{k-1} e^{-\lambda x}}{(k-1)!}, & x \geq 0, \lambda > 0, k \in \mathbb{Z}^+, \\ 0 & \text{otherwise.} \end{cases} \qquad (3.230)$$

Then, X is called the **Erlang random variable**, and its distribution is referred to as **Erlang probability distribution function**, denoted by $X \sim \text{Erlang}(k, \lambda)$.

Note 3.63

We leave it as an exercise to prove that the function defined by (3.230) is a pdf.

Note 3.64

The pdf defined in (3.230) reduces to the exponential pdf when $k = 1$. In that case, $1/\lambda$ is the mean, as expected. When $k = 2$, (3.230) reduces to the X^2 pdf, which we will study later in this chapter. Also, it can be seen that Erlang distribution is a special case of the gamma distribution function.

Example 3.82

Figure 3.19 shows the Erlang pdf for various k and λ. The x- and y-axes can only show finitely many values, but rest assured that the area under all curves is 1.

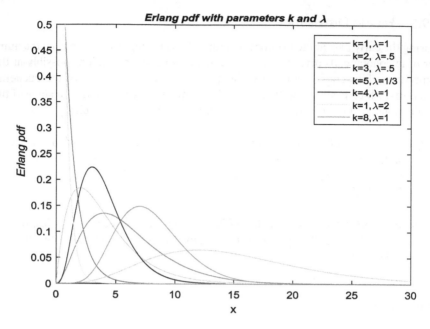

FIGURE 3.19   Presentation of the Erlang pdf with two parameters k and λ.

Remark 3.1

It can be shown that the solution of an initial-valued differential equation

$$xy' + (\lambda x + 1 - k)y = 0, \quad y(1) = \frac{e^{-\lambda}\lambda^k}{(k-1)!}, \quad x \ge 0, \lambda > 0 \tag{3.231}$$

is the Erlang pdf.

Proof:
We can rewrite the first part of (3.231) as:

$$y' + \left(\lambda + \frac{1-k}{x}\right)y = 0, \tag{3.232}$$

which is a homogeneous linear first-order ordinary differential equation with the initial value given in (3.231). Equation (3.232) can be solved by separable variable method or using integrating factor. We use the integrating factor as follows:

$$\mu = e^{\int\left(\lambda + \frac{1-k}{x}\right)dx} = e^{\lambda x}e^{(1-k)\ln x}$$

$$= e^{\lambda x}e^{\ln x^{1-k}} = \frac{e^{\lambda x}}{x^{1-k}}.$$

Thus, the general solution is:

$$y = c\frac{x^{1-k}}{e^{\lambda x}},$$

where c is a general integral constant. Now, to find the constant c, we will substitute the initial value and obtain:

$$\frac{e^{\lambda}\lambda^k}{(k-1)!} = \frac{1}{e^{\lambda}} \cdot c.$$

Thus,

$$c = \frac{\lambda^k}{(k-1)!}.$$

Hence, the particular solution is

$$y = \frac{\lambda^k}{(k-1)!} \cdot \frac{x^{k-1}}{e^{\lambda k}}$$

$$= \frac{\lambda^k x^{k-1} e^{-\lambda x}}{(k-1)!},$$

which is the same as in (3.230), that is, Erlang pdf.

Theorem 3.21

For an Erlang random variable, the cdf, mean, variance, and moment generating function are, respectively, as follows:

$$F_X(x;k,\lambda) = 1 - \sum_{j=0}^{k-1} \frac{(\lambda x)^j}{j!} e^{-\lambda x}, \quad x \ge 0, \lambda > 0. \tag{3.233}$$

$$E(X) = \frac{k}{\lambda}. \tag{3.234}$$

$$Var(X) = \frac{k}{\lambda^2}. \tag{3.235}$$

$$M(t) = \left(1 - \frac{t}{\lambda}\right)^{-k}, \quad t < \lambda. \tag{3.236}$$

Proof:
We prove the formula for cdf and leave the proof of the rest as an exercise.

We start the proof by definition of probability of X. Hence,

$$F_X(x) = P(X \le x) = \int_{0.}^{x} \frac{\lambda^k t^{k-1} e^{-\lambda t}}{(k-1)!} \, dt = \frac{\lambda^k}{(k-1)!} \int_0^x t^{k-1} e^{-\lambda t} \, dt$$

$$= \frac{\lambda^k}{(k-1)!} \left[\frac{-1}{\lambda} e^{-\lambda x} x^{k-1} + \frac{k-1}{\lambda} \int_0^x e^{-\lambda t} t^{k-2} \, dt \right]$$

$$= -\frac{(\lambda x)^{k-1}}{(k-1)!} e^{-\lambda x} + \frac{(k-1)\lambda^{k-1}}{(k-1)!} \left[\left(-\frac{1}{\lambda} e^{-\lambda x} x^{k-2} \right) + \frac{k-2}{\lambda} \int_0^x e^{-\lambda t} t^{k-3} dt \right]$$

$$= -\frac{(\lambda x)^{k-1}}{(k-1)!} e^{-\lambda x} - \frac{(\lambda x)^{k-2}}{(k-2)!} e^{-\lambda x} - \cdots - (\lambda x) e^{-\lambda x} + \lambda \left[-\frac{1}{\lambda} e^{-\lambda t} \right]_0^x$$

$$= -\frac{(\lambda x)^{k-1}}{(k-1)!} e^{-\lambda x} - \frac{(\lambda x)^{k-2}}{(k-2)!} e^{-\lambda x} - \cdots - (\lambda x) e^{-\lambda x} - e^{-\lambda x} + 1$$

$$= 1 - \sum_{j=0}^{k-1} \frac{(\lambda x)^j}{j!} e^{-\lambda x}.$$

Example 3.83

Figure 3.20 shows the Erlang cdf for various k and λ.

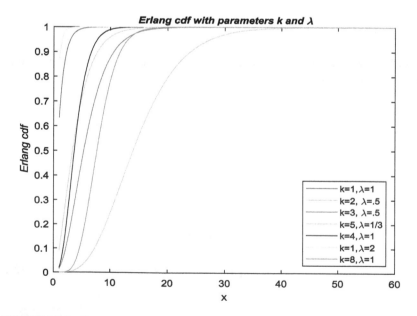

FIGURE 3.20   Presentation of the Erlang cdf with two parameters k and λ.

Example 3.84

Suppose you walk in a bank with one teller station open and a customer being severed. There are three people in line waiting for service. It is known that service times of customers in this branch are independent of each other and they are identically exponentially distributed with the mean time of 3 minutes. The question is: What is the probability that your waiting time in line before your service starts be more than 6 minutes? How about less than 4 minutes?

Answer

Let us represent the service times of the three customers present at the time of your arrival by T_1, T_2, T_3, T_4 with $T = T_1 + T_2 + T_3 + T_4$. Thus, based on Note 3.55, T is an Erlang random variable with parameters $k = 4$ and $\lambda = 1/3$. So, let us use the convolution property for the sum of independent exponential random variables with parameter λ given in Theorem 3.15. Hence, for $n = 4$, the pdf for T is

$$f_4(x) \equiv (f * f * f * f)(x) = \begin{cases} \dfrac{3^4 x^3}{3!} e^{-\lambda x}, & \lambda > 0, x \geq 0, \\ 0, & x < 0. \end{cases}$$

Thus,

1. $P(T > 6) = \dfrac{3^4}{6} \displaystyle\int_6^\infty x^3 e^{-3x} dx = 1.7560 e^{-5}$.

2. $P(T \leq 4) = \dfrac{3^4}{6} \displaystyle\int_0^4 x^3 e^{-3x} dx = 0.9977$.

To evaluate these integrals, we use a very simple four-line MATLAB program that gives answers for both.

```
syms x
f4 = (3^4/6)*(x^3)*exp(-3*x);
P6 = eval(int(f4,6,inf))
P4 = eval(int(f4,0,4))
```

Note 3.65

In a sequence of operations, when the completion of each operation has exponential pdf, the time of completion of n task will have the Erlang distribution. Proof is left as an exercise.

Note 3.66

If tasks arrive to a service location according to Poisson pmf, the inter-arrival times will have exponential pdf with the parameter as the arrival rate. The sum of inter-arrival times has an Erlang distribution. In other words, the Erlang distribution is the distribution of a sum of k iid random variables with the mean of $1/\lambda$ each. Proof is left as an exercise.

3.9.6 Normal Distribution

One of the very well-known and widely used continuous random variables is **normal**. Its distribution is very important in statistics, and its applications are in medicine, psychology, and social sciences. It was discovered by a French mathematician Abraham De Moivre in 1733 and later by another French mathematician Carl Friedrich Gauss in the nineteenth century. Because of this, normal distribution is referred to as the **Gaussian distribution**. All normal distribution curves are symmetric and have bell-shaped pdfs with a single peak. Hence, it is also referred to as the **bell-shaped curve** (Figures 3.21 and 3.22).

It is interesting that Google celebrated the 241st birthday of Gauss by showing logo on its site.

FIGURE 3.21   Abraham de Moivre (1667–1754).

FIGURE 3.22   Carl Friedrich Gauss 1840 by Jens.

Definition 3.47

Let X be a continuous random variable with its pdf, denoted by $\phi(x)$ with two parameters μ, a real number, and σ, a positive real number, defined as:

$$\phi(x) = \frac{1}{\sigma\sqrt{2\pi}} e^{-\frac{(x-\mu)^2}{2\sigma^2}}, \quad x, \mu \in (-\infty, +\infty), \sigma > 0, \tag{3.237}$$

which is called a **normal, Gaussian**, or **bell-shaped curve**, random variable, and (3.237) is referred to as the **normal or Gaussian pdf**.

Theorem 3.22

$\phi(x)$ defined by (3.237) is a pdf.

Proof:
Since $\sigma > 0$, the right hand of (3.237) is positive. Hence, we just have to show that integral of $\phi(x)$ over the entire real line is 1. So, first let

$$I \equiv \int_{-\infty}^{\infty} \phi(x)\,dx = \frac{1}{\sigma\sqrt{2\pi}} \int_{-\infty}^{\infty} e^{-\frac{(x-\mu)^2}{2\sigma^2}}\,dx.$$

Now, let $z = \dfrac{x-\mu}{\sigma}$. Then, $dz = \dfrac{dx}{\sigma}$ and

$$I = \frac{1}{\sqrt{2\pi}} \int_{-\infty}^{\infty} e^{-\frac{z^2}{2}}\,dz.$$

Hence,

$$I^2 = \left[\frac{1}{\sqrt{2\pi}} \int_{-\infty}^{\infty} e^{-\frac{z^2}{2}}\,dz \right]^2$$

$$= \left[\frac{1}{\sqrt{2\pi}} \int_{-\infty}^{\infty} e^{-\frac{z^2}{2}}dz \right]\left[\frac{1}{\sqrt{2\pi}} \int_{-\infty}^{\infty} e^{-\frac{y^2}{2}}\,dy \right]$$

$$= \frac{1}{2\pi} \int_{-\infty}^{\infty} e^{-\frac{(z+y)^2}{2}}\,dz\,dy. \tag{3.238}$$

Applying the polar coordinates: $z = r\cos\theta$, $y = r\sin\theta$, $r^2 = z^2 + y^2$, $\theta = \arctan\dfrac{y}{z}$, and $dzdy = rdrd\theta$, $0 < r < \infty$ and $0 \le \theta \le 2\pi$, from (3.238), we will have:

$$I^2 = \frac{1}{2\pi}\int_0^{2\pi}\int_0^{\infty} e^{-\frac{r^2}{2}} r\, dr\, d\theta = \frac{1}{2\pi}\int_0^{2\pi}\left(-\frac{e^{-r^2}}{2}\right)_0^{\infty} d\theta.$$

$$= \frac{1}{2\pi}\int_0^{2\pi} d\theta = \frac{2\pi}{2\pi} = 1.$$

Therefore, $I = 1$ and $\phi(x)$ is a pdf.

In order to give the cdf for normal random variable, we need the following definition.

Definition 3.48

An error function, denoted by **erf**, is defined as:

$$erf(x) = \frac{1}{\sqrt{\pi}}\int_{-x}^{x} e^{-t^2}\, dt$$

$$= \frac{2}{\sqrt{\pi}}\int_0^{x} e^{-t^2}\, dt. \qquad (3.239)$$

Figure 3.23 shows the graph of erf for values of x between -5 and 5.

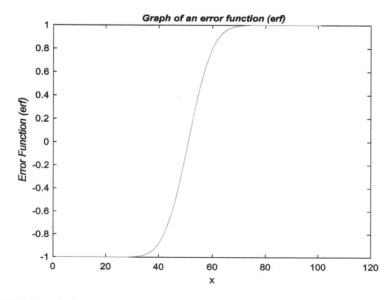

FIGURE 3.23   Graph of erf.

Theorem 3.23

For a normal or Gaussian random variable with pdf as defined in (3.237), the cdf, denoted by $\Phi(x;\mu,\sigma^2)$ or $\Phi(\mu,\sigma^2)$ or $\Phi(x)$ or any other notation such as $N(x;\mu,\sigma^2)$ or $F_X(x)$ or simply $F(x)$; mean, denoted by μ; variance, denoted by σ^2; and moment generating function, $M(t)$, are, respectively, as follows:

$$P(X \le x) \equiv \Phi(x;\mu,\sigma^2) = \int_{-\infty}^{x} \phi(t)\,dt,$$

$$= \int_{-\infty}^{x} \frac{1}{\sigma\sqrt{2\pi}} e^{-\frac{(t-\mu)^2}{2\sigma^2}}\,dt \qquad (3.240)$$

$$= \frac{1}{2}\left[1 + erf\left(\frac{x-\mu}{\sigma\sqrt{2}}\right)\right], \qquad (3.241)$$

where **erf** is given in (3.239).

$$E(X) = \mu, \; Var(X) = \sigma^2, \; \text{and } M(t) = e^{\mu t + \frac{\sigma^2 t^2}{2}}. \qquad (3.242)$$

Proof:
Proof is left as an exercise.

Note 3.67

As mentioned, the normal cdf is denoted by $N(\mu,\sigma^2)$ or $\Phi(\mu,\sigma^2)$; in other words, the random variable X having normal cdf is denoted by $X \sim N(\mu,\sigma^2)$ or $X \sim \Phi(\mu,\sigma^2)$.

Note 3.68

The smaller the value of σ, the higher the peak of the "bell curve", as is shown in Figures 3.18–3.20. The peak of normal pdf defined in (3.237) occurs at $x = \mu$ with its value equal to $\dfrac{1}{\sigma\sqrt{2\pi}}$, which is approximately equal to $\dfrac{0.399}{\sigma}$.

Note 3.69

It is interesting to note that a normal pdf is, indeed, an exponential pdf of a quadratic function, say $f(x)$, defined as

$$f(x) = e^{ax^2 + bx + c},$$

where

$$a < 0, c = \frac{b^2}{4a} + \frac{1}{2}\ln\left(\frac{-a}{\pi}\right).$$

In this case, the mean and standard deviations are

$$\mu = \frac{-b}{2a} \text{ and } \sigma = \sqrt{\frac{-1}{2a}}.$$

Definition 3.49

A normal random variable Z with mean $\mu = 0$ and variance standard deviation $\sigma = 1$ is called the **standard normal random variable**. The value z is a transformation of a regular normal random variable defined as:

$$z = \frac{x - \mu}{\sigma}. \tag{3.243}$$

The transformation (3.243) changes (3.238) to

$$\phi(z) = \frac{1}{\sqrt{2\pi}} e^{-\frac{z^2}{2}}, \quad -\infty < z < \infty, \tag{3.244}$$

which is referred to as the **standard normal** or **z-pdf**. Accordingly, this normal pdf has mean 0 and standard deviation 1. In this case, the cdf of the standard normal variable, $P(Z \le z)$, would be:

$$\Phi(z) = \int_{-\infty}^{z} \frac{1}{\sqrt{2\pi}} e^{-\frac{t^2}{2}} dt \tag{3.245}$$

$$= \frac{1}{2}\left[1 + erf\left(\frac{z}{\sqrt{2}}\right)\right]. \tag{3.246}$$

The number of standard deviations from the mean is sometimes referred to as **standard score** and sometimes **z-score**. The z-score is the transform defined in (3.242).

Note 3.70

The graph of normal pdf is also **asymptotic**, or the tails of the graph from both sides get very close to the horizontal axis without touching it. Hence, for standard normal cdf, we have:

$$\Phi(-z) = P(Z \le -z) = P(Z > z) = 1 - P(Z \le z) = 1 - \Phi(z), \quad z > 0. \tag{3.247}$$

Note 3.71

The approximate values of $\Phi(z)$ have been calculated and tabulated. However, the available technologies these days make the large tables of normal cdf useless. For instance, using some available tables, one can find $\Phi(0.9) = 0.8159$, $\Phi(0.09) = 0.5359$, and $\Phi(3.35) = 0.9996$. These values can easily be obtained using MATLAB. With tables, to find negative of each value, one needs $\Phi(-z) = 1 - \Phi(z)$. With MATLAB, for instance, one can obtain the negative values the same way as positive ones.

Note 3.72

We leave it as an exercise to verify that

$$\Phi(-\infty) = \lim_{z \to -\infty} = 0 \ \text{ and } \ \Phi(0) = \frac{1}{2}. \tag{3.248}$$

Note 3.73

We show that

$$\Phi(+\infty) = \lim_{z \to +\infty} = 1. \tag{3.249}$$

Proof:
Using polar coordinates, we will have:

$$[\Phi(\infty)]^2 = \int_{-\infty}^{\infty} \frac{e^{\frac{-x^2}{2}}}{\sqrt{2\pi}} \, dx \cdot \int_{-\infty}^{\infty} \frac{e^{\frac{-y^2}{2}}}{\sqrt{2\pi}} \, dy$$

$$= \int_{-\infty}^{\infty} \frac{e^{\frac{-(x^2+y^2)}{2}}}{2\pi} \, dx \, dy$$

$$= \int_{0}^{2\pi} \int_{0}^{\infty} \frac{e^{\frac{-r^2}{2}}}{2\pi} r \, dr \, d\theta.$$

Now, letting $u = \dfrac{r^2}{2}$, we have:

$$\int_{0}^{\infty} \frac{e^{\frac{-r^2}{2}}}{2\pi} r \, dr = \int_{0}^{\infty} \frac{e^{-u}}{2\pi} \, du = \frac{1}{2\pi}.$$

Thus,

$$\int_0^{2\pi} \int_0^\infty \frac{e^{\frac{-r^2}{2}}}{2\pi} r\, dr\, d\theta = \int_0^{2\pi} \frac{1}{2\pi} d\theta = 1.$$

We could see this also since the $\Phi(z)$ is symmetric and $\Phi(0) = \frac{1}{2}$, then

$$\Phi(\infty) = 2\Phi(0) = 2\left(\frac{1}{2}\right) = 1.$$

Some properties of normal distribution

1. An interesting property of the normal distribution arises from the following convolution property: Let X_1 and X_2 be two normal random variables with mean and variance for each as μ_1 and σ_1^2, and μ_2 and σ_2^2, respectively. That is,

If $X_1 \sim N\left(\mu_1, \sigma_1^2\right)$ and $X_2 \sim N\left(\mu_2, \sigma_2^2\right)$, then $X_1 + X_2 \sim N\left(\mu_1 + \mu_2, \sigma_1^2 + \sigma_2^2\right)$.   (3.250)

2. Let X be a normal random variable with mean and variance μ and σ^2, respectively. Then, for a constant c, we have:

$$X \sim N\left(\mu, \sigma^2\right) \rightarrow cX \sim N\left(c\mu, c^2\sigma^2\right).$$ (3.251)

3. Sum of a standard normal by itself is a scaled standard. Precisely,

$$N(0,1) + N(0,1) = \sqrt{(2)}N(0,1).$$ (3.252)

4. The mgf of the standard normal random variable is

$$M_X(t) = e^{\frac{t^2}{2}}.$$ (3.253)

This is because, for $X \sim N(0,1)$, we have:

$$M_X(t) = E\left(e^{tX}\right) = \int_{-\infty}^\infty e^{tx} \frac{1}{\sqrt{2\pi}} e^{-\frac{x^2}{2}} dx$$

$$= \int_{-\infty}^\infty \frac{1}{\sqrt{2\pi}} e^{-\left(\frac{x^2 - 2tx}{2}\right)} dx = \int_{-\infty}^\infty \frac{1}{\sqrt{2\pi}} e^{-\left[\frac{(x-t)^2 - t^2}{2}\right]} dx$$

$$= e^{\frac{t^2}{2}} \int_{-\infty}^\infty \frac{1}{\sqrt{2\pi}} e^{-\frac{(x-t)^2}{2}} dx,$$

where the integrant is the pdf of normal with parameters t and 1. Thus, the proof is complete.

Note 3.74

The empirical rule of statistics, sometimes denoted by (68, 85, 99), for normal pdf states that nearly all of the data will fall within three standard deviations of the mean, although some statisticians go to 4 standard deviations to cover the area completely. This general rule is broken down into three parts: (1) It is well known that about 68% (68.26%) of area under a normal pdf curve falls within 1 standard deviation around the mean, that is, $\mu \pm 1\sigma$; (2) about 95% (95.44%) falls within plus and minus 2 standard deviations of the mean, that is, $\mu \pm 2\sigma$; and (3) about 99.7% (99.74%) of data falls within 3 standard deviations of the mean, that is, $\mu \pm 3\sigma$.

Note 3.75

A practical way of finding normal probabilities is first to find the value of z from (3.250). Before progress in technology, tables were primarily used to find probabilities. However, with much development of technology and availability of several computer software packages for mathematics and statistics, there is no need for such long and tedious tables. We constantly use MATLAB, R, and others, as we see fit, for our calculations.

Example 3.85

Let us suppose that in a small company, the average height of staff is 5 feet and 7 inches with a standard deviation of 2 inches. Assume that the heights are normally distributed. What is the z-value for a staff member who is 5 feet and 5 inches?

Answer

It is easy to calculate the z-value of an $x = 5.42$ from (3.241) with $\mu = 5.583$ and $\sigma = 2$. That is,

$$z = \frac{x - \mu}{\sigma} = \frac{5.42 - 5.583}{2} = -0.0815.$$

Thus, this person is almost 2 inches below the average.

Example 3.86

Here is an example of standard normal random variable. Suppose final scores (out of 100 points) of a probability and statistics class with 15 students look like the following:

57, 84, 69, 74, 38, 55, 26, 44, 49, 70, 92, 18, 89, 24, 18.

Suppose that minimum passing score is 70. Looking at the scores, only 5 students, that is, 1/3 of the class, will pass this examination. The professor is unhappy with the failure of 2/3 of the class. He is not sure of the reason or reasons. He guesses the reason or reasons as follows: (1) Students did not study enough, (2) students did not understand the concepts, (3) questions on the test were too difficult, (4) the professor did not do a good job of teaching, or (5) a combination of these. The professor is looking to find a way to curve the scores to have a better passing rate. How could z-score be used?

Answer

To apply the z-score (3.241), we consider the students' scores a sample of independent observations. Although the sample size is supposed to be large (about at least 30–40 in practice) to apply normal distribution, we accept the number 15 for our approximation.

The professor calculates the mean and standard deviation as $\mu = 53.8$ and

$$\sigma = \sqrt{\frac{\sum\limits_{i=1}^{15} (x_i - \mu)^2}{14}},$$ respectively. These values could be obtained by MATLAB

or Microsoft Excel. Procedurally, and for the sake of ease of discussion, we sort the scores that are the values of our random variable and denote them by $x_i, i = 1, 2, \ldots, 15$. We now calculate $x_i - \mu, i = 1, 2, \ldots, 15$, and sum them. Then, calculate the standard deviation. Finally, calculate $z = \dfrac{x - \mu}{\sigma}$. Summary of these steps is in Table 3.32.

In MATLAB,

```
n = 15;
X = [57,84,69,74,38,55,26,44,49,70,92,18,89,24,18];
X1 = sort(X);
X3 = [18,18,24,26,38,44,49,55,57,69,70,74,84,89,92];
Sum_X3 = sum(X3);
mu = Sum_X3/n;
Diff = X3-mu;
Diff_2 = Diff.^2;
SS = sum(Diff_2);
s = sqrt(sum(Diff_2)/(n-1));
z = Diff/s;
```

Hence, we want students with z-scores of at least −1 to pass. That is, we want $z = \dfrac{x - 53.8}{25.5181} \geq -1$, or $x \geq 28.9899$. In other words, all original scores of 28.9899 or higher should pass. The minimum score satisfying this criterion is 38. Now, if passing score is 70/100, then we need to multiply all scores by $70/38 = 1.84211$ and pass all those with a score of 70 and higher. This is summarized in the right column of Table 3.32.

Example 3.87

Discuss Figures 3.18–3.20 of normal pdf with regard to the means.

Answer

Figures 3.24 and 3.25 show the normal pdf graphs for various μ and σ. In Figure 3.24, $\mu = 0$, while in Figure 3.24 $\mu = 2$, and in Figure 3.25, $\mu = -4$. Comparing

TABLE 3.32
Calculation of Parameters and z-Scores

i	x_i	$\mu = \dfrac{\sum_{i=1}^{15} x_i}{15}$	$x_i - \mu$	$(x_i - \mu)^2$	$\sum_{i=1}^{15}(x_i - \mu)^2$	$\sqrt{\dfrac{\sum_{i=1}^{15}(x_i - \mu)^2}{14}}$	$z = \dfrac{x - \mu}{\sigma}$	$(1.84211)x_i$
1	18		$18 - 53.8 = -35.8$	1281.64			-1.4029	33.15798
2	18		$18 - 53.8 = -35.8$	1281.64			-1.4029	33.15798
3	24		$24 - 53.8 = -29.8$	888.04			-1.1678	44.21064
4	26		$26 - 53.8 = -27.8$	772.64			-1.0894	47.89486
5	38		$38 - 53.8 = -15.8$	249.64			-0.6192	70.00018
6	44		$44 - 53.8 = -9.8$	96.04			-0.3840	81.05284
7	49		$49 - 53.8 = -4.8$	230.64			-0.1881	90.26339
8	55		$55 - 53.8 = 1.2$	1.44			0.0470	101.3161
9	57		$57 - 53.8 = 3.2$	10.24			0.1254	105.0003
10	69		$69 - 53.8 = 15.2$	231.04			0.5957	127.1056
11	70		$70 - 53.8 = 16.2$	262.44			0.6348	128.9477
12	74		$74 - 53.8 = 20.2$	408.04			0.7916	136.3161
13	84		$84 - 53.8 = 30.2$	912.04			1.1835	154.7372
14	89		$89 - 53.8 = 35.2$	1239.04			1.3794	163.9478
15	92		$92 - 53.8 = 38.2$	1459.24			1.4970	169.4741
		53.8			9116.4	25.5181		

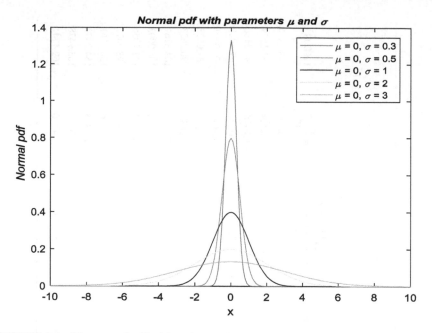

FIGURE 3.24   The normal pdf with various values of σ; each curve is shown by a different color. Curves in this figure are centered at $\mu = 0$. This figure is similar to the Google figure shown above.

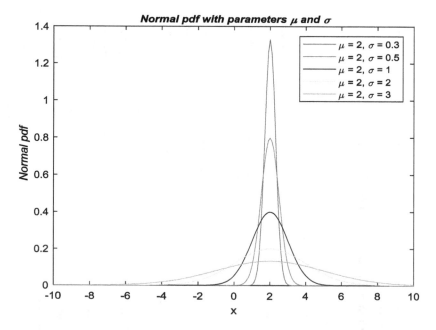

FIGURE 3.25   The normal pdf with various values of σ; each curve is shown by a different color. Curves in this figure are centered at $\mu = 2$. A shift of 2 units compared to Figure 3.18 can be seen.

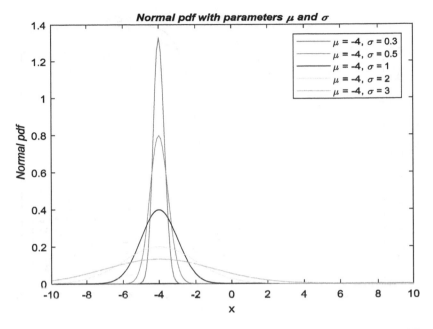

FIGURE 3.26 The normal pdf with various values of σ; each curve is shown by a different color. Curves in this figure are centered at $\mu = -4$. A shift of -4 units compared to Figure 3.18 can be seen.

these graphs, it can be seen that the change of value of μ only causes a shift. However, the shape of curves do not change. We have chosen the values of x within the interval $[-10,10]$ so that the curves are better seen (Figure 3.26).

Example 3.88

Discuss Figures 3.27–3.29 of normal cdf with regard to the means.
    Figures 3.27–3.29 show the normal cdf for various means, μ, and standard deviations, σ. We have chosen the values of x within the interval $[-10,10]$ so that the curves are better seen.

Example 3.89

A statistician is to find the price of an electronic item. She checks the price at several electronic stores locally and online. She finds the average price as $100 with the variance of $25. She decides to walk to a store and buy the item. Find the following probabilities:

    i. The price tag be between $95 and $105?
    ii. The price tag be greater than $105?
    iii. The price tag be less than $90?

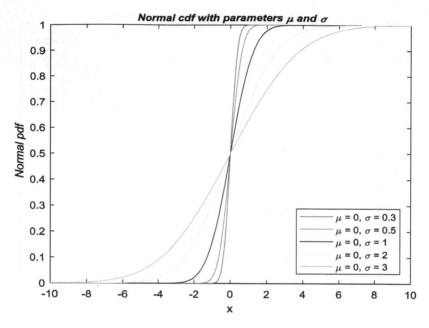

FIGURE 3.27　The normal cdf with various values of σ; each curve is shown by a different color. Curves in this figure are centered at $\mu = 0$.

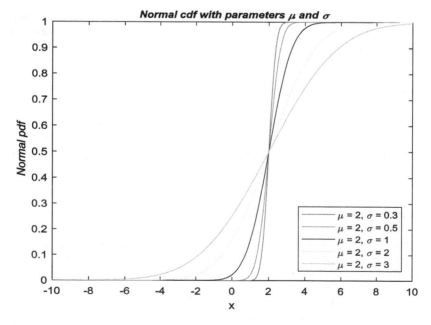

FIGURE 3.28　The normal cdf with various values of σ; each curve is shown by a different color. Curves in this figure are centered at $\mu = 2$.

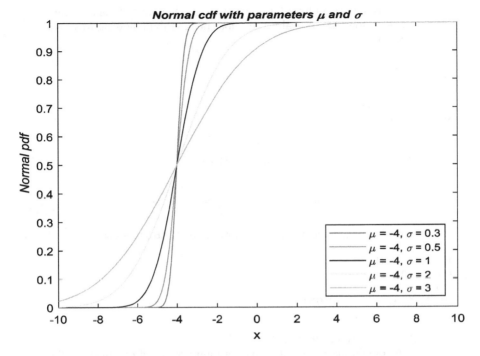

FIGURE 3.29 The normal cdf with various values of σ; each curve is shown by a different color. Curves in this figure are centered at $\mu = -4$.

Answer

   i. Using the cdf (230), we have:

$$P(95 \le X \le 105) = \int_{95}^{105} \frac{1}{5\sqrt{2\pi}} e^{-\frac{(t-100)^2}{2.25}} dt,$$

To evaluate this integral, use the following MATLAB program:

```
x = [95 105];
mu = 100;
sigma = 5;
p1 = normcdf(x,mu,sigma)
pr1 = p1(2)- p1(1)
```

The answer is 0.6827, or 68.27%.

   ii. Again, using the cdf (3.245), we have:

$$P(x > 105) = \int_{105}^{\infty} \frac{1}{5\sqrt{2\pi}} e^{-\frac{(t-100)^2}{2.25}} dt$$

$$= 1 - \int_{0}^{105} \frac{1}{5\sqrt{2\pi}} e^{-\frac{(t-100)^2}{2.25}} dt.$$

The reason we start from 0 is that we assume the price of the item cannot be below $0. To evaluate the integral, use the following MATLAB program:

```
x = [0 105];
mu = 100;
sigma = 5;
p2 = normcdf(x,mu,sigma)
pr2 = 1-(p2(2)- p2(1))
```

The answer is 0.1587, or 15.87%.

iii. Once more, using the cdf (3.245), we have:

$$P(0 \le X \le 90) = \int_0^{90} \frac{1}{5\sqrt{2\pi}} e^{-\frac{(t-100)^2}{2.25}} dt,$$

To evaluate the integral, use the following MATLAB program:

```
x = [95 105];
mu = 100;
sigma = 5;
p3 = normcdf(x,mu,sigma)
pr3 = p3(2)- p3(1)
```

The answer is 0.0228, or 2.28%.

To obtain the results of (i)–(iii), use the normalcdf command in the calculator. To activate this command, select 2nd and VARS and then select the normalcdf. The syntax of this command is as follows:

$$P(Lower\_Bound \le X \le Upper\_Bound)$$

$$= normalcdf(Lower\_Bound, Upper\_Bound, \mu, \sigma)$$

i. $P(95 \le X \le 105) = normalcdf(95, 105, 5) = 0.6827.$

ii. In this example, the upper bound is not given. As we know, the normal distribution can go toward the infinity. Hence, we assume the upper bound be very large. Thus,

$$P(X > 105) = normalcdf(105, 999999999, 100, 5) = 0.1587.$$

iii. In this part, we do not know about the lower bound. Similar to the part (ii), normal distribution can, theoretically, reach up to the negative infinity. Therefore, we consider the negative infinity as the lower bound. Hence, we have:

$$P(X < 90) = normalcdf(-999999999, 90, 100, 5) = 0.0228.$$

Example 3.90

In this example, for normal pdf, in Figure 3.30, we will show the probabilities $P(-8 < x < -4)$, $P(-2 < x < 2)$, and $P(3 < x < 5)$ as the areas under the curve of pdf. The areas colored in blue, purple, and green, respectively, represent each probability.

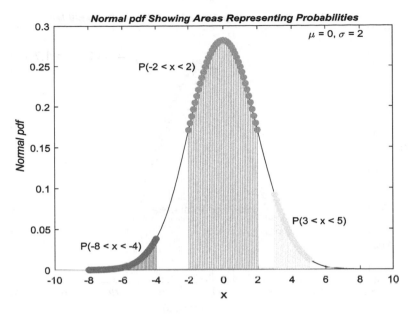

FIGURE 3.30 Presentation of normal probabilities as areas under the normal pdf curve with parameters $\mu = 0$ and $\sigma = 2$.

Example 3.91

Assume that the height of students in a class is normally distributed. Also, assume that 95% of students in the class are between 5 feet and 5 inches and 6 feet tall. Let us calculate the mean and standard deviation for the assumed normal distribution.

Answer

The bounds of the students' height in the class is 65 inches and 72 inches. Hence, the mean, μ, in this case, which is halfway between the given values, is:

$$\mu = \frac{65+72}{2} = 68.5 \text{ inches, or 5 feet and 8.5 inches.}$$

The assumption of 95% implies that there is a deviation of 2 units = each side of the mean, a total of 4 standard deviation that covers the 95% of the area under the normal pdf. See Figure 3.31. Hence,

$$\sigma = \frac{72-65}{4} = 1.75 \text{ inches, or } \approx 0.14583 \text{ feet.}$$

Example 3.92

Suppose that a periodical maintenance cost of a computer is known to have a normal distribution with parameters $\mu = \$60$ and $\sigma = \$3$. What is the probability that the actual cost exceeds $65?

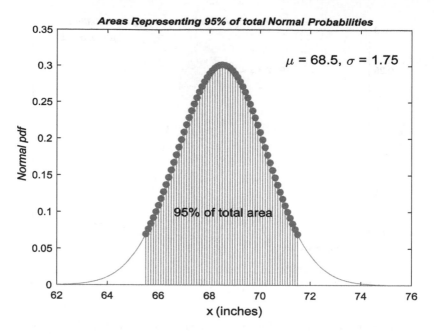

FIGURE 3.31 Presentation of 95% area under normal pdf as areas within 2 standard deviation around the mean. $\mu = 68.5$ inches and $\sigma = 1.75$ inches.

Answer

Let X represent the cost of the computer in a period. Using (317) and (318), and the MATLAB code "1-normcdf(65,60, 5)", we have:

$$P(X > 65) = P\left(X > \frac{65 - 60}{3}\right) = P\left(X > \frac{5}{3}\right)$$

$$= 1 - \frac{1}{2}\left[1 + erf\left(\frac{5}{3\sqrt{2}}\right)\right]$$

$$= 0.1587.$$

Example 3.93

For a standard normal random variable Z, let us find the following probabilities:

    i. $P(-0.25 < z \leq 0.46)$.
   ii. $P(0.87 \leq z \leq 7.3)$.
  iii. $P(z < 6)$.
  iv. $P(z > -0.40)$.

Answer

Formulate each question in terms of (3.232), and then use MATLAB. Hence,

i. $P(-0.25 < z \leq 0.46) = \Phi(0.46) - \Phi(-0.25) = 0.6772 - 0.4013 = 0.2759.$
ii. $P(0.87 \leq z \leq 7.3) = \Phi(7.3) - \Phi(0.87) = 1.0000 - 0.8078 = 0.1922.$
iii. $P(z < 6) = \Phi(6) = 1.0000.$
iv. $P(z > -0.40) = 1 - \Phi(-0.40) = 0.6554.$

$$P(X > 65) = P\left(X > \frac{-0.4}{1}\right) = P(X > -0.4)$$

$$= \frac{1}{2}\left[1 + erf\left(\frac{-0.4}{2}\right)\right]$$

$$= 0.6554.$$

Using calculator, we obtain the same results as follows:

i. $P(-0.25 < z \leq 0.46) = normalcdf(-0.25, 0.46, 0, 1) = 0.2759.$
ii. $P(0.87 \leq z \leq 7.3) = normalcdf(0.87, 7.3, 0, 1) = 0.1921.$
iii. $P(z < 6) = normalcdf(-99999, 6, 0, 1) = 1.$
iv. $P(z > -0.4) = normalcdf(-0.4, 99999, 0, 1) = 0.6554.$

The Normal Approximation to the Binomial Distribution
Previously, we saw an approximation of Poisson pmf to the binomial pmf when the product np remained fixed.

Note 3.76

We should point out that in this case, a pmf of a discrete random variable was approximated by the pmf of another discrete random variable.

Note 3.77

Consider tossing a coin with the probability of occurrence of a head as p. Then, as the number of tosses increases, the probability of occurrence of a head decreases so that product does not change; then, the distribution of the number of heads during the number of trials will be Poisson.

A more interesting idea is approximating the distribution of a discrete random variable by the distribution of a continuous random variable, namely, the normal distribution.

Example 3.94

Justifying the need of the normal approximation to the binomial for large sample size, by some, is because "solving problems using the binomial distribution might seem daunting". This is mainly due to the probabilities that have to be calculated by the binomial coefficient through the use of "factorial". For instance, suppose a fair coin is tossed 100 times. Let X be the number of heads. What is the probability that X is greater than 60?

Answer

The binomial table runs out of numbers for large enough trial counts. As another example, to find the probability that a binomial variable X is between 4 and 12 exclusive, we have to find the probability that X equals 5, 6, 7, 8, 9, 10, and 11, then add all of these probabilities together. But factorials may be large and some do not like to deal with them. However, the normal approximation allows one to bypass such a problem. Using the z-scores corresponding to 5 and 11 with a z-score table, one would find the probabilities. But with today's technology, those arguments may be obsolete. Mathematics software like MATLAB can calculate the values in a matter of seconds.

Note 3.78

In statistical practices, in order to use the normal approximation to a binomial, as one needs a large sample size, but what is "large"? Consider n as the number of observations and p as the probability of success in each trial. Here are some "rules of thumb":

i. Both np and $n(1-p) \geq 10$.
ii. Both np and $n(1-p) \geq 5$.
iii. $np(1-p) \geq 3$.

Example 3.95

In a binomial experiment, let $n = 100$ and $p = 0.5$. That is, we toss a unbiased coin with the probability of success $p = 0.5$. What is the probability of the number of heads to be at least 70?

Answer

This example could be a translation of a multiple-choice test with 100 two-choice questions, and a student needs at least 70% on the test to pass.

Let X be a random variable representing the number of heads occurring for the 100 trials. Then, X is a binomial random variable. In fact, the normal cdf could be used because $np = 50$, $n(1-p) = 50$, and $np(1-p) = 25$; that is, all cases (i), (ii), and (iii) are satisfied. Hence, we may find $P(X \geq 70)$ using the normal distribution.

Using (3.237) with

$$\mu = np = (100)(0.5) = 50 \text{ and } \sigma = \sqrt{np(1-p)} = \sqrt{(100)(0.5)(0.5)} = 5,$$

and

$$z_{70} = \frac{x - \mu}{\sigma} = \frac{70 - 50}{5} = 4,$$

since the maximum value of X is 100,

$$z_{100} = \frac{x - \mu}{\sigma} = \frac{100 - 50}{5} = 10.$$

Thus, the answer is

$$P(X \geq 70) = \Phi(10) - \Phi(4) = 0.000031671.$$

Using MATLAB code

```
normcdf(10) - normcdf(4)
```

or

```
normcdf(100,50,5) - normcdf(70,50,5)
```

we will have the same answer.
   Using the calculator

$$P(z \geq 70) = normalcdf(70,99999,50,5) = 0.00003167.$$

For the test question mentioned above, there is a near zero chance for passing a student desire to pass, if by completely guessing.
   We now summarize the approximation of normal to the binomial through the following theorem attributed to two French mathematicians Abraham de Moivre (1667–1754) and Pierre-Simon Laplace (1749–1827). De Moivre wrote a book on probability theory, *The Doctrine of Chances*, first published in 1718.

ABRAHAM DE MOIVRE (1667–1754)

Theorem 3.24 (de Moivre–Laplace)

Let X be a binomial random variable with pmf $\mathcal{B}_j(j;n,p)$, where j, n, and p are the number of desired successes, the number of trials, and the probability of success in each trial, respectively. If $n \to \infty$ (i.e., n becomes large), while p remains unchanged, then cdf of

$$\frac{X-np}{\sqrt{np(1-p)}}$$

is the standard normal cdf, that is,

$$\lim_{\substack{n\to\infty \\ p\to 0}} \mathcal{B}_j(j;n,p) = \lim_{\substack{n\to\infty \\ p\to 0}} cdf\left(\frac{X-np}{\sqrt{np(1-p)}}\right) = \Phi(0,1). \tag{3.254}$$

Proof:
Left as an exercise.
 We have presented a photo of Laplace in Chapter 2.

Theorem 3.25 (de Moivre–Laplace Central Limit Theorem)

Let X be a binomial random variable, with pmf $\mathcal{B}_j(j;n,p)$, where j, n, and p, $0 < p < 1$, and fixed, are the number of desired successes, the number of trials, and the probability of success in each trial, respectively. If $n \to \infty$ (i.e., n becomes large), while p remains unchanged, then for two values of X as a and b, we have:

$$P(a < X < b) \approx \Phi\left(\frac{b-np}{\sqrt{np(1-p)}}\right) - \Phi\left(\frac{a-np}{\sqrt{np(1-p)}}\right), \tag{3.255}$$

where Φ is the standard normal cdf.

Proof:
Left as an exercise.

Note 3.79

 The de Moivre–Laplace central limit theorem is a special case of the general central limit theorem that we will study in a later chapter, and it is only for binomial pmf (last phrase unclear). It states that if n is large, then the probability of having between a and b successes is approximately equal to the area between a and b under the normal curve with parameters $\mu = np$ and $\sigma^2 = np(1-p)$.

Example 3.96

Let us now consider a multiple-choice test that consists of 100 equal-weighted questions, where each question has one correct answer out of four choices. (1) A student wants to know his chance of getting at least 80% on the test and (2) another needs a chance of receiving at least 30%.

Answer

Let X be a random variable representing the student's number of correct answers. Hence, X has a binomial pmf with total observation $n = 100$, $p = 0.25$, mean $\mu = (100)(0.25) = 25$, and standard deviation $\sigma = \sqrt{(100)(0.25)(0.75)} = 4.3301$.

Now, as we did in Example 3.95, the normal distribution approximation to binomial, using (3.237), (3.240), and MATLAB, we have two cases:

$$z_{80} = \frac{x - \mu}{\sigma} = \frac{80 - 25}{4.3301} = 12.7018.$$

$$z_{30} = \frac{x - \mu}{\sigma} = \frac{30 - 25}{4.3301} = 1.1547.$$

$$z_{100} = \frac{x - \mu}{\sigma} = \frac{100 - 25}{4.3301} = 17.3206.$$

Thus, the answer is

  i. $P(X \geq 80) = \Phi(17.3206) - \Phi(12.7018) \approx 0$

  and

 ii. $P(X \geq 30) = \Phi(17.3206) - \Phi(1.1547) = 0.1241.$
     In MATLAB codes:

```
P_80_1 = normcdf(17.3206)  -  normcdf(12.7018);
```

or

```
P_80_2 = normcdf(100,25,4.3301)  -  normcdf(80,25,4.3301);
```

and

```
P_30_3 = normcdf(17.3206)  -  normcdf(1.1547)
```

or

```
P_30_4 = normcdf(100,25,4.3301)  -  normcdf(30,25,4.3301)
```

Example 3.97

We toss a fair coin 30 times. What is the probability of having up to 18 heads?

Answer

We will present the answer with two different methods:

   i. As we have seen before, this is a binomial problem with $n = 30$, $j = 0,1,$
      $2,\dots 18$, and $p = 0.5$ with pmf $\mathcal{B}_j(j;30,0.5)$, $j = 0,1,2,\dots,18$. Thus, letting
      X to represent the number of heads, we have:

$$P(0 \le X \le 18) = \sum_{i=0}^{j} \mathcal{B}_j(j;30,0.5) = \sum_{i=0}^{18} \binom{30}{i}(0.5)^i (0.5)^{30-i}$$

$$= 0.8998.$$

  ii. Alternatively, we use normal cdf. That is,

$$\mu = np = (30)(0.5) = 15,$$

$$\sigma = \sqrt{np(1-p)} = \sqrt{(30)(0.5)(0.5)} = 2.7386,$$

and

$$P(0 \le X \le 18) = \sum_{i=0}^{18} \mathcal{B}_j(j;30,0.5) = \Phi(15, 2.7386).$$

Example 3.98

Suppose we are interested in testing approval rating of a high-ranking political figure in a country, like the president of the United States. For this purpose, a random sample of size $n = 100$ is taken from the population of eligible voters in a part of the country. What is the probability that more than 60 of the 100 approve the politician?

Answer

Let X be a Bernoulli random variable representing a voter. Let the approving of his job be denoted by 1 with probability p and disapproving of his job be denoted by 0 with probability $1 - p$. Taking a sample of $n = 100$, it means repeating the trial 100 times independently. Hence, we are lead to a new random variable, say Y, which is the sum of $n = 100$ independent random variables, that is, $Y = X_1 + X_2 + \dots + X_{100}$, with parameters mean

$$\mu = np = (100)(0.5) = 50$$

and standard deviation

$$\sigma = \sqrt{np(1-p)} = \sqrt{(100)(0.50)(0.50)} = 5.$$

The question is finding the probability of approval rating of higher than 60%. Thus, referring to the de Moivre–Laplace theorem and using MATLAB, we have:

$$P(X > 60) = P(60 < X < 100) \approx \Phi\left(\frac{100-50}{5}\right) - \Phi\left(\frac{60-50}{5}\right)$$

$$= \Phi(10) - \Phi(2) = 1 - 0.9772 = 0.0228.$$

Example 3.99

Suppose that a certain population is comprised of half men and half women. A random sample of 10,000 from this population is selected. What is the probability that the percentage of the men in the sample is between 49% and 51%?

Answer

We use binomial pmf of a random variable X, representing the number of men in the sample, with $n = 10,000$ and $p = 0.5$. Hence, we are to find the probability of number of men in the sample to be between 4,900 and 5,100. Thus,

$$P(4,900 \le X \le 5,100) = \sum_{i=4,900}^{5,100} \binom{10,000}{i}(0.5)^i (0.5)^{10,000-i}$$

Now, since $np = 5,000$ and $\sigma = \sqrt{np(1-p)} = 50$, the normal approximation of the probability in question will be

$$P(4,900 \le X \le 5,100) \approx \Phi\left(\frac{5,100-5,000}{50}\right) - \Phi\left(\frac{4,900-5,000}{50}\right)$$

$$= \Phi(2) - \Phi(-2) = 0.9544$$

With the probability of approximately 95.44%, the percent of number of men in the sample is between 49% and 51%.

Note 3.80

The result obtained in the example above is a confirmation of what we have seen in the law of large numbers. The error of approximation, in this case, is a bit less than 5%.

3.9.7 χ^2, CHI-SQUARED, DISTRIBUTION

As an application of the normal distribution, we now discuss a very useful and widely used distribution, namely, χ^2 (read as chi-squared) distribution. It is an interesting distribution in testing statistical hypotheses that we will see in Chapter 4. Particular applications of χ^2 include the theoretical and observed values of a quantity, as well as about population variances and standard deviations.

Definition 3.50

Let $X_1, X_2, \ldots X_r$ be r independent standard normal random variables. Then, the cdf of the sum of square of these random variables is called χ^2, **chi-squared, distribution with r degrees of freedom**, denoted by

$$\sum_{i=1}^{r} X_i^2 \sim \chi^2_{(r)}. \tag{3.256}$$

Theorem 3.26

The pdf, cdf, mean, and variance for a χ^2 random variable denoted by $f_X(x,r)$, $F_X(x,r)$, μ, and σ^2 are, respectively, as follows:

$$f_X(x,r) = \begin{cases} \dfrac{x^{\frac{r}{2}-1} e^{-\frac{x}{2}}}{2^{\frac{r}{2}} \Gamma\left(\dfrac{r}{2}\right)}, & x > 0, \\ 0, & \text{otherwise,} \end{cases} \qquad F_X(x,r) = \begin{cases} \dfrac{\gamma\left(\dfrac{x}{2}, \dfrac{r}{2}\right)}{\Gamma\left(\dfrac{r}{2}\right)}, & x > 0, \\ 0, & \text{otherwise,} \end{cases} \tag{3.257}$$

where $\gamma(x,r)$ is called the **lower incomplete gamma function**, defined as:

$$\gamma(x,\alpha) = \int_0^x t^{\alpha-1} e^{-t}\, dt, \tag{3.258}$$

where α is a complex parameter, and

$$\mu = r, \text{ and } \sigma^2 = 2r. \tag{3.259}$$

Proof:
We leave the proof as an exercise.

Note 3.81

We leave it as an exercise to show that a particular case of the χ^2 cdf defined in (3.257) when $r = 2$ is an exponential of the form:

$$F_X(x,2) = \begin{cases} 1 - e^{-\frac{x}{2}}, & x > 0, \\ 0, & \text{otherwise.} \end{cases} \tag{3.260}$$

Example 3.100

Below is the graph of a χ^2 cdf with 6 degrees of freedom. MATLAB code for this graph is as follows (Figure 3.32):

```
x = 0:0.2:20;
y = chi2pdf(x,1);
plot(x,y, 'k')
hold on
y = chi2pdf(x,3);
plot(x,y, 'b')
hold on
y = chi2pdf(x,5);
plot(x,y, 'r')
hold on
y = chi2pdf(x,6);
plot(x,y, 'b')
hold on
y = chi2pdf(x,8);
plot(x,y, 'k')
hold off
ylabel('\chi^2 cdf');
xlabel('x');
title('\chi^2 cdf');
axis ([0 20 0 .4])
text(.9, 0.35, 'r = 1');
text(2.5, 0.2, 'r = 3');
text(3.8, 0.16, 'r = 5');
text(4.95, 0.14, 'r = 6');
text(9.5, 0.09, 'r = 8');
```

We utilize a similar program for calculating the cdf. See Figure 3.33.

There are many applications of X^2, such as **goodness-of-fit** and **test of independence** that we will discuss later.

3.9.8   THE *F*-DISTRIBUTION

A distribution that has a natural relationship with the χ^2 distribution is the *F*-distribution.

Definition 3.51

Let χ_1 and χ_2 each be a χ^2 random variable with r_1 and r_2 degrees of freedom, respectively. Then, the random variable χ denoted by

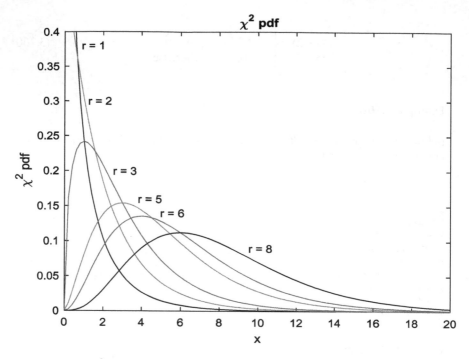

FIGURE 3.32 χ^2 pdf with $r = 1, 2, 3, 5, 6,$ and 8 degrees of freedom.

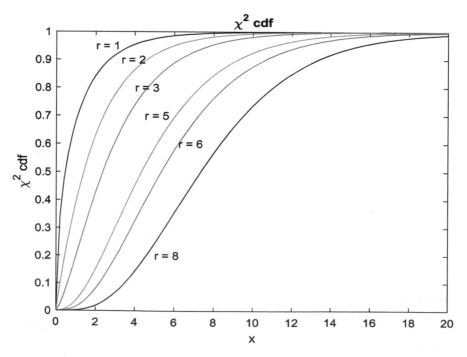

FIGURE 3.33 χ^2 cdf with $r = 1, 2, 3, 5, 6,$ and 8 degrees of freedom.

$$\chi = \frac{\dfrac{\chi_1}{r_1}}{\dfrac{\chi_2}{r_2}} \qquad (3.261)$$

has a pdf, denoted by $f_\chi(r_1, r_2)$ and defined as:

$$f_\chi(r_1, r_2) = \frac{\Gamma\left[\dfrac{r_1 + r_2}{2}\right]}{\Gamma\left(\dfrac{r_1}{2}\right)\Gamma\left(\dfrac{r_2}{2}\right)} \left(\frac{r_1}{r_2}\right)^{\frac{r_1}{2}} \frac{x^{\frac{r_1-2}{2}}}{\left[1 + \left(\dfrac{r_1}{r_2}\right)x\right]^{\frac{r_1+r_2}{2}}}, \qquad (3.262)$$

where $\Gamma(\cdot)$ has been defined before. χ is called **F random variable of F statistics.**

Note 3.82

The parameters r_1 and r_2 represent the number of independent pieces of information used to calculate χ_1 and χ_2, respectively.

Example 3.101

Calculate the pdf of an F-distribution with $r_1 = 7$ and $r_2 = 5$.

Answer

We use the MATLAB to calculate and plot the pdf value in question. The codes we use are as follows.

It can be seen from Figure 3.34 that the F-distribution exists on positive real axis and is skewed to the right. In other words, the F-distribution is an asymmetric distribution with a minimum value of 0. The peaks are close to the zero x-value. After the peak, a curve quickly approaches the x-axis. Also, from Figure 3.34, we see that the smaller the degrees of freedoms, the closer the peak to zero.

The MATLAB program for Figure 3.31 is as follows:

```
x = 0:0.1:10;
y = fpdf(x,5,3);
plot(x,y, 'r')
annotation('textarrow', [0.2,0.3], [0.5,0.6]);
text(2, .5,'(r_1=5,r_2=3)')
hold on
y = fpdf(x,7,5);
plot(x,y, 'k')
annotation('textarrow', [0.2,0.3], [0.6,0.7]);
text(2.2, .6,'(r_1=7,r_2=5)')
hold on
y = fpdf(x,7,9);
plot(x,y, 'b')
```

Figure 3.35 shows graphs of F cdf for a variety of degrees of freedoms for the numerator and the denominator.

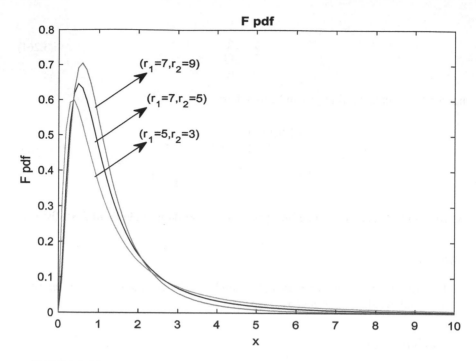

FIGURE 3.34 F pdf with $(r_1, r_2) = (7, 9)$, (7,5), (5, 3), two degrees of freedoms, one for the numerator and one for the denominator.

Note 3.83

To find numerical probabilities, we can remove the ":" and obtain a list of values.

Definition 3.52

A sample point that falls far away from the sample points is referred to as an **outlier**. The median is less sensitive to outliers than the mean. Hence, when a sample contains outliers, the median is used. Such samples are sometimes referred to as **skewed samples**. An outlier on the far left will cause a **negative or left skew**, while an outlier on the far right will cause a **positive or right skew**. See Wilks (1962), p.265, for the calculation of skewness.

Not all data sets are distributions the same way not all pdfs or pmfs are of the same shape. As we have seen in Chapter 2, some are symmetric, while others are asymmetric. Some are even skewed. We have even seen distributions that are multimodal.

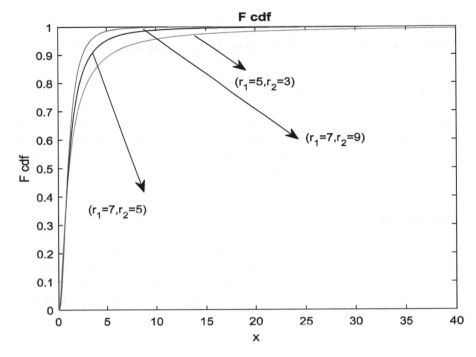

FIGURE 3.35 F cdf with $(r_1, r_2) = (7, 9), (7, 5), (5, 3)$, two degrees of freedoms, one for the numerator, and one for the denominator.

Definition 3.53

Kurtosis ("kurtos" means "arched" or "bulging") measures the sharpness of peak of a distribution. It also measures the thickness or heaviness of the tails of a distribution.

Subjectively, kurtosis measures are obtained by comparing with normal distribution. However, to be precise, the ratio $\dfrac{\mu_4}{\sigma^4}$ is used, where μ_4 is **Pearson's fourth moment about the mean** and σ is the standard deviation. The **kth moment about the mean**, μ_k, is defined as follows: Let X be a random variable with n values $x_1, x_2, \ldots, x_n$. Let us also denote the mean of X by μ. Then,

$$\mu_k = \frac{\displaystyle\sum_{i=1}^{n}(x_i - \mu)^k}{n}. \tag{3.263}$$

Note 3.84

We leave it as an exercise to show that the kurtosis of the normal distribution is 3.

Definition 3.54

Similar to skewness, positive or negative values of kurtosis will cause flatter than or sharper than the peak of the normal curve. For the calculation of kurtosis, the reader may see Wilks (1962, p. 265). A distribution that has tails similar to a normal distribution is referred to as **mesokurtic**.

Definition 3.55

If a distribution has kurtosis greater than a mesokurtic distribution, it is referred to as a **leptokurtic** ("lepto" means "skinny") distribution. Such distributions are sometimes identified by peaks that are thin and tall. The tails of such distributions are thick and heavy. "Student's *t*-distribution" is an example of a leptokurtic distribution.

Definition 3.56

Distributions with slender tails are referred to as **platykurtic** ("platy" means "broad") distributions. Often their peaks are lower than those of mesokurtic distributions. Uniform distributions are the examples of platykurtic distributions. Bernoulli pmf is another example of a platykurtic distribution.

Essentially, a kurtosis distribution is used to be a baseline for mesokurtic, leptokurtic, and platykurtic distributions. Thus, we can subtract 3 (the kurtosis for normal distribution) from the standard calculation formula, $\frac{\mu_4}{\sigma^4}$, that is, $\frac{\mu_4}{\sigma^4} - 3$, to measure **excess kurtosis**. Hence, the excess kurtosis for mesokurtic, leptokurtic, and platykurtic distributions are 0, negative excess kurtosis, and positive excess kurtosis, respectively.

3.9.9 STUDENT'S *t*-DISTRIBUTION

Historically, the Irish chemist William Gosset used the opportunity of "staff leave for study" at the place he was working to continue his study enrolling in "Karl Pearson's Biometric Laboratory" at University College London. He developed and published a paper in 1908 regarding a continuous random variable with a distribution he called, under the pen name, **Student**, referring to the opportunity's name he used. Later, the distribution was also referred to as **Student's *t***, or simply ***t*-distribution.** Here is how it is defined (Figure 3.36).

FIGURE 3.36 William Sealy Gosset 1876–1937.

Definition 3.57

Let X represent a random sample $X_1, X_2, \ldots, X_n$ of size n (i.e., $X_1, X_2, \ldots, X_n$ are iid random variables) from a normal population with mean and standard deviation μ and σ, respectively, that is, $X \sim N(\mu, \sigma^2)$. Let $\overline{X}$ and S^2 denote the sample mean and the corrected variance of the sample, respectively, that is (Figure 3.37),

$$\overline{X} = \frac{1}{n}\sum_{i=1}^{n} X_i \text{ and } S^2 = \frac{1}{n-1}\sum_{i=1}^{n}\left(X_i - \overline{X}\right)^2.$$

FIGURE 3.37 Karl Pearson 1857–1936.

Then, as we have seen earlier, $Z = \dfrac{\overline{X} - \mu}{\dfrac{\sigma}{\sqrt{n}}} \sim N(0,1)$.

Now, if we define the random variable t by

$$t = \frac{\overline{X} - \mu}{\dfrac{S}{\sqrt{n}}}, \tag{3.264}$$

then t **has t-distribution with** $n - 1$ **degrees of freedom**, denoted by "*df*".

Definition 3.58

Let $f(t,v)$ be a function defined as

$$f(t,v) = \frac{\Gamma\left(\dfrac{v+1}{2}\right)}{\sqrt{v\pi}\,\Gamma\left(\dfrac{v}{2}\right)} \frac{1}{\left(1 + \dfrac{t^2}{v}\right)^{-\frac{v+1}{2}}}, \tag{3.265}$$

where v is the fixed number of *df* and $\Gamma(\cdot)$ is the gamma function defined earlier. Then, $f(t,v)$ is a pdf (proof left as an exercise), given the value of v, and is referred to as **Student t's pdf.**

Note 3.85

Using beta function, (3.265) can be written as

$$f(t,v) = \frac{1}{\sqrt{v}\,\mathbf{B}\left(\dfrac{1}{2},\dfrac{v}{2},\right)} \left(1 + \frac{t^2}{v}\right)^{-\frac{v+1}{2}}, \tag{3.266}$$

where $\mathbf{B}$ is the beta function defined earlier.

Note 3.86

In case v **is an integer**, then there are two cases for (3.265) as follows:

Case 1, for $v > 1$ and **even,** the first term on the right-hand side of (3.265) becomes

$$\frac{\Gamma\left(\dfrac{v+1}{2}\right)}{\sqrt{v\pi}\,\Gamma\left(\dfrac{v}{2}\right)} = \frac{(v-1)(v-3)\cdots 5\cdot 3}{2\sqrt{v}\,(v-2)(v-4)\cdots 4\cdot 2}. \tag{3.267}$$

Case 2, for $v > 1$ and odd, the first term on the right-hand side of (3.266) becomes

$$\frac{\Gamma\left(\dfrac{v+1}{2}\right)}{\sqrt{v\pi}\,\Gamma\left(\dfrac{v}{2}\right)} = \frac{(v-1)(v-3)\cdots 4\cdot 2}{\pi\sqrt{v}\,(v-2)(v-4)\cdots 5\cdot 3}. \tag{3.268}$$

Note 3.87

Noting that $v = n - 1$, it can be seen from (3.268) that a t-distribution depends upon the sample size n. That is, there is a different t-distribution for each sample size n with different degrees of freedom. Hence, (3.268) defines a class of distributions.

Student's t-distribution is a symmetric bell-shaped distribution with peak at 0. It looks like a standard normal curve, but with lower height and wider spread. The reason of more spread than the standard normal curve is because the standard deviation for a t-curve with, say $v, v > 2$, df is $\sqrt{\dfrac{v}{v-2}} > 1$, which is the standard deviation of the standard normal distribution curve; the spread is thus larger for a t-curve. However, as the number of df increases, the t-distribution curve approaches the standard normal curve.

Note 3.88

We should note that as the degrees of freedom v becomes large and approaches infinity, the t-distribution would approach the standard normal distribution.

Theorem 3.27

The cdf of Student's t-random variable, $F(x,v)$, can be obtained by integrating (3.265) from $-\infty$ to x. That is, $F(x,v)$ is as follows:

$$F(x,v) = \frac{1}{2} + x\Gamma\left(\frac{v+1}{2}\right) \cdot \frac{{}_2F_1\left(\dfrac{1}{2}, \dfrac{v+1}{2}, \dfrac{3}{2}; -\dfrac{x^2}{v}\right)}{\sqrt{\pi v}\,\Gamma\dfrac{v}{2}}, \tag{3.269}$$

where ${}_2F_1(\cdot)$ is the **hypergeometric function** defined as

$${}_2F_1(a,b,c;z) = \begin{cases} \displaystyle\sum_{j=0}^{\infty} \frac{(a)_j (b)_j}{(c)_j} \frac{z^j}{j!}, & c > 0, |z| < 1, \\[2ex] \text{undefined or } \infty, & C \le 0, \end{cases} \tag{3.270}$$

where $(r)_i$ is defined as

$$(r)_i = \begin{cases} 1, & i = 0, \\ r(r+1)\cdots(r+i-1), & i > 0. \end{cases} \tag{3.271}$$

Proof:
Left as an exercise.

Note 3.89

If either a or b is a nonpositive integer, the series in (3.270) will become finite. In this case, for $c > 0$ and $|z| < 1$, (3.269) becomes a polynomial as follows:

$$_2F_1(-k,b,c;z) = \sum_{j=0}^{k} (-1)^j \binom{k}{j} \frac{(b)_j}{(c)_j} z^j, \quad c > 0, |z| < 1. \tag{3.272}$$

We leave it as an exercise to show that the mean and variance of t-random variable are, respectively,

$$\text{mean}(t) = \begin{cases} 0, & v > 1, \\ \text{undefined}, & v \le 0, \end{cases} \quad \text{and} \quad Var(t) = \begin{cases} \dfrac{v}{v-2}, & v > 2, \\ \infty, & 1 < v \le 2, \\ \text{undefined}, & \text{otherwise}. \end{cases} \tag{3.273}$$

Code for finding the mean and variance of a t-distribution, given *df* and say v, is " [m,v] = tstat(nu)".

Example 3.102

Suppose a t-random variable with $v = 24$. Then, using MATLAB code "[m,v] = tstat(nu)", the mean and variance of this t are 0 and 1.0909, respectively.

Example 3.103

Let X represent a random variable with Student's t-pdf with v df. We let X take values between 0 and 7 and $v = 4,99,999,4999$. Using MATLAB, we want to see how the shape of the pdf and cdf curves will change as the value of v changes. Do the same for X taking values between –7 and 7.

Answer

Using MATLAB, the following programs (the first 5 lines of pdf should be repeated for cdf) records all four plots for pdfs, Figure 3.38, and cdfs, Figure 3.39, for values of x between 0 and 7 and a visual comparison.

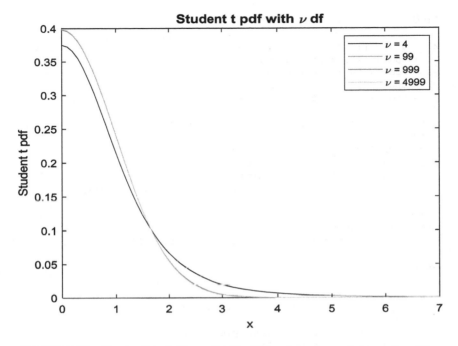

FIGURE 3.38 Graph of Student's t-pdf with different v values, x between 0 and 7.

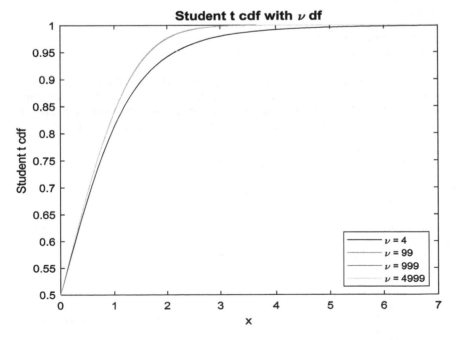

FIGURE 3.39 Graph of Student's t-cdf with different v values, x between 0 and 7.

t-pdf

```
x = [0:.1:7];
y1 = tpdf(x,4); % For nu = 4
y2 = tpdf(x,99); % For nu = 99
y3 = tpdf(x,999); % For nu = 999
y4 = tpdf(x,4999); % For nu = 4999
figure;
plot(x,y1,'Color','black','LineStyle','-')
hold on
plot(x,y2,'Color','red','LineStyle','-')
plot(x,y3,'Color','blue','LineStyle','-')
plot(x,y4,'Color','green','LineStyle','-')
legend({'\nu = 4','\nu = 99','\nu = 999', '\nu = 4999'})
hold off
xlabel('x')
ylabel('Student t pdf')
title('Student t pdf with \nu df','fontsize',12)
```

t-cdf

```
figure;
plot(x,y1,'Color','black','LineStyle','-')
hold on
plot(x,y2,'Color','red','LineStyle','-')
plot(x,y3,'Color','blue','LineStyle','-')
plot(x,y4,'Color','green','LineStyle','-')
legend({'\nu = 4','\nu = 99','\nu = 999', '\nu = 4999'},
'Location',
'southeast')
hold off
xlabel('x')
ylabel('Student t cdf')
title('Student t cdf with \nu df','fontsize',12)
```

Figures 3.40 and 3.41 show the closeness of t-pdf and t-cdf to the standard normal pdf and cdf, respectively, for values of x between –7 and 7 and a visual comparison. MATLAB codes are similar to the other figures with the exception of the addition of the normal plot. Here is the t-cdf with the standard normal cdf:

```
x = [-7:.1:7];
y1 = tcdf(x,4); % For nu = 4
y2 = tcdf(x,99); % For nu = 99
y3 = tcdf(x,999); % For nu = 999
y4 = tcdf(x,4999); % For nu = 4999
figure;
plot(x,y1,'Color','black','LineStyle','-')
hold on
plot(x,y2,'Color','red','LineStyle','-')
plot(x,y3,'Color','blue','LineStyle','-')
plot(x,y4,'Color','green','LineStyle','--')
z = normcdf(x,0,1);
plot(x,z,'-.m*')
legend({'\nu=4','\nu=99','\nu=999','\nu=4999','N(0,1)'},'Location
','southeas')
hold off
```

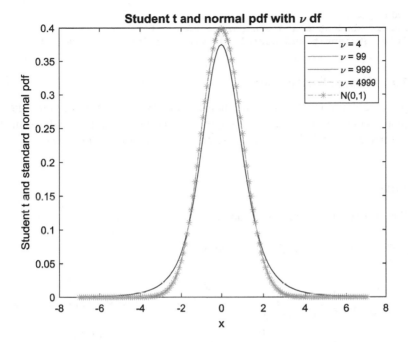

FIGURE 3.40   Graph of Student's t-pdf with different ν values, x between -7 and 7.

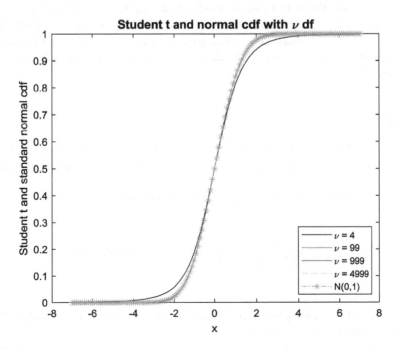

FIGURE 3.41   Graph of Student's t-cdf with different ν values, x between -7 and 7.

```
xlabel('x')
ylabel('Student t and standard normal cdf')
title('Student t and normal cdf with \nu df','fontsize',12)
```

3.9.10 WEIBULL DISTRIBUTION

We give the definition for Weibull (named after Waloddi Weibull (1887–1979), a Swedish engineer for his work on strength of materials and fatigue analysis, in 1939) pdf in three cases of 1, 2, and 3 parameters. We start with 3 parameters.

Definition 3.59

Let X be a continuous random variable.

1. The function $f(x,\alpha,\beta,\gamma)$, defined by

$$f(x,\alpha,\beta,\gamma) = \begin{cases} \dfrac{\beta}{\alpha}\left(\dfrac{x-\gamma}{\alpha}\right)^{\beta-1} e^{\left(-\frac{x-\gamma}{\alpha}\right)^{\beta}}, & x > 0, x \geq \gamma; \beta,\alpha > 0; -\infty < \gamma < \infty, \\ 0, & x < 0, \end{cases}$$

(3.274)

is called the **three-parameter Weibull pdf**, that is, a $X \sim$ Weibull(α,β,γ). α,β, and γ are the parameters. We leave it as an exercise to prove that $f(x,\alpha,\beta,\gamma)$ is, indeed, a pdf. The parameter α is referred to as the **shape parameter**, β is called the **scale parameter**, and γ is referred to as the **location parameter**.

Note 3.90

Changing the value of α will cause the Weibull pdf to change its shape and can model a wide variety of data. This is why it is referred to as the **shape parameter**.

2. In (3.274), if $\gamma = 0$, the function $f(x;\alpha,\beta)$, defined by

$$f(x;\alpha,\beta) = \begin{cases} \dfrac{\beta}{\alpha}\left(\dfrac{x}{\alpha}\right)^{\beta-1} e^{-\left(\frac{x}{\alpha}\right)^{\beta}}, & x > 0, \alpha,\beta > 0, \\ 0, & x \leq 0, \end{cases}$$

(3.275)

is called the **two-parameter Weibull pdf**, that is, a $X \sim$ Wibull(α,β). α and β are the parameters.

We leave it as an exercise to prove that $f(x;\alpha,\beta)$ is, indeed, a pdf. The parameter α is referred to as the **shape parameter** and β is called the **scale parameter.**

We leave it as an exercise to prove that the cdf in this case, denoted by $F(x;\alpha,\beta)$, is:

$$F(x;\alpha,\beta) = \int_0^\infty f(x;\alpha,\beta)\,dx = \begin{cases} 1 - e^{-\left(\frac{x}{\alpha}\right)^\beta}, & x \geq 0, \alpha, \beta > 0 \\ 0, & X < 0. \end{cases} \qquad (3.276)$$

Note 3.91

Changing α and β will change the shape of a variety of distribution. If β is between 3 and 4, the Weibull distribution approximates the normal distribution.

Note 3.92

Changing the values of β stretches or compresses the curves in the x-direction and so it is called a **scale parameter.**

Example 3.104

Letting X vary from 1 to 10 with 0.1 incriminate, $\alpha=1,2,3$, $\beta=1,2,3$, and MATLAB codes,

```
x = 1:.1:10;
alpha = 1:3;
beta = 1:3;
p1 = wblpdf(x,1,1);
plot(x,p1,'r')
text(1.5, 0.1, '\alpha=1, \beta = 1')
hold on
p2 = wblpdf(x,2,3);
plot(x,p2,'b')
text(4.2, 0.15, '\alpha=3, \beta = 2')
hold on
p3 = wblpdf(x,3,2);
plot(x,p3,'k')
text(2.5, 0.4, '\alpha=2, \beta = 3')
hold off
xlabel('x', 'fontsize', 12)
ylabel('2-Parameters Weibull pdf', 'fontsize', 10)
title('2-parameters Weibull pdf','fontsize', 12)
```

we will have Figure 3.42. Coding for cdf is the same except instead of pdf, we should use cdf. See Figure 3.43.

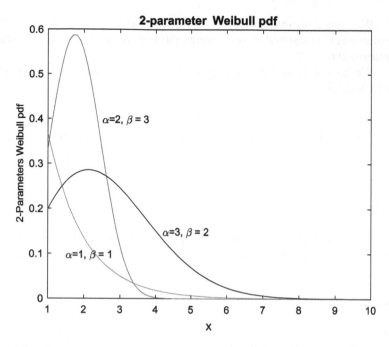

FIGURE 3.42 Graph of a two-parameter Weibull pdf with a variety of values for α and β, Formula (3.276).

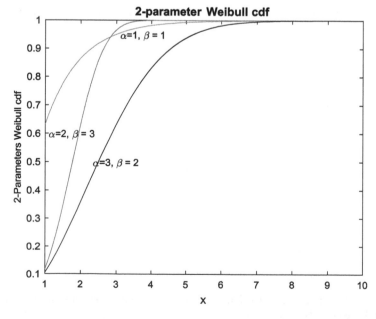

FIGURE 3.43 Graph of a two-parameter Weibull cdf with a variety of values for α and β, Formula (3.277).

3. In (3.274), if $\gamma = 0$ and $\beta = 1$, the function $f(x;\alpha)$, defined by

$$f(x,\alpha) = \begin{cases} \dfrac{1}{\alpha}e^{-\frac{x}{a}}, & x \geq 0, \alpha > 0, \\ 0, & x < 0, \end{cases} \qquad (3.277)$$

is called the **one-parameter Weibull pdf**, or **exponential pdf**. α is the parameter. We leave it as an exercise to prove that $f(x;\alpha)$, indeed, is a pdf. The parameter β is called the **scale parameter.**

We leave it as an exercise to prove that the cdf in this case, denoted by $F(x;\beta)$, is:

$$F(x;\beta) = \int_0^\infty f(x;\beta)\,dx = \begin{cases} 1 - e^{-\frac{x}{a}}, & x \geq 0, \alpha > 0, \\ 0, & x < 0. \end{cases} \qquad (3.278)$$

We also leave it as an exercise to show that the mean, μ, and standard deviation, σ, of the Weibull random variable are:

1. For three-parameter Weibull:

$$\mu = \gamma + \alpha \cdot \Gamma\left(\frac{1}{\beta}+1\right) \text{ and } \sigma = \alpha \cdot \sqrt{\Gamma\left(\frac{2}{\beta}+1\right) - \Gamma\left(\frac{1}{\beta}+1\right)^2}, \qquad (3.279)$$

   where $\Gamma(\cdot)$ is the gamma function.
2. For two-parameter Weibull:

$$\mu = \alpha \cdot \Gamma\left(\frac{1}{\beta}+1\right) \text{ and } \sigma = \alpha \cdot \sqrt{\Gamma\left[\left(\frac{2}{\beta}+1\right) - \Gamma\left(\frac{1}{\beta}+1\right)\right]}. \qquad (3.280)$$

3. For one-parameter Weibull:

$$\mu = \alpha \text{ and } \sigma = \alpha^2. \qquad (3.281)$$

Example 3.105

Let X be a two-parameter random variable with Weibull distribution function, that is, $X \sim \text{Wibull}(\alpha,\beta)$. Let $\alpha = 0.4$ and $\beta = 0.8$. We want to answer the following items:

   i. Probability of $X < 0.6$,
  ii. Probability of $0.8 < X < 2$,
 iii. Probability of $2 < X < 6$,
 iv. The mean for this distribution,
  v. The variance for this distribution,
 vi. Graph of this distribution for $0 \leq X \leq 10$.

Answer

Using the following MATLAB codes, we answer the questions:

```
x = [0.6, 0.8, 2, 6];
alpha = 0.4;
beta = 0.8;
p1 = wblcdf(0.6,0.4,0.8) % P(X < 0.6)
p2 = wblcdf(0.8,0.4,0.8) % P(X < 0.8)
p3 = wblcdf(2,0.4,0.8) % P(0.8 < X < 2)
p4 = p3 - p2 % P(0.8 X < 2)
p5 = wblcdf(6,0.4,0.8) % P(X < 6)
p6 = p5 - p3 % P(2 X < 26)
x = 0:0.1:10;
p7 = wblcdf(x,0.4,0.8);
plot(x,p7,'k')
xlabel('x', 'fontsize', 12)
ylabel('2-Parameters Weibull cdf', 'fontsize', 10)
title('2-parameters Weibull cdf, \alpha=0.4, \
beta=0.8','fontsize', 12)
[M,V]=wblstat(0.4,0.8)
```

   i. $P(X < 0.6) = 0.7492$.
  ii. $P(X < 0.8) = 0.8247$, $P(X < 2) = 0.9733$, $P(0.8 < X < 2) = 0.1486$.
 iii. $P(X < 6) = 0.9998$, $P(2 < X < 6) = 0.0265$.
  iv. $\mu = 0.4532$.
   v. $\sigma^2 = 0.3263$
  vi. For $0 \le X \le 10$, see Figure 3.44.

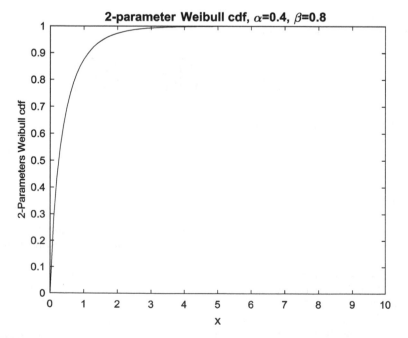

FIGURE 3.44   Graph of a two-parameter Weibull cdf, Example 3.105, (vi).

3.9.11 Lognormal Distribution

Definition 3.60

A random variable X (with positive real values) is called **lognormally distributed** if its logarithm, say $Y = \ln X$, is normally distributed. Conversely, if Y is normally distributed, then $X = e^Y$ is lognormally distributed.

Lognormal distribution sometimes is referred to as the **Galton distribution**, referencing to Francis Galton (1822–1911), a British statistician. There are other names in the literature that this distribution is associated with, such as **Galton–McAlister, Gibrat, and Cobb–Douglas (Figure 3.45)**.

Note 3.93

To see the meaning of the definition, let $Y = e^X$. Hence, $X = \ln Y$. Thus, if X is a normally distributed random variable, then Y is a lognormally distributed random variable.

Theorem 3.28

Let $X \sim N(x; \mu, \sigma^2)$. Let Y be lognormally distributed random variable with pdf denoted by $f_Y(x)$. Then,

$$f_Y(y) = \frac{1}{x\sigma\sqrt{2\pi}} e^{-\frac{(\ln x - \mu)^2}{2\sigma^2}}, \quad 0 < x < \infty, \quad -\infty < \mu < \infty, \quad \sigma > 0. \quad (3.282)$$

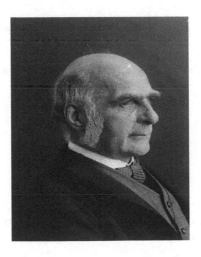

FIGURE 3.45 Francis Galton 1822–1911.

Proof:

$$f_Y(y) = \frac{d}{dy} P(Y \le y)$$

$$= \frac{d}{dx} P(Y \le \ln x) = \frac{d}{dx} F_Y(x) = \frac{d}{dx} F_{\ln X}(x)$$

$$= \frac{d}{dx} F_X\left(\frac{\ln x - \mu}{\sigma}\right) = f_X\left(\frac{\ln x - \mu}{\sigma}\right) \frac{d}{dx}\left(\frac{\ln x - \mu}{\sigma}\right)$$

$$= f_X\left(\frac{\ln x - \mu}{\sigma}\right) \frac{1}{x\sigma} = \frac{1}{x} \frac{1}{\sigma\sqrt{2\pi}} e^{-\frac{(\ln x - \mu)^2}{2\sigma^2}}.$$

Theorem 3.29

Let Y be a lognormally distributed random variable. Denote the cdf of Y by $F_Y(x)$, where X is normally distributed with mean and variance as μ and σ^2, respectively. Then,

$$P(Y \le x) \equiv F_Y(x) = \int_0^\infty \frac{1}{t\sigma\sqrt{2\pi}} e^{-\frac{(\ln t - \mu)^2}{2\sigma^2}} dt$$

$$= \frac{1}{2}\left[1 + erf\left(\frac{\ln x - \mu}{\sigma\sqrt{2}}\right)\right] = \frac{1}{2} erfc\left(-\frac{\ln x - \mu}{\sigma\sqrt{2}}\right), \qquad (3.283)$$

where *erf* is the **error function** given in (3.262) and *erfc* is the **complimentary error function**. The mean and variance of Y are given by:

$$E(Y) = e^{\mu + \frac{\sigma^2}{2}} \text{ and } Var(Y) = \left[e^{\sigma^2} - 1\right] e^{2\mu + \sigma^2}. \qquad (3.284)$$

Also, the median and mode of Y denoted by $Md(Y)$ and $Mo(Y)$ are, respectively, as:

$$Md(Y) = e^\mu \text{ and } Mo(Y) = e^{\mu - \sigma^2}. \qquad (3.285)$$

Proof:
Left as an exercise.

Example 3.106

Let X be a normally distributed random variable with $\mu = 0.6$, $\sigma = 1$. Let Y be log-normally distributed random variable defined by $Y = \ln X$. Answer the following items:

    i. Probability of $1 < Y < 3$,
   ii. Probability of $Y > 4$,
  iii. The mean of Y,
  iv. The variance of Y,
   v. For values of $0 < X \le 25$, and two sets of means and variances as: $\mu = 0.6$, $\sigma = 1$ and $\mu = 0.5$, $\sigma^2 = 0.9$, display the graphs of pdf of Y.
  vi. For values of $0 < X \le 2$, and two sets of means and variances as: $\mu = 0.6$, $\sigma = 1$ and $\mu = 0.5$, $\sigma^2 = 0.9$, display the graphs of cdf of Y.

Answer

    i. $P(1 < Y < 3) = 0.4167$.
   ii. $P(Y > 4) = 1 - P(Y \le 4) = 0.2158$.
  iii. $\mu_Y = 3.0042$.
  iv. $\sigma_Y^2 = 15.5075$.
   v. For graph of lognormal pdf, see Figure 3.46.
  vi. For graph of lognormal cdf, see Figure 3.47.

Note 3.94

It should be observed that the lognormal distribution is positively skewed with many small values and just a few large values. Consequently, the mean is greater than the mode in most cases.

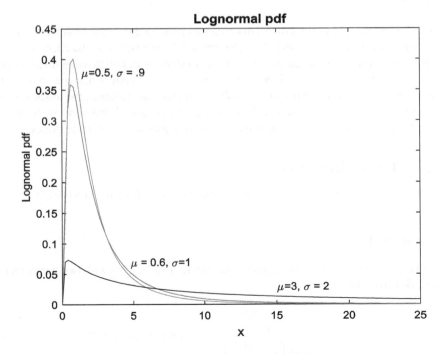

FIGURE 3.46   Graph of a lognormal pdf, Example 3.106, (v).

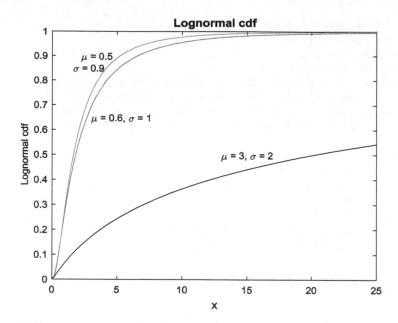

FIGURE 3.47   Graph of a lognormal cdf, Example 3.106, (vi).

Note 3.95

In applications, the normal distribution cannot be used due to the possible negative values of the random variable. Thus, lognormal distribution is an excellent alternative. For example, for modeling asset prices in stock prices, lognormal distribution can be used since it is bound by zero on the lower side. Thus, while the returns on a stock (continuously compounded) follow a normal distribution, the stock prices follow a lognormal distribution. However, the lognormal distribution is the most appropriate for stock prices, regardless of the distribution for the asset prices.

3.9.12 Logistic Distribution

One of the widely used distributions in engineering and clinical trials is logistic.

Definition 3.61

Let X be a continuous random variable. The function $f_X(x; \mu, \gamma)$ with two parameters α and β defined by

$$f_X(x; \mu, \gamma) = \frac{e^{\frac{x-\mu}{\gamma}}}{\gamma \left(1 + e^{-\frac{x-\mu}{\lambda}}\right)^2}, \quad -\infty < x, \mu < \infty, \gamma > 0, \qquad (3.286)$$

where the parameters μ and γ are referred to as the **location** and **scale parameter**, respectively, is a pdf called the **logistic pdf** for X.

Theorem 3.30

1. The function $f_X(x; \mu, \gamma)$ defined in (3.286) is, indeed, a pdf.
2. The function $F_X(x, \mu, \gamma)$ defined as:

$$F_X(x; \mu, \gamma) = \frac{1}{1 + e^{\frac{x-\mu}{\gamma}}}, \quad -\infty < x, \mu < \infty, \gamma > 0, \tag{3.287}$$

is the cdf of the logistic pdf.
3. The mean, median, and mode for the logistic random variable, denoted by $E(X)$, $Md(X)$, and $Mo(X)$, respectively, are equal, and they are equal to μ. That is,

$$E(X) = Md(X) = Mo(X) = \mu. \tag{3.288}$$

4. Also, the variance of the logistic random variable is:

$$Var(X) = \frac{\gamma^2 \pi^2}{3}. \tag{3.289}$$

Proof:
Left as an exercise.

Note 3.96

It can be seen that:

$$f_X(x; \mu, \gamma) = \frac{1}{\gamma \left(e^{\frac{x-\mu}{2\gamma}} + e^{-\frac{x-\mu}{2\gamma}} \right)^2} = \frac{1}{4\gamma} \operatorname{sech}^2 \left(\frac{x-\mu}{2\gamma} \right), \tag{3.290}$$

and

$$F_X(x, \mu, \gamma) = \frac{1}{2} \left[1 + \tanh \left(\frac{\pi(x-\mu)}{2\sqrt{3}\gamma} \right) \right]. \tag{3.291}$$

Example 3.107

Let $\mu = 0, 2$. We graph the pdf and cdf of logistic for values of $\gamma = 1, 2, 3$ in Figures 3.48 and 3.49, respectively.

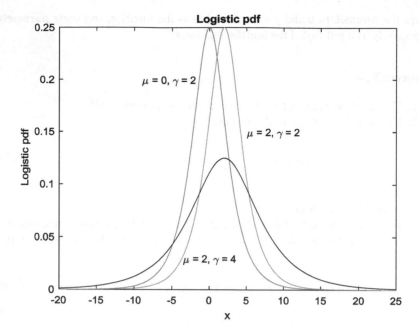

FIGURE 3.48 Graph of logistic *pdf* for $\mu = 0$ and 2 and $\gamma = 1, 2, 3$.

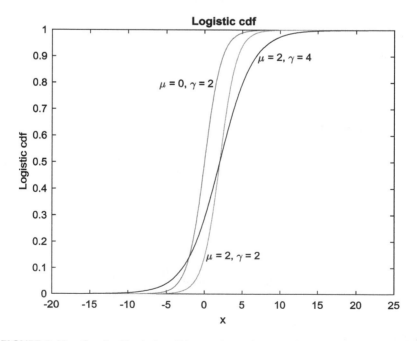

FIGURE 3.49 Graph of logistic *cdf* for $\mu = 0$ and 2 and $\gamma = 1, 2, 3$.

We show how MATLAB programs can be written for the logistic pdf and cdf to obtain Figures 3.40 and 3.41.

MATLAB Logistic pdf

```
x = -20:0.1:25;
mu_1 = 0;
gamma_1=2;
p_1 = (exp((x-mu_1)/gamma_1))/gamma_1./(1+exp((x-mu_1)/
gamma_1).^2);
plot(x,p_1,'b')
hold on
mu_2 = 2;
gamma_2=2;
p_2 = (exp((x-mu_2)/gamma_2))/gamma_2./(1+exp((x-mu_2)/
gamma_2).^2);
plot(x,p_2,'r')
hold on
mu_3 = 2;
gamma_3=4;
p_3 = (exp((x-mu_3)/gamma_3))/gamma_3./(1+exp((x-mu_3)/
gamma_3).^2);
plot(x,p_3,'k')
hold off
ylabel('Logistic pdf')
xlabel('x')
text(-9,0.2,'\mu = 0, \gamma = 2')
text(5,0.15,'\mu = 2, \gamma = 2')
text(-3.5,0.03,'\mu = 2, \gamma = 4')
title('Logistic pdf')
```

MATLAB Logistic cdf

```
x = -20:0.1:25;
mu_1 = 0;
gamma_1=2;
c_1 = (1/2)*(1+tanh((pi*(x-mu_1))/(2*(sqrt(3))*gamma_1)));
plot(x,c_1,'b')
hold on
mu_2 = 2;
gamma_2=2;
c_2 = (1/2)*(1+tanh((pi*(x-mu_2))/(2*(sqrt(3))*gamma_2)));
plot(x,c_2,'r')
hold on
mu_3 = 2;
gamma_3=4;
c_3 = (1/2)*(1+tanh((pi*(x-mu_3))/(2*(sqrt(3))*gamma_3)));
plot(x,c_3,'k')
hold off
ylabel('Logistic cdf')
xlabel('x')
text(-6,0.8,'\mu = 0, \gamma = 2')
text(0.2,0.15,'\mu = 2, \gamma = 2')
text(7,0.9,'\mu = 2, \gamma = 4')
title('Logistic cdf')
```

3.9.13   Extreme Value Distribution

Another distribution that is widely used these days in engineering and clinical trials is **extreme value**. The distribution was called the **extreme value** because it was used for probabilistic modeling of extreme or rare events arising in areas, where such events could have very negative consequences. The examples of rare events included events such as storms with high wind speeds, extreme high or low temperatures, extreme floods and snowfalls, large fluctuations in exchange rates, and market crashes. The area of study is referred to as the **extreme value theory**, pioneered by the British researcher, in cotton industry, Leonard Tippett (1902–1985). He was trying to make cotton thread stronger. R. A. Fisher helped him finding the necessary theory; see Fisher and Tippett (1928). The German mathematician Emil Julius Gumbel (1891–1966) focused on the applications of extreme value to engineering problems, in particular, modeling of meteorological phenomena like annual flood flows. He was the first to bring engineers and statisticians' attentions to possible application of *extreme value theory* to some empirical distributions they were using such as radioactive emission, strength of materials, flood analysis, and rainfall. He presented the theory in his book titled *Statistics of Extremes* published in 1958, see *Gumbel (2004, reprint of 1958)*. One of the three types of extreme value distribution is included in his book and bears his name.

Definition 3.62

Let X be a continuous random variable. There are three types of pdf (and cdf) for X called the **extreme value**.

These types are **one-parameter, two-parameter, and three-parameter**. Here are there definitions:

1. The random variable X has **extreme value type 1** (or **Gumbel-type extreme value** or **double exponential**) pdf with local parameter μ and scale parameter θ, denoted by $f_X(x; \mu, \theta)$ defined as

$$f(X; \mu, \theta) = \frac{1}{\theta} e^{-\frac{x-\mu}{\theta} - e^{-\frac{x-\mu}{\theta}}}, \quad -\infty < x < \infty, \theta > 0. \tag{3.292}$$

2. The random variable X has **extreme value type 2** (or **Fréchet**, named after a French mathematician Maurice Fréchet (1878–1973)) pdf with shape parameter α and scale parameter β, denoted by $f_X(x; \alpha, \beta)$, defined as

$$f(X; \alpha, \beta) = \frac{\alpha}{\beta} \left(\frac{\beta}{x} \right)^{\alpha+1} e^{-\left(\frac{\beta}{x} \right)^{\alpha}}, \quad 0 < x < \infty, \alpha, \beta > 0. \tag{3.293}$$

3. The **extreme value type 3** pdf is the Weibull pdf with two and three parameters defined in (3.274) and (3.275).

A combination of the three types of extreme value distributions, that is, Gumbel, Fréchet, and Weibull, is referred to as the **generalized extreme value distribution (GEV)** (or **von Mises-type** or **von Mises–Jenkinson-type distribution**), with its pdf denoted by $f(x,\alpha,\beta,\gamma)$, defined as:

$f(x,\alpha,\beta,\gamma)$

$$
= \begin{cases}
\dfrac{1}{\alpha}\left[1+\beta\left(\dfrac{x-\gamma}{\alpha}\right)\right]^{-1-\frac{1}{\beta}} e^{-\left[1+\beta\left(\frac{x-\gamma}{\alpha}\right)\right]^{\frac{1}{\beta}}}, & \text{when} \quad \left[1+\dfrac{\beta(x-\gamma)}{\alpha}\right] > 0, \quad \text{for } \beta \neq 0, \\[4mm]
\dfrac{1}{\alpha} e^{-\frac{x-\gamma}{\alpha}-e^{-\frac{x-\gamma}{\alpha}}}, & \text{when} \quad -\infty < x < \infty, \quad \text{for } \beta = 0,
\end{cases}
$$

$$(3.294)$$

which is left to be proved. In other words, each one of the three types can be obtained from (3.294).

Note 3.97

Extreme value type 1 distribution is an appropriate distribution to use when the variable of interest is the minimum of many real-valued random factors. However, Weibull, the extreme value type 3 distribution, would be a candidate distribution when lifetime random phenomena are being considered.

Note 3.98

We leave it as an exercise to show the following:

i. If X is a Weibull random variable, then $X = \ln X$ is a type 1 extreme value random variable.
ii. Type 2 and type 3 extreme value distributions can be obtained from each other by changing the sign of the random variable.
iii. A type 2 extreme value random variable X can be transformed to a type 1 extreme value Y by the following transformation: $Y = \ln(X - \mu)$.
iv. A type 3 extreme value random variable X can be transformed to a type 1 extreme value Z by the following transformation: $Z = -\ln(\mu - X)$.

Theorem 3.31

1. Prove the following:
   i. $f_X(x; \mu, \theta)$ for type 1 extreme value random variable, defined by (3.299), is, indeed, a pdf.

  ii. $f_X(x;\alpha,\beta)$ for type 2 extreme value random variable, defined by (3.293), is, indeed, a pdf.

  iii. cdf for type 1 extreme value random variable, denoted by $F_X(x;\mu,\theta)$, is given as

$$F_X(x;\mu,\theta) \equiv P(X \le x) = e^{-e^{-\frac{x-\mu}{\theta}}}, \quad -\infty < X <, \theta > 0. \tag{3.295}$$

  iv. cdf for type 2 extreme value random variable, denoted by $F_X(x;\mu,\theta)$, is given as

$$F_X(x;\alpha,\beta) \equiv P(X \le x) = e^{-\left(\frac{\beta}{x}\right)^{\alpha}}, \quad x > 0, \alpha,\beta > 0. \tag{3.296}$$

2. For a type 1 extreme value random variable X, show that the moment generating function $M(t)$, mean, and variance are as follows:

$$M(t) = E\left(e^{tX}\right) = e^{t\mu}\Gamma(1-\theta t), \quad \theta|t| < 1. \tag{3.297}$$

$$E(X) = \mu - \theta\phi(1) = \mu + \eta\theta = \mu + 0.57722\theta, \tag{3.298}$$

where $\phi(\cdot)$ is the **digamma function** defined as $\phi(t) = \mu t - \ln[\Gamma(1-\theta)]$ and η is Euler's constant equal to 0.57722.

$$Var(X) = \frac{\pi^2\theta^2}{6} = 1.64493\,\theta^2. \tag{3.299}$$

3. Find the mean and variance of type 2 extreme value distributions.

Example 3.108

Let $\mu=2$. We graph the pdf and cdf of extreme value type 1 for values of $\theta=1,3,5$, with x running from −15 to 25, in Figures 3.50 and 3.51, respectively.

MATLAB Extreme Value Type 1, pdf

```
mu = 2;
theta = [1,3,5];
for i = 1:length(theta)
 x = -15:0.1:25;
 f = (1/theta(i))*exp(-(x-mu)/theta(i)).*exp(-exp(-(x-mu)/
theta(i)));
 plot(x,f)
 hold on
end
ylabel('Extreme-value type 1, pdf')
xlabel('x')
text(-5,0.30,'\mu = 2, \theta = 1')
text(-7,0.08,'\mu = 2, \theta = 3')
text(-12,0.02,'\mu = 2, \theta = 5')
```

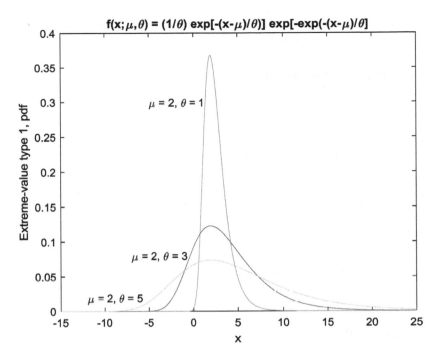

FIGURE 3.50   Graph of extreme value type 1 *pdf* for $\alpha = 2$ and $\beta = 1,3,5$.

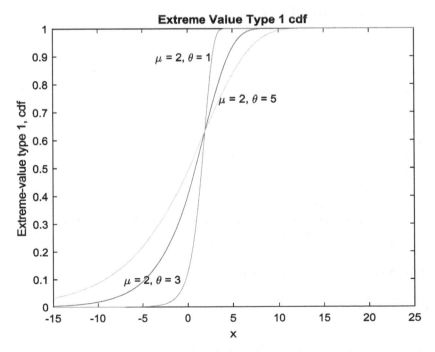

FIGURE 3.51   Graph of extreme value type 1 *cdf* for $\alpha = 2$ and $\beta = 1,3,5$.

```
title('f(x;\mu,\theta) = (1/\theta) exp[-(x-\mu)/\theta)]
exp[-exp(-(x-\mu)/\theta]')
hold off
```

MATLAB Extreme Value Type 1, cdf

```
mu = 2;
theta = [1,3,5];
for i = 1:length(theta)
 x = -15:0.1:25;
 evcdf_1=evcdf(x,mu, theta(i));
 plot(x,evcdf_1)
 hold on
end
ylabel('Extreme-value type 1, cdf')
xlabel('x')
text(-3.5,0.9,'\mu = 2, \theta = 1')
text(-7.1,0.1,'\mu = 2, \theta = 3')
text(3.5,0.75,'\mu = 2, \theta = 5')
title('Extreme Value Type 1 cdf')
hold off
```

Example 3.109

Let $\alpha = 2$. We graph the pdf and cdf of extreme value type 2 for values of $\beta = 1,3,5$, with x running from 0.1 to 25, in Figures 3.52 and 3.53, respectively.

MATLAB Extreme Value Type 2, pdf

```
alpha = 2;
beta = [1,3,5];
for i = 1:length(beta)
 x = 0.1:.1:25;
 f = (alpha/beta(i))*((beta(i)./x).^(alpha+1)).*exp(-beta(i)./x).
^alpha;
 plot(x,f)
 hold on
end
ylabel('Extreme Value Type 2, pdf')
xlabel('x')
text(1.1,0.25,'\alpha = 2, \beta = 1')
text(3,0.1,'\alpha = 2, \beta = 3')
text(8.4,0.03,'\alpha = 2, \beta = 5')
title('f(x;\alpha,\beta) = (\alpha/\beta)(\beta/x)^\alpha
exp(-\beta/x)^\alpha)')
hold off
```

MATLAB Extreme Value Type 2, cdf

```
alpha = 2;
beta = [1,3,5];
for i = 1:length(beta)
 x = .1:.1:25;
 EVcdf_2 = exp(-(beta(i)./x).^alpha);
 plot(x,EVcdf_2)
```

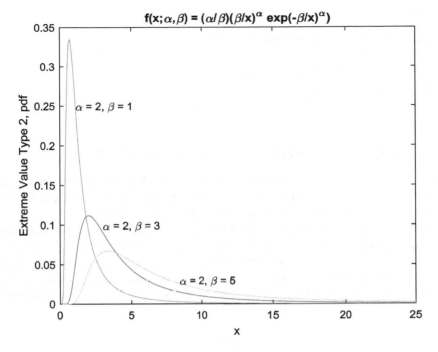

FIGURE 3.52   Graph of extreme value type 2 *pdf* for $\alpha = 2$ and $\beta = 1, 3, 5$.

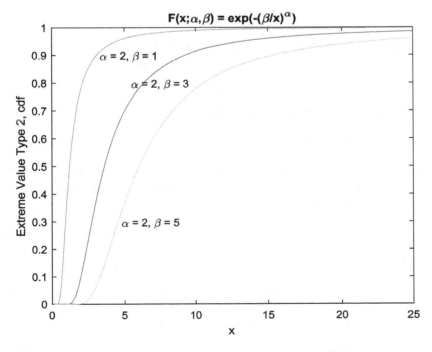

FIGURE 3.53   Graph of extreme value type 2 *cdf* for $\alpha = 2$ and $\beta = 1, 3, 5$.

```
        hold on
        end
        ylabel('Extreme Value Type 2, cdf')
        xlabel('x')
        text(3.3,0.9,'\alpha = 2, \beta = 1')
        text(5.5,0.8,'\alpha = 2, \beta = 3')
        text(4.8,0.3,'\alpha = 2, \beta = 5')
        title('F(x;\alpha,\beta) = exp(-(\beta/x)^\alpha)')
        hold off
```

3.10 ASYMPTOTIC PROBABILISTIC CONVERGENCE

For a sequence of random variables and for samples when the sizes become large, there are properties, some of which we will discuss in this section. These properties are applicable in both probability theory and inferential statistics.

Before we start discussing the main purpose of the section, we remind the readers that by a **sequence**, it is meant a function whose domain is the set of positive integers. Also, in a classical sense, by **convergence of a sequence** $\{X_n\}$ to X as n approaches infinity is meant that the difference of X_n and X, in absolute value, gets closer to 0 as n becomes larger and larger. Formally, a **sequence**, $S_n = \{a_n\}$, $n = 1, 2, \ldots$, **converges** to a number (limit) S, denoted by

$$\lim_{n \to \infty} S_n = S,$$

if for every positive number $\epsilon > 0$, there exists a number N such that

$$|S_n - S| < \epsilon, \quad \forall n > N.$$

If the sequence does not converge, it is said to **diverge**.

We now discuss the asymptotic probabilistic convergence.

Definition 3.63 Convergence in Probability

We offer two versions of this definition.

Version a: The sequence of random variables $\{X_n, n \in \mathbb{N}\}$, not necessarily all defined on the same sample space Ω, **converges in probability** to a real number c, denoted by

$$X_n \xrightarrow{\text{prob}} c, \tag{3.300}$$

if

$$\lim_{n \to \infty} P\big(|X_n - c| \ge \epsilon\big) = 0, \tag{3.301}$$

for any arbitrary positive number ϵ. Or equivalently,

$$P\big(|X_n - c| > \epsilon\big) \to 0, \quad \text{as } n \to \infty, \quad \text{for a fixed } \epsilon > 0. \tag{3.302}$$

Or equivalently,

$$P\big(|X_n - c| \le \epsilon\big) \to 1, \quad \text{as } n \to \infty, \quad \text{for a fixed } \epsilon > 0. \tag{3.303}$$

Version b: The sequence of random variables $\{X_n, n \in \mathbb{N}\}$ defined on a sample space Ω **converges in probability** to a random variable X, defined on a sample space Ω, denoted by

$$X_n \xrightarrow{\text{prob}} X, \tag{3.304}$$

if and only if

$$\lim_{n \to \infty} P\big(|X_n - X| \ge \epsilon\big) = 0, \tag{3.305}$$

for any arbitrary positive number ϵ. In other words, as n increases, $P\big(|X_n - X| > \epsilon\big)$ decreases, where X is called the **probability limit of the sequence**.

Note 3.99

In Definition 3.64, if X is a constant, say c, rather than a random variable, then both versions are the same.

Note 3.100

Suppose each of the two sequences $\{X_n, n \in \mathbb{N}\}$ and $\{Y_n, n \in \mathbb{N}\}$ converges in probability to X and Y, respectively, with all random variables defined on the same sample space Ω. Then, we have the following implication:

$$\text{If } X_n \xrightarrow{\text{prob}} X, \text{ and } Y_n \xrightarrow{\text{prob}} Y, \text{ then } \big(X_n + Y_n\big) \xrightarrow{\text{prob}} \big(X + Y\big). \tag{3.306}$$

Proof:
Left as an exercise.

Note 3.101

The concept of convergence in probability can be extended to a sequence of random vectors.

Definition 3.64 Convergence in Distribution

The sequence of random variables $\{X_n, n \ge 1\}$ is said to **converge in distribution** to a random variable X, denoted by

$$X_n \xrightarrow{\text{dist}} X, \tag{3.307}$$

if

$$\lim_{n \to \infty} F_{X_n}(x) = F_X(x), \qquad (3.308)$$

for every $x \in \mathbb{R}$ at which $F_X(x)$ is continuous.

Definition 3.65 Almost Surely Convergence

The sequence of random variables $\{X_n, n \in \mathbb{N}\}$, defined on the sample space Ω, **converges almost surely** (or **with probability one**) to a random variable X, denoted by

$$X_n \xrightarrow{\;a.s.\;} X \left(\text{or } X_n \to X \text{ with probability one} \right), \qquad (3.309)$$

if there is a set $A \subset \Omega$ such that

$$\lim_{n \to \infty} X_n(\omega) = X(\omega), \quad \text{for all } \omega \in A, \qquad (3.310)$$

and $P(A) = 1$ or

$$P\left(\left\{ \lim_{n \to +\infty} X_n(\omega) = X(\omega), \quad \forall \omega \in \Omega \right\} \right) = 1. \qquad (3.311)$$

Note 3.102

The almost sure convergence implies convergence in probability, but the converse is not true.

Proof:
Proof is left as an exercise.

Note 3.103

The convergence in probability implies that there exists a subsequence that converges almost surely to the same limit.

Proof:
We leave the proof as an exercise.

Note 3.104

We leave it as an exercise to prove that the almost sure convergence reduces the idea to convergence of deterministic sequences.

Note 3.105

We leave it as an exercise to prove that convergence almost surely is the probabilistic version of pointwise convergence in real analysis.

Note 3.106

The random variables X_n, $n \geq 1$ mentioned in Theorem 3.32 are generally highly dependent.

When a chance experiment is repeated a large number of times, the average of occurrences of an event will approach the expected value of that event. For instance, if we rolling a fair die once, the average points obtained would be 3.5, which is the sum of numbers from 1 to 6 divided by 6. Now if we repeat the rolling of the same die many times, we expect the average value of numbers obtained will be the same as 3.5; that is, the sample mean will approach the population mean.

Definition 3.66

The mean of a random sample is referred to as the **sample mean**.

Now, let us consider a random sample from a population. Then, the idea of a sample mean approaching the population mean is the essence of what is referred to as the **law of large numbers (LLN)**. There are two versions of the LLN, simply, weak and **strong**. **The first** is referred to as **Khinchin's law**.

Theorem 3.32 The Weak Law of Large Numbers, Khinchin's Law

The mean of a random sample referred to as the **sample mean** approaches the population mean (or the expected value) as the sample size approaches infinity. Symbolically, consider a random sample of size n and denote the sample mean by $\bar{X}_n$, that is,

$$\bar{X}_n \equiv \frac{X_1 + \cdots + X_n}{n}. \tag{3.312}$$

Then,

$$\bar{X}_n \xrightarrow{\text{P}} \mu, \text{ as } n \to \infty. \tag{3.313}$$

That is, $\bar{X}_n$ approaches μ in probability. In other words, given a small positive number ϵ, we have:

$$\lim_{n \to \infty} P\left(\left|\bar{X}_n - \mu\right| > \epsilon\right) = 0. \tag{3.314}$$

Proof:
Left as an exercise.

Theorem 3.33 The Strong Law of Large Numbers, Kolmogorov's Strong Law

The sample mean defined in (3.43) converges almost surely (*a.s.*) to the population mean as the sample size, n, approaches infinity. Symbolically,

$$\bar{X}_n \xrightarrow{a.s} \mu, \text{ as } n \to \infty. \tag{3.315}$$

In other words,

$$P\left(\lim_{n\to\infty} \bar{X}_n = \mu\right) = 1. \tag{3.316}$$

Yet, another way of stating this theorem is the following: Suppose $\{X_n, n \geq 1\}$ is a sequence of iid random variables, sometimes referred to as a **random sample**, such that $\mu = E\left[|(X_1)|\right] = E\left[|(X_2)|\right] = \cdots = E\left[|(X_n)|\right]$ is finite. Then,

$$\bar{X}_n \xrightarrow{a.s.} \mu. \tag{3.317}$$

Proof:
Left as an exercise.

Note 3.107

Relation (3.313) can be generalized replacing random variables by functions of random variables. Thus, let g be a function from reals to reals, that is, $g : \mathbb{R} \to \mathbb{R}$ such that $\mu \equiv E\left[g(X_1)\right] = E\left[g(X_2)\right] = \cdots = E\left[|(X_n)|\right]$ exits and

$$\frac{g(X_1) + \cdots + g(X_n)}{n} \to \mu = \int_{-\infty}^{\infty} g(x) f_{X_1}(x) dx, \quad \text{as } n \to \infty. \tag{3.318}$$

Example 3.110

Consider tossing a fair coin n times, independently. Observing each occurrence of heads in a row. Let H_n be the row with longest number of heads. That is, H_n represents the longest sequence of consecutive tosses that produce heads. Show that as n becomes too large, the value of H_n approaches $\log_2 n$. In other words, show that

$$\frac{H_n}{\log_2 n} \to 1, \quad \text{in probability,} \quad \text{as } n \to \infty. \tag{3.319}$$

Answer

To see the problem better, let, for example, $n = 25$ and the outcome of tosses of a fair coin is:

$$\underbrace{TTHTHTHHHHHTHHHHTHTHTHHTTH}_{25 \text{ in total}}.$$

This shows $H_n = 5$. From (3.319), it can be seen that to have the limit to be 1, the length of the highest row should be about 20 (in fact, it is 1.0034) and that happens in after a million tosses.

We now prove (3.319) that we used in this example. For the number of heads in a sequence of tosses, we will calculate the upper and lower bounds on probabilities that H_n is large; that is, for an arbitrary number j, we will calculate $P(H_n \geq j)$ and $P(H_n \leq j)$ as follows.

For the upper bound, the number of heads in a row starts at some point, say i, where $0 \leq i \leq n - j + 1$; that is, i is the size of the first row of heads. Hence,

$$P(H_n \geq j) = P(\text{number of heads in a row start at some point } i, 0 \leq i \leq n - j + 1)$$

$$= P\left(\bigcup_{j=1}^{n-j+1} i \text{ is the first heads in a row of at least } j \text{ heads} \right) \leq n \cdot \frac{1}{2^j}. \qquad (3.320)$$

For the lower bound, after the n tosses of the coin, there is a string of size n of heads and tails. We divide this string into disjoint blocks of the same size j. Thus, there will be $\left[\dfrac{n}{j}\right]$ (where $[x]$ or "bracket x" is the greatest integer less than or equal x) blocks, such that if n is not divisible by j, the leftover smaller block at the end will be ignored. Hence, as soon as one of the blocks consists of heads only, $H_n \geq j$ and different blocks are independent. Thus,

$$P(H_n < j) = \left(1 - \frac{1}{2^i} \right)^{\left[\frac{n}{j}\right]} \leq e^{-\frac{1}{2^j}\left[\frac{n}{j}\right]}. \qquad (3.321)$$

The Maclaurin expansion of a function $f(x) = e^{-x}$ is:

$$e^{-x} = 1 - x + \frac{x^2}{2} + \cdots + \frac{(-1)^n x^n}{n!}.$$

Hence, we have:

$$1 - x \leq e^{-x}, \forall x.$$

Also, if $x \geq 2$, the following inequalities are known:

$$x + 1 \leq 2x, x - 1 \geq \frac{x}{2}, [x] \leq x + 1 \text{ and } [x] \geq x - 1,$$

where $[x]$ or "bracket x" is the greatest integer less than or equal x (we leave the proofs as exercises).

Now, to show (3.321), we have to show that for any fixed $\epsilon > 0$, the following are true:

$$P(H_n \geq (1 + \epsilon) \log_2 n) \to 0 \qquad (3.322)$$

and

$$P(H_n \leq (1 - \epsilon) \log_2 n) \to 0, \qquad (3.323)$$

because

$$P\left(\left|\frac{H_n}{\log_2 n} - 1\right| \geq \epsilon\right) = P\left(\frac{H_n}{\log_2 n} \geq 1 + \epsilon \text{ or } \frac{H_n}{\log_2 n} \leq 1 - \epsilon\right)$$

$$= P\left(\frac{H_n}{\log_2 n} \geq 1 + \epsilon\right) + P\left(\frac{H_n}{\log_2 n} \leq 1 - \epsilon\right)$$

$$= P\left(H_n \geq (1+\epsilon)\log_2 n\right) + P\left(H_n \leq (1-\epsilon)\log_2 n\right). \quad (3.324)$$

Of course, neither $(1+\epsilon)\log_2 n$ nor $(1-\epsilon)\log_2 n$ is an integer, and these may cause some concerns regarding the proof. However, this type of concern is common in problems such as the one under consideration.

Now to prove (3.323), in (3.324), we let $j = \left[(1+\epsilon)\log_2 n\right]$ and obtain:

$$P\left(H_n \geq (1+\epsilon)\log_2 n\right) = n \cdot \frac{1}{2^{(1+\epsilon)\log_2 n - 1}}$$

$$= n \cdot \frac{2}{n^{1+\epsilon}} = \frac{2}{n^\epsilon} \to 0, \quad n \to \infty. \quad (3.325)$$

To prove (3.318), in (3.319), we let $j = \left[(1-\epsilon)\log_2 n\right] + 1$ that yields:

$$P\left(H_n \leq (1-\epsilon)\log_2 n\right) \leq P\left(H_n < j\right)$$

$$\leq e^{-\frac{1}{2^j}\left[\frac{n}{j}\right]}$$

$$\leq e^{-\frac{1}{2^j}\left(\frac{n}{j}-1\right)}$$

$$\leq e^{-\frac{1}{32} \cdot \frac{1}{n^{1-\epsilon}} \cdot \frac{n}{(1-\epsilon)\log_2 n}}$$

$$\leq e^{-\frac{1}{32} \cdot \frac{n^\epsilon}{(1-\epsilon)\log_2 n}},$$

which approaches 0, as $n \to \infty$, because n^ϵ is much larger than $\log_2 n$. This completes the proof of (3.325) and the answer to Example 3.102.

Example 3.111

Let X be a random variable. Let also $Y_n, n = 1, 2, \ldots$, be a sequence of iid random variables such that

$$E(Y_n) = \frac{1}{n} \text{ and } Var(Y_n) = \frac{v}{n}, v > 0, n = 1, 2, \ldots. \quad (3.326)$$

We further suppose that

$$X_n = X + Y_n. \quad (3.327)$$

We want to show that

$$X_n \xrightarrow{\text{prob}} X. \tag{3.328}$$

Answer

In order to apply the fact that $|x+y| \le |x|+|y|$, we write Y_n as $Y_n = Y_n - E(Y_n) + E(Y_n)$. Hence,

$$|Y_n| \le |Y_n - E(Y_n)| + \frac{1}{n}.$$

Thus, for $\epsilon > 0$, we have:

$$P\left(|Y_n - E(Y_n)| \ge \epsilon\right) = P\left(|Y_n| \ge \epsilon\right) \le P\left(|Y_n - E(Y_n)| + \frac{1}{n} \ge \epsilon\right)$$

$$= P\left(|Y_n - E(Y_n)| - \frac{1}{n} \ge \epsilon\right).$$

Using Chebyshev's inequality, we will have:

$$= P\left(|Y_n - E(Y_n)| - \frac{1}{n} \ge \epsilon\right) \le \frac{Var(Y_n)}{\left(\epsilon - \frac{1}{n}\right)^2} = \frac{v/n}{\left(\epsilon - \frac{1}{n}\right)^2} \to 0, n \to \infty. \tag{3.329}$$

Relation (3.329) implies (3.328).

Example 3.112

Let $X_n, n = 1, 2, \ldots$, be a sequence of iid exponential random variables. Let also the random variable $X = 0$ show that

$$X_n \xrightarrow{\text{prob}} 0. \tag{3.330}$$

Answer

Since $X_n \ge 0$ and X_n is exponentially distributed, we have:

$$\lim_{n \to \infty} P\left(|X_n - 0| \ge \epsilon\right) = \lim_{n \to \infty} e^{n\epsilon} = 0. \quad \forall \epsilon > 0. \tag{3.331}$$

Relation (3.331) implies (3.330).

Example 3.113

Suppose the distribution of the grades of an examination is skewed with mean and standard deviation of 75 and 5, respectively. What percent of students have received grade between 50 and 100?

Answer

Using Chebyshev's theorem, we know that $1 - \dfrac{1}{k^2}$ proportion of data is in the interval of $(\mu - ks, \mu + ks)$. Therefore, $\mu + ks = 100$, or $75 + k(5) = 100$. Hence, $k = 5$. Therefore, the proportion of grades between 50 and 100 is $1 - \dfrac{1}{k^2} = 1 - \dfrac{1}{25} = 0.96$.

Theorem 3.34 Central Limit Theorem

Let $X_1, X_2, \ldots,$ be a sequence of iid random variables with mean μ and variance σ^2. Let $\{X_1, X_2, \ldots, X_n\}$ be a sample of size n. Define

$$S_n = \sum_{k=1}^{n} x_k. \tag{3.332}$$

Then, S_n is the total of n observations of the sample. Hence, the sample mean $\bar{X}$ will be defined as in (3.238), that is,

$$\bar{X} = \frac{S_n}{n}. \tag{3.333}$$

For large n, $S_n \sim N\left(n\mu, \sqrt{n}\sigma\right)$, that is,

$$\lim_{n \to \infty} P\left(\frac{S_n - n\mu}{\sigma\sqrt{n}} \leq x\right) = \Phi(x), \quad -\infty < x < \infty. \tag{3.334}$$

Thus, using (3.333), for large n, $\bar{X} \sim N\left(\mu, \dfrac{\sigma}{\sqrt{n}}\right)$, that is, for $-\infty < a < b < \infty$,

$$\lim_{n \to \infty} P\left(a \leq \frac{\bar{X} - \mu}{\dfrac{\sigma}{\sqrt{n}}} \leq b\right) = \Phi(b) - \Phi(a). \tag{3.335}$$

If $\bar{X}$ is transferred to the standard normal random variable, that is, with $\mu = 0$ and $\sigma^2 = 1$, then

$$\lim_{n \to \infty} P\left(a \leq \frac{S_n}{\sqrt{n}} \leq b\right) = \Phi(b) - \Phi(a), \quad -\infty < a < b < \infty. \tag{3.336}$$

Proof:
There are different versions of the proof for this theorem. We offer that from Rice (2007). It suffices to prove the theorem for $\mu = 0$. This is because, if $\mu \neq 0$, then for each k, we let $Y_k = X_k - \mu$. We also let $T_n = Y_1 + Y_2 + \cdots + Y_n$. Then, for $-\infty < x < \infty$, we have:

$$\lim_{n\to\infty} P\left(\frac{S_n - n\mu}{\sigma\sqrt{n}} \le x\right) = P\left(\frac{T_n}{\sigma\sqrt{n}} \le x\right). \tag{3.337}$$

Thus, we start the proof with $\mu = 0$, using the moment generating function, mgf, denoted by $M_x(t)$. We show that the mgf of

$$Z_n = \frac{S_n}{\sigma\sqrt{n}} \tag{3.338}$$

approaches the mgf of the standard normal pdf.

Since S_n is the sum of n independent random variables, from a property of the mgf, we have (3.338) and

$$M_{S_n}(t) = [M(t)]^n = M_{Z_n}(t) = \left[M\left(\frac{t}{\sigma\sqrt{n}}\right)\right]^n. \tag{3.339}$$

To complete the proof, we need to show that

$$L \equiv \lim_{n\to\infty} n \ln\left[\frac{t}{\sigma\sqrt{n}}\right] = \frac{t^2}{2}, \tag{3.340}$$

by taking the exponential of this and using one of the properties of the mgf mentioned in an Chapter 2. The essence of that property is that if $M_n(t) \to M(t)$, $\forall t$ in an open interval containing zero, then $F_n(x) \to F(x)$, for all x at which F is continuous, where F_n is a sequence of cdfs with corresponding mgfs $M_n(t)$. A similar property would let $X_1, X_2, \ldots$, be a sequence of random variables with cdf $F_1, F_2, \ldots$, respectively, and also, X be a random variable with cdf $F_X(x)$. Then, X_n will **converge in distribution** to X, denoted by $X_n \xrightarrow{d} X$, if $\lim_{n\to\infty} F_n(x) = F(x)$ at each point of continuity of $F(x)$. For an interval $[a,b]$, this property states that

$$P(a \le F_n \le b) \xrightarrow{d} P(a \le F \le b). \tag{3.341}$$

Now, changing $1/\sqrt{2}$ to x, the unknown L will become:

$$L = \lim_{x\to\infty} \frac{\ln M\left(\frac{tx}{\sigma}\right)}{x^2}. \tag{3.342}$$

However, at $x = 0$, $M(0) = 1$. Hence, (3.342) becomes $\dfrac{0}{0}$, which is indeterminate. Using L'Hôpital's rule, we will have:

$$L = \lim_{x\to\infty} \frac{\dfrac{M'\left(\dfrac{tx}{\sigma}t\right)}{M\left(\dfrac{tx}{\sigma}\right)}}{2x} = \frac{t}{2\sigma} \lim_{x\to\infty} \frac{M'\left(\dfrac{tx}{\sigma}\right)}{xM\left(\dfrac{tx}{\sigma}\right)}. \tag{3.343}$$

Once again, at $x = 0$, (3.343) is indeterminate. Using L'Hôpital's rule again, we will have:

$$L = \frac{t}{2\sigma} \lim_{x \to 0} \frac{M''\left(\frac{tx}{\sigma}\right)\frac{t}{\sigma}}{M\left(\frac{tx}{\sigma}\right) + xM'\left(\frac{tx}{\sigma}\right)\frac{t}{\sigma}}.$$

$$= \frac{t^2}{2\sigma^2} \frac{\lim\limits_{x \to 0} M''\left(\frac{tx}{\sigma}\right)}{\lim\limits_{x \to 0} M\left(\frac{tx}{\sigma}\right) + \frac{t}{\sigma}\lim\limits_{x \to 0} xM'\left(\frac{tx}{\sigma}\right)}$$

$$= \frac{t^2}{2\sigma^2} \frac{M''(0)}{M(0) + \frac{t}{\sigma}0M'(0)}.$$

Further, using the properties of the mgf, from $M^r(0) = E\left(X^r\right)$, we will have:

$$M(0) = E(1) = 1, \quad M'(0) = E(X) = 0$$

$$M''(0) = E\left(X^2\right) = \left[E(X)\right]^2 + Var(x) = 0 + \sigma = \sigma.$$

Thus,

$$L = \frac{t^2}{2\sigma^2} \frac{\sigma^2}{1+0} = \frac{t^2}{2},$$

as desired.

Note 3.108

Essentially, the central limit theorem says that the average **of the sample means will be the population mean.**

Note 3.109

The central limit theorem is sometimes stated differently. For instance, Hogg and Tanis (1993) state and prove the result as follows: As n becomes very large $(n \to \infty)$, the cdf of

$$\frac{\bar{X} - \mu}{\sigma/\sqrt{n}} = \frac{\sum\limits_{i=1}^{n} X_i - n\mu}{\sigma\sqrt{n}}$$

is the standard normal cdf, that is,

$$\lim_{n\to\infty} cdf\left(\frac{X-\mu}{\sigma/\sqrt{n}}\right) = \lim_{n\to\infty} cdf \frac{\sum_{i=1}^{n} X_i - n\mu}{\sigma\sqrt{n}} = \Phi(0,1).$$

Note 3.110

The central limit theorem, is stated for a "large sample size", as it was mentioned earlier. The question is how large is "large"? The answer may be "the larger the better". But this is not specific enough. For practical purposes, some statisticians accept standard large samples as $n \geq 30$ or $n \geq 40$, when the population distribution is roughly normal and much larger if the population distribution is not normal or extremely skewed.

Note 3.111

As an extension of a property of normal distribution mentioned in Chapter 2, we now state a property that is referred to as **weak form of the central limit theorem**.

Theorem 3.35 The Weak Central Limit Theorem

Let $\{X_1, X_2, ..., X_n\}$ be a sample of size n; that is, let $X_1, X_2, ..., X_n$ be n iid random variables with mean 0 and variance σ^2. Then,

$$\frac{X_1 + X_2 + \cdots + X_n}{\sqrt{n}} \text{ approaches } N(0, \sigma^2) \text{ as } n \text{ becomes large.}$$

Of course, in case the variables are with mean different from 0, subtraction of the mean can normalize them.

Proof:
Left as an exercise.

Note 3.112

The central limit theorem applies to almost all types of probability distributions. For instance, it can be proved for independent variables with bounded moments, and even more general versions; even limited dependency can be tolerated. Moreover, random variables not having finite moments, that is, $E(X_n)$ doesn't converge for all n, like a Cauchy distribution, are sometimes sufficiently well behaved to induce convergence.

Theorem 3.36

Let $X_1, X_2, \ldots, X_n$ be a random sample of size n, that is, n iid random variables with mean and finite variance μ and $\sigma^2 < \infty$, respectively. Define the sample mean as

$$\bar{X} = \frac{X_1 + X_2 + \cdots + X_n}{n}. \tag{3.344}$$

Also, define S_n as a random variable representing the sum $X_1 + X_2 + \cdots + X_n$, that is,

$$S_n = \sum_{i=1}^{n} X_i. \tag{3.345}$$

Then,

i. $E(\bar{X}) = \mu,$ (3.346)

ii. $Var(\bar{X}) = \dfrac{\sigma^2}{n},$ (3.347)

iii. $E(S^2) = \sigma^2,$ (3.348)

iv. $E(S_n) = n\mu,$ (3.349)

v. $Var(S_n) = n\sigma^2.$ (3.350)

Proof:
We prove part (i) and leave the rest for the readers to refer to Casella and Berger (1990, p. 208).

From the definition of $\bar{X}$, (3.344), for a sample of size n, we have:

$$E(\bar{X}_n) = E\left(\frac{1}{n}\sum_{i=1}^{n} X_i\right) = \frac{1}{n}\left(\sum_{i=1}^{n} X_i\right) = \frac{1}{n}\left[nE(X_i)\right] = \frac{1}{n}(n\mu) = \mu.$$

What (3.345) says is that the mean of the distribution of the sample mean is the same as the mean of the population. In other words,

As the sample size increases and approaches infinity, we will have:

$$E(\bar{X}) = \lim_{n \to \infty}\left(\frac{1}{n}\sum_{k=1}^{n} X_k\right) = \mu. \tag{3.351}$$

In Chapter 2, we discussed the **de Moivre–Laplace central limit theorem.** There it was mentioned as a special case of the general central limit theorem. Here, we present the general central limit theorem.

Theorem 3.37 Central Limit Theorem (For Sample Mean)

Let $X_1, X_2, \ldots, X_n$ be a random sample of size n (a natural number, i.e., $n \in \mathbb{N}$) from a population, that is, n iid random variables, with mean and finite variance, μ and $\sigma^2 < \infty$, respectively. Then,

$$\lim_{n \to \infty} \left(\frac{\bar{X} - \mu}{\frac{\sigma}{\sqrt{n}}} \right) = N(0,1). \tag{3.352}$$

In other words,

$$\bar{X} \to N\left(\mu, \frac{\sigma}{\sqrt{n}} \right). \tag{3.353}$$

That means the sample mean can be approximated with a normal variable with mean μ and standard deviation $\sigma \sqrt{n}$.

Example 3.114

Let X be a random variable with $\mu = 6$ and $\sigma = 5$. Suppose a sample of size 100 is taken from this population. We are looking for the probability that the sample mean is less than 5.

Answer

Applying the central limit theorem, we will have:

$$P(X < 5) = P\left(z < \frac{5-6}{\frac{5}{\sqrt{100}}} \right) = P(z < -2) = \Phi(-2) = 0.0228.$$

Example 3.115

Suppose the population of female secretaries with college degrees has normal pdf with a mean salary of \$39,523 and a standard deviation of \$3,125. A sample of 200 female secretaries is taken. What is the probability that their mean salaries will be less than \$38,500?

Answer

Since $n = 200$, it is accepted as large enough. Hence,

$$z = \frac{x - u}{\sigma} = \frac{38,500 - 39,523}{3,125} = -0.3274.$$

$$P(X < 38,500) = \Phi(-0.3274) = 0.3717 \approx 37.17\%.$$

Yet another way of stating the central limit theorem is the following. This is similar to the de Moivre–Laplace central limit theorem. We leave the proof as an exercise.

Theorem 3.38

Let $X_1, X_2, \ldots, X_n$ be a random sample of size n (a natural number, i.e., $n \in \mathbb{N}$) from a population, that is, n iid random variables, with mean and finite variance, μ and $\sigma^2 < \infty$, respectively. Also, let a and b be two fixed numbers. Define Z_n as the random variable

$$Z_n = \frac{\bar{X} - \mu}{\frac{\sigma}{\sqrt{n}}} = \frac{\frac{1}{n}\sum_{k=1}^{n} X_k - U}{\frac{\sigma}{\sqrt{n}}}, \qquad (3.354)$$

that is, the standard normal random variable with $\mu = 0$ and $\sigma^2 = 1$. Then, for large n, Z_n can be approximated by the standard normal distribution $N(0,1)$ or $\Phi(x)$, that is,

$$\lim_{n \to \infty} P\left(a \leq \frac{\bar{X} - \mu}{\frac{\sigma}{\sqrt{n}}} \leq b \right) = \Phi(b) - \Phi(a). \qquad (3.355)$$

Proof:
Left as an exercise.

Example 3.116

Let us consider $X_1, X_2, \ldots, X_{10}$ be 10 uniformly iid random variables distributed on the half-open interval $[0,16)$. What is the probability that the sample mean falls in this interval?

Answer

From what is given, and from Chapter 2, we know that $a = 0$, $b = 16$, and

$$\mu_X = E(X) = \frac{b+a}{2} = \frac{16+0}{2} = 8$$

and

$$\sigma_{\bar{X}}^2 = \frac{(b-a)^2}{12} = \frac{(16-0)^2}{12} = 21.33.$$

Hence, from (3.314), we have:

$$P\left(0 \le \bar{X} < 16\right) = P\left(\frac{0-8}{\frac{\sqrt{21.33}}{\sqrt{10}}} \le \frac{\bar{X}-8}{\frac{\sqrt{21.33}}{10}} \le \frac{16-8}{\frac{\sqrt{21.33}}{\sqrt{10}}}\right)$$

$$= P(-5.48 \le Z \le 5.48) = \Phi(5.48) - \Phi(-5.48)$$

$$= 1 - \left(2.1266e^{-08}\right) = 1 - 0 = 1.$$

One additional central limit theorem for the sum of random variables is as follows.

Theorem 3.39 Central Limit Theorem (For Sum of n iid Random Variables)

Let $X_1, X_2, \ldots, X_n$ be a random sample of size n (a natural number, i.e., $n \in \mathbb{N}$) from a population, that is, n iid random variables, with mean and finite variance, μ and $\sigma^2 < \infty$, respectively. Consider S_n as a random variable representing the sum $X_1 + X_2 + \cdots + X_n$. Then,

$$\lim_{n \to \infty} \left(\frac{S_n - n\mu}{\sigma\sqrt{n}}\right) = N(0,1). \tag{3.356}$$

Proof:
Left as an exercise.

Example 3.117

Let X be a random variable with $\mu = 10$ and $\sigma = 4$. A sample of size 100 is selected from this population. We want to find the probability that the sum of these 100 observations is less than 950.

Answer

Let the sum be denoted by S_n. Then, we will have (See Table 3.33)

TABLE 3.33
The (Student) t-Distribution

$$F(x) = P(X \leq x) = \frac{\Gamma\left(\dfrac{d+1}{2}\right)}{\Gamma\left(\dfrac{d}{2}\right)\sqrt{d\pi}} \int_{-\infty}^{x} \left(1 + \frac{t^2}{d}\right)^{-(d+1)/2} dt$$

$$0.0005 \leq P(X \leq x) \leq 0.30$$

$P(X \leq x)$

d	0.30	0.20	0.10	0.05	0.025	0.01	0.005	0.001	0.0005
1	0.726543	1.376382	3.077684	6.313752	12.706205	31.820516	63.656741	318.308839	636.619249
2	0.617213	1.060660	1.885618	2.919986	4.302653	6.964557	9.924843	22.327125	31.599055
3	0.584390	0.978472	1.637744	2.353363	3.182446	4.540703	5.840909	10.214532	12.923979
4	0.568649	0.940965	1.533206	2.131847	2.776445	3.746947	4.604095	7.173182	8.610302
5	0.559430	0.919544	1.475884	2.015048	2.570582	3.364930	4.032143	5.893430	6.866827
6	0.553381	0.905703	1.439756	1.943180	2.446912	3.142668	3.707428	5.207626	5.958816
7	0.549110	0.896030	1.414924	1.894579	2.364624	2.997952	3.499483	4.785290	5.407883
8	0.545934	0.888890	1.396815	1.859548	2.306004	2.896459	3.355387	4.500791	5.041305
9	0.543480	0.883404	1.383029	1.833113	2.262157	2.821438	3.249836	4.296806	4.780913
10	0.541528	0.879058	1.372184	1.812461	2.228139	2.763769	3.169273	4.143700	4.586894
11	0.539938	0.875530	1.363430	1.795885	2.200985	2.718079	3.105807	4.024701	4.436979
12	0.538618	0.872609	1.356217	1.782288	2.178813	2.680998	3.054540	3.929633	4.317791
13	0.537504	0.870152	1.350171	1.770933	2.160369	2.650309	3.012276	3.851982	4.220832
14	0.536552	0.868055	1.345030	1.761310	2.144787	2.624494	2.976843	3.787390	4.140454
15	0.535729	0.866245	1.340606	1.753050	2.131450	2.602480	2.946713	3.732834	4.072765
16	0.535010	0.864667	1.336757	1.745884	2.119905	2.583487	2.920782	3.686155	4.014996
17	0.534377	0.863279	1.333379	1.739607	2.109816	2.566934	2.898231	3.645767	3.965126
18	0.533816	0.862049	1.330391	1.734064	2.100922	2.552380	2.878440	3.610485	3.921646
19	0.533314	0.860951	1.327728	1.729133	2.093024	2.539483	2.860935	3.579400	3.883406

(Continued)

TABLE 3.33 (Continued)
The (Student) t-Distribution

$$F(x) = P(X \le x) = \frac{\Gamma\left(\dfrac{d+1}{2}\right)}{\Gamma\left(\dfrac{d}{2}\right)\sqrt{d\pi}} \int_{-\infty}^{x} \left(1 + \frac{t^2}{d}\right)^{-(d+1)/2} dt$$

$$0.0005 \le P(X \le x) \le 0.30$$

d	0.30	0.20	0.10	0.05	0.025	0.01	0.005	0.001	0.0005
20	0.532863	0.859964	1.325341	1.724718	2.085963	2.527977	2.845340	3.551808	3.849516
21	0.532455	0.859074	1.323188	1.720743	2.079614	2.517648	2.831360	3.527154	3.819277
22	0.532085	0.858266	1.321237	1.717144	2.073873	2.508325	2.818756	3.504992	3.792131
23	0.531747	0.857530	1.319460	1.713872	2.068658	2.499367	2.807336	3.484964	3.767627
24	0.531438	0.856855	1.317836	1.710882	2.063899	2.492159	2.796940	3.466777	3.745399
25	0.531154	0.856236	1.316345	1.708141	2.059539	2.485107	2.787436	3.450189	3.725144
26	0.530892	0.855665	1.314972	1.705618	2.055529	2.478530	2.778715	3.434997	3.706612
27	0.530649	0.855137	1.313703	1.703288	2.051831	2.472560	2.770683	3.421034	3.689592
28	0.530424	0.854647	1.312527	1.701131	2.048407	2.467140	2.763262	3.408155	3.673906
29	0.530214	0.854192	1.311434	1.699127	2.045230	2.462021	2.756386	3.396240	3.659405
30	0.530019	0.853767	1.310415	1.697261	2.042272	2.457262	2.749996	3.385185	3.645959
32	0.529665	0.852998	1.308573	1.693889	2.036933	2.448678	2.738481	3.365306	3.621802
34	0.529353	0.852321	1.306952	1.690924	2.032244	2.441150	2.728394	3.347934	3.600716
36	0.529076	0.851720	1.305514	1.688298	2.028094	2.434494	2.719485	3.332624	3.582150
38	0.528828	0.851183	1.304230	1.685954	2.024394	2.428568	2.711558	3.319030	3.565678
40	0.528606	0.850700	1.303077	1.683851	2.021075	2.423257	2.704459	3.306878	3.550966
50	0.527760	0.848869	1.298714	1.675905	2.008559	2.405272	2.677793	3.261409	3.496013
60	0.527198	0.847653	1.295821	1.670649	2.000298	2.390119	2.660283	3.231709	3.460200
120	0.526797	0.846786	1.293763	1.666914	1.994437	2.380807	2.647905	3.210789	3.435015

$$P(X \le x)$$

EXERCISES

3.1. Consider the experiment of rolling two dice and the following random variables.

Let $W =$ The sum of the two numbers that appear on two dice.

$X =$ The absolute value of the difference of two numbers that appear on two dice.

$Y =$ Both numbers that appear are the same.

  i.   State the possible values of each of the above random variable.
  ii. Construct the probability mass of X.
  iii. Represent the cumulative distribution of X in a graphical form.
  iv. Calculate the expected value of X.

3.2. Assume you examine three bicycles reaching to a junction. You are going to observe whether the bicycles are going straight (0), turn to left (−1), or turn to right (1) from the junction. Let X be the random variable that represents the sum of the values assigned to each direction of the above three bicycles. List the values of X.

  i. Construct the probability mass values of X in a graphical form.
  ii. Calculate the expected value of X.

3.3. Consider the probability distribution of the random variable X.

X	0	1	2	3	4
$p(x)$	0.2	0.3	0.1	0.2	

  i.   Complete the above table.
  ii.   What is the most likely occurring value of the random variable X?
  iii. Compute $P(X < 3)$.
  iv.   Compute $P(X \leq 2)$.
  v.   Compute $P(X > 3)$.
  vi.   Compute $P(1 < X < 3)$.
  vii. Compute $P(1 < X \leq 3)$.

3.4. A computer vender knows the following information by his experience. 20% of the time he does not sell any computer, 35% of the time he sells one computer, 25% of the time he sells two computers, and 20% of the time he sells three computers. Taking the number of computers the vender sells as the random variable X, answer the following questions:

  i.   Construct the probability distribution of X in the tabular form.
  ii.   Construct the cumulative distribution function (cdf) of X.
  iii. Calculate the expected number of computers the vender sells.
  iv.   Calculate the variance of the number of computers the vender sells.

3.5. Consider the probability distribution of the random variable X.

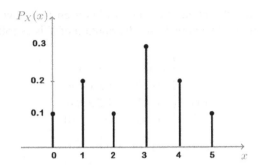

i. Construct the probability distribution X in the tabular form.

ii. Calculate the expected value of X.

3.6. A particular gas station has five gas pumps. Let Y denote the number of gas pumps that are in use at a specified time. Consider the following pmf of Y.

y	0	1	2	3	4	5
$P(y)$	0.10	0.15	0.20	0.25	0.22	0.08

Calculate the probability of each of the following events.

i. {at most 3 pumps are in use},

ii. {fewer than 3 pumps are in use},

iii. {at least 3 pumps are in use},

iv. {between 2 and 5 pumps, inclusive, are in use},

v. {between 2 and 4 pumps, inclusive, are not in use},

vi. {at least 4 pumps are not in use}.

3.7. A sample of 100 married women was taken. Each was asked how many children she had. Responses were tallied as follows:

Number of children	0	1	2	3	4	5 or more
Number of women	25	22	29	13	7	4

i. Find the average number of children in the sample.

ii. Find the sample standard deviation of the number of children.

iii. Find the sample median.

iv. What is the first quartile of the number of children?

v. What proportion of the women had more than the average number of children?

vi. For what proportion of women was the number of children more than one standard deviation greater than the mean?

vii. For what proportion of the women was the number of children within one standard deviation of the mean?

3.8. Let X represent the number of months between two power outages that occur. The cumulative probability distribution of X is as follows:

$$F(x) = \begin{cases} 0 & x < 1 \\ 0.20 & 1 \le x < 4 \\ 0.45 & 4 \le x < 6 \\ 0.60 & 6 \le x < 10 \\ 1 & 10 \le x \end{cases}$$

   i. Construct the pmf of X.

  ii. What is the expected number of months between two successive power outages?

 iii. What is the probability of the number of months between two successive powered outages at least 5 months?

3.9. Suppose the number of years of experience (X) an employee possesses when joining to a particular type of job is given by the following pmf:

$$P(X = x) = p(x) = \begin{cases} \dfrac{x}{10}; & x = 1,2,3,4 \\ 0; & \text{else} \end{cases}$$

   i. Calculate the expected number of years of experience a new employee possesses when joining to the above job.

  ii. What is the probability that a new employee has at least 2 years of experience when joining to the above job?

3.10. Consider a random variable with the following probability mass function

$$(\text{pmf}): \ P(x) = \begin{cases} \dfrac{k}{x^2}; & x = 1,2,3 \\ 0; & \text{else} \end{cases}, \text{ where } k \text{ is a positive constant.}$$

   i. Calculate the constant k.

  ii. Construct the cumulative distribution function (cdf) of the random variable.

3.11. Suppose an individual plays a gambling game where it is possible to lose $1.00, break even, win $3.00, or win $10.00 each time she plays. The probability distribution for each outcome is provided by the following table:

Outcome	−$1.00	$0.00	$3.00	$5.00
Probability	0.30	0.40	0.20	0.10

Calculate the expected value and the variance of the amount an individual going to receive at the end.

3.12. Consider a bicycle seller, who purchases a batch of five bicycles from a company and sells to customers. He buys these bicycles from the company for $100 per bicycle and sells one for $150. According to his experience, he knows that he has 1% chance not to sell any bicycle, 5% chance to sell one bicycle, 10% chance to sell two bicycles, 14% to sell three bicycles, 40% to sell four bicycles, and 30% chance to sell all the five bicycles.

    i.   What is the expected number of bicycles he expects to sell?

    ii. What is the expected profit he gets at the end of this business from one batch of bicycles?

3.13. An examination consists of 20 multiple-choice questions. Suppose each question has four possible choices, but only one of them is the correct solution. If a student who did not study for the examination is going to guess at the examination, what is the probability that he guesses

    i.   Exactly five questions correctly?

    ii. At most five questions correctly?

    iii. At least five questions correctly?

    iv. Between five and ten questions (inclusively) correctly?

    v. How many questions do you expect him to guess correctly?

3.14. A laboratory network consisting of 25 computers was attacked by a computer virus. This virus enters each computer with a probability of 0.4, independently of other computers. What is the probability that

    i.   At least ten computers are affected by the virus?

    ii. More than ten computers are affected by the virus?

3.15. A clothing store has determined that 30% of the customers who enter the store will use a Visa credit card. Consider a collection of 15 customers who have entered to the store.

    i.   How many of them you expect to use Visa card?

    ii. What is the probability that more than five customers use Visa card?

3.16. A quality controller knows by his experience that 10% of the items he inspects are defective. He randomly selects 30 items from the production and wants to inspect them.

    i.   Calculate the expected number of defective items he is going to find.

    ii. What are the variance and the standard deviation of the number of defective items that the quality controller is going to find?

    iii. What is the probability that the number of defective items he is going to find is more than one standard deviation of the mean value?

3.17. Suppose 10% of the drivers in a particular city do not possess a valid driver's license. A traffic inspector is inspecting all the drivers coming to a particular junction in this city. What is the probability that the inspector needs to inspect five drivers until he finds the first driver without a valid license?

3.18. Usually a particular type of candy bags contains ten candies in each bag. Only 20% of the candy bags contains 15 candies in the candy bag. Suppose a kid is trying to find a bag with 15 candies. How many candy bags does he needs to open until he finds the first bag with 15 candies?

3.19. Consider a particular type of lottery with 2% chance of winning any kind of prize. Suppose a person wants to buy this lottery until he wins a prize. What is the probability that he gets his first win from the fifth lottery he purchases?

3.20. Assume that there is a 10% chance a certain baseball player hits a home run in a game. What is the probability that the player completes his second home run in his fifth game?

3.21. The number of telephone calls receiving per 5 minutes to a hotel's reservation center follows a Poisson random variable with a mean of 3. Find the probability that no call is received by the hotel in a given 5-minute period.

3.22. Let us suppose that sale of a flood insurance by a salesperson is according to a Poisson pmf, with an average of 3 per week. Use Poisson pmf to calculate the probability that in a given week the salesperson will sell

    i. Some policies.

    ii. Two or more policies but less than five policies.

    iii. Assuming that there are five working days per week, what is the probability that in a given day he will sell one policy?

3.23. A service station experiences ten vehicles arriving in an hour. Assume that the arrival of vehicles is independent and no more than one vehicle arrives at the same time. Considering a particular hour at the service station,

    i. What is the probability that there are at most five vehicles arriving to the service station?

    ii. If the number of vehicles arrived exceeds ten vehicles, the service station needs to arrange additional parking space. What is the probability that the service station needs to arrange?

3.24. Assume that the number of bacterial colonies of a certain type is grown at a rate of 2 colonies per minute. Assume that two or more colonies do not grow at the same time, and they are independent. Considering a 5-minute period,

    i. What is the probability that no more than 15 colonies are grown during this 5-minute period?

    ii. How many colonies do you expect to grow during this 5-minute period?

3.25. Consider the following joint pmf of X and Y.

	$Y = 0$	$Y = 1$	$Y = 2$
$X = 0$	0.3	0.2	0.1
$X = 1$	0.2	0.1	0.1

    i. Construct the marginal pmf of X and Y.

    ii. Calculate the expected values and the variances of X and Y.

iii. Calculate the covariance of X and Y.
iv. Calculate the coefficient of correlation.

3.26. Note that the pH value of a soil in a certain area has the following distributed function:

$$f(x) = k\left(x^2 - 20x + 100\right); 3 \le X \le 10 \text{ and } k > 0$$

i. Calculate the constant k.
ii. What is the mean pH value of this soil?
iii. If a soil sample from this area is selected, what is the probability that the pH value of the soil is more than 5?

3.27. The error in the reaction time X (in minutes) of a certain machine has the following distribution:

$$f(x) = \frac{2}{3}x^2; \quad -1 \le x \le 1$$

i. Construct the cdf of X.
ii. What is the probability that the error in the reaction time is over 0.5?

3.28. Consider a layer of a particular paint applied on metal surfaces to protect from corrosion. Assume that the thickness of the above paint has a uniform distribution, which is distributed between 10 and 20 μm. If a metal surface with this paint is randomly selected,

i. What is the expected value of the thickness of it?
ii. What is the probability that the thickness of the paint is less than 18 μm?

3.29. Suppose a random variable X has a uniform distribution over $[1, \beta]$ for some $\beta > 1$. Calculate the value of β if $P(2 < X < 5) = 0.5$.

3.30. Suppose a random variable X has a uniform distribution over $[a, b]$, where $0 \le a < b$. If the mean value of X is 5 and variance is 3/4, what are the values of a and b?

3.31. Let X be a continuous random variable with pdf $f(x) = 1 - |x - 1|, 0 \le x \le 2$.
i. Find cdf of X and graph it.
ii. Find the first decile.
iii. Find 20th and 95th percentiles, and show them on the graph of cdf of X.
iv. Find the IQR.

3.32. Suppose a particular examination that is needed to enter to a graduate school has a normal distribution with a mean of 150 and a standard deviation of 10. Considering a student is going to take this examination,
i. What is the probability that his score is between 145 and 160?
ii. What is the probability that he scores at least 165?
iii. What is the probability that he scores at most 145?
iv. A reputed graduate school admits students whose score is in top 10%. In order to get the admissions to this school, what is the minimum score a student needs to have?

3.33. Assume the monthly income of a family in an identified city is approximately normally distributed with a mean of $2,000 and a standard deviation of $100.

    i. If a family from this city is randomly selected, what is the probability that the income of the family is over $2250 per month?

    ii. The mayor of this city decided to give financial assistance for the families whose income is in the bottom 1%. What is the highest monthly income of a family who qualifies for this financial assistance?

3.34. The weight of a bag of chips of a certain brand sold in vending machines is normally distributed with a mean of 16 ounce and a standard deviation of 0.3 ounce. If a customer buys two bags of chips, what is the probability that weights of both of them exceed 16.5 ounce?

3.35. A fisherman knows by experience the average weight and the standard deviation of a catfish are 3.1 pounds and 0.8 pounds, respectively. Assuming the distribution of catfish is normally distributed, what is the probability that the weight of a randomly caught catfish is not more than 4 pounds?

3.36. The lifetime of a certain type of light bulbs is approximately exponentially distributed with an expected lifetime of 250 hours. If a light bulb of the above type is randomly selected, what is the probability that the lifetime of the light bulb is

    i. Less than 300 hours?

    ii. Between 200 and 300 hours?

    iii. Suppose a light bulb has lasted more than 300 hours, what is the probability that it will last more than 400 hours?

3.37. A bicycle tire manufacturer states that the average lifetime of his tire is 3 years. If a customer purchases two bicycle tires, then calculate the probabilities.

    i. Both will survive more than 4 years?

    ii. At least one will survive more than 4 years?

    iii. None of them will survive more than 4 years?

    iv. At least one of them will survive more than 4 years?

3.38. Let X be the time (in minutes) a doctor spends to talk to a patient. Assume that this time is exponentially distributed with an average of 5 minutes.

    i. Construct the cumulative distribution function (cdf) of X.

    ii. Using the above cdf, calculate the following probabilities.

        a. What is the probability that the doctor spends less than 4 minutes with a patient?

        b. What is the probability that the doctor spends less between 3 and 7 minutes with a patient?

        c. What is the probability that the doctor spends more than 8 minutes with a patient?

3.39. Assume that the height (n inches) of a certain plant can be approximated with a gamma distribution with shape parameter 2 and the scale parameter 1. If a plant of this type is selected randomly, what is the probability the height of the plant is

     i.   More than 3 inches?

    ii.   Not more than 4 inches?

3.40. Consider a random variable that follows a gamma distribution. If its expected value is 4 and variance is 2, calculate the two parameters of the random variable.

3.41. Using the moment generating function of the chi-square distribution, calculate the expected value and the variance.

3.42. The gamma pdf with two parameters k and θ for a random variable X is defined as:

$$f_X(x;\theta;k) = \begin{cases} \dfrac{1}{\theta^k \Gamma(k)} x^{k-1} e^{-\frac{x}{\theta}}, & k, \theta > 0, x > 0, \\[2ex] 0 & x \leq 0, \end{cases}$$

where $\Gamma(\alpha)$ is the gamma function. Prove that $E(X) = k\theta$ and $Var(X) = k\theta^2$

4 Descriptive Statistics

4.1 INTRODUCTION AND HISTORY OF STATISTICS

It seems the word "statistics" is based on "data" and "analysis of data". In 1749, *Statistik*, a book in German, was published describing the analysis of demographic and economic data about the state (political arithmetic in English). However, the word "statistics" originates from the Latin word "statisticum collegium", which means "council of state". "Statista" is the Italian word for statistics that means "statesman" or "politician". In 1800s, the word "statistics" expanded its meaning to cover summarizing and analyzing data. It has further widened its scope to include the concept of probability for the purpose of statistical inference.

4.2 BASIC STATISTICAL CONCEPTS

Statistics is essentially linked to the theory of probability, and probability theory is a branch of mathematics. Hence, it could be said that statistics is a branch of mathematics. On the other hand, since statistics deals with gathering, analyzing, and interpreting data, there are lots of human judgments involved in statistical analysis. This idea seems to separate statistics from mathematics, to the extent that it is becoming difficult for pure mathematicians to accept statistics as part of mathematics. However, the second part of statistics, inferential statistics, is mainly mathematics with less human judgment.

Thus, statistics is sometimes considered as a branch of mathematics, and at other times, a discipline in its own right. Regardless of how it is looked at, it is a very important concept now that its applications are so vast and diverse that no area of science can do without it. In fact, it has been so spread out that humanity, psychology, social sciences, and even communication cannot do without gathering and analysis of data.

We start this chapter by defining some important terms that are widely used throughout this chapter. Of course, concepts of probability theory discussed in Chapter 2 will help us provide necessary basis.

Definition 4.1

A data point can be quantitative that is, numerical that can be measured by counting and qualitative, that is, a trait, characteristic, or category.

Definition 4.2

A data set can be classified according to the level of measurement as follows:

i. **Nominal data points** are qualitative data that are categorized by names, labels, color, favorite political party, gender, and brand or features. No mathematical operations can be applied on these data points.

ii. **Ordinal data points** are both qualitative and quantitative. Most importantly, these data can be arranged according to an **order or a rank**. Examples of this category are classification of students as freshman, sophomore, junior, and seniors; also, level of pain (say 1 through 10, 10 being highest) and level of satisfaction (such as very bad, bad, okay, good, very good); and rank of a movie (such as one star, two stars, three stars, four stars, five stars). Mathematical operations on these points do not make sense as well.

iii. **Interval data** are continuous sample data, which are quantified. Some mathematical operations on these intervals are quite possible, for instance, the differences between these data intervals. However, multiplication of these intervals does not make sense. If the value of an interval is zero, it does not mean that the interval is "empty" or "nonexistent".

iv. **Ratio** as data points represent quantitative data such as weight and height. Most importantly, mathematical operations may be applied on such data points. Value zero of a data represents the "empty" or "nonexistent".

Example 4.1

Let us consider the following variables: favorite food, goodness of the instructor (very poor, poor, okay, good, very good), and standardized examination score (like z-score). We summarize these variables as follows:

Variable	Quantitative/ Qualitative	Level of Measurement
Favorite food	Qualitative	Nominal
Goodness of the instructor (very poor, poor, okay, good, very good)	Qualitative	Ordinal
Standardized examination score (z-score)	Quantitative	Interval
Time you spend on exercise	Quantitative	Ratio

4.2.1 DATA COLLECTION

In practice, there are three ways to collect data, which are as follows:

i. Observational studies,
ii. Designed experiments,
iii. Published sources.

We will now discuss these methods and offer some examples.

i. Observational Method

In this case, the values of the variables (measurements) are observed and recorded without manipulating them. Surveys and interviews are examples of this method.

Example 4.2

To study the relationship between smoking and lung cancer, from a population, we collect a sample of those who state they have been smoking for years and another sample of those who state they do not smoke. Then, we use the observed values as they are, conduct a statistical analysis, and make an appropriate comparison between the two groups for the purpose of investigating the correlation between the smoking and lung cancer.

In this example, we just use the observations as they are. A problem with this method is that we are ignoring other possibilities of the causes of lung cancer such as family history, type of occupation, diet, age, and the living places of the elements in the samples.

ii. Designed Experiments (Experimental Design) Method

Unlike the observational method, in the designed experiments, we control the observed values of the variables in order to find the relationship among them. This is the method mostly used in the laboratory experiments.

Example 4.3

In order to find out which level of fertilizer (0 oz, 5 oz, or 10 oz) of a certain brand is the best, we can run a designed experiment as follows:

1. Select 60 plants of a certain type and of the same size.
2. Place plants individually, each in a similar pot with same type of soil.
3. Measure the length of each plant.
4. Place all plants in a location so that they all receive the same amount of sunlight.
5. Water all plant the same amount and at the same time.
6. Now select 20 of these plants randomly and apply 10 oz of the fertilizer and label them as such.
7. Randomly select another set of 20 plants from the remaining 40 and apply 5 oz of the fertilizer and label them accordingly.
8. Do not fertilize the remaining 20 plants.
9. Let a certain time period elapse and measure the lengths of all these 60 plants.
10. Compare the growth of plants in each of the three categories (i.e., items 6, 7, and 8), that is, plants fertilized with 10 oz, 5 oz, and none.

Note 4.1

It can be seen that the observed measures from the experimental design method are much more accurate for comparison than those from the observational method. However, the experimental design method is more time-consuming and more costly than the observational method. On the other hand, researchers who are concerned about the ethics of research would rather go with the observational method rather than the experimental design method.

iii. **Published Sources**

Standard published sources for the sake of conducting research and references are reliable research journals, reliable newspapers with reliable sources, and reliable and public databases.

4.3   SAMPLING TECHNIQUES

Now we know how to collect data, a sample from the population. The process of selecting a sample from the population is called the **sampling**. The essence of sampling is to generalize one or more properties of a small sample to a large population, that is, obtaining information of a large population from a sample. We should caution that it is not always easy to select a random sample, particularly, when the population size is very large and the sample size is small. For instance, to inspect thousands of cartons of canned food in the storage for safety concerns, selecting a sample of size ten cartons will be very difficult. This is because it is almost impossible to number all these cartons in the storage and then choose ten at random. Hence, in cases like this, we do not have many choices; we have to do the best we can and hope that we are not seriously violating the randomness property of the sample.

Sampling needs to be under certain conditions to take the place of the entire population. One very important condition is that a sample must be **representative** of the entire population. One way to guarantee this property is that the **sample be random**. There are some popular methods of selecting a sample: **simple random sample, cluster sample, stratified sample**, and **systematic sample**. Here are brief definitions of each.

Definition 4.3

i. A **simple random sample** is a set of n objects in a population of size N where all possible samples are equally likely to happen. Most importantly, in a simple random sample, each object has the same probability to represent the sample.

ii. **Cluster sampling** means dividing the population into separate groups, referred to as the **clusters**. Then, a simple random sample of clusters is selected from the population.

iii. Stratified **sampling** is a sampling technique wherein the entire population is divided into different subgroups or strata and then the final subjects are randomly selected from the different **strata**.

4.4 TABULAR AND GRAPHICAL TECHNIQUES IN DESCRIPTIVE STATISTICS

To extract more information from a sample, we first organize the data points somehow such as grouping, table, or graphical display. However, as it was noted for the measures of central tendencies, individual characteristics of data points may be lost when they are grouped. That is, the data is no longer "raw data". This is because the information obtained is for a group rather than individuals in the group. We explain these methods below.

A data set is grouped for reasons. Hence, data points in a grouped data are related in some sense. For instance, a collected set of data from people may be grouped by age, race, color of skin, nationality, income, and education. Even a group may be further grouped into other subgroups. For instance, the age group may be further grouped by teens, twenties, thirties, forties, middle age, and aged. In education, for instance, the subgrouping may be by high-school dropouts, high-school diploma, undergraduate degree, graduate degree, doctorate degree (such as Ph. D., Ed. D., J. D., D. of Music, MD), etc. In such cases, each subgroup is referred to as the **class interval** or **class size**.

Below, we will discuss the following grouping and displaying of data points:

i. Frequency distribution for qualitative data
ii. Bar chart (graph),
iii. Pie chart.
iv. Frequency distribution for quantitative data
v. Histogram,
vi. Stem-and-leaf plot,
vii. Dot plot.

4.4.1 FREQUENCY DISTRIBUTION FOR QUALITATIVE DATA

Definition 4.4

Grouping a data set of observations according to the number of repeated data points is referred to as a **frequency distribution**.

Note 4.2

The terms "relative frequency" and "percent relative frequency" defined in Definition 2.6b are consequences of frequency distribution defined in Definition 4.4.

Let us first consider a set of qualitative data and organize the data points using a frequency distribution as in the example below.

Example 4.4

Final grades of students in a class of 30 are as follows:

B	C	A	D	C	B	C	A	F	C
C	B	F	C	B	A	A	C	B	F
D	C	D	A	D	C	F	C	B	D

Using frequency distribution, we summarize the grades in Table 4.1.

From Table 4.1, 5 out of 30 students have made A. According to the third column, relative frequencies, the probability of making A is approximately 0.17; that is, approximately 17% of the grades are A. Similarly, percentage of other grades may be interpreted.

4.4.2 BAR GRAPH

Definition 4.5

A bar **chart** or a **bar graph** is a chart or graph that presents a categorical data set or a discrete random variable. The height or a length of a bar is proportional to the frequency. The bars can be plotted vertically or horizontally. A bar graph shows comparisons among discrete categories. One axis of the chart shows the specific categories being compared, and the other represents the frequency of it.

Note 4.3

William Playfair (1759–1824) is given credit for the first to develop and use the bar chart for commercial and political uses in Scotland during Christmas in 1780 and 1781.

Comparing the definition of a bar graph with that of a histogram, we can see that the main difference between them is that a histogram displays quantitative

TABLE 4.1
Frequency and Relative Frequencies

Value of the Qualitative Variable	Frequency	Relative Frequency (Frequency/Total)	Percent Relative Frequency (Relative Frequency * 100)
A	5	$\frac{5}{30} \approx 0.17$	17
B	6	$\frac{6}{30} = 0.20$	20
C	10	$\frac{10}{30} \approx 0.33$	33
D	5	$\frac{5}{30} \approx 0.17$	17
F	4	$\frac{4}{30} \approx 0.13$	13
Total	30	1.00	

(or nominal) data, while a bar graph displays qualitative (or ordinal) data. However, another distinction between the bar graph and the histogram is their arrangements. Bar graph has gaps between bars, while histogram, usually, has connected bins. When bars in a bar graph are arranged in decreasing order of their heights, it is referred to as **Pareto charts**. On the other hand, rectangles in a histogram are arranged in the order of their classes (bins) occurrences.

Example 4.5 (Data Using Excel)

Given the following frequency table for bird and rat counts in a particular neighborhood, we wish to draw a bar chart displaying this information (Figure 4.1).

Category	Count
Chicken	350
Duck	210
Cat	84
Dog	283
Rat	56

We use the Microsoft Excel 2010 (although 365 is available these days) and take the following steps that are for Excel 2016 and 365 with some less steps for 2010:

Step 1. Open an Excel file.
Step 2. Enter the categories in a column.
Step 3. Insert the counts (frequencies) in another column.
Step 4. Highlight both columns.
Step 5. From the Header, press Insert.
Step 6. Select "Bar" or "Column", as you wish.
Step 7. Choose "Chart Tools", then "Layout".
Step 8. Select "Chart Title".
Step 9. Choose other options from the Header.

You may see: https://projects.ncsu.edu/labwrite/res/gt/gt-bar-home.html.

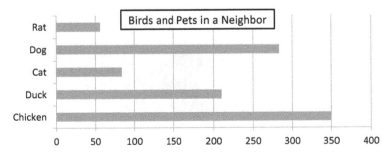

FIGURE 4.1 Bar graph using Excel for Example 4.5.

Note 4.4

When constructing bar graphs, we usually leave a gap between each bar. The reason is here x-axis represents a qualitative variable and usually we cannot compare the values of it. This is different when constructing histograms for quantitative data.

4.4.3 PIE CHART

Another common display of data points is the **pie chart**. Its basic use is the immediate visibility of comparison of slices. The slices are similar to the rectangles in a histogram. Each slice of the pie represents a value of a group of the data set. Similar to bar graphs, pie charts are also used to represent categorical (qualitative) data.

Example 4.6

Let us create a pie chart from the scratch. We first complete Table 4.2.
Now construct the pie chart with angles of the corresponding sectors, which are also given in Figure 4.2.

TABLE 4.2
Frequency, Relative Frequency, and Angle

Category	Count (Frequency)	Relative Frequency	Angle = Relative Frequency * 360
Chicken	350	0.36	128°
Duck	210	0.21	77°
Cat	84	0.09	31°
Dog	283	0.29	104°
Rat	56	0.06	21°
Total	**983**	**1.00**	**360°**

Animal count

■ Chicken ■ Duck ■ Cat ■ Dog ■ Rat

FIGURE 4.2 Pie chart for Example 4.6.

Example 4.7

Consider household income consisting of the income from the wife, husband, daughter, and the son. Then, each of the four slices of the pie represents a portion of the total household income. Considering each portion as a percent of the total area of the pie, the sum of all portions should be 100%.

Example 4.8

Consider the primary in an election with eight candidates. A pie chart can represent the percent votes for each candidate as a slice of a pie. By just looking at the size of each slice, one may predict electability of each candidate in the actual election time (Figure 4.3).

A pie chart has its own message, and it simply displays parts of a set of observations. In industry, for example, it is often a standard method to display the relationship of parts to the whole. However, because it is not as narrow as other displays of observations, for instance, when slices are almost of the same size or there are too many slices of narrow areas, it is not quite popular among statisticians. Pie charts are not good to show changes of observations over time, for example.

If, nonetheless, one wants to display the data set with the pie chart, here are the steps to create it in Excel. We may need some adjustment going from one version to another on different versions of Windows. In our case, we use Office 2010 on Windows 10. We use Example 4.6 for our illustration.

In our example, we want to show the relationship between the different levels of donors that give to our charity as compared to total giving. A pie chart is perfect for this illustration. We will start with a summary of the total giving by level.

Step 1. Insert the names of categories, for instance, incomes of husband, wife, daughter, and son.
Step 2. Insert the amount of income for each member of the family, say, $70,000, $50,000, $25,000, and $10,000 for husband, wife, son, and daughter, respectively.
Step 3. From the menu, select "Insert", "Charts", and then "Pie".

For more details, see: https://www.pryor.com/blog/create-outstanding-pie-charts-in-excel/.

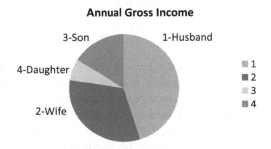

FIGURE 4.3 Pie chart using Excel for Example 4.8.

Note 4.5

In addition to the bar graph and the pie charts, there is another name called "Pareto Graph", which is basically the bar graph ordering from highest frequency to the lowest frequency. As this is a different arrangement of the bar graph, we do not discuss it as a separate graph.

4.4.4 FREQUENCY DISTRIBUTION FOR QUANTITATIVE DATA

As discussed in Section 4.3.1, we can use the frequency distributions for quantitative data, as well. Quantitative data can be in both the discrete and continuous forms. In the discrete case, we have a finite or a countable collection of values, whereas in the continuous case, we have an infinite collection of values. In the latter case, we can organize the data in intervals or groups.

Example 4.9

Suppose 25 rational numbers have been randomly chosen and listed as below:

A Set of 25 Observations (Raw Data)				
12.4	14.8	13.6	15.9	12.4
13.6	13.6	15.9	11.3	11.3
11.3	14.8	15.9	11.3	12.4
12.4	15.9	14.8	15.9	14.8
14.8	11.3	14.8	15.9	15.9

We want to do the following:

a. Group the given data points,
b. Find the frequencies,
c. Find the relative frequencies,
d. Find the percent relative frequencies.

Answer

Table 4.3 contains answers to all the questions.

Definition 4.6

Grouping a data set of observations that are real numbers is called a **frequency distribution with intervals**. Otherwise, it is referred to as **frequency distribution with classes** although "class" can cover both. The **length** (or **size**) **of an interval or a class** is defined as the difference between the upper bounds and the lower bounds of the interval. To group a real-valued data by intervals, one may (1) calculate the range by subtracting the minimum from the maximum and (2) decide the number of intervals (not too many and not too few). Classes may be with the same length,

TABLE 4.3
Grouping Data from Example 4.9

i	Data Point x_i	Frequency f_1	Relative Frequency	Percent Relative Frequency
1	11.3	5	1/5	20
2	12.4	4	4/25	16
3	13.6	3	3/25	12
4	14.8	6	6/25	24
5	15.9	7	7/25	28
Total		25	1	100

different lengths, or mixture of both kinds. In this case, we may divide the range by the number of desired intervals and round up the number obtained to have a natural number (or a whole number) as the size of classes. The endpoints of the intervals are referred to as the **class limits**. The left and right limits of an interval are referred to as the **lower class limit** and the **upper class limit**, respectively. Exact class limits are referred to as the **class boundaries**, and the left boundary and the right boundary are referred to as the **class lower boundary** and the **class upper boundary**, respectively.

It is customary to avoid overlaps in obtaining class boundaries, by adding 0.5 to the upper limit and subtracting 0.5 from the lower limit, respectively. The addition of ±0.5 to the limits is when the limits are integers; otherwise, ±0.05 if it is with no digits and starts with the tenth decimal place, ±0.005 if it starts with the hundredth decimal place, and so on, down and up from each end of the interval's limits. In such cases, the interval will be chosen half-open on the right end.

In case of grouping by classes and frequency distribution, the set of sample points is redefined. We may choose the midpoint of each class (sometimes referred to as the **class mark**) as an individual revised data point for the sake of finding the mean, variance, and standard deviation using frequencies.

Example 4.10

Let us suppose that we are given a set of data points with minimum and maximum 67 and 100, respectively. Hence, the range of the set is 33. Now let us assume that we want to group the data point into six equal-sized classes. Based on what we have already discussed, the class sizes will be $\frac{33}{6} \approx 6$. Therefore, the nonoverlapping intervals are $[66.5, 72.5)$, $[72.5, 78.5)$, $[78.5, 84.5)$, $[84.5, 90.5)$, $[90.5, 96.5)$, and $[96.5, 102.5)$.

Note 4.6

In constructing a histogram, it is important to decide on the number of intervals to select. As a convention, we select the number of interval close to $\sqrt{\text{the number observations}}$.

Example 4.11

As an example for equal-sized interval classes, suppose that the passing score of a test is from 70 to 100 with three grades C, B, and A with scores 70 to 79, 80 to 89, and 90 to 100, respectively, for the particular test that perfect is almost impossible. To find the boundaries, the range is $99 - 70 = 29$. Hence with three classes, the class size is $\frac{29}{3} \approx 10$ Frequency distribution for this test is given in Table 4.4.

Example 4.12

For a given sample size of 250, Table 4.5 is an example of a group organization of data points with six unequal-sized intervals (0-age indicates a not-born child).

Example 4.13

This is an example of a mixed equal-sized and an unequal-sized interval classes. We refer to the Example 4.12 and include other sections' grades to have a larger sample. So, suppose that scores of a test range from 0 to 100 with five grades F, D, C, B, and A with corresponding scores 0 to 59, 60 to 69, 70 to 79, 80 to 89, and 90 to 100. Frequency distribution for this test is given in Table 4.6.

TABLE 4.4
Grouping Data from Example 4.11

Interval No.	Test Score	Classes with Boundaries	Grade	x_i	Frequency (No. of Students Earning the Score) f_i	Interval Size
1	70–79	(69.5–79.5)	C	74.5	35	10
2	80–89	(79.5–89.5)	B	84.5	18	10
3	90–99	(89.5–99.5)	A	94.5	12	10
Total					60	

TABLE 4.5
Grouping Data from Example 4.12

Interval No.	Age Interval (Year)	x_i	Frequency f_i	Interval Size
1	0–11	5.5	35	11
2	12–17	14.5	28	16
3	18–25	21.5	47	8
4	26–40	33	46	15
5	41–64	52.5	69	24
6	65–99	82	25	35
Total			250	

TABLE 4.6
Grouping Data from Example 4.13

Interval No.	Test Score	Grade	x_i	Frequency (No. of Students Earning the Score) f_i	Interval Size
1	0–59	F	29.5	8	60
2	60–69	D	64.5	13	10
3	70–79	C	74.5	35	10
4	80–89	B	84.5	18	10
5	90–100	A	95	12	11
Total				81	

To graphically summarize qualitative data, the bar graph, Pareto chart, and pie charts may be used. For a quantitative data, however, the rest of the graphical options such as histogram, dot plot, stem-and-leaf plot, and box plot may be used.

4.4.5 Histogram

We are distinguishing two cases, namely, values of discrete and continuous random variables.

Definition 4.7

A very common graphical display or presentations of distribution of a numeric data set (observations, as values of a continuous random variable) is referred to as a **histogram**. A histogram displays a grouped data set into **connected, nonoverlapping intervals** (or **bins**). It also shows the **shape** of the data set (observations) as a group. It can assess the **symmetry** or **skewness** of the data set.

The histogram was first introduced by Carl Pearson in 1895. It was constructed from a simple frequency distribution as adjacent rectangles (**bars**).

The bars may be vertical or horizontal. Representations of sides remain the same as mentioned. One side of a bar represents the distance between the endpoints of the bin, referred to as the **width of a bin**.

Carl Pearson
1857 - 1936

Widths may be uniform (equal size) or of different sizes. Let n be the number of data points (observations) and k the number of desired bins. Let us, also, denote the boundaries of a bin (an interval) by b_i and b_{i+1}, $i = 1, 2, ..., k$, for the lower and higher limits, respectively. Then, the width of a bin is $b_{i+1} - b_i$, $i = 1, 2, \ldots, k$. In case the uniformity of widths is desirable, suppose that the range of data set is r. Then, if k uniform bins are desired, the width of each bin will be r/k.

The other side of a bar, that is, the **height of a bar**, represents the frequency of a data point that fall within the bin, say $f_i, i = 1, 2, \ldots, k$, of the ith bin, $i = 1, 2, \ldots, k$, in the general case, and relative frequency (frequency divided by the total number of the data points in the set of data points), $f_i/n, i = 1, 2, \ldots k$, of the ith bin, $i = 1, 2, \ldots, k$ in a particular case. Of course,

$$\sum_{k=1}^{k} f_i = n \text{ and } \sum_{k=1}^{k} \frac{f_i}{n} = 1. \tag{4.1}$$

Definition 4.8

In a sense, a histogram is an approximation of the probability density function (pdf) of the underlying distribution of the data set, when the height of each bin is the relative frequency of the data points in the bin. A histogram created with relative frequencies is referred to as a **relative frequency histogram**.

Steps to **construct** a histogram (vertical or horizontal) or relative frequency histogram are as follows:

Step 1. Determine the number of data points in the data set, say n.
Step 2. Mark the x-axis to some values above the maximum value of the data points.
Step 3. Decide the number of bins you desire, say k (usually, taken as $k \approx \sqrt{n}$).
Step 4. Determine the minimum and maximum values of the set data points.
Step 5. Calculate the range r.
Step 6. Decide what type of bins you desire: with different widths or uniform widths.
Step 7. Calculate the width of bins.
   i.   With desired different widths, take the difference between the upper and lower boundaries: $b_{i+1} - b_i, i = 1, 2, \ldots, k$, or
   ii. With uniform widths, divide the range into equal intervals, that is, r/k.
Step 8. Mark the vertical axis from 0, 1, or the minimum data point value, all the way to the maximum data point value or higher.
Step 9. Determine the frequency for each bin, that is, data points belonging to each bin, $f_i, i = 1, 2, \ldots, k$, of the ith bin, $i = 1, 2, \ldots, k$, or the relative frequency for each bin, $f_i/n, i = 1, 2, \ldots k$, of the ith bin, $i = 1, 2, \ldots, k$
Step 10. Draw a bar (a vertical or horizontal rectangle or a strip) over each bin. The height of a bin is either frequency $f_i, i = 1, 2, \ldots, k$ or relative frequency $f_i/n, i = 1, 2, \ldots, k$.

Note 4.7

The area of a bin is calculated as follows:

i. For a relative frequency histogram with different bar widths:

$$\text{Area of a bin} = \left(b_{i+1} - b_i\right)\frac{f_i}{n}, \quad i = 1, 2, \ldots, k, \qquad (4.2)$$

and

ii. For a relative frequency histogram with uniform widths,

$$\text{Area of a bin} = \frac{rf_i}{k}, \quad i = 1, 2, \ldots, k. \qquad (4.3)$$

Example 4.14 (Histogram Descriptively)

Let the range of a data set be $r = 80$. We divide the set into $k = 5$ uniform bins. That is, the width of each bin will be $\frac{80}{5} = 16$. Finally, suppose the number of data point in each of the interval is 3, 18, 5, 35, and 39, respectively, with a total of $n = 100$ data points. Then, the areas of the rectangles are $3 \cdot 16 = 48, 18 \cdot 16 = 288, 5 \cdot 16 = 80, 35 \cdot 16 = 560,$ *and* $39 \cdot 16 = 624,$ respectively. In this case, the relative frequencies are 0.03, 0.18, 0.05, 0.35, and 0.39, respectively. These relative frequencies represent the probabilities of each bin, sum of which is 1, of course.

Example 4.15   (MATLAB® Simulated)

Using MATLAB codes:

```
x = randn(1000,1);(choosing 1000 natural numbers, randomly, the
    observations)
h = histogram(x)
```

We simulate a set of data points with the following information to create a histogram:

$n = 1{,}000$
Minimum value $= -3.0$
Maximum value $= 3.25$
Range $= 6.25$
Bin width: 0.30 (arbitrarily chosen)
Number of bins $= \dfrac{3:25 - (-0:3)}{0.3} \approx 21$, arranged in ascending order, 1–21
Frequency is the number of times each data point has reoccurred.
Relative frequencies:

$$\text{Relative frequency of a bin} = \frac{\text{Number of times a data point occurs in a bin}}{\text{Total number of data points in the set of data}}$$

$$\text{Height of a bar} \begin{cases} \text{(i) Frequency relative frequency of data points in that bar or} \\ \quad \text{the of a bin depending upon how the illustration is to} \\ \quad \text{be shown, for discrete random variable} \\ \text{(ii) } \dfrac{\text{relative frequency of a bin}}{\text{width of the bin}} \text{, for discrete random variable.} \end{cases}$$

Summary and calculations are illustrated in Table 4.7, and the graph is shown in Figure 4.4.

Note 4.8

A sample point will fall within one of the intervals $[b_{i-1}, b_i)$ or $(b_{i-1}, b_i]$, whether we want to take the half-open interval to be open from the left or from the right; we take it open from the left.

TABLE 4.7
Simulating Numerical Random Data Points and Grouping Them Using Histogram, Example 4.15

Bin No.	Bin (Interval) Boundaries	Frequency	Relative Frequency = Height	Bar's Area = Height * Width
1	[−3.00, −2.70)	1	0.001	(0.001)(0.3) = 0.0003
2	[−2.70, −2.40)	7	0.007	(0.007)(0.3) = 0.0021
3	[−2.40, −2.10)	8	0.008	(0.008)(0.3) = 0.0024
4	[−2.10, −1.80)	10	0.010	(0.010)(0.3) = 0.0010
5	[−1.80, −1.50)	31	0.031	(0.031)(0.3) = 0.0093
6	[−1.50, −1.20)	52	0.052	(0.052)(0.3) = 0.0156
7	[−1.20, −0.90)	87	0.087	(0.087)(0.3) = 0.0261
8	[−0.90, −0.60)	103	0.103	(0.103(0.3) = 0.0309
9	[−0.60, −0.30)	109	0.109	(0.109)(0.3) = 0.0327
10	[−0.30, 0.00)	122	0.122	(0.122)(0.3) = 0.0366
11	[0.00, 0.30)	104	0.104	(0.104)(0.3) = 0.0312
12	[0.30, 0.60)	109	0.109	(0.109)(0.3) = 0.0327
13	[0.60, 0.90)	79	0.079	(0.079)(0.3) = 0.0237
14	[0.90, 1.20)	71	0.071	(0.071)(0.3) = 0.0213
15	[1.20, 1.50)	51	0.051	(0.051)(0.3) = 0.0153
16	[1.50, 1.80)	23	0.023	(0.023)(0.3) = 0.0069
17	[1.80, 2.10)	25	0.025	(0.025)(0.3) = 0.0075
18	[2.10, 2.40)	3	0.003	(0.003)(0.3) = 0.0009
19	[2.40, 2.70)	3	0.003	(0.003)(0.3) = 0.0009
20	[2.70, 3.00)	1	0.001	(0.001)(0.3) = 0.0003
21	[3.00, 3.30]	1	0.001	(0.001)(0.3) = 0.0003
	Sum	1,000	1	0.3

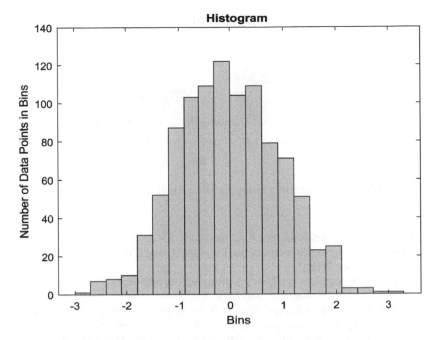

FIGURE 4.4 Creation of a histogram using MATLAB®, for Example 4.15.

Example 4.16 (Data Using MS Excel 2010)

Suppose we are given the following set of data points (Figure 4.5). It could be the daily observation of emission (in tons) of sulfur from an industrial plant about a decade ago in part of the United States, in an ordered data form as:

07.0	07.9	07.9	10.2	11.0
11.0	11.0	11.0	12.7	12.7
12.7	12.7	12.7	17.1	17.1
17.1	16.0	18.3	18.3	18.3
18.3	20.2	20.2	20.2	20.2
21.5	21.5	23.0	23.0	23.0

We want to group this data as a histogram.

Answer

We are to use the Microsoft Excel 2010 to create the requested histogram. Here are the steps:

Step 1. Open an Excel file.
Step 2. Go to the "File" tab, and open the "Option" tab.
Step 3. Then, click on "Add-ins" tab.

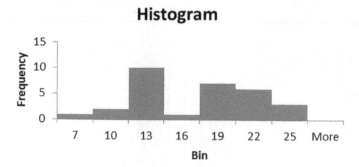

FIGURE 4.5   Construction of a histogram using Excel 2010, for Example 4.16.

Step 4. Choose "Analysis Tool Pak", and click "OK".

Step 5. Now, go to the menu of the Excel sheet, and choose "Insert".

Step 6. Choose "Data".

Step 7. Click on "Data Analysis". A new window appears.

Step 8. Choose "Histogram", and click "OK".

Step 9. In the "Input Range", type in $ (insert the column of the first data point, say A) $ (insert the row number of the first data point, say 5): $ (insert the column of the last data point, should be A) $ (insert the row number of the last data point, say 34).

Step 10. In the "Bin Range", type in $ (insert the column of the first data bin, say C) $ (insert the row number of the first data bin, say 5): $ (insert the column of the last data bin, should be C) $ (insert the row number of the last data bin, say 11).

Step 11. Check mark "Pareto (sorted histogram)", "Cumulative Percentages" (optional), and "Chart output", and then hit "OK".

Step 12. To delete the legend, right-click on a bar and press "Delete".

Step 13. To remove the space between the bars, right-click a bar, click "Format Data Series", and change the "Gap Width" to "0%".

For more details, see, for instance, https://www.excel-easy.com/examples/histogram.html.

Bin	Frequency
7	1
10	2
13	10
16	1
19	9
22	6
25	3
More	0

Example 4.17 (Data Using R)

Let us consider the following data set.

55	57	61	59	55
59	63	63	63	63
61	55	57	55	59
65	59	61	63	61
55	63	63	57	55
59	55	55	65	59

We want to group this data as a histogram (Figure 4.6)

Answer

We follow the same steps as mentioned in Example 4.16.

Bin	Frequency	Cumulative %	Bin	Frequency	Cumulative %
54–56	8	26.67%	55	8	26.67%
56–58	3	36.67%	63	7	50.00%
58–60	6	56.67%	59	6	70.00%
60–62	4	70.00%	61	4	83.33%
62–64	7	93.33%	57	3	93.33%
64–66	2	100.00%	65	2	100.00%

Example 4.18 (Data Using Minitab)

Again let us consider the following data set:

6	6	4	3	6
6	5	5	6	7
7	7	7	4	2
4	4	6	3	1
4	2	5	4	1
3	3	6	5	1

We want to group this data as a histogram using Minitab.

Answer

We use the following steps on the Minitab statistical software. We note, however, that for our example, we will be using Minitab 14, although its current version is 18.

Histogram

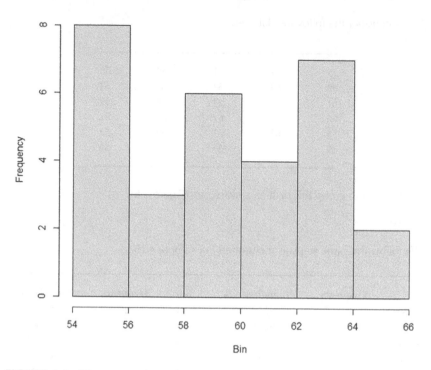

FIGURE 4.6   Histogram using R for Example 4.17.

Steps to create a histogram using Minitab are as follows:

Step 1. Insert the data points in a Minitab worksheet's column, and name this column in the gray box just above the data points, say "Data Points".

Step 2. On the menu bar, click on "Graph", click on "Histogram", "Simple", and then, "OK".

Step 3. Choose the type of graph you wish, "Simple" or any other, and then, click on "OK". We for our example have chosen with normal curve. The normal pdf is an approximating curve to fit the histogram.

Step 4. Choose the variable you want to graph, that is, "Grouped Data", and click "Select".

Step 5. Click on the "Labels…" box, and type in a description of the data in the box next to "Title". For example, you may want to insert a title, like "Histogram".

Step 6. If you have a Microsoft Word document open, you can right-click on the graph and choose "Send Graph to Microsoft Word", and the histogram will appear in your Word document.

For more details, the reader may Google for steps. For instance, https://www2.southeastern.edu/Academics/Faculty/dgurney/Math241/StatTopics/Minitab%20Stats/MntbHistogram.pdf.

4.4.6 STEM-AND-LEAF PLOT

As mentioned before, grouping data points like the histogram will cause losing the individual characteristics of data points. To avoid this problem, in 1977, John W. Tukey in his book *Exploratory Data Analysis* introduced **stem plots**, or **stem-and-leaf plot**, as it is known, nowadays, as an alternative method of graphically organizing numerical data that will not lose the individual characteristics of data points. This method is based on Arthur Bowl's method developed in 1900s. It is mainly used for large data sets such as tracking series of scores on sports teams, temperatures over a period of time, and series of test scores.

Interest in using stem-and-leaf plot in 1980s, in addition to its capability of holding on to the individual data point's characteristics, was its capability using the typewriter those days that leads to using computer later. However, these days, there are computer software packages that can produce graphical presentation of data that the popularity of stem-and-leaf plot has almost vanished.

In a sense, a stem-and-leaf plot can be thought of as a special case of a histogram in such a way that the frequency tallies in each bin as in a histogram and each individual value in the bin. This is how the information is preserved. Hence, in creating the plot stems take the place of bins and leaves take the place of the bars.

Note 4.9

We have to caution the reader that the term "stem plots" is used only by some computer software packages like MATLAB. Hence, the use of the stem-and-leaf plot is recommended, if at all necessary.

To display a stem-and-leaf plot, each data point is considered as a combination of a "stem" and a "leaf".

Step 1. Sort the data points according to their first digits in ascending (or descending) order.

Step 2. Create two vertical column tables: the left column for the stems and the right column for the leaves.

Step 3. Split each number into two parts. The first part is the first digit on the left, which will be the stem, and the second part contains all other digits of the number, which will be the leaves. The leaves do not have to be sorted.

Step 4. Put the stem in the first column and the leaves with space in between in the second column. In other words, the "stem" values are listed down, and the "leaf" values go right from the stem values.

Example 4.19

Consider the following test scores of 30 participants of a test (the scores are 1, 2, 3, 4, and 5, with 1 as the lowest and 5 as the highest):

2.2	1.7	3.3	2.4	4.1
5.0	3.7	2.2	3.8	3.2
4.7	3.7	1.9	3.5	3.0
1.5	4.6	4.7	3.3	3.8
2.9	2.9	3.9	2.3	2.0
4.3	4.6	2.8	3.8	3.2

We want to create a stem-and-leaf plot.

Answer

Stem							Leaf					
1	7	9	5									
2	2	4	2	9	9	3	0	8				
3	3	7	8	2	7	5	0	3	8	9	8	2
4	1	7	6	7	3	6						
5	0											

Hence, the value of "stem 3 and leaf 7" is 3.7. In this case, the leaves are decimal numbers. At a glance, the plot shows that 11 participants received scores below 3 that may not be a passing score, if the scores are related to letter grades, denoted by g, as A, B, C, D, and F as follows:

A	5
B	$4 \leq g < 5$
C	$3 \leq g < 4$
D	$2 \leq g < 3$
F	$g < 2$

4.4.7 DOT PLOT

Another simple way to display a grouped data is called the **dot plot** (or in R programming language terminology, **strip plot or strip chat**). This plot can be displayed similar to the bar graph with dots replacing the bars and is usually good for categorical data set. However, dot plots can include outliers. It can be displayed without bars. The method is a "had-drawn" before technology appeared in practice. It is used for small sample sizes. For large sample sizes, like 20 or more, perhaps other methods like the histogram are preferred. However, we should be cautious that dot plots may not be spaced uniformly along the horizontal axis.

Steps to create dot plots with Excel (we are using Excel 2010 for our illustration), that is, making dot plots using Minitab, are as follows:

Step 1. Put your data in one of the columns of the Minitab worksheet.
Step 2. Add a variable name in the gray box just above the data values.

Step 3. Click on "Graph", and then click on "Dotplot...".

Step 4. Make sure "Simple" is highlighted under "One Y", and click on "OK".

Step 5. Choose the variable you want to graph, and click on "Select".

Step 6. Click on the "Labels" box.

Step 7. Under "Title:", type in a description for your dot plot.

Step 8. Under "Subtitle 1:", state that you created the graph.

Step 9. Click on "OK" in that window, and click on "OK" in the next window.

Step 10. Double-click on the axis title. A window will appear. The axis title will be in a box near the bottom of the window under the word "Text".

You may edit the axis title in this box, and then if you click on "OK", the axis title on the graph will show your changes.

Step 11. If you have a Microsoft Word document open, you can right-click on the graph and choose "Send Graph to Microsoft Word" and the histogram will appear in your Word document. Otherwise, click in the gray area outside the graph.

Choose the "Edit" tab at the very top of the window, and then choose "Copy Graph".

Example 4.20

We consider the weather temperature, in Fahrenheit (maximum and minimum), of the city of Houston in Texas, the United States, for a 30-day period from November 20, 2018, through December 28, 2018, chosen from https://www.accuweather.com/en/us/houston-tx/77002/december-weather/351197 (Table 4.8; Figure 4.7).

It can be seen that graphical techniques are helpful to have an overview of data and the shape of its distribution. Graphs sometimes do not display the exact data points. They will rather show only intervals, and if exact information is needed, the theory can assist.

4.5   MEASURES OF CENTRAL TENDENCY

Definition 4.9

When a set of observations is in hand, the most trivial question that comes to mind is finding the midpoint or the **center** of the values. But how do we find the center? The center is found in different ways, each with a separate property. These centers are called the **measures of central tendency**. The ones we will discuss are the most popular ones, and they are

  i. The arithmetic mean or the balance point or center of sample points, or sample mean,
  ii. The proportion,
  iii. The median,
  iv. The mode.

TABLE 4.8

Data Regarding Weather Temperature of Houston, Texas, during the Period of 11/29/2018, Example 4.20

	Date	High	Low
1	Nov. 29, 2018	76	66
2	Nov. 30, 2018	81	68
3	Dec. 01, 2018	81	66
4	Dec. 02, 2018	79	59
5	Dec. 03, 2018	68	52
6	Dec. 04, 2018	61	45
7	Dec. 05, 2018	63	43
8	Dec. 06, 2018	64	52
9	Dec. 07, 2018	75	55
10	Dec. 08, 2018	55	45
11	Dec. 09, 2018	48	43
12	Dec. 10, 2018	61	41
13	Dec. 11, 2018	64	45
14	Dec. 12, 2018	68	54
15	Dec. 13, 2018	75	50
16	Dec. 14, 2018	50	45
17	Dec. 15, 2018	64	48
18	Dec. 16, 2018	68	50
19	Dec. 17, 2018	72	55
20	Dec. 18, 2018	63	50
21	Dec. 19, 2018	61	55
22	Dec. 20, 2018	63	54
23	Dec. 21, 2018	64	45
24	Dec. 22, 2018	77	54
25	Dec. 23, 2018	70	55
26	Dec. 24, 2018	66	48
27	Dec. 25, 2018	73	61
28	Dec. 26, 2018	70	63
29	Dec. 27, 2018	73	54
30	Dec. 28, 2018	60	47

Note 4.10

Although different types of measures of central tendencies inform us about a data set by a number or two, individual information is lost, which is the price to pay.

We define each of these statistics for a sample as below.

i. Sample Arithmetic Mean

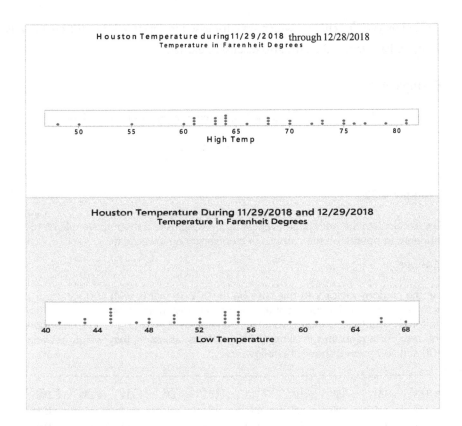

FIGURE 4.7   Bar pie chart using Excel for Example 4.20.

Definition 4.10

Consider a random sample $X_1, X_2, ..., X_n$. Then, the statistic **sample mean** or **arithmetic mean** or **arithmetic average**, denoted by $\bar{x}$ (read as: x-bar), is defined as

$$\bar{x} = \frac{1}{n} \sum_{k=1}^{n} x_k. \tag{4.4}$$

The essence of Definition 4.6 is based on sampling from a discrete uniform probability mass function (pmf) and finding its expected value. It states that the arithmetic average of n items is obtained by adding the n sample values and dividing the total by n.

Note 4.11

The sample mean may be extended to population mean, which is the expected value of the underlying distribution describing the population. The population mean is usually denoted by μ. For a discrete pmf, $p_X(x)$, the population mean is defined by $\mu = \sum xp_X(x)$, which was discussed in Chapter 3. For the continuous case, we have to use the pdf, also discussed earlier.

Example 4.21

In rolling a fair die once, the arithmetic average of number of dots will be

$$\frac{1+2+3+4+5+6}{6} = 3.5.$$

Example 4.22

Let us calculate the arithmetic average of the weights of a random sample of 100 students, in pound, on the campus of a university on a weekday.

Answer

Let the random vector **x** in this case be the 10£ 10 matrix with 100 values, that is, a vector of 100 independent and identically distributed random variable (iid) uniformly distributed random variables denoted by $X = (x_1, x_2, \ldots, x_{100})$. Suppose the result of a random sampling of 100 students (assuming they weigh between 100 and 200 pound) shows the following:

163	161	183	102	118	147	136	111	183	186
114	119	200	159	195	159	198	125	131	191
157	114	138	173	140	108	114	115	160	184
140	177	119	165	114	150	178	182	147	175
105	165	200	106	154	178	136	163	166	126
113	134	168	115	103	113	128	166	126	195
139	105	194	176	155	170	154	167	128	185
146	142	164	150	140	183	141	155	169	185
133	146	165	138	176	151	192	128	198	178
186	154	138	129	116	100	136	109	128	147

Each $x_i, i = 1,2,\ldots,100$, is one of the numbers in the set above. For the sake of example, we created the matrix by the MATLAB program using the code: "x = randi([90 200],10,10)". To find the arithmetic average, $\bar{x}$, we use MATLAB code "x _ bar = mean(x,'all')". Hence, the arithmetic average (4.4), for this example, is $\bar{x} = 149.87$ pounds.

Definition 4.11

The sample mean may be obtained by grouping the data. That is, for a sample of size n, the sample points x_i, $i = 1,2,\ldots n$, may be repeated values. In such a case, the data may be grouped according to the values. The number of times a data point is repeated is referred to as the **frequency**. The ratio of a frequency to the total sample points is called the **relative frequency**. Multiplying a relative frequency by 100 yields what is called a **percent relative frequency**. For a simple frequency table, the sample mean (4.4) becomes:

$$\bar{x} = \frac{1}{n}\sum_{i=1}^{k} f_i x_i,$$ (4.5)

where x_i $i = 1,2,\ldots,n$, represent the data points and $f_i, i = 1,2,\ldots,k$, $k \le n$, are their corresponding frequencies.

Example 4.23

Let us use MATLAB to generate 25 numbers between 1 and 10, randomly (data points), that is, $n = 25$, with repetition and call the matrix obtained as **X**, whose elements are x_i, $i = 1,2,\ldots,25$. Then, we want to find

  i. The frequencies,
  ii. The relative frequencies,
  iii. The percent relative frequency,
  iv. The mean using the frequencies.

Answer

We try to answer all four questions, using the following MATLAB code:

```
x = randi(10,5,5)
N = unique(x);
Freq_Table = [N,histc(x(:),N)]
Freq_Table_C1=Freq_Table(:,1);
Freq_Table_C1'
Freq_Table_C2=Freq_Table(:,2);
Freq_Table_C2'
```

Thus, we obtain the following:

X = A Set of Observations (Raw Data)				
7	8	5	4	1
7	3	7	5	7
2	9	9	8	7
5	9	7	9	5

As we see from the data above, only numbers from 1 to 9 were called. We now tabulate the results in Table 4.9.

Hence, from (4.5), we have $\bar{x} = \dfrac{146}{25} = 5.84$.

In case grouped data is available, the mean value may be calculated.

TABLE 4.9
Data from Example 4.23

											Sum
x_i = randomly chosen numbers	1	2	3	4	5	6	7	8	9	10	
f_i = frequencies of numbers	1	1	3	2	5	1	5	3	40		25
Relative frequencies	0.04	0.04	0.12	0.08	0.2	0.04	0.2	0.12	0.16	0	1
% Relative frequencies	4	4	12	8	20	4	20	12	16	0	100
$f_i \cdot x_i$	1	2	9	8	25	6	35	24	36	0	146

Example 4.24

We return to an example we earlier considered. For a randomly chosen 25 rational numbers, we obtain Table 4.10, from which we can calculate the sample mean.

Thus, $\bar{x} = \dfrac{346}{25} = 13.86$.

Existence of outliers in real data is a fact. The sample mean value is changed with the existence **of** outliers. If there is an outlier, which is extremely larger than the rest of data points, then the sample mean will be large, too. Similarly, if the outlier is an extremely smaller value, this will make the sample mean very small. To obtain a more consistent analysis of data, it is customary to **trim or truncate the data.** If there are a larger number of data points, the trimmed mean will be a better measurement of the center of the data than the sample mean. Data **trimming** means sorting the data in ascending order and cutting or discarding it by 5%–25% from both ends. This process will affect some measures of the data set. For instance, the mean of the trimmed mean is referred to as the **trimmed mean** or **truncated mean.** The statistical software Minitab uses the code **TrMean** to calculate the trimmed mean.

Usually, the trimmed mean is used if there are enough data points. Otherwise, it is not used for a small sample of data points. For instance, suppose we have 100 observations. Then, 5% trimmed mean means that we sort the data and remove the lowest 5 data points and the highest 5 data points, and then average the remaining of 90 data points.

TABLE 4.10
Grouping Data from Example 4.24

i	Data Point x_i	Frequency f_i	Relative Frequency	Percent Relative Frequency	$f_i \cdot x_i$
1	11.3	5	1/5	20	56.5
2	12.4	4	4/25	16	49.6
3	13.6	3	3/25	12	40.3
4	14.8	6	6/25	24	88.8
5	15.9	7	7/25	28	111.3
Total		25	1	100	346.5

Example 4.25

Consider an ambulance dispatching service. It is possible that mistakes are made and response times may be with long delays; the dispatcher forgets to record some calls, that is, response times 0, or the dispatcher forgets to stop the response times, that is, extremely delayed response times. In such cases for a measure of tendency, the trimmed mean is obtained from the sorted data by discarding possible 10% from each end, which is called the **10% trimmed mean.**

ii. Sample Proportion

In statistics, proportion is a percentage of a total in which a certain characteristic is observed. Essentially, a proportion of a characteristic that is observed j times in a sample of size n is defined as $\frac{j}{n}$. Similarly, for a finite population of size N, the population proportion is defined as $\frac{k}{n}$, where k is the number of characteristics observed in the population. We will discuss the sample proportion in detail in Chapter 5.

iii. Sample Median

In order to define the median, we need to define some terms.

Occasionally, it is helpful to **order the data points** (in this case, the leaves in each row) from small to large. Generally, we order the data points as in the following definition:

Definition 4.12

Suppose we have a sample with a set of n data points (observations) $x_1, x_2, \ldots, x_n$. We order these data points from small to large. The resulting set of ordered data points is referred to as the **ordered statistics** of the sample. We may rank these ordered data points and denote them by a symbol.

Example 4.26

Suppose we have a sample that has 5 points 23, 18, 12, 44, and 15. We can order them as 12, 15, 18, 23, and 44. Then, we can rank them as $y_1 = 12, y_2 = 15, y_3 = 18, y_4 = 23, y_5 = 44$. Hence, we will have $y_1 < y_2 < y_3 < y_4 < y_5$.

Definition 4.13

The median of a sample is the point of the random variable, which splits the distribution into two parts. Let us consider a random sample of size n, numbered and arranged (sorted) in ascending or descending order, say $X_1, X_2, \ldots, X_n$, whose values, that is, observations, are, respectively, as $x_1, x_2, \ldots, x_n$. If n is odd, the **median** of the sample points, denoted by m, is the sample data right in the middle of the observations, that is,

$$m = \frac{n+1}{2}. \tag{4.6}$$

However, if n is an even natural number, the median is the average of the middle two data points, that is,

$$m = \frac{1}{2}\left(\frac{n}{2} + \frac{n+2}{2}\right). \tag{4.7}$$

iv. Sample Mode

Definition 4.14

The most frequent data point is referred to as the **mode**. However, there might be more than one mode within the set of observations. In such a case, for two modes, the set is referred to as the **bimodal**, and for more than two modes, the set is referred to as the **multimodal**.

Example 4.27

Consider the data set 3, 7, 6, 8, 6, 3, 3, 4, 2. Let us find its mode.

Answer

Looking at the data, we see that the value 3 has been repeated three times. Hence, it is the mode of the data.

As mentioned before, grouping will cause losing the individual characteristics, and hence, finding mode is not possible any longer. For this reason, the class with the highest frequency is taken, and it is referred to as the **modal class**. Of course, it is possible that more than one class is with the same highest frequency. In that case, the sample will have a **multiple modal class**.

Example 4.28

For a given sample of size 250, Table 4.11 is an example of a group organization of data points with 6 unequal-sized intervals (0-age indicates a not-born child). Modal class in this example is the interval [41, 69). From Table 4.11,

$$\bar{x} = \frac{10849.5}{250} = 43.40.$$

Example 4.29

This is an example of a mixed equal-sized and unequal-sized interval classes. We refer to the Example 4.13 and include other sections' grades to have a larger sample. So, suppose that scores of a test range from 0 to 100 with five grades F, D, C, B, and A with corresponding scores 0 to 59, 60 to 69, 70 to 79, 80 to 89, and 90 to 100. Frequency distribution for this test is given in Table 4.12.

TABLE 4.11
Grouping Data from Example 4.28

Interval No.	Age Interval (Year)	x_i	Frequency f_i	Interval Size	$f_i \cdot x_i$
1	0–11	5.5	35	11	192.5
2	12–17	14.5	28	16	406.0
3	18–25	21.5	47	8	1,010.5
4	26–40	33	46	15	1,518.0
5	41–64	52.5	69	24	3,622.5
6	65–99	82	25	35	2,050.0
Total			250		10,849.5

TABLE 4.12
Grouping Data from Example 4.29

Interval No.	Test Score	Grade	x_i	Frequency (No. of Students earning the score) f_i	Interval Size	$f_i \cdot x_i$
1	0–59	F	29.5	8	60	236.0
2	60–69	D	64.5	13	10	838.5
3	70–79	C	74.5	35	10	2,607.5
4	80–89	B	84.5	18	10	1,521.0
5	90–100	A	95	12	11	1,134.0
Total				81		6,337.0

Modal class in this example is [70, 79). From Table 4.12, $\bar{x} = \dfrac{6337}{81} = 78.23$

4.6 MEASURE OF RELATIVE STANDING

So far we discussed the measures of central tendency that help us understand the location of the center of a set of data. In practice, when we have a large collection of data sets, we may need to compare two data points that may be in the same data set or in two different data sets. In that case, we may use measurements such as **percentile**, **quartile**, or **z-scores**. Such measures are referred to as the **measure of relative standing**. We will define these terms below.

4.6.1 PERCENTILE

Definition 4.15

A **percentile** is a good application of ordering. Let us consider a sample of size n ordered in ascending order, denoted by S. Let also p be a number between 0 and 1, exclusive, that is, $0 < p < 1$. Then,

a. **If $(n+1)p$ is an integer,** then the **(100p)th sample percentile** is a sample point at which there are approximately np sample points that fall below (left of) it and $n(1-p)$ sample points that fall above (right of) it. In other words, the area of the left side of the distribution is p and the area of the right side is $(1-p)$, as shown in Figure 4.8.

b. **If $(n+1)p$ is not an integer,** but can be written as a whole number plus a proper fraction, of the form $r+a$, where r is the whole part and a is the proper fraction part, then the **(100p)th sample percentile** takes the weighted average of the rth and $(r+1)$st ordered data points.

$$(1-a)\mathrm{D}_{r+a} \cdot \mathrm{D}_{(r+1)}$$

c. **If $(n+1)p < 1$,** the **sample percentile is not defined.**
In **other** words, the points that divide the set of data points (observations), S, into one hundred equal parts, say C_i $i=1,2,\ldots,99$, are called the **percentiles**. $C_1, C_2, \ldots, C_{99}$ are called the **first, second, ..., ninety-ninth percentile**, respectively.

The percentiles are usually used for a very large number of data points. In other words, $C_1 = $ the value of $\left(\dfrac{n}{100}\right)$th data point, $C_2 = $ the value of $\left[2\left(\dfrac{n}{100}\right)\right]$th data point, $C_3 = $ the value of $\left[3\left(\dfrac{n}{100}\right)\right]$th data point, $\ldots$, $C_{99} = $ the value of $\left[99\left(\dfrac{n}{100}\right)\right]$th data point.

As a special case of percentile, we define the decile. If the sorted sample data, S, is divided into 10 equal-sized parts, each of the 9 parts, say D_i, $i=1,2,\ldots,9$, is called a **decile**. Thus, each decile contains 10% of the data points (observations) contained in S.

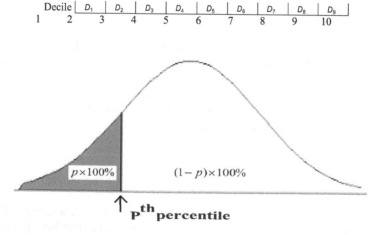

FIGURE 4.8   (100p)th sample percentile, when $(n+1)$ is an integer.

In other words, D_1 = the value of $1\left(\dfrac{n+1}{10}\right)$ th data point, D_2 = the value of $\left[2\left(\dfrac{n+1}{10}\right)\right]$ th data point, D_3 = the value of $\left[3\left(\dfrac{n+1}{10}\right)\right]$ th data point, ..., D_9 = the value of $\left[9\left(\dfrac{n+1}{10}\right)\right]$ th data point.

4.6.2 QUARTILE

Yet another special case of a percentile is the **quartile**. If the sorted sample data, S, is divided into 4 equal-sized parts, each of the 3 parts is called a **quartile**. Thus, the **first quartile**, denoted by Q_1 or the **lower quartile**, is the number that 25% of the lowest data points fall below. The **second quartile** (also the **50th percentile**) of the sample, denoted by Q_2, is also called the **median**. This is the number that 50% of the lowest data points fall below. Similarly, the **third quartile**, denoted by Q_3, also referred to as the **higher quartile**, is the number that 75% of the lowest data points fall below and 25% fall above.

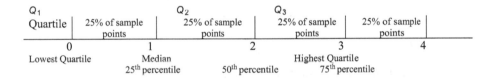

In a set of **data points, S, the interquartile range, denoted by IQR, is a measure of where the "middle** fifty" is. In other words, IQR is the range of the middle 50% of the data points. It is calculated as the difference between Q_3 and Q_1, that is, $Q_3 - Q_1$.

As seen above, quartiles are a special case of percentiles. We will illustrate how to calculate the percentiles and quartiles in the following examples.

Example 4.30

Let us return to the Example 4.23 with data set, S, as:

S = A Set of Data Points				
7	8	5	4	1
7	3	7	5	7
2	9	9	8	7
5	9	7	9	5

We sort S as follows:
1, 2, 3, 4, 5, 5, 5, 5, 5, 7, 7, 7, 7, 7, 8, 8, 9, 9, 9, 9

As we can see:

a. The minimum = 1,
b. The maximum = 9,
c. The range = 9 − 1 = 8,
d. $Q_1 = 5$,
e. $Q_2 = 7$,
f. $Q_3 = 8$,
g. $Q_3 − Q_1 = 8 − 5 = 3$.

Example 4.31

Let us consider another data set:

77	72	83	66	59
97	73	43	61	91
54	47	70	65	90
69	76	60	38	74
76	58	73	75	93

Suppose we wish to:

1. Create the stem-and-leaf plot display.
2. Find the first, second, and third quartiles, Q_1, Q_2, and Q_3, respectively.
3. Find the 60th and 90th percentiles.
4. Find the *IQR*.

Answer

For (1), the ordered stems and leaves (which are referred to as the **ranked stem-and-leaf display**) are shown in Table 4.13.

TABLE 4.13
Stem-and-Leaf Plot, Example 4.31

Stem			Leaf							Frequency	Cumulative Frequency
3	8									1	1
4	3	7								2	3
5	4	8	9							3	6
6	0	1	5	6	9					5	11
7	0	2	3	3	4	5	6	6	7	9	20
8	3									1	21
9	0	1	3	7						4	25

For (2a), we find the first quartile, Q_1, and the percentiles as follows:
Q_1 is the 25th percentile. Thus, $p = 0.25$ and $(n+1)p = (26)(0.25) = 6.50 = 6 + 0.50$. Hence, $r = 6$ and $a = 0.50$. Therefore,

$$\tilde{\pi}_{0.25} = (1 - 0.5)D_6 + 0.5D_7 = (0.5)(59) + (0.5)(60) = 59.5.$$

For (2b), Q_2 is the 50th percentile. Thus, $p = 0.50$ and $(n+1)p = (26)(0.50) = 13$. Hence,

$$\text{the 13th ordered data point is 72 and } Q_2 = \tilde{\pi}_{50} = 72.$$

For (2c), Q_3 is the 75th percentile. Thus, $p = 0.75$ and $(n+1)p = (26)(0.75) = 19.5 = 19 + 0.50$. Thus, $r = 19$ and $a = 0.50$. Hence,

$$\tilde{\pi}_{0.75} = (1 - 0.5)D_{19} + 0.5D_{20} = (0.5)(76) + (0.5)(77) = 76.5.$$

For (3a), the 60th percentile, $p = 0.60$ and $(n+1)p = (26)(0.60) = 15.60 = 15 + 0.60$. Thus, $r = 15$ and $a = 0.60$. Hence,

$$\tilde{\pi}_{0.60} = (1 - 0.6)D_{15} + 0.6D_{16} = (0.4)(73) + (0.6)(74) = 73.6.$$

For (3b), the 90th percentile $p = 0.90$ and $(n+1)p = (26)(0.90) = 23.40 = 23 + 0.40$. Thus, $r = 23$ and $a = 0.40$. Hence,

$$\tilde{\pi}_{0.90} = (1 - 0.6)D_{23} + 0.6D_{24} = (0.4)(91) + (0.6)(93) = 92.2.$$

For (4), the $IQR = 76 - 59.5 = 16.5$.

We may interpret (a) through (e) as "approximately 25%, 50%, 75%, 60%, and 90% of the sample points are less than 59.5, 72, 76.5, 73.6, and 92.2, respectively". The IQR in (f) tells us that about 16.5% of data points fall between the first and the third quartiles.

Note 4.12 Calculating Percentile

To calculate a sample percentile, whether the population from which the sample is taken is known or not known. If it is known and easy to work with, the calculation is easy. However, if it is known, but difficult to work with, the sample points may be used. "Numerical approach" is another method to use. In this case, there are a variety of software methods such as Tables, MATLAB, Minitab, R, and the Calculator.

Example 4.32

Let X be a continuous random variable with pdf as follows:

$$f_X(x) = 2x, \quad 0 < x < 1.$$

To calculate the 90th percentile of this pdf, we denote the 90th percentile by x_0. Then, we have to find $P(X < x_0) = 0.90$. Using the given pdf, we will have:

$$P(X < x_0) = \int_0^{x_0} f(x)dx = \int_0^{x_0} 2x\,dx = x^2\Big|_0^{x_0} = 0.9.$$

Hence,

$$x_0^2 = 0.90 \rightarrow x_0 = 0.95.$$

Therefore, the 90th percentile is 0.95. That is, 90% of the data points are less than or equal to 0.95.

Example 4.33

For a community at random, it is determined that the family income can be approximated by a normal distribution with mean and variance as $2,000 and $10,000, respectively. Calculate the 95th percentile of the income of a family in that community and interpret it.

Calculation
Let X represent the income of a family in the community. Hence, X is normally distributed with $v = 2000$ and $\sigma = 100$. Let us also denote the 95th percentile by x_0.

a. **Using the Calculator**
   Since X is normally distributed, we use the inverse normal command "invNorm" on a "graphing calculator" with statistics feature, to calculate the wanted percentile. Here is how the command is used:

$$x_0 = invNorm(p, \mu, \sigma),$$

where p represents the decimal value of the percentile. This will lead to

$$x_0 = invNorm(0.95, 2000, 100) \approx \$2,164.50.$$

b. **Using the Z-Table**
   Based on the desired 95% percentile, we have $P(X < x_0) = 0.5$. We may convert the normal distribution into standard normal distribution as follows:

$$P(X < x_0) = 0.95,$$

$$P\left(\frac{X - \mu}{\sigma} < \frac{x_0 - \mu}{\sigma}\right) = 0.95,$$

$$P\left(Z < \frac{x_0 - 2,000}{100}\right) = 0.95,$$

$$P(Z < z_0) = 0.95,$$

where

$$z_0 = \frac{x_0 - 2,000}{100}.$$

According to the Z-table, z_0 value occurs exactly between $z = 1.64$ and $z = 1.65$. Hence, the middle value will be taken as the z_0 value. Having the z_0 value as 1.645,

from

$$Z_0 = \frac{x_0 - 2,000}{100} = 1.645, \text{we have } x_0 = 2,000 + (1.645)(100) = 2,164.5.$$

z	0.00	0.01	0.02	0.03	0.04	0.05	0.06
0.0	0.5000	0.5040	0.5080	0.5120	0.5160	0.5199	0.5239
0.1	0.5398	0.5438	0.5478	0.5517	0.5557	0.5596	0.5636
0.2	0.5793	0.5832	0.5871	0.5910	0.5948	0.5987	0.6026
0.3	0.6179	0.6217	0.6255	0.6293	0.6331	0.6368	0.6406
0.4	0.6554	0.6591	0.6628	0.6664	0.6700	0.6736	0.6772
0.5	0.6915	0.6950	0.6985	0.7019	0.7054	0.7088	0.7123
0.6	0.7257	0.7291	0.7324	0.7357	0.7389	0.7422	0.7454
0.7	0.7580	0.7611	0.7642	0.7673	0.7704	0.7734	0.7764
0.8	0.7881	0.7910	0.7939	0.7967	0.7995	0.8023	0.8051
0.9	0.8159	0.8186	0.8212	0.8238	0.8264	0.8289	0.8315
1.0	0.8413	0.8438	0.8461	0.8485	0.8508	0.8531	0.8554
1.1	0.8643	0.8665	0.8686	0.8708	0.8729	0.8749	0.8770
1.2	0.8849	0.8869	0.8888	0.8907	0.8925	0.8944	0.8962
1.3	0.9032	0.9049	0.9066	0.9082	0.9099	0.9115	0.9131
1.4	0.9192	0.9207	0.9222	0.9236	0.9251	0.9265	0.9279
1.5	0.9332	0.9345	0.9357	0.9370	0.9382	0.9394	0.9406
1.6	0.9452	0.9463	0.9474	0.9484	0.9495	0.9505	0.9515
1.7	0.9554	0.9564	0.9573	0.9582	0.9591	0.9599	0.9608
1.8	0.9641	0.9649	0.9656	0.9664	0.9671	0.9678	0.9686
1.9	0.9713	0.9719	0.9726	0.9732	0.9738	0.9744	0.9750

That is, the 95th percentile is $2,164.5. This means that 95% of the incomes of families in this community are less than or equal to $2,164.5.

Example 4.34 Calculating Quartile

Suppose the lifetime (in years) of a car tire can be approximated as an exponential distribution with parameter $\frac{1}{3}$. Calculate the third quartile and interpret the result.

Calculation
Let X be the lifetime of the tire. Then, the pdf of X is:

$$f_X(x) = \frac{1}{3} e^{-\frac{1}{3}x}, \quad x \geq 0.$$

Also, let x_0 be the third quartile.

$$P(X < x_0) = \int_0^{x_0} f_X(x)\,dx = \int_0^{x_0} \frac{1}{3} e^{-\frac{1}{3}x}\,dx = \frac{1}{3}\left[\frac{e^{-\frac{1}{3}x}}{-\frac{1}{3}}\right]_0^{x_0} = e^{-\frac{1}{3}x_0} = 0.75.$$

Thus,

$$-\frac{1}{3}x_0 = \ln(0.75), \text{ giving } x_0 = -\ln(0.75) = 0.8630.$$

So, under the assumptions of the problem, the lifetime of 75% of the car tires will be less than or equal to 0.8630 years, that is, about 10 months and 10 days.

Example 4.35 Calculating Quartile

Suppose the thickness or depth of a protective paint applied on a metal surface is uniformly distributed on [20, 30] microns. Calculate the lower quartile for the thickness of the paint. What does this value mean?

Calculation
Let X be the random variable representing the thickness of the paint. Since X is uniformly distributed on the interval [20, 30], the pdf of X is:

$$f_X(x) = \frac{1}{30-20} = \frac{1}{10}, \quad 20 < X < 30.$$

Let x_0 be the first quartile, Q_1. Then,

$$P(X < x_0) = \int_{20}^{x_0} f_X(x)\,dx = \int_{20}^{x_0} 0.1\,dx = 0.1x\big|_{20}^{x_0} = 0.25.$$

That is,

$$10(x_0 - 20) = 25, \text{ giving } x_0 = 22.5.$$

Therefore, the 25th percentile is 22.5. That is, 25% of the data points are less than or equal to 22.5 μm.

The fifth **decile** of the sample is also called the **median** (this is because there are five parts on each side of the fifth mark).

The second **quartile**, also the **50th percentile**, of the sample is also called the **median** (this is the number that 50% of the lowest data points fall below it).

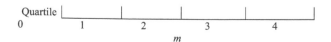

Example 4.36

Suppose we are given two sets of the data as:

1. 3, 7, 6, 8, 6, 3, 3, 4, 2,
   and
2. 4, 1, 4, 6, 9, 7, 1, 9, 2, 7.

For each data set, calculate the median, the fifth decile, and the second quartile.

Answer

We first sort each of the data set in ascending order. The sorted samples are as follows:

2, 3, 3, 3, 4, 6, 6, 7, 8,
   and
1, 1, 2, 4, 4, 6, 7, 7, 9, 9.

For set (1):
   Having 9 data points, the middle point, that is, the median, is 4. Deciles start from 0.9 and increasingly move by 0.9, as follows.

		0.9	1.8	2.7		3.6		4.5		5.4		6.3		7.2		8.1		
Decile		1	2	3		4		5		6		7		8		9		10
Data				2		3, 3, 3		4				6, 6		7		8		

Median *m* (fifth decile)

For deciles, dividing 9 by 4 gives the first decile as 2.25. Then, adding 2.25, we obtain the median and the higher decile.

		2.25		4.5		6.75		8
Quartile		1		2		3		4
Data	2		3, 3, 3, 4		6, 6		7, 8	

Median *m* (second quartile)

For set (2):
   Having 10 data points, the middle 2 points are 4 and 6, and the average of those points, that is, the median, is 5. Deciles start from 1 and increasingly move by 1, as follows.

		1	2	3		4		5		6		7		8		9	10
Decile		1	2	3		4		5		6		7		8		9	10
Data	1, 1	2			4, 4				6		7, 7				9, 9		

Median *m* (fifth decile)

For deciles, dividing 10 by 4 gives the first decile as 2.5. Then, adding 2.5, we obtain the median and the higher decile.

	2.5	5	7.5	10
Quartile	1	2	3	4
Data	1, 1, 2	4, 4	6, 7, 7	9, 9

Median m (second quartile)

4.6.3 Z-SCORE

Let us start with an example.

Example 4.37

Suppose a test in two versions, 1 and 2, is given in a statistics class. Two classmates S_1 and S_2 are given two different versions of the examinations. S_1 received a score of 65 and S_2 received 80. If we are to select the better performer out of S_1 and S_2, just by looking at the scores, we would select S_2 since S_2 received a better score. Now the question is, did we make a correct decision?

Answer

To give the correct answer, we should have asked the following questions of ourselves before making a decision. Questions such as:

What is the average for each version of the examination?
Is the S_1 's score better or worse than the class' average for version S_1?
Is S_2 's score better or worse than the class' average for version S_2?

In a case similar to the above example, the best way to select the best performer is to calculate the z-score, which is the difference between an observation and the average value in terms of the standard deviation, that is,

$$z = \frac{\text{Observation} - \text{Mean}}{\text{Stdev}} = \frac{x - \mu}{\sigma}. \tag{4.8}$$

So, let us assume the following statistics for the examination in the example above:

Student	Score	Version Average	Version Standard Deviation
S_1	65	50	5
S_2	80	85	10

We now calculate the z-scores of S_1 and S_2, using (4.8), denoted by z_1 and z_2, respectively.

$$z_1 = \frac{x_1 - \mu_1}{\sigma} = \frac{65 - 50}{5} = 3 \text{ and } z_2 = \frac{x_2 - \mu_2}{\sigma} = \frac{80 - 85}{10} = -0.5.$$

Thus, based on the z-scores, S_1 has a z-score of 3, which means that S_1's grade is higher than 3 standard deviations of the average value for those taking the first version of the examination. S_2's z-score, on the other hand, is -0.5, which means that S_2's grade is lower than the average value of the second version of the test by 0.5 standard deviation. Thus, comparing the z-scores, it is clear that S_1 performed better than S_2.

4.7 MORE PLOTS

4.7.1 BOX-AND-WHISKER PLOT

Yet another graphical display of a data set is the **box-and-whisker plot** or **box plot**. This is also called the **five number summary**. It shows the minimum, Q_1, Q_2, Q_3, and the maximum value of a given data set. It involves a center box containing 50% of the data and two **whiskers**, each of which represents 25% of the data. In other words, the plot divides the distribution of a data set into four parts:

   i. The lower "whisker" that contains the minimum value;
   ii. The lower portion of the box that contains the first quartile, Q_1, and the middle line of the box that contains the second quartile, Q_2;
   iii. The top part of the box, above the median line, which is the third quartile, Q_3;
   iv. The upper "whisker" that represents the maximum value.

In summary, the box plot is based on five numbers:

   1. The minimum,
   2. The first quartile, Q_1,
   3. The median, the second quartile, Q_2,
   4. The third quartile, Q_3,
   5. The maximum of the data set.

The rectangular shape of the box plot graph is the basis for its name since its shape looks like a box.

Steps to create the box-and-whisker plot are as follows:

Step 1. Use the equal interval scale.
Step 2. Draw a rectangular box such that:
   i. One end at Q_1 and the other end at Q_3,
   ii. Draw a vertical segment at the median value,
   iii. Draw two horizontal segments on each side of the box, one down to the minimum value and the other one up to the maximum value; these segments are the **whiskers**.

Example 4.38

We want to draw a box-and-whisker plot for the following data set S:

$$S = \{5,7,9,10,14,15,16,20,23\}$$

Answer

From the given data points, we had the following necessary five numbers:
Minimum: 5, Q_1: 8, median: 14, Q_3: 18, and maximum: 23.
Using MATLAB 2018b, we can write the following simple codes to obtain Figure 4.9.

```
S=[5 7 9 10 14 15 16 20 23];
boxplot(S)
ylabel('Observations')
title('Box Plot, MATLAB')
```

As it can be seen from Figure 4.8, the following information describes the figure (plot):

i. The vertical spacing between the labels indicates the values of the variable in proportion;
ii. A vertical line extending from the top of the rectangle indicates the maximum;
iii. The top part of the rectangle indicates the third quartile, Q_3;

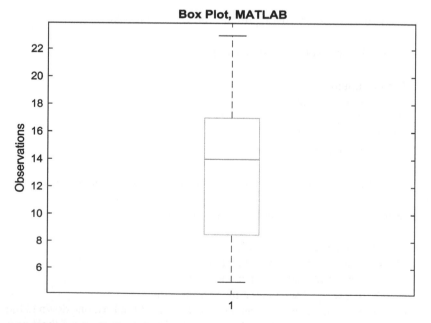

FIGURE 4.9   Box plot using MATLAB for Example 4.38.

iv. A horizontal line near the middle of the rectangle indicates the median, "Md", Q_2;

v. The bottom of the rectangle indicates the first quartile, Q_1;

vi. The vertical line extending from the bottom of the rectangle indicates the minimum;

vii. The vertical dotted lines above and below the box are the whiskers.

Now, using Minitab 18-1 for the same example, we will have Figure 4.10. Descriptions are the same as for MATLAB figure.

The box plot can be used to illustrate multivariables.

Example 4.39

Referring to Example 4.20, use MINTAB to do a box plot to show both high and low temperatures during the period of November 29, 2018, and December 28, 2018, in Houston, Texas, the United States (Figure 4.11).

Answer

Note 4.13

This is how we can find outliers, using the box plot. Box plot can be used to identify the outliers. We consider any data point as an outlier if its value is less than the lower fence, denoted by *LF*, defined by

$$LF = Q_1 - (1.5)(IQR) \tag{4.9}$$

or higher than the upper fence, denoted as *UF*, defined by

$$UF = Q_3 + (1.5)(IQR). \tag{4.10}$$

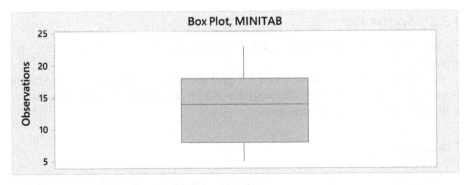

FIGURE 4.10   Box plot using Minitab for Example 4.38.

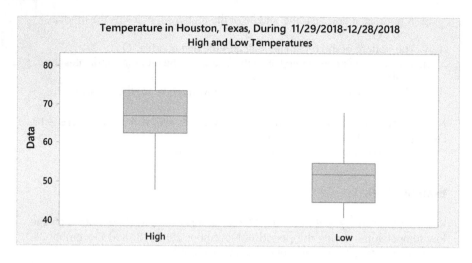

FIGURE 4.11   Box plot using Minitab for Example 4.39.

Example 4.40

Let us consider the following data points:
   8, 7, 10, 8, 6, 9, 5, 5, 7, 8, 14

   Check for outliers.

Answer

Here, $Q_1 = 6, Q_3 = 9$. and $IQR = 9 - 6 = 3$. Then,

$$LF = Q_1 - (15)(IQR) = 1.5 \text{ and } UF = Q_3 + (15)(IQR) = 13.5.$$

Clearly, only the data point 14 is beyond the *LF* and *UF*. Therefore, 14 is the outlier.

4.7.2   SCATTER PLOT

Finally, the **scatter plot** or an **XY plot** illustrates the relationship between two variables (if any).

   A scatter plot is also known as an **XY plot** since the variables are plotted on the *x*- and *y*-axis. The strength of the linear correlation on a scatter plot can be measured using a linear correlation coefficient.

Example 4.41

Referring to Example 4.20, use MINTAB to do a scatter plot to show both high and low temperatures during the period of November 29, 2018, through December 28, 2018, in Houston, Texas, the United States (Figure 4.12).

Answer

Note 4.1

Scatter plots help to identify the linear relationship between two quantitative variables (X and Y).

4.8   MEASURES OF VARIABILITY

So far, we have discussed the measures of central tendency. When studying a data set, it is not enough to know only the measures of the central tendency.

Example 4.42

Suppose after an examination is over, the instructor goes to the class and mentions the class average of the examination. By knowing only the average, can you guess your grade?

Answer

The answer is not easy. It is because you are not sure about the distribution of the grades. Are all grades very close to the average, or are grades far away from the average value?

In a situation like this, you need not only the measures of the central tendency, but also measurements that represent how far the rest of the data is away from the mean value. These are called the **measures of variability**. Measures of variability

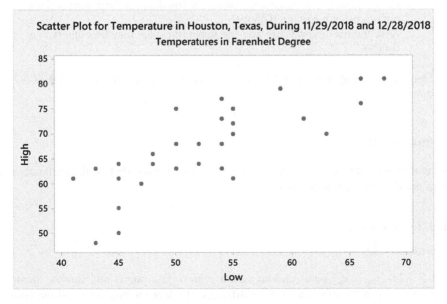

FIGURE 4.12   Scatter plot using Minitab for Example 4.41.

are also called the **measures of dispersions**, that is, the distance of a data point from the sample mean. Under the measures of variability, we will talk about the range, the variance, and the standard deviation.

4.8.1 RANGE

The range is the difference between the maximum and the minimum of data values.

Example 4.43

Consider the following sample of data that shows the average gas price of cities.

$3.00	$3.25	$2.75	$3.20	$2.95
$2.00	$3.50	$3.10	$2.50	$2.90

Calculate the range of the above data.

Calculation

Range = Maximum Value − Minimum Value = $3.50 − $2.00 = $1.50.

As we saw, it is very easy to calculate the range. In practice, we would like to look for a better measurement to represent the variability of the data than the range, due to the following reasons:

   i. Range is based only on the two values in the data set. Therefore, it does not represent the entire data set;
   ii. Also, the maximum value and the minimum value may be outliers in a data set.

Hence, it is an important and better measure to represent the variability of the data.

Definition 4.16

Measures of variability means measures of dispersions, that is, the distance of a data point from the sample mean.

4.8.2 VARIANCE

Definition 4.17

Let $\mathbf{X} = (x_1, x_2, \ldots, x_n)$ be a random vector with n data points and sample mean $\bar{x}$. **Deviation from the mean** means the difference between a data point and the sample mean, that is, $x_k - \bar{x}$, $k = 1, 2, \ldots, n$.

Since the sample mean is fixed, a deviation from the mean can be positive, negative, or 0, depending upon the value of x_i whether it is greater than, less than, or equal to the value of the sample mean; however, the sum of the deviations from the sample mean is zero, that is,

$$\sum_{k=1}^{n}(x_k - \bar{x}) = 0. \tag{4.11}$$

This is because substituting (4.4) in (4.11), we will have:

$$\sum_{k=1}^{n}(x_k - \bar{x}) = \sum_{k=1}^{n} x_k - \sum_{k=1}^{n} \bar{x} = \sum_{k=1}^{n} x_k - n\bar{x}$$

$$= \sum_{k=1}^{n} x_k - n\left(\frac{1}{n}\sum_{k=1}^{n} x_k\right) = \sum_{k=1}^{n} x_k - \sum_{k=1}^{n} x_k = 0.$$

$$\sum_{k=1}^{n} \frac{|x_k - \bar{x}|}{n}, \quad n = 1, 2, \dots$$

As we saw, the sum of the deviations from the sample mean is 0. However, this is not a desirable information since we are interested in the average deviation from the sample mean. Hence, one option is the average absolute value, that is,

$$\sum_{k=1}^{n} \frac{|x_k - \bar{x}|}{n}, \quad n = 1, 2, \dots$$

But absolute value is not an interesting concept in mathematical analysis since operation with it is difficult. The next option is the average of squares of deviations. Therefore, we define the deviation from the mean as follows.

Definition 4.18

Consider a finite population of size N with elements as $x_1, x_2, \dots, x_N$. Let us also denote the population mean and variance by μ and σ^2, respectively. Then, given the population mean, the **population variance** is defined as

$$\sigma^2 = \frac{1}{N}\sum_{i=1}^{N}(x_i - \mu)^2. \tag{4.12}$$

Since the calculation of the population variance using (4.12) is a difficult task, we approximate it by the sample variance.

Definition 4.19

The **sample variance** or **mean square**, denoted by s^2, is a statistic defined by

$$s^2 = \frac{1}{n-1}\sum_{k=1}^{n}(x_k - \bar{x})^2. \tag{4.13}$$

The denominator of the fraction on the right-hand side of (4.12) should logically be n, as was used up to a couple of decades ago. However, it was found that how $1/n$ causes under estimating the population variance, σ^2, as the role of s^2 is. Thus, replacing n by $n-1$ enlarges s^2, which is a better estimate for σ^2.

Theorem 4.1

Another form of the sample variance is:

$$s^2 = \frac{1}{n-1}\left[\sum_{k=1}^{n}x_k^2 - \frac{1}{2}\left(\sum_{k=1}^{n}x_k\right)^2\right]. \tag{4.14}$$

Proof:
Using the definition of $\bar{x}$ (4.4), and the definition of s^2 (4.14), we will have the following argument that leads to (4.13):

$$s^2 = \frac{1}{n-1}\sum_{k=1}^{n}(x_k - \bar{x})^2 = \frac{1}{n-1}\left[\sum_{k=1}^{n}x_k^2 - 2\bar{x}\sum_{k=1}^{n}x_k + \sum_{k=1}^{n}\bar{x}^2\right]$$

$$= \frac{1}{n-1}\left[\sum_{k=1}^{n}x_k^2 - 2\bar{x}\cdot n\bar{x} + n\bar{x}^2\right]$$

$$= \frac{1}{n-1}\left[\sum_{k=1}^{n}x_k^2 - n\bar{x}^2\right].$$

Relation (4.13) can be used in case the data is grouped in a simple frequency form $f_k, \quad k=1,2,\ldots,n$. The sample variance will also be of the following form:

$$s^2 = \frac{1}{n-1}\left[\sum_{k=1}^{n}f_k\cdot x_k^2 - \frac{1}{n}\left(\sum_{k=1}^{n}f_k\cdot x_k\right)^2\right]. \tag{4.15}$$

Example 4.44

Let us go back to Example 4.20, and consider the data sets of low and high city temperatures during the given period. We want to calculate the sample variances and sample standard deviations for both.

Calculation

To compute this, we use the Minitab 18 with the following steps:

Step 1: Type the data into a column in a Minitab worksheet.
Step 2: Click "Stat", then click "Basic Statistics", then click "Display Descriptive Statistics".
Step 3: Click the variables you want to find the variance for ("High", in our case), and then click "Select" to move the variable names to the right window.
Step 4: Click "Statistics".
Step 5: Check the "Variance" box, and then click "OK" **twice**.

Here are the results:

Statistics: Low				Statistics: High			
Variable	**Mean**	**StDev**	**Variance**	**Variable**	**Mean**	**StDev**	**Variance**
Low	52.27	7.30	53.31	High	67.10	8.30	68.92

Using Excel program, following (4.4) and (4.3), we calculate each statistic in Table 4.14.

High Temperature	**Low Temperature**
$\bar{x} = 67.1$	$\bar{x} = 52.27$
$\sum_{k=1}^{30}(x_k - 67.1)^2 = 2,016.7$	$\sum_{k=1}^{30}(x_k - 52.27)^2 = 1,545.87$
$s^2 = \dfrac{1}{29}\sum_{k=1}^{30}(x_k - 67.1)^2 = 69.54$	$s^2 = \dfrac{1}{29}\sum_{k=1}^{30}(x_k - 52.27)^2 = 53.31$
$s = \sqrt{69.54} = 8.34$	$s = \sqrt{53.31} = 8.34$

4.8.3 STANDARD DEVIATION

Standard deviation is considered as the square root of the variance. So, we can state that σ is the population standard deviation, and S is the sample standard deviation.

TABLE 4.14

Data Regarding Weather Temperature of Houston, Texas, during the Period of 11/29/2018, Example 4.44

Day No.	Date	High	$x_k - \bar{x}$	$(x_k - \bar{x})^2$	Low	$x_k - \bar{x}$	$(x_k - \bar{x})^2$
1	Nov. 29, 2018	76	8.9	79.21	66	13.73	188.60
2	Nov. 30, 2018	81	13.9	193.21	68	15.73	247.54
3	Dec. 01, 2018	81	13.9	193.21	66	13.73	188.60
4	Dec. 02, 2018	79	11.9	141.61	59	6.73	45.34
5	Dec. 03, 2018	68	0.9	0.81	52	−0.27	0.07
6	Dec. 04, 2018	61	−6.1	37.21	45	−7.27	52.80
7	Dec. 05, 2018	63	−4.1	16.81	43	−9.27	85.87
8	Dec. 06, 2018	64	−3.1	9.61	52	−0.27	0.07
9	Dec. 07, 2018	75	7.9	62.41	55	2.73	7.47
10	Dec. 08, 2018	55	−12.1	146.41	45	−7.27	52.80
11	Dec. 09, 2018	48	−19.1	364.81	43	−9.27	85.87
12	Dec. 10, 2018	61	−6.1	37.21	41	−11.27	126.94
13	Dec. 11, 2018	64	−3.1	9.61	45	−7.27	52.80
14	Dec. 12, 2018	68	0.9	0.81	54	1.73	3.00
15	Dec. 13, 2018	75	7.9	62.41	50	−2.27	5.14
16	Dec. 14, 2018	50	−17.1	292.41	45	−7.27	52.80
17	Dec. 15, 2018	64	−3.1	9.61	48	−4.27	18.20
18	Dec. 16, 2018	68	0.9	0.81	50	−2.27	5.14
19	Dec. 17, 2018	72	4.9	24.01	55	2.73	7.47
20	Dec. 18, 2018	63	−4.1	16.81	50	−2.27	5.14
21	Dec. 19, 2018	61	−6.1	37.21	55	2.73	7.47
22	Dec. 20, 2018	63	−4.1	16.81	54	1.73	3.00
23	Dec. 21, 2018	64	−3.1	9.61	45	−7.27	52.80
24	Dec. 22, 2018	77	9.9	98.01	54	1.73	3.00
25	Dec. 23, 2018	70	2.9	8.41	55	2.73	7.47
26	Dec. 24, 2018	66	−6.1	37.21	48	−4.27	18.20
27	Dec. 25, 2018	73	−4.1	16.81	61	8.73	76.27
28	Dec. 26, 2018	70	2.9	8.41	63	10.73	115.20
29	Dec. 27, 2018	73	5.9	34.81	54	1.73	3.00
30	Dec. 28, 2018	60	−7.1	50.41	47	−5.27	27.74

The population standard deviation, $\sigma = \sqrt{\dfrac{\sum\limits_{i=1}^{N}(x_i - \mu)^2}{N}}$.

The sample standard deviation, $S = \sqrt{\dfrac{\sum\limits_{i=1}^{n}(x_i - \bar{x})^2}{n-1}}$.

Note 4.15

The unit of the sample standard deviation is the same as the unit of each data point unlike the variance. The unit of the variance is always squared units.

4.9 UNDERSTANDING THE STANDARD DEVIATION

We have discussed both the variance and the standard deviation before. Both represent how far a data point is away from the mean value on average. The standard deviation can be used to understand the data with the help of following rules.

4.9.1 THE EMPIRICAL RULE

The empirical rule applies to a distribution, which is symmetric and mound-shaped. According to this rule, approximately 68% of the data lies between one standard deviation of the mean value. Approximately 95% of the data lies between two standard deviations of the mean value. Furthermore, approximately 99.7% of the data lies between three standard deviations of the mean value.

This is illustrated in Figure 4.13.

Example 4.45

Suppose the monthly income of a family in a certain city is approximately normally distributed with a mean of $2,000 and a standard deviation of $100. Using the empirical rule, we obtain the results which are given in Table 4.15.

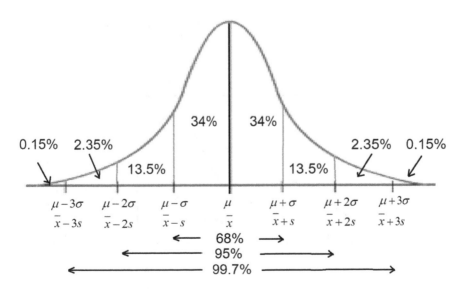

FIGURE 4.13 Empirical rule illustration.

TABLE 4.15
Intervals and Proportions

k	$(\mu - k\sigma, \mu + k\sigma)$	Proportion of Data	Interpretation
1	$(2{,}000 - 100, 2{,}000 + 100)$	68%	68% of the families have an income between $1,900 and $2,100
2	$(2{,}000 - 2 * 100, 2{,}000 + 2 * 100)$	95%	95% of the families have an income between $1,800 and $2,200.
3	$(2{,}000 - 3 * 100, 2{,}000 + 3 * 100)$	99.7%	99.7% of the families have an income between $1,700 and $2,300.

4.9.2 CHEBYSHEV'S RULE

This rule can be applied to any distribution, unlike the empirical rule. According to Chebyshev's rule, the proportion of data between k standard deviations of the mean is approximately $\left(1 - \dfrac{1}{k^2}\right) \times 100\%$, where $k > 1$. This is illustrated accordingly in Figure 4.14.

For various $k > 1$ values, we can construct the following table.

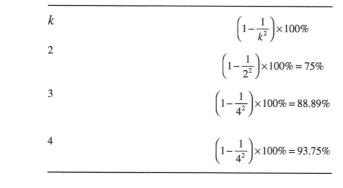

k	$\left(1 - \dfrac{1}{k^2}\right) \times 100\%$
2	$\left(1 - \dfrac{1}{2^2}\right) \times 100\% = 75\%$
3	$\left(1 - \dfrac{1}{4^2}\right) \times 100\% = 88.89\%$
4	$\left(1 - \dfrac{1}{4^2}\right) \times 100\% = 93.75\%$

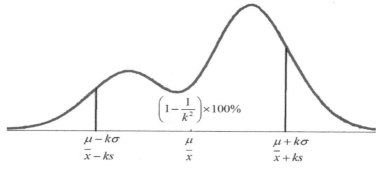

FIGURE 4.14 Chebyshev's rule illustration.

Example 4.46

Let's consider Example 4.44, and assume that the shape of the distribution is skewed. Therefore, so we can't apply the empirical rule. According to Chebyshev's rule, we can conclude as follows.

Approximately 75% of the families have an income between $1,800 and $2,200. Similarly, 88.89% of the families have an income between $1,700 and $2,300. Furthermore, approximately 93.75% of the families have an income between $1,600 and $2,400.

EXERCISES

4.1. Consider the following variables related to a student. Identify them as quantitative or qualitative.
   a. Major.
   b. Number of credits the student takes.
   c. Student's classification (freshman/sophomore/junior/senior).
   d. Student's GPA.
   e. Student number.

4.2. Classify the following variables as quantitative or qualitative.
   a. Brand of a car.
   b. Car mileage.
   c. Price of a car.
   d. Year of manufacture of a car.
   e. Car's VIN number.

4.3. Identify the following variables of a car engine, according to the level of the measurements.
   a. Country of manufacture.
   b. Year of manufacture.
   c. Type of the engine (two cylinders, three cylinders, four cylinders, five cylinders, six cylinders, and eight-plus cylinders).
   d. Gas mileage.
   e. Average temperature.

4.4. Consider the following grades of a course in elementary statistics.

A	A	C	B	F
B	C	C	B	C
F	C	B	C	A
C	C	B	F	F
A	B	B	C	C

   a. Construct an appropriate graph to summarize the above data.
   b. What percentage of students have received a grade "B"?

4.5. A survey was conducted to find out the means of transportation to a school.

Means of Transportation	Number of Students
Walking	150
Bicycle	145
Bus	300
Car	405

    a. Represent the above information using an appropriate graph.
    b. What percentage of students use buses to come to school?
    c. What percentage of students use either a bicycle or a car to reach the school?

4.6. Consider the set of numbers 2, 5, 6, 4, 3. Calculate the following:

    a. $\displaystyle\sum_i x_i$

    b. $\displaystyle\sum_i x_i^2$

    c. $\displaystyle\sum_i x_i - 4$

    d. $\displaystyle\sum_i (x_i - 4)$

4.7. Using the data sets 4, 1, 2, and 3, compute the following:

    a. $\displaystyle\sum_i x_i^2 - 5$

    b. $\displaystyle\sum_i (x_i - 5)^2$

    c. $\displaystyle\sum_i (2x_i - 5)^2$

    d. $\displaystyle\sum_i 2(x_i - 5)^2$

4.8. The average grade of 10 students for an examination is 70. If the grades of nine students are 75, 80, 65, 90, 68, 40, 75, 60, and 70.
    Calculate the other students' examination grade.

4.9. Consider the following car mileages of randomly selected 10 cars:
    33.5   33.0   35.630.7 36.3   31.5   25.0   33.2   35.0   40.5
    a. Calculate the average car mileage using the above data.
    b. Calculate the 10% trimmed mean car mileage for the above data.

4.10. Consider the following grades of an examination:

92	95	72	85	50	60
88	73	70	87	75	83
40	71	70	75	97	40
70	77	88	45	56	90
91	83	87	71	74	86

a. Construct a stem-and-leaf plot for the above data.
b. Comment on the shape of the above plot.
c. Calculate the mean, median, and the mode of the above grades.

4.11. An instructor constructed a histogram for the grades students got for one of his classes. The shape of the grades distribution is as follows.

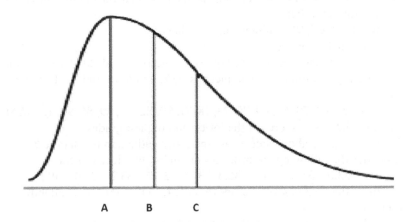

A B C

Out of mean, median, and mode, what are suitable for A, B, and C?

4.12. Consider the following amount of money spent by several companies for advertising during a month:

Name	Amount ($)
A	20,000
B	60,000
C	55,000
D	70,000
E	25,000

a. Calculate the range of above expenditures
b. Calculate the average value of the above expenditures.
c. Calculate the variance of the expenditures.

4.13. John and Sam sat for two different examinations. Suppose John's grade was 75 and the average and the standard deviation of his examination were 60 and 10, respectively. Similarly, assume that Sam's grade was 85 and the average and the standard deviation of his examination were 90 and 0.5, respectively.
a. Whose grade is better?
b. Calculate z-scores for both John's and Sam's grades?
c. Interpret both z-scores.
d. What do you think about their grades now?

4.14. Consider gas price in several gas stations of a big city.

| $2.54 | $2.60 | $2.35 | $2.45 | $2.65 |
| $2.30 | $2.40 | $2.55 | $2.56 | $2.62 |

a. Calculate the median of the above data.
b. Calculate the first, the second, and the third quartiles of the above data and interpret them.
c. What is the IQR and interpret the value of it?
d. Construct a box plot.

4.15. The following sample of data represents the scores of 20 students on an engineering examination. The mean is 70.2, and the standard deviation is 18.5.

70, 39, 51, 63, 59, 68, 69, 75, 73, 78, 77, 80, 82, 88, 89, 85, 90, 15, 67, 86.

a. Calculate the z-scores of the lowest and highest grades.
b. Based on the above values, are there any outliers in the data set?

4.16. Consider the following car mileages of randomly selected 9 cars.

33.5   33.0   35.6   30.7   36.3   31.5   25.0   33.2   35.0   40.5

Using the z-scores of the above values, check for the availability of outliers.

4.17. Consider the following sample of data, which represents the pitching speed (miles per hour) of pitchers in 20 games of a major baseball league.

85, 90, 95, 88, 92, 91, 87, 89, 86, 92, 80, 70, 75, 68, 69, 74, 77, 80, 75, 73.

a. Construct a stem-and-leaf plot for the above data.
b. Calculate the 70th and 80th percentiles for the above data.
c. Interpret the above percentiles.

4.18. Consider a particular type of car engine, which has an average gas mileage of 35 miles per gallon and a standard deviation of 5 miles per gallon. Calculate the 95th percentile of the gas mileage and interpret the result.

4.19. Suppose the waiting time for a bus in a particular bus routine follows a uniform random variable between 0 and 10 minutes. Calculate the 80th percentile for the waiting time.

4.20. Assume that the lifetime of a car engine follows an exponential distribution with an average of 10 years.
a. Calculate the median lifetime of the car engine.
b. What is the 90th percentile of the car engine and interpret the result?

4.21. Suppose a set of observations of daily emission (in tons) of sulfur from an industrial plant is given below.

A Set of Observations (Raw Data)				
20.2	12.7	18.3	18.3	23.0
21.5	10.2	17.1	07.3	11.0
18.3	20.2	20.2	20.2	18.3
17.1	11.0	12.7	11.0	23.0
23.0	18.3	12.7	07.9	12.7
17.1	21.5	07.9	11.0	12.7

a. Calculate the mean and median for the above data.
b. Calculate the first, the second, and the third percentiles.
c. Construct the box plot for the above data.
d. Calculate the lower fence and the upper fence for the above data and identify any outlier(s) if there are any.

4.22. Suppose an engineering firm has five engineers who are paid annual salaries as $70,000, $80,000, $85,000, $85,000, and $90,000, respectively.
a. Calculate the average annual salary based on the above data.
b. What is the median salary above?
c. Suppose the chief executive officer (CEO), who is also an engineer, earns $180,000. How does the inclusion of the CEO's salary in to the above data change both calculations in parts (a) and (b)?

4.23. Suppose an electronic company sells three types of top of the line high-tech TVs at prices $4,000, $2,000, and $1,500 each, respectively. The company's data shows that the percentages of sales of these items are 10, 60, and 30, respectively. We want to find the distribution of the sample mean and also the average revenue of selling a unit of TV across the nation.

4.24. Suppose a set of observations of daily emission (in tons) of sulfur from an industrial plant is given below.

20.2	12.7	18.3	18.3	23.0
21.5	10.2	17.1	07.3	11.0
18.3	20.2	20.2	20.2	18.3
17.1	11.0	12.7	11.0	23.0
23.0	18.3	12.7	07.9	12.7
17.1	21.5	07.9	11.0	12.7

a. Construct a relative frequency distribution for the above data.
b. Construct a histogram using the intervals [4,8), [8,12), ..., [20,24).
c. Discuss the shape of the above distribution.

4.25. The lifetime of a certain type of light bulb has an exponential distribution with an average value of 200 hours.
a. Calculate the first median value of the lifetime of this type of bulb and interpret this value.
b. What is the 90th percentile of a bulb of the above type?

4.26. A certain brand of car has an average gas mileage of 35 miles per hour and a standard deviation of 5 miles per hour. If you select a car of the above brand, what is the third quartile of the gas mileage? What does this third quartile represent?

4.27. The error in the reaction time (in minutes) of a certain machine has the following distribution:

$$f(x) = \frac{3}{2}x^2; \quad -1 \le x \le 1.$$

Compute the 95th percentile of the error in reaction.

4.28. Suppose a random variable X has a uniform distribution over $[a,b]$, where $0 \le a < b$. Compute the first, the second, and the third quartiles in terms of values a and b.

4.29. Suppose the GPA of 20 randomly selected students is as follows:

   2.50, 3.00, 2.75, 3.80, 3.60, 2.95, 3.00, 1.70, 2.80, 2.00
   3.99, 2.10, 2.65, 2.85, 2.30, 3.45, 2.00, 2.70, 1.85, 3.00

   a. Calculate the mean and the standard deviation of the above data.
   b. Calculate the number of GPA values contained in the intervals of $(\bar{x} - s, \bar{x} + s)$, $(\bar{x} - 2s, \bar{x} + 2s)$, and $(\bar{x} - 3s, \bar{x} + 3s)$.
   c. Calculate the proportion of data points in the above intervals, and compare them with the empirical rule.

4.30. The grades of one of the national examinations follow a skewed distribution with a mean value of 70 and a standard deviation of 5.
   a. What proportion of students have scored one standard deviation of the mean value?
   b. What proportion of students have scored within three standard deviations of the mean value?
   c. Interpret the above values.

5 Inferential Statistics

5.1 INTRODUCTION

Already familiar with basics of statistics, the vocabularies, and properties of the measures of central tendency and dispersion, we are now ready for data analysis. The essence of data analysis is to choose a sample, collect data, and address questions regarding the data. The idea is to reach conclusions about the population from the sample and predict the possible future. In other words, the purpose of the most statistical investigations is to generalize information contained in samples of the populations and can tell what the future data may look like. This is referred to as **estimating parameters by the statistics,** which is the basis for statistical inference or **inferential statistics**. The term "inference" means a conclusion or a deduction, that is, essentially, a **statistical decision-making process**. In other words, inferential statistics are derived via complex mathematical theory allowing scientists to infer trends about a larger population based on results from the sample. Often scientists use inferential statistics to examine the relationships between variables within a sample and then generalize or predict how the variables may relate to a larger population. It seems that inferential statistics started in the fifth century BC about the time Athenians needed to estimate the height of ladders necessary to scale walls.

Note 5.1

There are ways that decision-making could be deterministic such as in linear optimization. However, here in this book, we mainly consider stochastic decision-making.

Now, before we go into details of this chapter, we are to refresh the readers' memories. Hence, we re-state some definitions from Chapter 4 and give some examples.

In a study, the **population** represents all the observations under the investigation. Usually, the population is very large and consists of an enormous number of observations, which make it practically impossible to use it all as a collected data set. Thus, a subset of the population, called the **sample**, is used instead. Practically, this subset consists of a finite collection of observations that is easy to gather. Consequently, unlike the population, the sample is an easier collection to work with.

A numerical value that summarizes the population is referred to as a **population parameter**. The mean, median, mode, variance, and standard deviation are some examples of population parameters. Usually, these values are unknown or difficult to calculate. Often, population parameters are denoted, symbolically, by Greek letters such as μ (mu), mostly used for mean, and σ (sigma), mostly used for standard deviation. In the case of samples, they are referred to as **the sample statistics** and are denoted by English letters such as $\bar{X}$ (X-bar) for sample mean and S for sample standard deviation. Using the terms "population parameter" and "sample statistic", we can state that the **inferential statistics** is studying about unknown population parameters using sample statistics.

Example 5.1

In studying the monthly income of a family in the United States, the population mean, that is, the average income of a family, is very difficult to compute. The reason is that the number of families in the country is huge and it is not easy to reach each and every one of them to obtain their monthly income as needed to average. However, selecting a random sample is a very easy and practical way of doing it. So, this sample statistics summarizes the entire sample into a single numerical value.

Example 5.2

For the purpose of statistical study of the average family income in the United States, suppose a sample of 1,000 families is selected. It can be seen that the calculation of the sample mean in this case is easily obtained since the number of families is finite and possibly very small.

Note 5.2

In inferential statistics, selecting a good sample is very crucial. Since the calculation of the population parameters is based on the sample, it should be truly representative of the population.

Example 5.3

A large technology company, like Apple, is deciding to develop a new version of its smart phone. Before start developing the product, it tried to collect opinions of potential users in some large cities' malls in the United States to see how the public will react to such a development before the product is developed and marketed. Hence, a questionnaire (a survey) is developed and the data collected will help the company decide. In this example, the population will be all people who may use this new smart phone. As the company cannot reach everyone in the population, they can get an idea about people's opinions in the population, by using the collected sample.

There are two ways of making inference about population parameters. One way is to make a good guess about the unknown population parameters. This is referred to as the **estimation**. The unknown parameter may be estimated by a single value, in which case the estimation is referred to as the **point estimation**. Instead of considering point estimation, it is more practical to consider finding an interval for estimating the parameter, in which case the estimation is referred to as an **interval estimation**. The second way to make an inference about the population parameters is by testing the parameter. This method is referred to as the **hypothesis testing**.

Essentially, inferential statistics is classified into four major areas:

    i. Point estimation,
   ii. Interval estimation,
  iii. Hypotheses testing,
  iv. Nonparametric estimation (which will be discussed in Chapter 6).
      There are other methods and areas of inferential statistics that are used
in special cases. For instance,
   v. **Sparse adaptive channel estimation** (under assumption of normal distribution),
  vi. **The statistical model of source localization based on range difference measurements** (under assumption of normal distribution),
 vii. Localization.

However, they will not be discussed in this book.

5.2 ESTIMATION AND HYPOTHESIS TESTING

One may ask, what is estimation? The answer is: Estimation is predicting or stating some information about the population parameters from the sample statistics, for instance, generalizing the obtained value of the sample mean to the mean of the population. We usually ask the question, how far or how close is the population mean from the sample mean? To answer this question, we need the following definition.

Definition 5.1

A statistic is said to be an **unbiased estimate** of a given parameter if the mean of the sampling distribution of that statistic can be shown to be equal to the parameter which is being estimated; that is, it is *neither* **underestimated** *nor* **overestimated**; otherwise, it is **biased**. An **unbiased estimator** is an accurate statistic that is used to approximate a population parameter.

Note 5.3

Definition 5.1 states that since estimation is an educated guess, it is expected to have the estimated value not exact. Hence, the word "bias" is to state that there is some distance from actuality.

Remark 5.1

The word "bias" and its opposite "unbias" are used to confirm that an estimate is not as the real value or an estimate and the actual value are the same, respectively. Thus, the **sample mean**, $\bar{X}_n$ (or if there is no ambiguity, simply, $\bar{X}$), is an **unbiased estimate of the population mean**, $E(X) = \mu$, if **the estimator** (the sample mean) equals the parameter (the population mean). In other words, **if the statistic equals the parameter, then the estimation is unbiased.**

5.2.1   POINT ESTIMATION

Definition 5.2

A **point estimation** is an estimation of a population parameter such as the mean, variance, or standard deviation, by a single value of a statistic.

Since the most popular parameters of a population are the mean, variance, and standard deviation, which often need to be estimated by random samples, we start this section with some properties of sample parameters, that is, statistics.

Let X be a random variable representing a **random sample** of size n denoted by $X_1, X_2, \ldots, X_n$, from a population; that is, $X_1, X_2, \ldots, X_n$ are n identically independent random variables. Also, let $\bar{X}_n$ represent the mean of the random sample $X_1, X_2, \ldots, X_n$. We further let S^2 denote the variance of this random sample. Thus, we are considering a sample mean as a random variable with its own distribution. Hence, it has its own properties such as the mean and variance. We summarize some of these properties in the following theorem.

Theorem 5.1

Let $X_1, X_2, \ldots, X_n$ be a random sample of size n from a population with mean $\bar{X}_n$ and variance S^2. We also suppose the population has a mean and finite variance denoted by μ and σ^2, respectively. Then,

i. $E(\bar{X}_n) = \mu.$ (5.1)

   Thus, the sample mean is an unbiased estimate of the population mean. Hence, if the statistic equals the parameter, then the estimation is unbiased.

ii. $\mathrm{Var}(\bar{X}_n) = \dfrac{\sigma^2}{n},$ (5.2)

   So the variance of the sample mean is a biased estimate of the population variance. But the variance of the sample mean is an unbiased estimate of the population variance divided by n.

iii. $E(S^2) = \sigma^2,$ (5.3)

iv. $E(S) \leq \sigma,$ (5.4)

v. $S^2 \to \sigma^2$, as $n \to \infty$, with probability 1. (5.5)

Proof:

i. Since the random variables are independent and identically distributed random variables (iid), we have:

$$E(\bar{X}_n) = E\left(\frac{1}{n}\sum_{k=1}^{n} x_k\right) = \frac{1}{n}\sum_{k=1}^{n} E(x_k) = \frac{1}{n}\left[E(x_1) + E(x_2) + \cdots + E(x_n)\right]$$

$$= \frac{1}{n}(\mu + \mu + \cdots + \mu) = \frac{1}{n} \cdot n\mu = \mu.$$

We might look at the sample mean approaching the population mean from a different viewpoint.

ii. To prove (5.2), again, since the random variables are iid, we have:

$$\text{Var}(\bar{X}_n) = \text{Var}\left(\frac{1}{n}\sum_{k=1}^{n} x_k\right) = \frac{1}{n^2}\sum_{k=1}^{n}\text{Var}(x_k)$$

$$= \frac{1}{n^2}\left[\text{Var}(x_1) + \text{Var}(x_2) + \cdots + \text{Var}(x_n)\right]$$

$$= \frac{1}{n^2}\left(\sigma^2 + \sigma^2 + \cdots + \sigma^2\right) = \frac{1}{n^2}\cdot n\sigma^2 = \frac{\sigma^2}{n}.$$

iii. To prove (5.3), we note that not only are the random variables iid, but also if X is a random variable from a general population with mean $\bar{X}$, from a property of variance in Chapter 4, we have:

$$\sigma^2(X) = E(X^2) - \mu^2 \text{ and } S^2 = \frac{1}{n-1}\sum_{k=1}^{n}(X_k - x)^2.$$

Hence,

$$\sum_{k=1}^{n}(x_k - \bar{x})^2 = \sum_{k=1}^{n}\left(x_k^2 - 2x_k\bar{x} + \bar{x}^2\right) = \sum_{k=1}^{n}x_k^2 - 2n\bar{x}\sum_{k=1}^{n}x_k + n\bar{x}^2 = \sum_{k=1}^{n}x_k^2 - n\bar{x}^2.$$

Thus,

$$E(S^2) = \frac{1}{n-1}E\left[\sum_{k=1}^{n}(xk - \bar{x})^2\right] = \frac{1}{n-1}\left[\sum_{k=1}^{n}(\sigma^2 + \mu^2) - n\left(\frac{\sigma^2}{n} + \mu^2\right)\right]$$

$$= \frac{1}{n-1}\left[n(\sigma^2 + \mu^2) - n\left(\frac{\sigma^2}{n} + \mu^2\right)\right] = \frac{1}{n-1}\left[(n-1)\sigma^2\right] = \sigma^2.$$

iv. Proof of (5.4) is left as an exercise.

v. To prove (5.5), we use the strong law of large numbers stated above. From the definition of the sample variance, we have:

$$S^2 = \frac{1}{n-1}\sum_{k=1}^{n}x_k^2 - \frac{n}{n-1}\bar{x}^2 = \frac{n}{n-1}\left[E(X^2) - [E(X)]^2\right].$$

Now, with probability 1, $E(X^2) \to (\sigma^2 + \mu^2)$ as $n \to \infty$ and $[E(X)]^2 \to \mu^2$ as $n \to \infty$.

Note 5.4

Based on (5.2), the variance of the sample mean, S^2, decreases as the sample size, n, increases.

Going back to (5.1), one might ask the following question: With a random sample of size n, how close the population mean, without knowing the distribution of the random variable X representing it, could be to the sample mean? We answer this question using Chebyshev's inequality, which we have seen before defined as

$$P(|X - \mu| \ge \epsilon) \le \frac{\sigma^2}{\epsilon^2}, \forall_\epsilon \in \mathbb{R}^+.$$

Definition 5.3

The standard deviation of the sample mean, denoted by SE($\bar{X}$), is called the **standard error of the mean**, that is,

$$\mathrm{SE}(\bar{X}_n) = \frac{\sigma}{\sqrt{n}}. \tag{5.6}$$

It is a commonly used index of the error involved in estimating a population mean based on the information from a random sample of size n.

Note 5.5

Usually, it is unlikely for a value of a random variable to be more than two standard deviations away from the mean. This is particularly true when the random variable is almost normal. This is the case of the sample mean when the sample size is large. That gives us assurance that the sample mean is about ±2 standard error (SE) away from the population mean.

Note 5.6

Due to the square root of the sample size in the denominator of the SE, when the sample size increases, the SE will decrease. For instance, if the sample size quadruples, the SE will be cut in half.

Example 5.4

A very destructive Hurricane Katrina originated over the Bahamas on August 23, 2005, as Category 5 hurricane over the warm waters of the Gulf of Mexico and made its first landfall in the United States of America on Florida and Louisiana. However, it weakened to Category 5 hurricane and made a second landfall on

August 29, 2005, over Southeastern Louisiana and Mississippi causing catastrophic and devastating fatal and financial damages, partially as a result of fatal engineering flaws in the flood protection system known as **levees** around the city of New Orleans. The total property damage was estimated as $125 billion. The damages included oil spills estimated over 10,000 gallons (38,000 L). The oil spills, in gallons, reported by nine company's locations are as follows:

991,000; 3,780,000; 13,440; 819,000; 53,000; 1,050,000; 25,000; 461,000; 13,000.

We want to find the estimate of the mean and SE of all oil spills as a result of Hurricane Katrina in August 2005.

Answer

From Equation (5.1), the mean of oil spill is $\overline{X}$, that is,

$$\overline{X} = \frac{991,000 + 3,780,000 + 13,440 + 819,000 + 53,000 + 1,050,000 + 25,000 + 461,000 + 13,000}{9}$$

$$= \frac{7,205,440}{9} \approx 800,604.$$

From Theorems 5.1 and (5.2), the variance of mean of oil spills is $\frac{\sigma^2}{n}$, and hence, the standard deviation of the sample mean is $\frac{\sigma}{\sqrt{n}}$. Using Minitab 18, we have the following information:

				Descriptive Statistics				
N	Mean	SE Mean	StDev	Minimum	Q1	Median	Q3	Maximum
9	800,604	399,502	1,198,506	13,000	19,220	461,000	1,020,500	3,780,000

Therefore, the standard deviation of the sample mean is $\frac{\sigma}{\sqrt{n}} = 399,502$. That is, the SE of $\overline{X}$ is 399,502 gallons.

Example 5.5

It is well known that blood sugar (glucose) monitoring is the main tool to check one's diabetes control. If the sugar level goes too low, one can lose the ability to think and function normally. On the other hand, if the level goes too high and stay high, it can cause damage or complications to the body. A1C is a tool for blood test that tells us what the blood sugar levels have been over the last two to three months. However, a blood sugar meter tells us what the blood sugar level is at the moment of reading. It is important for the blood sugar levels to stay in a healthy range. There are other devices like GluCall to measure the blood sugar level in a body. Although the National Institute for Clinical Excellence (NICE) has determined the standard glucose ranges, each individual's target range is supposed to be agreed by their doctor or diabetic consultant. The NICE recommended that the target blood glucose level ranges are as follows:

Target Levels by Type	Upon Waking	Before Meals (Preprandial)	At Least 90 Minutes after Meals (Postprandial)
Nondiabetic		4.0–5.9 mmol/L	under 7.8 mmol/L
Type 2 diabetes		4–7 mmol/L	under 8.5 mmol/L
Type 1 diabetes	5–7 mmol/L	4–7 mmol/L	5–9 mmol/L
Children w/ type 1 diabetes	4–7 mmol/L	4–7 mmol/L	5–9 mmol/L

Now, we consider 30 adult persons with type 1 diabetes. Each individual gets the reading of his/her blood sugar level on a morning day in a certain week after waking up. The following are the readings:

6	6	5	8	6	4	9	6	5	7
5	5	5	7	8	9	5	4	6	8
7	7	5	9	9	8	5	6	7	9

We want to find the estimate of the mean and SE of all of these readings so that a decision can be made nationwide.

Answer

Here again, we use Minitab18 to get general information on the given data set.

Descriptive Statistics								
N	Mean	SE Mean	StDev	Minimum	Q1	Median	Q3	Maximum
30	6.533	0.291	1.592	4.000	5.000	6.000	8.000	9.000

Thus, from the Minitab results and (5.1), we have:

$$\mu = \bar{X}_{30} = 6.533$$

and from the Minitab results and (5.6), we have:

$$\text{STD}\left(\bar{X}_{30}\right) = \text{SE}\left(\bar{X}_{30}\right) = \frac{1.592}{\sqrt{30}} = \frac{1.592}{5.477} \approx 0.291.$$

Example 5.6

Consider a random variable X representing a population with a mean of 20 and a variance of 144. We choose a sample of size 225 from this population. What is the probability that the

    i. Sample mean is <18?
    ii. Total result of all 225 observations is not <5,000?

Answer

From the problem, it is given that $n = 225$, $\mu = 20$, $\sigma^2 = 144$, $\sigma = 12$, and $T = 5{,}000$.

i. The first question is to find $P(\bar{X} < 18)$.

$$P(\bar{X} < 18) = P\left(z < \dfrac{18 - 20}{\dfrac{12}{\sqrt{225}}}\right) = P(z < -2.5) = 0.0062.$$

ii. The second question is to find $1 - P(T < 4{,}950)$.

$$P(\bar{X} < 5{,}000) = P\left(z < \dfrac{5{,}000 - (225)(20)}{(\sqrt{225})(12)}\right) = P(z < 2.78) = 0.9973.$$

Thus,

$$P(\bar{X} \geq 5{,}000) = 1 - 0.9973 = 0.0027.$$

Example 5.7

Suppose we want to estimate the usage of Internet activities. The following information is in a recent report at https://www.att.com/esupport/data-calculator/index.jsp regarding the amount of data used in an Internet activity (estimates based on typical file sizes; approximately 1 MB = 1,000 KB; 1 GB = 1,000 MB; 1 TB = 1,000 GB):

1 **email** (no attachments)	**20 KB**
1 **email** (with standard attachments)	**300 KB**
1 min. of surfing the **web**	**250 KB** (15 MB/h)
1 **song** downloaded	**4 MB**
1 **photo** upload to social media	**5 MB**
1 minute of streaming **standard-definition video**	**11.7 MB** (700 MB/h)
1 minute of streaming **high-definition video**	**41.7 MB** (2,500 MB/h)
1 minute of streaming **4K video**	**97.5 MB** (5,850 MB/h)
1 min. of **online games**	**200 KB** (12 MB/h)

From (5.5), we see that S^2, calculated on a sample, is an unbiased estimate of the variance of the population from which the sample was drawn. In other words, the calculation from a statistical sample of an estimated value of the variance (and, hence, standard deviation, a measure of statistical dispersion) of a population of data points is such that the expected value of the calculation equals the true value. We will discuss this concept in more detail later.

Ignoring the unit of each activity and considering MB as the units used, answer the following questions:

1. Using a mathematical software such as Excel, Minitab, or MATLAB, find the
     i. Mean,
    ii. Standard deviation,
   iii. Mean standard error (SE),
    iv. Minimum,
     v. Maximum,
    vi. Range,
   vii. First quartile,
  viii. Third quartile,
    ix. Median.
2. Display the data by histogram.
3. Find the probability that the sample mean of the usage of the Internet is >20.

Answer

Using Minitab 18, to answer questions 1(i) through 1(viii) as (Figure 5.1)

Variable	N	N*	Mean	SE Mean	StDev	Minimum	Q1	Median	Q3	Maximum
MB Usage	9	0	17.9	10.9	32.7	0.0	0.2	4.0	26.7	97.5

     i. Mean = 17.9,
    ii. Standard deviation = 32.7,
   iii. Mean standard error (SE) = 10.9,
    iv. Minimum = 0 (i.e., rounding 0.02),
     v. Maximum = 97.5,
    vi. Range = 97.5 (rounded),
   vii. First quartile = 0.2,

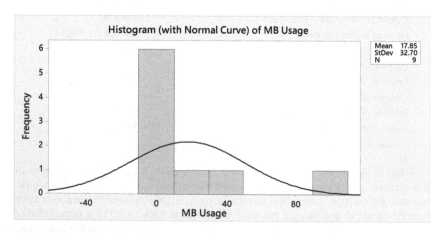

FIGURE 5.1   Display of Internet use, Example 5.13.

      viii. Third quartile = 26.7,
       ix. Median = 4.0.
   4. Histogram display is shown in Figure 5.1.
   5. The third question is to find $P(\bar{X} > 20)$.

$$P(\bar{X} > 20) = 1 - P\left(z < \frac{20 - 17.9}{\frac{32.7}{\sqrt{9}}} \right) = 1 - P(z < 0.1927) = 1 - 0.5764 = 0.4236.$$

In Chapter 4, we briefly mentioned "proportion". We now discuss this term in more detail. Although often a mathematical proportion is taken as an equality of two ratios, in statistics the proportion is a percentage of a total in which a certain characteristic is observed. Thus, we start with the definition of this term.

Definition 5.4

Let us now consider a finite population of size N. There are elements in this population with a certain characteristic of our interest, say x. We denote the **proportion** of the population with this characteristic by $\tilde{p}$. Then, of the total population, the proportion will be

$$\tilde{p} = \frac{x}{N}. \tag{5.7}$$

That is, a total of $N\tilde{p}$ elements have the characteristic x and $N(1 - \tilde{p})$ do not.

Example 5.8

Suppose an electronic manufacturer has developed a chip to improve the quality of the current television products. The new chip has shown to be effective. To see the proportion of the improved televisions out of new 1,000 televisions, a random sample has been chosen and each element has been tested. The tested items show that seven out of each ten items in the sample are improved televisions. Thus, the proportion of the improved televisions is $p = \dfrac{700}{1,000} = 0.7$. Clearly, there are $Np = (1,000)(0.7) = 700$ improved items and $N(1 - P) = (1,000)(0.3) = 300$ unimproved items.

    Let us now consider choosing a random sample of size n from a finite population of size N. Let us also define a random variable $X_k, k = 1,2,\ldots,n$, representing the elements $x_k, k = 1,2,\ldots,n$, of the sample with a specific feature. In other words, X is a binary random variable described as follows: For $k = 1,2,\ldots,n$.

$$X_k = \begin{cases} 1, & \text{if } k\text{th element of the sample has the feature,} \\ 0, & \text{otherwise.} \end{cases} \tag{5.8}$$

Now, let X represent the total number of elements with the feature, that is,

$$X = X_1 + X_2 + \ldots + X_k = \sum_{k=1}^{n} X_k, \qquad (5.9)$$

by (5.8). Thus,

$$p = \frac{\sum_{k=1}^{n} X_k}{n} = \overline{X}. \qquad (5.10)$$

Note 5.7

From (5.10), we see that the proportion of a sample having a certain feature is the sample mean. In other words,

$$P(X_k = 1) = \frac{N_p}{N} = p, \quad k = 1, 2, \ldots, n, \qquad (5.11)$$

and

$$P(X_k = 0) = 1 - P(X_k = 1) = 1 - p, \quad k = 1, 2, \ldots, n. \qquad (5.12)$$

From (5.11) and (5.12), it is found that each $X_k, k = 1, 2, \ldots, n$, has either 1 or 0 with a probability of p or $1 - p$, respectively.

Let us consider (5.8) through (5.11) from a different viewpoint. So, we consider a random sample of size n in which member points have or don't have a certain feature. This is like considering the members as male or female and our interest is the female members. Thus, $\overline{X}$ is the proportion of elements having the feature. Hence, we can write the random variable X in terms of $\overline{X}$ as $X = n\overline{X}$. Thus, from (5.8) through (5.11), we can see that X is a binomial random variable with parameters n and p. In this case, X will be the random variable representing the number of success in n independent Bernoulli trials (i.e., (5.9)) with the probability of success in each trial as p. Thus, (5.8) for $k = 1, 2, \ldots, n$ can be rewritten as follows: For $k = 1, 2, \ldots, n$

$$X_k = \begin{cases} 1, & \text{if the } k\text{th trial is a success,} \\ 0, & \text{if the } k\text{th trial is a failure.} \end{cases} \qquad (5.13)$$

Hence, as we have seen in Chapter 4 for a binomial random variable, for $k = 1, 2, \ldots, n$,

$$E(X_k) = np \text{ and } \mathrm{Var}(X_k) = np(1 - p).$$

Thus, X being the number of success in a sample of size n from a population with mean p and variance $p(1 - p)$, $\frac{X}{n}$ can be thought of as the mean of the sample. Therefore, from the central limit theorem, when n is large ($n \geq 30$, for instance, or both $np > 5$ and $np(1 - P) > 5$), the expression $\dfrac{\frac{X}{n} - p}{\sqrt{\dfrac{p(1-p)}{n}}}$ or $\dfrac{X - np}{\sqrt{np(1-p)}}$ will approximately possess a standard normal probability density function (pdf).

Example 5.9

Suppose a manufacture producing cellphone covers produces 55% of its products for a particular location in pink, aiming at female customers in that location. To test the market, the marketing department selects a random sample of size 300 ladies from that location. We want to find the probability that more than 60% of the sample elements favor the color pink for their cellphone covers.

Answer

Let X represent the number of those females in the sample who favor the pink color for their cellphone covers. As discussed above, X will be a binomial random variable with parameters $n = 300$ and $p = 0.55$. We are looking for $P(X > 180)$. Thus, the standard normal random variable in this case will be:

$$\frac{X - np}{\sqrt{np(1-p)}} = \frac{X - (300)(0.55)}{\sqrt{(300)(0.55)(0.44)}} = \frac{X - 165}{8.521}.$$

Note 5.8

To approximate a binomial random variable X by a standard normal distribution, since binomial is discrete and normal is continuous, it is customary to create an interval such as

$$P(X = k) \approx P(k - 0.5 \leq X \leq k + 0.5) \tag{5.14}$$

to approximate a discrete probability with a continuous one. This relationship (i.e., (5.14)) is referred to as a **continuity correction**.

Hence, using MATLAB 2019a, we will have the following:

$$P(X > 180) = P(X > 179.5) = P\left(\frac{X - 165}{8.521} > \frac{179.5 - 165}{8.521}\right)$$

$$= P(Z > 1.702) = 0.0444.$$

That is, the chance that more than 180 females purchase the pink color is more than 40%. In fact, had we not approximated the probability with normal and gone with actual binomial, we would gotten

$$P(X > 180) = B(180, 300, 0.55) = 0.0356.$$

That is, the exact probability of the event under question is about 3.56%.

Definition 5.5

As we have already seen, an **unbiased point estimate** of a parameter μ is the most plausible value of μ. The statistic mean, $\bar{x}$, that estimates this parameter is called the **point estimator** of μ and is denoted by $\hat{\mu}$ (reads as "mu hat").

Definition 5.6

The absolute value of the difference between the population mean and the sample mean, that is, $|\mu - \bar{x}|$, is referred to as the **sampling error**. A probability statement about the magnitude of the sampling error is incorporated in the **interval estimation**.

From Definition 5.5, if the *pdf* is known but the parameter, μ, is unknown, first take a random sample n times independently. Then, using the information obtained from the samples taken, try to estimate (guess) the value of the parameter μ. This means that we want to find a number $\hat{\mu}$ as a function of observations $(x_1, x_2, \ldots, x_n)$ for instance, $\hat{\mu} = G(x_1, x_2, \ldots, x_n)$. The function G is a statistic estimating μ (and, hence, a random variable with a pdf or probability mass function (pmf) of its own), and it is referred to as a **point estimator** for μ. The idea is to calculate $\hat{\mu} = G(x_1, x_2, \ldots, x_n)$ as close to the actual value of the parameter μ as possible. But how close is a question to be answered. Here are questions:

   i. How close should an estimate be to the parameter in question? In other words, what should the properties of an estimator be?

   ii. How one would know that there is not a better estimator than the one chosen?

The following definition answers question (i).

Definition 5.7

A **good estimator** (a statistic) is the one that is **unbiased** (i.e., *neither* **underestimated** *nor* **overestimated**), that is, the one that on the average is **equal to the parameter** to be estimated, and its variance is as small as possible, that is, with **minimum variance**. The **bias of an estimator** is the difference between the expected value of the estimator and the parameter.

In other words, if we are to select an estimator for the population parameter, we select an unbiased estimator. When there is more than one unbiased estimator, we select the one with the minimum variance. This estimator is referred to as the **minimum variance unbiased estimator (MVUE)**.

As mentioned earlier, the main idea of sampling, in addition to finding information about the population from which the sample is taken from, which is the most fundamental idea, is estimating the population mean. We saw in Chapter 4 that $E(\bar{X}) = \mu$. This suggests that $\bar{X}$ is an estimator of μ, that is, $\hat{\mu}$. In other words, we can point-estimate the population mean by the sample mean. However, we have to accept that in the estimation process, some errors may occur. This is due to the fact that it is the entire population that is used for the calculation of the desired statistics, rather a subset of it; that is, a sample has been used to point-estimate the desired parameters. The **error** is referred to as the **bias of the estimator**, $\hat{\mu}$, denoted by $B(\hat{\mu})$, and is defined as

$$B(\hat{\mu}) = \hat{\mu} - \mu. \tag{5.15}$$

We also saw in Chapter 4 that S^2/n is an unbiased estimate of the variance of the sampling distribution of mean for a random sample of size n, and the square root of this quantity is called the **standard error of the mean.** It is a commonly used index of the error involved in estimating a population mean based on the information in a random sample of size n.

Let $E(X) = \mu$ be the expected value of X. We want to investigate the biasedness or unbiasedness of the sample mean, $\bar{X}_n$, and sample variance, S^2. It is easy to see that

$$E\left(\frac{1}{n}\sum_{i=1}^{n} X_i\right) = \frac{1}{n}E\left(\sum_{i=1}^{n} X_i\right). \tag{5.16}$$

Hence, from (5.15) and (5.16), we have:

$$E\left(\bar{X}_n\right) = \frac{1}{n}(n\mu) = \mu. \tag{5.17}$$

Equation (5.17) shows that the sample mean is unbiased. In other words, the sample mean is centered about the actual mean.

Now let us denote the sample variance S^2 by $\mathrm{Var}\left(\bar{X}_n\right)$ and the population variance by σ_X^2. Then,

$$\mathrm{Var}\left(\bar{X}_n\right) = \mathrm{Var}\left(\frac{1}{n}\sum_{i=1}^{n} X_i\right) = \frac{1}{n^2}\mathrm{Var}\left(\sum_{i=1}^{n} X_i\right) = \frac{\sigma_X^2}{n^2}. \tag{5.18}$$

Equation (5.18) shows that the sample variance is not unbiased.

Note 5.9

For the sake of comparative studies, the estimation procedures may be extended to more than one population (often, 2). For instance, a study may be to compare GPA (grade point average) of a group of students during their sophomore and junior years to see if there are any changes and if so what the factors may be. This is to look into the general student populations of sophomore and junior. Hence, two independent random samples will be selected, one from the population of sophomore and the other from the population of juniors, with means $\bar{x}_1$ and $\bar{x}_2$, respectively. Hence, the difference between the two means may be used as a point estimator for the two means. It may be a biased or unbiased estimator, as it should be investigated. However, since, $E\left(\bar{X}_1 - \bar{X}_2\right) = E\left(\bar{X}_1\right) - E\left(\bar{X}_2\right) = \mu_1 - \mu_2$, $\bar{X}_1 - \bar{X}_2$ is an unbiased estimator of $\mu_1 - \mu_2$.

Definition 5.8

For a population, let μ and $\hat{\mu}$ be a parameter and an estimator of it, respectively. Then, the **mean square error** of $\hat{\mu}$, denoted by $MSE(\hat{\mu})$, is defined as

$$\text{MSE}(\hat{\mu}) = \text{Var}(\hat{\mu}) + \left[E(\hat{\mu}) - \mu\right]^2. \tag{5.19}$$

We remind the reader that the last term of (5.19) is the square of the bias. Thus, if the bias is zero, such as the sample mean that is unbiased estimator of the population mean, that is, $\hat{\mu} = E(\bar{X})$, then we can have the following definition.

Definition 5.9

The variance of the sample mean, $\text{Var}(\bar{X})$, is the **mean square error of the sample mean**, which is a measure of the quality of the estimator.

Note 5.10

From Definitions (5.4) and (5.2), for a sample of size n, we have:

$$\text{MSE}(\bar{X}) = \frac{\sigma^2}{n}, \tag{5.20}$$

where σ^2 is the population variance.

Note 5.11

The case $\text{MSE}(\bar{X}) = 0$ means that $\bar{X}$ is a perfect estimator of μ.

Example 5.10

Consider 16 uniformly iid random variables $X_1, X_2, \ldots, X_{16}$ distributed on the open interval [0,16). Let X represent this sample. What is the mean and variance of X?

Answer

From what is given, and from Chapter 4, we know that $a = 0$, $b = 16$, and

$$\mu x = E(X) = \frac{b+a}{2} = \frac{16+0}{2} = 8$$

and

$$\sigma_X^2 = \frac{(b-a)^2}{12} = \frac{(16-0)^2}{12} = 21.33.$$

Example 5.11

Consider a binomial random variable X with parameters n and p, that is, sample of size n and unknown probability of success in each trial p. Choose $\hat{p} = \dfrac{X}{n}$ as an estimator of p.

i. Show that $\hat{p}$ is unbiased,
ii. Find its *MSE*.

Answer

i. As we have already shown before, $E(X) = np$ and $\text{Var}(X) = np(1-p)$. Thus, we have:

$$E(\hat{p}) = E\left(\frac{X}{n}\right) = \frac{E(\bar{X})}{n} = \frac{np}{n} = p.$$

Thus,

$$E(\hat{p}) - p = p - p = 0.$$

Hence, $\hat{p}$ is unbiased.

ii. $\text{Var}(\hat{p}) = \text{Var}\left(\frac{X}{n}\right) = \frac{\text{Var}(X)}{n^2} = \frac{np(1-P)}{n^2} = \frac{p(1-P)}{n}.$

Thus,

$$\text{MSE}(\hat{p}) = 0 + \frac{p(1-p)}{n} = \frac{p(1-p)}{n}.$$

This result could have been anticipated from (5.19).

Definition 5.10

If an estimate of a parameter approaches the parameter with an arbitrary accuracy when the sample size increases without bound, the estimate is referred to as **consistent**.

Example 5.12

Consider a sample of size n and an unknown parameter α with an estimator $\hat{\alpha}$. Then, for an arbitrary $\epsilon > 0$, α is consistent if:

$$P\{|\hat{\alpha} - \alpha| < \epsilon\} \rightarrow 1, \text{ as } n \rightarrow \infty.$$

Example 5.13

Show that the sample mean is a consistent estimator of the population mean.

Answer

Let $X_k, k = 1,2,\ldots,n$ represent the sample points of a sample of size n with $\bar{X}_n$ as its sample mean. To answer the question, it suffices to show that the variance of $\bar{X}_n$ is 0. But this is easy since $\text{Var}(\bar{X}_n) = \sigma^2/n$, where σ^2 is the population variance. Hence,

$$\text{Var}(\bar{X}_n) \rightarrow 0, \text{ as } n \rightarrow \infty.$$

Definition 5.11

If an estimate of a parameter approaches the parameter without much wide fluctuations, the estimate is referred to as **efficient**. In other words, an estimator is efficient if its variance is minimal. Let α be an unknown parameter of a population. Let also $\hat{\alpha}_j$ be different estimators of α, j being finite. Then, an estimator $\hat{\alpha}$ is efficient if among all $\hat{\alpha}_j$, we have:

$$\text{Var}(\hat{\alpha}) = \min_j \text{Var}(\hat{\alpha}_k).$$

Let us now discuss the median of a sample. However, before doing that, we pause and go back to order statistics that we discussed in Chapter 4 to study it a bit more in detail so that we can analyze **the sample median**. The use of order statistics has increased in recent years, particularly, in nonparametric that we will study in Chapter 6. The difference between the largest data points and the smallest data points is referred to as the **sample range**.

To develop the distribution of order statistic, consider a sample of size n with elements as n independent random variables $X_1, X_2, \ldots, X_n$ from a continuous distribution with cumulative distribution function (cdf) and pdf, $F_X(x)$ and $f_X(x)$, respectively, such that, for constants a and b, $a < x < b$ with the possibility of $a = -\infty$ and $b = +\infty$, $0 < F_X(x) < 1$, $F_X(x) = 0$, and $F_X(b) = 1$. Let $Y_1 < Y_2 \ldots < Y_n$ be the order statistics for this sample. Each Y_i of $Y_1 < Y_2 \ldots < Y_n$ is interpreted as follows: Y_1 is the smallest of $X_1, X_2, \ldots, X_n$; Y_2 is the second smallest of $X_1, X_2, \ldots, X_n$; $\ldots$ and Y_n is the largest of $X_1, X_2, \ldots, X_n$.

Now, consider the event of the *rth* order statistic Y_r be less than or equal to y, that is, $Y_r \leq y$. This event can occur if and only if at least r observations out of n are less than or equal to y. The order statistic of the sample is, in fact, a set of n Bernoulli trials with probability of success in each trial as $F_X(y)$. Thus, for the event Y_r, there should be at least r successes. Hence, we are dealing with a binomial pmf. Thus, denoting the pmf of Y_r as $G_r(y) = P(Y_r \leq y)$, we will have the following:

$$G_r(y) = P(Y_r \leq y) = \sum_{j=r}^{n} \binom{n}{j} [F_X(y)]^j [1 - F_X(y)]^{n-j}. \tag{5.21}$$

Or

$$G_r(y) = P(Y_r \leq y) = \sum_{j=r}^{n-1} \binom{n}{j} [F_X(y)]^j [1 - F_X(y)]^{n-j} + [F_X(y)]^n. \tag{5.22}$$

Taking the derivative of (5.22) with respect to y, we will have the pdf of Y, denoted by $g_r(y)$, as follows:

$$g_r(y) = P(Y_r \leq y) = \sum_{j=r}^{n-1} \binom{n}{j} j f_X(y) [1 - F_X(y)]^{n-j}$$

$$+ \sum_{j=r}^{n-1} \binom{n}{j} [F_X(y)]^j (n-j) [1 - F_X(y)]^{n-j-1} [-f_X(y)] + n [F_X(y)]^{n-1} f_X(y). \tag{5.23}$$

Expanding the combinations, we have the following:

$$\binom{n}{j}j = \frac{n!}{(j-1)!(n-j)!} \text{ and } \binom{n}{j}(n-) j = \frac{n!}{(j)!(n-j-1)!}. \tag{5.24}$$

Thus, from (5.23) and (5.24), for $a < y < b$, only the first term of the first sum in (5.23) has a nonzero value as follows:

$$g_r(y) = \frac{n!}{(r-1)!(n-r)!}[F_X(y)]^{r-1}[1-F_X(y)]^{n-r} f_X(y). \tag{5.25}$$

Note 5.12

It is left as an exercise to show that the rest of the terms of (5.23) add up to zero.

Note 5.13

It is left as an exercise to show that

1. The pdf of the smallest order statistic is

$$g_1(y) = n[1-F_X(y)]^{n-1} f_X(y), \qquad a < y < b. \tag{5.26}$$

2. The pdf of the largest order statistic is

$$g_n(y) = n[F_X(y)]^{n-1} f_X(y), \qquad a < y < b. \tag{5.27}$$

Note 5.14

Using the definition of sample median, in terms of ordered statistics, which is denoted by $\tilde{X}$, we can redefine it as follows:

$$\tilde{X} = \begin{cases} Y_{\frac{n+1}{2}}, & n \text{ is odd,} \\ \dfrac{Y_{\frac{n}{2}} + Y_{\frac{n}{2}+1}}{2}, & n \text{ is even} \end{cases} \tag{5.28}$$

Example 5.14

Let us consider a random sample of size 6 with elements as $X_1, X_2, X_3, X_4, X_5, X_6$ with pdf as:

$$f_X(x) = 3x^2, \qquad 0 < x < 1. \tag{5.29}$$

Also, let $Y_1 < Y_2 < Y_3 < Y_4 < Y_5 < Y_6$ be the order statistics associated with this random sample. Find

　　i. The pdf of Y_4,
　　ii. The probability that the value of Y_4 will be between $\dfrac{1}{4}$ and $\dfrac{3}{4}$.

Answer

First, we see from (5.29) that the cdf is:

$$F_X(x) = \int_0^x 3t^2\, dt = x^3, \quad 0 < x < 1. \tag{5.30}$$

　　i. Now, we see from the order statistics that Y_4 is the fourth smallest among the six observations. From (5.25), (5.29), and (5.30), we have:

$$g^4(y) = \frac{6!}{(4-1)!(6-4)!}\left(y^3\right)^{4-1}\left[1 - y^3\right]^{6-4}\left(3y^2\right) = \frac{3 \cdot 6!}{3!2!}y^9\left[1 - y^3\right]^2 \cdot y^2,$$

or

$$g_4(y) = 180 y^{11}\left[1 - y^3\right]^2, \quad 0 < y < 1.$$

　　ii. $P\left(\dfrac{1}{4} < Y_4 < \dfrac{3}{4}\right) = \displaystyle\int_{\frac{1}{4}}^{\frac{3}{4}} 180 y^{11}\left[1 - y^3\right]^2 = 180\left(\dfrac{y^{12}}{12} - \dfrac{2y^{15}}{15} + \dfrac{y^{18}}{18}\right)\Big|_{\frac{1}{4}}^{\frac{3}{4}}$

$$= 180\left[\frac{1}{12}\left(\frac{3}{4}\right)^{12} - \frac{2}{15}\left(\frac{3}{4}\right)^{15} + \frac{1}{18}\left(\frac{3}{4}\right)^{18}\right.$$

$$\left. -\left(\frac{1}{12}\left(\frac{1}{4}\right)^{12} - \frac{2}{15}\left(\frac{1}{4}\right)^{15} + \frac{1}{18}\left(\frac{1}{4}\right)^{18}\right)\right]$$

$$= 0.2108.$$

Example 5.15

Let us consider a random sample of size 5 with elements as X_1, X_2, X_3, X_4, X_5 with pdf and cdf as in Example 5.14 as (5.30):

$$f_X(x) = 3x^2 \text{ and } F_X(x) = x^3, 0 < x < 1. \tag{5.31}$$

Also, let $Y_1 < Y_2 < Y_3 < Y_4 < Y_5$ be the order statistics associated with this random sample. Find

　　i. The variance of mean,
　　ii. The pdf of median,
　　iii. The variance of median.

Answer

i. From (5.31), the mean and variance of the population, denoted by μ and σ^2, are calculated as follows:

$$\mu = E(X) = \int_0^1 x \cdot 3x^2 \, dx = \frac{3}{4}.$$

$$E\left(X^2\right) = \int_0^1 x^2 \cdot 3x^2 \, dx = \frac{3}{5}.$$

$$\sigma^2 = \text{Var}(X) = E\left(x^2\right) - \mu^2 = \frac{3}{5} - \frac{9}{16} = \frac{3}{80}.$$

Hence,

$$\text{Var}\left(\overline{X}\right) = \frac{\sigma^2}{n} = \frac{3}{400} = 0.0075.$$

ii. Now, from (5.28), since $n = 5$, which is odd, the median is:

$$\overline{X} = Y_{\frac{5+1}{2}} = Y_3.$$

Hence, the pdf of the median is:

$$g_3(y) = \frac{5!}{(3-1)!(5-3)!}\left(y^3\right)^{3-1}\left[1 - y^3\right]^{5-3}\left(3y^2\right) = \frac{3 \cdot 5}{2!2!}y^6\left[1 - y^3\right]^2 \cdot y^2,$$

or

$$g_3(y) = 90y^8\left[1 - y^3\right]^2, \quad 0 < y < 1.$$

iii. The variance of the median is calculated as follows:

$$E\left(\tilde{X}\right) = \int_0^1 y \cdot 90y^8\left[1 - y^3\right]^2 \, dy = \frac{81}{104}.$$

$$E\left(\tilde{X}^2\right) = \int_0^1 y^2 \cdot 90y^8\left[1 - y^3\right]^2 \, dy = \frac{810}{1,309}.$$

$$\text{Var}\left(\tilde{X}\right) = \frac{810}{1,309} - \left(\frac{81}{104}\right)^2 = 0.25.$$

Note 5.15

As we see in the example above, $\text{Var}\left(\overline{X}\right) < \text{Var}\left(\tilde{X}\right)$. We leave it as an exercise to show that this property may be generalized through examples and estimation, for both cases of median (odd and even n). Hence, the mean is a better efficient estimator for population mean than the median.

Example 5.16

Let us consider students' scores on a test in a statistics course as 35, 67, 89, 94, 38, 49, 76, 46, 58, and 90. Now we write these ten observed values of ten independent random values $X_1, X_2, ..., X_{10}$, as:

$$x_1 = 35, \quad x_2 = 67, \quad x_3 = 89, \quad x_4 = 94, \quad x_5 = 38,$$

$$x_6 = 49, \quad x_7 = 76, \quad x_8 = 46, \quad x_9 = 58, \quad x_{10} = 90.$$

We order these values from small to large and rewrite them as $Y_1, Y_2, ..., Y_{10}$ such that:

$$y_1 = 35 < y_2 = 38 < y_3 = 46 < y_4 < 49 < y_5 = 58$$

$$y_6 = 67 < y_7 = 76 < y_8 = 89 < y_9 < 90 < y_{10} = 94.$$

Hence, the sample median in this case, denoted by $\tilde{X}$, is:

$$\tilde{X} = \frac{Y_5 + Y_6}{2} = \frac{58 + 67}{2} = \frac{125}{2} = 62.5.$$

The sample range, in this case, denoted by D_s, is:

$$D_S = D_{10} - D_1 = 94 - 35 = 59.$$

There are various methods of point estimation. We discuss two of them here, namely, (i) **maximum-likelihood estimator (MLE)** and (ii) **method of moments.**

(i) Maximum-likelihood estimator (*MLE*)

One of the most used methods of point estimation is the **maximum-likelihood estimator, abbreviated as MLE.** Among statisticians, the most popular method of finding out how unbiased an estimator may be is the *MLE*. This is particularly the case when the sample size is large. The *MLE* is a technique used for estimating the parameters of a given distribution, using some observed data. The *MLE* method lets us choose a value for the unknown parameter that most likely is the closest value to the observed data.

Example 5.17

If a population is known to follow a normal distribution but the mean and variance are unknown, the *MLE* can be used to estimate them using a limited sample of the population, by finding particular values of the mean and variance so that the observation is the most likely result to have occurred.

Example 5.18

As another example, suppose we are to take a random sample $X_1, X_2, ..., X_n$, where the X_j, $j = 1, 2, ..., n$, are assumed to be normally distributed with mean μ and standard deviation σ; then among others, the aim may be to find a good estimate of μ, using the data points $x_1, x_2, ..., x_n$ obtained from the random sample.

Thus, the *MLE* method obtains the parameter estimates by finding the parameter values that maximize the **likelihood function** that we will define below. The estimates obtained are called the **maximum-likelihood estimates**, which are also abbreviated as *MLE*.

So, suppose we have a set of data points and want to estimate an unknown parameter or a set of parameters from a population. Then, what the *MLE* does is choosing the estimate that explains the data best, with **maximum probability (likelihood)**. That is, when the sample size increases, the estimate converges faster toward the population parameter(s). Generally speaking, the likelihood of a sample is the probability of getting that sample, given a specified probability distribution model. The **likelihood function** is a way to express that probability: The parameters that maximize the probability of getting that sample are the **maximum-likelihood estimators**.

We now formally define *MLE*.

Definition 5.12

Let us suppose that $\{X_1, X_2, ..., X_n\}$ is a set of iid random variables with observed values $\{x_1, x_2, ..., x_n\}$ taken from an unknown population distribution pmf or cdf with a pdf, say $f_X(X_j; \alpha)$ depending upon a parameter α, where f is the model, X_j is the set of random variables, and α is the unknown parameter. For the maximum-likelihood function, we want to know what the most likely value for α is, given the set of random variables X_j. The goal of the MLE is to maximize the **likelihood function:**

$$L(\alpha) = f_{X_1, X_2, ..., X_n}(x_1, x_2, ..., x_n | \alpha) = P(X_1 = x_1, X_2 = x_2, ..., X_n = x_n)$$

$$= f_X(x_1 | \alpha) \cdot f_X(x_2 | \alpha) ... f_X(x_n | \alpha) = \prod_{j-1}^{n} f_X(x_j | \alpha). \tag{5.32}$$

Often, the **log-likelihood** function

$$\ln L(\alpha) = \sum_{j=1}^{n} \ln f(x_j | \alpha) \tag{5.33}$$

is easier to work with.

Definition 5.13

We generalize Definition 5.12 for n parameters as follows. Consider a random sample (random vector) $X_1, X_2, ..., X_n$ with $x_1, x_2, ..., x_n$ as the observed values of the random sample, with a joint *pmf* or *pdf*

$$f_{X_1, X_2, ..., X_n}(x_1, x_2, ..., x_n; \alpha_1, \alpha_2, ..., \alpha_n), \tag{5.34}$$

depending on n unknown parameters. Then, the function f, defined in (5.33), is referred to as the **likelihood function**. The value of each one of the parameters $\alpha_1, \alpha_2, \ldots, \alpha_n$ that maximizes the likelihood function f, defined in (5.33), denoted by $\hat{\alpha}_1, \hat{\alpha}_2, \ldots, \hat{\alpha}_n$, respectively, is referred to as the **maximum-likelihood estimate** of each parameter $\alpha_1, \alpha_2, \ldots, \alpha_n$, respectively, that each is abbreviated as **MLE**. If each random variable $X_1, X_2, \ldots, X_n$ is taken instead of $x_1, x_2, \ldots, x_n$, then each value $\hat{\alpha}_1, \hat{\alpha}_2, \ldots, \hat{\alpha}_n$ is referred to as the **maximum-likelihood estimator** of each $\alpha_1, \alpha_2, \ldots, \alpha_n$, respectively. In summary, **the goal of MLE is to find a point estimator** $f(X_1, X_2, \ldots, X_n)$ **such that** $f(x_1, x_2, \ldots, x_n)$ **is a "good" point estimate of** α, **where** $x_1, x_2, \ldots, x_n$ **are the observed values of the random sample.**

Note 5.16

Here is a reason why the expression maximum likelihood is used rather than maximum probability: Although most people often use probability and likelihood interchangeably, statisticians and probabilists distinguish between the two. The reason for the confusion is if two expressions of "likelihood L (parameter 1, parameter 2; data)" are the same as those of "probability P (parameter 1, parameter 2; data)". The answer may be no because one is asking about the data and the other is asking about the parameter values. This is why the method is called the **maximum likelihood** and not the **maximum probability**.

Note 5.17

The question may arise if a maximum-likelihood estimation can always be solved in an exact manner. The answer is no. It's more likely that in a real-world scenario, the derivative of the log-likelihood function is still analytically intractable (i.e., it is way too hard or impossible to differentiate the function manually). Therefore, iterative methods like expectation–maximization (EM) algorithms are used to find numerical solutions for the parameter estimates.

An EM algorithm is an iterative method to find the MLE or maximum a posteriori (MAP) estimates of parameters in statistical models, where the model depends on unobserved latent variables. The EM iteration alternates between performing an expectation E-step, which creates a function for the expectation of the log-likelihood evaluated using the current estimate for the parameters, and a maximization M-step, which computes the parameters maximizing the expected log-likelihood found on the E-step.

Example 5.19

We present this example to show the **algorithmic method of finding the MLE.** Let us consider a Bernoulli random variable X with p as the probability of success and $1 - p$ as the probability of failure. The probability p, as a population parameter, can be thought of as a proportion of an element in a population with a special feature, say with higher education, for example. What is the MLE for p?

Answer

Step 1. Identify pmf or cdf describing the population characteristics
Let us denote the success and failure by 1 and 0, respectively. The pmf of X, being a Bernoulli random variable with parameter p, will be:

$$f_X(x) = p^x(1-p)^{1-x}, \quad x = 0,1. \tag{5.35}$$

Step 2. Choose a random sample and find the joint pmf or pdf
Now, choose a sample of size n and denote the random vector by $X_1, X_2, \ldots, X_n$ with observed values denoted by $x_1, x_2, \ldots, x_n$, respectively. Hence, the joint pmf of $X_1, X_2, \ldots, X_n$ with values $x_1, x_2, \ldots, x_n$ will be as follows:

$$P(X_1 = x_1, X_2 = x_2, \ldots, X_n = x_n) = \prod_{j=1}^{n} p^{x_j}(1-p)^{1-x_j} = \left[p^{\sum_{j=1}^{n} x_j} \right]\left[(1-p)^{\left(n - \sum_{j=1}^{n} x_j\right)} \right]. \tag{5.36}$$

Relation (5.36) is the likelihood function we need. Let us denote this function by $L_{(p)}$.

Step 3. Write the likelihood function
We now have to find the value of p that maximizes $L_{(p)}$. Since the Bernoulli trials are independent, we will have:

$$L(p) = f_{X_1, X_2, \ldots, X_n}(X_1, X_2, \ldots, X_n; p) = P(X_1 = x_1, X_2 = x_2, \ldots, X_n = x_n)$$

$$= f_{X_1}(x_1; p) f_{X_2}(x_3; p) \ldots f_{X_n}(x_n; p) \tag{5.37}$$

$$= \left[p^{\sum_{j=1}^{n} x_j} \right]\left[(1-p)^{\left(n - \sum_{j=1}^{n} x_j\right)} \right], \quad 0 \leq p \leq 1.$$

Step 4. Find the value of p that maximizes $L_{(p)}$
To maximize $L(p)$, as is the standard practice, we take the derivative of $L(p)$ with respect to p and set it equal to zero. Hence,

$$\frac{dL(p)}{dp} = \left(\sum_{j=1}^{n} x_j \right)\left[p^{\sum_{j=1}^{n} x_j - 1} \right]\left[(1-p)^{\left(n - \sum_{j=1}^{n} x_j\right)} \right]$$

$$- \left(p^{\sum_{j=1}^{n} x_j} \right)\left(n - \sum_{j=1}^{n} x_j \right)\left[(1-p)^{\left(n - \sum_{j=1}^{n} x_j - 1\right)} \right] = 0, \quad 0 \leq p \leq 1. \tag{5.38}$$

From (5.38), we will have:

$$p^{\sum_{j=1}^{n} x_j - 1}\left((1-p)^{\left(n - \sum_{j=1}^{n} x_j - 1\right)} \right)\left[\frac{(1-p)\sum_{j=1}^{n} x_j}{p} - \frac{p(n - \sum_{j=1}^{n} x_j)}{1-p} \right] = 0, \quad 0 < p < 1.$$

Or

Or

$$\sum_{j=1}^{n} x_j - np = 0,$$

from which we find:

$$p = \frac{\sum_{j=1}^{n} x_j}{n} = \bar{x}, \qquad (5.39)$$

where $\bar{x}$ is the value of the sample mean.

Step 5. Calculate the *MLE* of parameter *p*

Relation (5.39) shows that the statistic sample mean is the *MLE* of p. Hence, denoting the MLE of p by $\hat{p}$, we see that, for this case,

$$\hat{p} = \bar{X}. \qquad (5.40)$$

Note 5.18

At this point, we should verify that (5.29) indeed results in a maximum. This can be done by verifying that the second derivative of the log-likelihood with respect to p is negative. We leave it as an exercise to complete this verification.

Note 5.19

It is customary for the sake of ease of calculation to use the natural logarithm of the likelihood function for taking derivative. Hence, for Example 5.19, we would have to add $\ln L(p)$ in Step 3 to obtain the following:

$$\ln L(p) = \left(\sum_{j=1}^{n} x_j \right)(\ln p) + \left(n - \sum_{j=1}^{n} x_j \right)[\ln(1-p)], \quad 0 < p < 1. \qquad (5.41)$$

Then,

$$\frac{d \ln L(p)}{dp} = \left(\sum_{j=1}^{n} x_j \right)\left(\frac{1}{p} \right) + \left(n - \sum_{j=1}^{n} x_j \right)\left(\frac{-1}{1-p} \right) = 0, \quad 0 < p < 1. \qquad (5.42)$$

The other steps follow similarly. We leave it as an exercise to verify that the same end result will be obtained.

Example 5.20

For the sake of quality control, a small company selects a random sample of ten staff members from total workers to test the effectiveness of the jobs they are performing. The question is for what value of p (as the probability of success) a sample may most likely look like the one observed?

Answer

The sample staff members are numbered, and it shows the job performances of staff members 3, 5, and 8 are below expectation. Let us denote the performance effectiveness and below effectiveness expected by 1 and 0, respectively. Then, the sample elements will transfer to the set $\{1,1,0,1,0,1,1,0,1,1\}$. If X is a random variable representing the sample, then the elements of the sample may be denoted by $\{X_1, X_2, \ldots, X_{10}\}$. Based on the effectiveness and noneffectiveness of performances, X will have two values 0 and 1.

Now, let p denote the probability that the performance of the staff chosen is at the accepted effective ness level and $1 - p$ denote the probability if the performance of the staff is not at the level of expectation. Symbolically, we have:

$$p = P(X_j = 1) \text{ and } 1 - p = P(X_j = 0), \, j = 1, 2, \ldots, 10.$$

Thus, the joint *pmf* of $\{X_1, X_2, \ldots, X_{10}\}$ will be as follows:

$$f_{X_1, X_2, \ldots, X_{10}}(x_1, x_2, \ldots, x_{10}; p) = \underbrace{p}_{1}\underbrace{p}_{2}\underbrace{1-p}_{3}\underbrace{p}_{4}\underbrace{1-p}_{5}\underbrace{p}_{6}\underbrace{p}_{7}\underbrace{1-p}_{8}\underbrace{p}_{9}\underbrace{p}_{10}$$

$$= p^7(1-p)^3. \tag{5.43}$$

We now need to maximize the likelihood function found in (5.43) or, as we showed above, equivalently, its logarithm, that is,

$$\ln f_{X_1, X_2, \ldots, X_{10}}(x_1, x_2, \ldots, x_{10}; p) = 7\ln p + 3\ln(1 - p).$$

Hence,

$$\frac{d \ln f_{X_1, X_2, \ldots, X_{10}}(x_1, x_2, \ldots, x_{10}; p)}{dp} = \frac{7}{p} + \frac{-3}{1 - p} = 0. \tag{5.44}$$

Solving p from (5.44), we have:

$$\frac{7}{p} + \frac{-3}{1 - p} = \frac{7 - 10p}{p(1 - p)} = 0 \rightarrow p = \frac{7}{10}. \tag{5.45}$$

The number p in (5.44) may be looked at as the ratio of number of success over the sample size in a binomial pmf created by repeated independent Bernoulli trial with probability of success as p. Thus, $p = \dfrac{7}{10}$ is the probability of observing the particular data in hand (10 data points) as large as possible.

Note 5.20

It should be noted that to begin solving the problem, we specified the location of noneffective performances as 2, 5, and 8. However, this is not necessary. This is because just knowing the number of noneefficient staff is sufficient; since if we were told that the number of noneefficient staff was 3, using the binomial pmf, we could find the probability of having exactly three noneefficient staff members among ten staff members, in total, as:

$$p = P(X = 3) = \binom{10}{3} p^7 (1-p)^3 = \frac{7}{10},$$

which, indeed, is the relative frequency of the efficient staff members in the sample.

Example 5.21

In this example, we try to estimate the Poisson parameter. Let $\{X_1, X_2, \ldots, X_n\}$ represent a random sample (iid random variables) with size n, containing n independent observed values $x_1, x_2, \ldots, x_n$ from a Poisson random variable X with parameter μ. We want to find the *MLE* for λ.

Answer

Using the Poisson pmf, the probability of observing the data point $x_j, j = 1, 2, \ldots, n$, at the jth trial is:

$$f_X(x_j | \lambda) = P(X_j = (x_j | \lambda)) = \frac{\lambda^{x_j} e^{-\lambda}}{x_j!}, \quad x_j = 0, 1, 2, \ldots; j = 1, 2, \ldots, n. \qquad (5.46)$$

Hence, the log-likelihood function will be:

$$\ln L(\lambda) = \sum_{j=1}^{n} \ln P_X(x_j | \lambda) = \sum_{j=1}^{n} (x_j \ln \lambda - \lambda - \ln(x_j!)) = \ln \lambda \sum_{j=1}^{n} (x_j) - n\lambda - \sum x_i \ln(x_j!).$$
$$\qquad (5.47)$$

Now, we go through the rest of the steps. We take the derivative of (5.47) and set it equal to 0 to find λ. Hence,

$$\frac{d \ln L(\lambda)}{d\lambda} = \frac{1}{\lambda} \sum_{j=1}^{n} x_j - n = 0 \rightarrow \lambda = \bar{x}.$$

That is, the *MLE* for λ is $\bar{x}$, the sample mean.

Example 5.22

Using computer software like MATLAB, we simulate (generate) a set of normally distributed data points. How could we calculate the MLE of the parameters mean

μ and standard deviation σ, for the normal distribution used, if we randomly select three points from this set, say 5, 7.3, and 12.8?

Answer

To answer the question, we need to calculate the total probability of observing all of the data points under assumption of independence, that is, the joint probability distribution of all observed data points, assuming iid random variables. To do this, we would need to calculate some conditional probabilities. So it is here that we'll make our first assumption. Thus, the joint pdf for the three selected points 5, 7.3, and 12.8 is as follows:

$$P(5,7.3,12.8;\mu,\sigma) = \frac{1}{\sigma\sqrt{2\pi}}e^{-\frac{(5-\mu)^2}{2\sigma^2}} \cdot \frac{1}{\sigma\sqrt{2\pi}}e^{-\frac{(7.3-\mu)^2}{2\sigma^2}} \cdot \frac{1}{\sigma\sqrt{2\pi}}e^{-\frac{(12.8-\mu)^2}{2\sigma^2}}. \quad (5.48)$$

We now have to find the values of μ and σ that maximize (5.48). Hence, we have to differentiate (5.38) once with respect to μ and set it equal to 0 and once with respect to σ and set it equal to 0. To simplify the differentiation, we use the logarithm, that is, the log-likelihood function, and do the same. Hence, applying the natural log to both sides of (5.48), we obtain the following:

$$\ln[P(5,7.3,12.8;\mu,\sigma)] = -3\ln\sigma - \frac{3\ln 2}{2} - \frac{3\ln\pi}{2}$$

$$-\frac{1}{2\sigma^2}\left[(5-\mu)^2 + (7.3-\mu)^2 + (12.8-\mu)^2\right]. \quad (5.49)$$

We first take the derivative of (5.49) with respect to μ, as follows:

$$\frac{\partial\ln[P(5,7.3,12.8;\mu,\sigma)]}{\partial\mu} = \frac{1}{\sigma^2}\left[(5-\mu) + (7.3-\mu) + (12.8-\mu)\right],$$

$$\frac{1}{\sigma^2}\left[(5-\mu) + (7.3-\mu) + (12.8-\mu)\right] = 0,$$

$$(5-\mu) + (7.3-\mu) + (12.8-\mu) = 25.1 - 3\mu = 0,$$

and hence,

$$\mu = 8.357. \quad (5.50)$$

Now, we take the derivative of (5.49) with respect to σ, as follows:

$$\frac{\partial\ln[P(5,7.3,12.8;\mu,\sigma)]}{\partial\mu} = \frac{-3}{\sigma} + \frac{1}{\sigma^3}\left[(5-\mu)^2 + (7.3-\mu)^2 + (12.8-\mu)^2\right],$$

$$\frac{-3}{\sigma} + \frac{1}{\sigma^3}\left[(5-\mu)^2 + (7.3-\mu)^2 + (12.8-\mu)^2\right] = 0,$$

$$-3\sigma^2 + \left[(5-\mu)^2 + (7.3-\mu)^2 + (12.8-\mu)^2\right] = 0,$$

$$3\sigma^2 = (5-\mu)^2 + (7.3-\mu)^2 + (12.8-\mu)^2,$$

and hence,

$$\sigma = \frac{1}{3}\sqrt{(5-\mu)^2 + (7.3-\mu)^2 + (12.8-\mu)^2}. \tag{5.51}$$

Now from (5.50), using the value $\mu = 8.357$ in (5.51), we will have:

$$\sigma = \frac{1}{3}\sqrt{(5-8.357)^2 + (7.3-8.357)^2 + (12.8-8.357)^2},$$

$$= \frac{1}{3}\sqrt{(-3.357)^2 + (-1.057)^2 + (4.443)^2},$$

$$= \frac{1}{3}\sqrt{11.269 + 1.117 + 19.740} = \frac{1}{3}\sqrt{32.126} = \frac{5.668}{3},$$

and hence

$$\sigma = 1.889. \tag{5.52}$$

Finally, the *MLEs* for μ and σ are $\hat{\mu} = 8.357$ and $\hat{\sigma} = 1.889$, respectively.

(ii) Method of Moments (MoM)

The second common method of point estimation is the **method of moments (MoM)**. It seems this is the oldest method of point estimation that goes back at least to Carl Pearson, an English mathematician and biostatistician (1857–1936), in the mid-nineteenth century. One of the strengths of MoM is its simplicity in use, and that it almost always yields some sort of estimate. However, in many cases, this method is not the best and often needs improvement.

Steps for this method are as follows.

Step 1. Choose a random sample

Suppose $X = \{X_1, X_2, \ldots, X_n\}$ is the set of elements of a set of a random sample with size n, containing n independent observed values in the set $x = \{x_1, x_2, \ldots, x_n\}$ from a population with pdf or pmf $f(x|\mu_1, \mu_2, \ldots, \mu_j)$, where $\mu_1, \mu_2, \ldots, \mu_j$ are the parameters of the distribution.

Step 2. Obtain the MoM estimators

To obtain the MoM estimators, equate the first j sample moments to the corresponding j population moments. Now solve the system of j equations and j unknowns. To do that, define the sample and population moments, respectively, denoted by $M_{S_k}, k = 1, 2, \ldots, j$, and $M_{P_k}, k = 1, 2, \ldots, j$. Then, $M_{S_k}, k = 1, 2, \ldots, j$, are as follows:

$$\begin{cases} M_{S_1} = E(X) = \bar{X} = \dfrac{\sum_{k=1}^{n} X_k}{n}, \\[2ex] M_{S_2} = E(X^2) = \dfrac{\sum_{k=1}^{n} X_k^2}{n}, \\[2ex] \vdots \\[1ex] M_{S_j} = E(X^j) = \dfrac{\sum_{k=1}^{n} X_k^j}{n}. \end{cases} \tag{5.53}$$

Now let each of M_{P_k}, $k = 1, 2, \ldots, j$, be a function of $\mu_1, \mu_2, \ldots, \mu_j$, say $\eta(\mu_1, \mu_2, \ldots, \mu_j)$. Then, M_{P_k}, $k = 1, 2, \ldots, j$, are as follows:

$$\begin{cases} M_{P_1} = \eta_1(\mu_1, \mu_2, \ldots, \mu_j), \\ M_{P_2} = \eta_2(\mu_1, \mu_2, \ldots, \mu_j), \\ \vdots \\ M_{P_j} = \eta_j(\mu_1, \mu_2, \ldots, \mu_j). \end{cases} \tag{5.54}$$

From (5.54), we solve for $\mu_1, \mu_2, \ldots, \mu_j$. The solutions, denoted by $\hat{\mu}_1, \hat{\mu}_2, \ldots, \hat{\mu}_j$, are the MoM estimators of $\mu_1, \mu_2, \ldots, \mu_j$, respectively.

Example 5.23

Suppose $X - \{X_1, X_2, \ldots, X_n\}$ is the set of elements of a set of a random sample with size n containing n independent observed values in the set $x = \{x_1, x_2, \ldots, x_n\}$ from a population with a parameter α and pdf

$$f(x; \alpha) = \alpha x^{\alpha - 1}, \quad 0 < x < 1, 0 < \alpha < \infty. \tag{5.55}$$

Using MoM, we want to find a point estimate for α.

Answer

Here $M_{S_1} = \bar{X}$ and $M_{P_1} = E(X) = \int_0^1 xf(x)\,dx$. According to MoM, $M_{S_1} = M_{P_1}$, then

$$\bar{X} = E(X) = \int_0^1 x\alpha x^{\alpha - 1}\,dx = \alpha \int_0^1 x^\alpha\,dx = \left. \frac{\alpha x^{\alpha + 1}}{\alpha + 1} \right|_0^1 = \frac{\alpha}{\alpha + 1}. \tag{5.56}$$

Since $\bar{x}$ is an unbiased estimator of the population mean, solving for α from (5.42) will give us the estimator we are looking for. Hence,

$$\bar{X} = \frac{\alpha}{\alpha + 1} \rightarrow \alpha = \alpha\bar{X} + \bar{X} \rightarrow \alpha(1 - \bar{X}) = \bar{X}. \tag{5.57}$$

Therefore, from (5.57):

$$\hat{\alpha} = \frac{\bar{X}}{1 - \bar{X}}$$

is a MoM point estimator for α.

5.2.2 INTERVAL ESTIMATION

A point estimate, even the best one, of an unknown parameter, such as the mean, variance, or standard deviation, is not an accurate estimate since it is an estimate only based on a single random sample. It by itself provides no information about the precision and reliability of the estimation. For instance, we have no idea how close the value of the sample mean, $\bar{X}$, that we found as an estimate for the population mean, μ, is to μ, that is, a bit higher or little lower. To remedy this concern, an alternative way of estimating a parameter is **interval estimation**, recognized by Neyman (1937).

Definition 5.14

An **interval estimation** is a range of values obtained by repeated sampling and finding two values, say a and b, within which the true value of the unknown parameter is expected to fall. That is, the value of the parameter falls in the interval $a < x < b$ with some probability, say 90%, 95%, or 99%. Such number, that is, the **probability**, is referred to as the **confidence coefficient**, which is also called the **confidence level**, and the interval is called the **confidence interval** (sometimes abbreviated as **CI**). he endpoints of a CI are called the **upper and lower confidence limits**.

The confidence limits incorporate certain degrees of true values of parameters. Thus, if we determine a **90%** CI estimate, we understand that the probability that the interval contains the true parameter is **0.90**. That is, out of **100** possible intervals, **90** of the intervals are certain to contain the true parameter. Yet, in other words, **with a probability of 0.90, the actual value of the parameter will fall in the interval** $a < x < b$.

Hence, for interval estimation, we sample repeatedly. Then, the **confidence level** is an indication of the success of the process of construction of a CI. Thus, for the confidence level of 0.95, it means that if the same population is sampled numerous times and the interval estimates are calculated each time, the resulting intervals would lead to containing the true population parameter in approximately 95% of the cases. The inverse significance can be thought as $1 - 0.95 = 0.05$, or 5%. Also, a confidence level of 90% implies that 90% of all samples would give an interval that includes the unknown parameter, and $1 - 0.9 = 0.10$, or 10%, of all samples would yield an erroneous interval. Further, CIs with a confidence level of 80% will, in the long run, miss the true population parameter 20% of the times, that is, one out of every five times.

Note 5.21

Often there is a misunderstanding regarding CIs. A CI does not predict with a given probability that a parameter lies within the interval. The problem arises because the word "confidence" is misinterpreted as implying probability. Parameters are fixed, not random variables, and so a probability statement cannot be made about them. The fact is **that when a CI has been constructed, it either contains the parameter or it does not**.

Note 5.22

The higher confidence level implies that one can more strongly believe that the value of the parameter lies within the interval.

Note 5.23

If X is not actually continuous, it is sometimes argued that a **correction for continuity** should be applied by adding and subtracting 0.5 to x value. In case of normal pdf, the correction for continuity reduces the magnitude of z. That is, failing to correct for continuity will result in a z-score that is too high. In practice, especially when the sample size n is large, the correction for continuity tends to get ignored, but for small n or borderline cases, the correction can be important. However, the correction for continuity can sometimes make things worse, particularly, if it is a close decision. In that case, it is better to use a computer program that can make a more exact calculation, like Stata (with its **bitest** and **bitesti** routines).

Example 5.24 (Known σ)

Let us use the interval estimation to estimate a normal population's parameter mean, μ, assuming that the value of the standard deviation, σ, is known.

Answer

Let X represent the sample points of a sample chosen from a normal population with unknown mean and known standard deviation μ and σ, respectively. Then, $\bar{X}$ has a normal distribution, that is, $\bar{X} \sim N(\mu, \sigma^2/n)$, or $Z = \dfrac{\bar{X} - \mu}{\frac{\sigma}{\sqrt{n}}} \sim N(0,1)$. It is

known that probabilities of standard normal random variable within some intervals are shown in Table 5.1.

Hence,

$$P(-1.96 < Z < 1.96) = P\left(-1.96 < \frac{\bar{X} - \mu}{\frac{\sigma}{\sqrt{n}}} < 1.96\right)$$

$$= P\left(\bar{X} - 1.96\frac{\sigma}{\sqrt{n}} < \mu < \bar{X} + 1.96\frac{\sigma}{\sqrt{n}}\right) = 0.95.$$

Thus, we found an interval estimate for μ with a probability of 0.95. In other words, with the probability of 0.95, the interval $\left(\bar{X} - 1.96\frac{\sigma}{\sqrt{n}}, \bar{X} + 1.96\frac{\sigma}{\sqrt{n}}\right)$ will contain μ. Hence, for a given value of the sample mean, we can estimate the interval that the population mean will fall in with the probability of 0.95.

Example 5.25 (Known σ)

Suppose a statistics course with 10 sections, and 40 students enrolled in each section, is offered in a semester at a university. The same test is given simultaneously to all five sections. Results of this test are selected for a randomly chosen section

TABLE 5.1

Probabilities of Standard Normal Random Variable within Some Intervals

Interval for Z $-a < Z < a$	Probability $P(-a < Z < a)$
$(-1.282, 1.282)$	0.800
$(-1.440, 1.440)$	0.850
$(-1.645, 1.645)$	0.900
$(-1.960, 1.960)$	0.950
$(-2.576, 2.576)$	0.990
$(-2.807, 2.807)$	0.995
$(-3.291, 3.291)$	0.999

and its mean is found to be 78%. Assume that the population standard deviation is 3. What is the estimate of the mean of the test for the entire 400 students in this course with 95% CI?

Answer

We assume that the random variable X representing the test results of students enrolled in this statistics course is normally distributed with a standard deviation of 3. Then, the CI for the mean of the first test in this class is calculated as follows:

$$P\left(\bar{X} - 1.96\frac{\sigma}{\sqrt{n}} < \mu < \bar{X} + 1.96\frac{\sigma}{\sqrt{n}}\right) = P\left(78 - 1.96\frac{3}{\sqrt{40}} < \mu < 78 + 1.96\frac{3}{\sqrt{40}}\right)$$

$$= P(77.0703 < \mu < 78.9297) = 0.95.$$

In other words, with a 95% CI, the average test grade for the entire 400 students enrolled in this course will be between 77.0703% and 78.9292%. The CI in this case is $(77.0703\%, 78.9293\%)$.

Note 5.24

For the example above, the stated result is somewhat misleading. The correct way of stating it is that if the sample had been taken repeatedly infinitely many times, then the mean would fall within the founded interval.

Note 5.25

We note that intervals for Z may be created with different probabilities. From the interval, its length or the range of Z may be calculated, as well. For instance, see Table 5.2.

From Table 5.2, it is clear that the length of the interval becomes shortest when the interval is symmetric around the mean value of Z (i.e., 0). Therefore, when we

TABLE 5.2

The Same Probability of Standard Normal Random Variable within Some Intervals

Interval for Z $-a < Z < a$	Probability $P(-a < Z < a)$	Length of the Interval of the Range of Z
(−1.96, 1.96)	0.95	1.96 − (−1.96) = 3.92
(−1.80, 2.19)	0.95	2.19 − (−1.80) = 3.99
(−2.36, 1.74)	0.95	1.74 − (−2.36) = 4.10
(−2.69, 1.68)	0.95	1.68 − (−2.69) = 4.37

calculate the CIs, we consider a symmetric interval $\left(-Z_{\frac{\alpha}{2}}, Z_{\frac{\alpha}{2}}\right)$, which is around the zero as discussed in the following terms:

i. *z-interval* is used for large sample size (greater than or equal to 30) and population standard deviation, σ, is unknown. Furthermore, if the population is normally distributed and the population standard deviation, σ is known, z-interval can be used

ii. *t-interval* is used when the population standard deviation, σ, is unknown and the population is normally distributed. Furthermore, if the sample size is greater than or equal to 30 and the population standard deviation, σ, is unknown, t-interval can be used

Definition 5.15

For a given value α, a $100(1-\alpha)\%$ CI for the population mean μ of a normal population with known standard deviation σ is given as

$$\left(\bar{X} - Z_{\frac{\alpha}{2}}\left(\frac{\sigma}{\sqrt{n}}\right), \bar{X} + Z_{\frac{\sigma}{2}}\left(\frac{\sigma}{\sqrt{n}}\right)\right). \tag{5.58}$$

The interval in (5.58) is often written compactly as

$$\bar{X} \pm Z_{\frac{\alpha}{2}}\left(\frac{\sigma}{\sqrt{n}}\right). \tag{5.59}$$

The value α is referred to as the **confidence level**, $Z_{\frac{\sigma}{2}}$ is called the **confidence coefficient or critical value**, the term "$Z_{\frac{\alpha}{2}}\left(\frac{\sigma}{\sqrt{n}}\right)$" is referred to as the **margin of error of a confidence level**, and the term "$\frac{\sigma}{\sqrt{n}}$" is called the **standard error**.

Note 5.26

The center of the above interval is

$$\frac{\left(\bar{X} - Z_{\frac{\alpha}{2}} \frac{\sigma}{\sqrt{n}}, \bar{X} + Z_{\sigma} \frac{\sigma}{\sqrt{n}}\right)}{2} = \bar{X}$$

and the width of the interval is

$$\left(\bar{X} + Z_{\frac{\alpha}{2}} \frac{\sigma}{\sqrt{n}}\right) - \left(\bar{X} - Z_{\sigma} \frac{\sigma}{\sqrt{n}}\right) = 2 \cdot Z_{\frac{\alpha}{2}} \frac{\sigma}{\sqrt{n}}.$$

According to the above width, it is clear that when the confidence level (α) increases, the width of the CI also increases. Practically, we like to have a higher CI and lower width, as such a CI has higher confidence that the unknown parameter will be in the smaller interval. Unfortunately, it is impossible to achieve both of these at the same time. In practice, we fix the confidence level and the width of the interval we require, before conducting the study. Then, we collect the required sample size in order to achieve the above fixed values. As far as the confidence levels are concerned, 90%, 95%, and 99% are frequently used and 95% is the default in most of the situations. For these levels of significance, using the normal Z-table, available online, we can construct the following table:

$100(1-\alpha)\%$	α	$\frac{\alpha}{2}$	$Z_{\frac{\alpha}{2}}$
90	0.10	0.05	1.645
95	0.05	0.025	1.96
99	0.01	0.005	2.575

In practice, 95% confidence is preferred over 99% because, for normal distribution the interval for which data points fall in is wider, that is, $(-2.58, 2.58)$, and hence, less precise. If high CI is desired, the confidence limits may be fixed and the sample size be calculated accordingly, as discussed above (Figure 5.2).

Using TI83 calculator, the Z_α value for any given α value can be calculated. For instance, here is for 80% CI?

$$100(1-\alpha)\% = 80 \rightarrow 1-\alpha = 0.8 \rightarrow \alpha = 0.2.$$

Hence, $\frac{\alpha}{2} = 0.10$. Now we need to look at the standard normal table to match the right tail area of 0.1. Then, we can find the $Z_{\frac{\alpha}{2}} = 1.28$. This can be calculated using the TI-83/84 calculator with the following command:

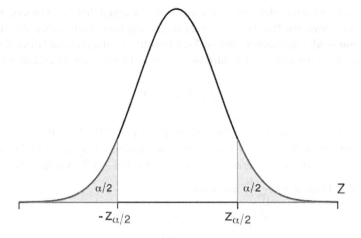

FIGURE 5.2   Display of normal distribution.

$$Z_{\frac{\alpha}{2}} = \text{invNorm}\left(1 - \frac{\alpha}{2}, \text{mean}, \text{stdev}\right) = \text{invNorm}(0.9, 0, 1).$$

See Figure 5.3 for illustration.

Note 5.27

Since the central limit theorem assures us that the sample mean $\bar{X}$ is approximately normally distributed, that is, $\bar{X} \sim N\left(\mu, \sigma^2/n\right)$, $\bar{X} \pm Z_{\frac{\alpha}{2}}\left(\frac{\sigma}{\sqrt{n}}\right)$ is a CI for μ with a confidence level of $\sim 100(1 - \alpha)\%$.

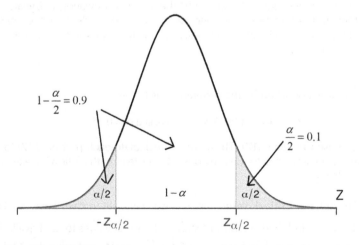

FIGURE 5.3   Display of CIs.

In the above examples and discussions, we assumed that the standard deviation is known. However, this is not a realistic assumption. Thus, when the **standard deviation is also unknown,** we will try to estimate the standard deviation of the population by the sample standard deviation, s. In this case, the CI for μ becomes

$$\bar{X} \pm Z_{\frac{\alpha}{2}} \left(\frac{s}{\sqrt{n}} \right) \tag{5.60}$$

with yet the same confidence level of approximately $100(1-\alpha)\%$.

In Definition 5.7, if σ is not known, we replace it by its sample estimates. Thus, to maintain the coverage probability of $100(1-\alpha)\%$, we need to adjust the factor $Z_{\frac{\alpha}{2}}$. Thus, $Z = \dfrac{\bar{X} - \mu}{\dfrac{\sigma}{\sqrt{n}}}$ becomes

$$T = \frac{\bar{X} - \mu}{\dfrac{s}{\sqrt{n}}}, \tag{5.61}$$

which has the Student t-distribution with $v = n - 1$ degrees of freedom. Thus, (5.17) becomes:

$$\bar{X} \pm t_{\frac{\alpha}{2}, v} \left(\frac{s}{\sqrt{n}} \right), \tag{5.62}$$

which is a $100(1-\alpha)\%$ CI for μ, where $t_{\frac{\alpha}{2}, v}$ is the $(1-\alpha)$ quantile of the distribution.

Example 5.26 (Unknown σ)

Consider a study that aims to estimate the age of junior students of a very large university in the United States. For this, a sample of 2,000 junior students is selected, and the sample average is 22 years and the standard deviation is 8 years. Using this information, estimate the average age of all junior students of the university, the accuracy of this estimate within a given CI with confidence level of 95%, and the margin of error.

Answer

The information given by the problem is as follows:

$$n = 2,000, \bar{X} = 22, s = 8, \text{ and } \alpha = 0.95.$$

We are to use formula (5.17). Thus, we will calculate each part of (5.17) to reach the conclusion. Referring to the previous table instead of the following explanation to calculate $Z_{\frac{\sigma}{2}}$.

    i. To find the **critical value**, or $Z_{\frac{\sigma}{2}}$, we have alternative ways: (a) Use Table 5.1 and (b) use tables of standard normal values for z, if available. Hence, the confidence level α is 0.05, $\dfrac{\alpha}{2} = 0.025$, and using the Z-table.

(c) Use MATLAB to find this value. Thus, from any of the suggested ways, we will find the level as 1.96.

ii. To calculate the **SE**, we have $\dfrac{s}{\sqrt{n}} = \dfrac{8}{\sqrt{2,000}} = \dfrac{8}{44.72} = 0.179$.

iii. Multiply 1.96 by 0.179 (your critical value by your SE), that is, $Z_{\frac{\alpha}{2}}\left(\dfrac{s}{\sqrt{n}}\right) = (1.96)(0.179) = 0.351$, which is the **margin of error**.

iv. Thus, to find the CI, we have to use $\bar{X} \pm Z_{\frac{\alpha}{2}}\left(\dfrac{s}{\sqrt{n}}\right)$, that is, the mean plus or minus the margin of error. Thus, we have:

$$\textbf{Confidence interval} = \bar{X} \pm Z_{\frac{\alpha}{2}}\left(\frac{s}{\sqrt{n}}\right) = 22 \pm 0.351 = (21.649, 22.351).$$

Therefore, given the aforementioned information, the average age of male junior students in the college will be approximately between 21.649 years (or 21 years, 7 months, and 24 days) to 22.351 years (or 22 years, 4 months, and 6 days) with 95% confidence level. In other words, we are 95% confident that the average age of junior students of the college is between 21.649 years and 22.351 years.

To calculate a CI for a small sample size, n, that is, the size of <30, $n < 30$, choose what is called the **degrees of freedom**, denoted by df, which is defined as $n - 1$.

For instance, let us assume that the sample size is 11 and we want to calculate the 95% CI. Then, here $df = 11 - 1 = 10$ and

$$100(1 - \alpha)\% = 95 \rightarrow 1 - \alpha = 0.05 \rightarrow \frac{\alpha}{2} = 0.025.$$

Therefore, we have one-tailed level of significance value $\dfrac{\sigma}{2} = 0.025$ and two-tailed level of significance value $\alpha = 0.05$. Now you can use the following t-table using both $df = 10$ and either one-tailed level of significance value $\dfrac{\sigma}{2} = 0.025$ or two-tailed level of significance value $\alpha = 0.05$ to find the critical value, $t = 2:228$ (Figure 5.4).

Note 5.28

Unlike the Z-table, using the t-table, one cannot find the critical values for any given α value. We can use only the given α values in the table. Therefore, a statistical software is used to calculate the t-value for any given level of significance.

5.2.3 HYPOTHESIS TESTING

CIs and **hypothesis testing** are closely related. Some textbooks discuss how these two are equivalent. A reason for this idea is that since the true values of the most population parameters are unknown, finding them would be possible by estimation, that is, hypothesize them.

Critical Values of the t Distribution

	Level of significance for one-tailed test					
	.10	.05	.025	.01	.005	.0005
	Level of significance for two-tailed test					
df	.20	.10	.05	.02	.01	.001
1	3.078	6.314	12.706	31.821	63.657	636.619
2	1.886	2.920	4.303	6.965	9.925	31.598
3	1.638	2.353	3.182	4.541	5.841	12.941
4	1.533	2.132	2.776	3.747	4.604	8.610
5	1.476	2.015	2.571	3.365	4.032	6.859
6	1.440	1.943	2.447	3.143	3.707	5.959
7	1.415	1.895	2.365	2.998	3.499	5.405
8	1.397	1.860	2.306	2.896	3.355	5.041
9	1.383	1.833	2.262	2.821	3.250	4.781
10	1.372	1.812	2.228	2.764	3.169	4.587
11	1.363	1.796	2.201	2.718	3.106	4.437
12	1.356	1.782	2.179	2.681	3.055	4.318

FIGURE 5.4   Part of the t-table.

In hypothesis testing, the main aim is to make inference about the unknown population parameters instead of estimating them. In fact, the major purpose of hypothesis testing is to choose between two competing hypotheses about the value of a population parameter. In other words, it is to determine whether there is enough statistical evidence in favor of the hypothesized value of the parameter. Hypothesis testing is like raising a question of truth of a statement and a possible error answering such question.

Statistically speaking, hypothesis testing is the process of using statistics in determining the probability that a given hypothesis is true. In other words, hypothesis testing is a method of statistical inference, that is, finding out if an assumption is true or false based on some evidence or the lack of it.

Example 5.27

Hypothesis testing is for the acquisition of information about the population parameters based on the sample statistic; that is, no generalization is involved. For instance, in assigning dose level to a cancer treatment, a doctor needs to know if the average dose level assigned was effective or noneffective in order to make a decision about the next dose level. So, if the doctor wants to change the dose level, he needs to make some assumption, test, and then decide.

Example 5.28

Suppose a pharmaceutical company has developed a new drug that will treat at least 75% of the patients to recover from the disease. As a standard practice, the company needs to choose a random sample of identically conditioned patients, splits them into two subgroups, tests the drug on one subgroup, applies placebo to the other, and compares the results. The company is to see if there is enough statistical evidence in favor of the minimum 75% capability of treatment of the first subgroup as hypothesized.

In hypothesis testing, there are four essential terms. We briefly state them and then subsequently discuss them in some detail.

i. **Null hypothesis, H_0** (reads "H subzero"). The symbol H_0 is used for a hypothesis to be tested and is commonly referred to as the **null hypothesis**. H_0 is assumed to be true unless there is strong evidence to the contrary.

ii. **Alternative hypothesis, H_1 or H_A** (reads "H sub-one or H sub-A"). The other hypothesis, which is assumed to be true when the null hypothesis is false, is referred to as the **alternative hypothesis** and is often symbolized by H_1 or H_A. In fact, the alternative hypothesis is the complement of the null hypothesis.

iii. **Test statistic.** In statistics, the evidence we mentioned in (i) is to provide proof or disproof from a data set that the null hypothesis is true or false. This means, in order to process a hypothesis testing, we would need a **test statistic**, which is a random variable that determines how close a specific sample result falls to one of the hypotheses being tested. Hence, a test statistics is a numerical quantity calculated using the attributes of a sample.

iv. **p-value.** The strength of evidence supporting the truth of H_0 is measured by what is referred to as the **p-value**.

For instance, if the parameter tested is the population mean, we may choose the sample mean as our random variable. Hence, the question we might ask is, how do we assess our evidence to be manifested in the value of the sample mean? To answer this question, a standard practice is to choose the z-score. Hence, as a general practice, one should **quantify** the evidences and use the central limit theorem. As mentioned, the strength of evidence supporting the truth of H_0 is measured by what is referred to as the **p-value**. It originated in 1700s by a physician from Scotland John Arbuthnot. The p-value evaluates how well the sample data support the devil's advocate argument that the null hypothesis is true (Figure 5.5).

In hypothesis testing, before testing starts, both the null and alternative hypotheses should be stated. Although it is commonly practiced to state the null hypothesis first with an " = " sign, or "true", it could contain statements such as "=", ≤, or ≥. The alternative hypothesis is stated second with the symbol "<", or ">", or "not true" or ≠, that is, "different". Thus, if the parameter to be tested is "x" and it is to be tested for the value "m", possible choices are as follows:

FIGURE 5.5   John Arbuthnot 1667–1735.

$$H_0 : x \geq m \quad H_0 : x > m \quad H_0 : x = m$$

$$H_1 : x < m \quad H_1 : x \leq m \quad H_1 : x \neq m$$

$$(5.63)$$

$$H_0 : x \leq m \quad H_0 : x < m \quad H_0 : x \neq m$$

$$H_1 : x > m \quad H_1 : x \geq m \quad H_1 : x = m.$$

Three of choices are seen often for means, say μ_1 and μ_2:

$$H_0 : \mu_1 = \mu_2 \quad H_0 : \mu_1 \geq \mu_2 \quad \text{and} \quad H_0 : \mu_1 \leq \mu_2$$

$$H_1 : \mu_1 \neq \mu_2 \quad H_1 : \mu_1 < \mu_2 \quad\quad\quad H_1 : \mu_1 > \mu_2.$$

$$(5.64)$$

Other choices are possible. However, since the idea behind H_0 is assumption of truth and hope for its opposite, the last choice is not appropriate; it is similar for some other choices. Perhaps, this is why it is standard practice to use "=" for H_0 and other choices will be stated for the case H_1. Choices with "unequal" sign are the **"two-tailed"** cases, and others are **"one-sided"** or one-tailed cases. In other words, the first case of (5.64) is a two-sided case and the other two cases are one-sided.

Note 5.29

The terms "two-tailed" and "two-sided" are often used interchangeably. The term "side" refers to the hypothesis on which the side of 0 the difference $\mu_1 = \mu_2$ lies (negative or positive). Since this is a statement about the hypothesis, it is independent of the choice of test statistic.

We use H_0: $\mu_1 = \mu_2$ hypothesis testing when we compare two population means. These two means are not from the same population. For instance, suppose we want to know whether the average GPA of a college is higher than that of another college. In this case, μ_1 is the average GPA of one college and μ_2 of the other.

One might ask why do we have to test the equality of two means or proportions? The answer is because two means are always unequal if we keep all decimal places. Then, one may ask if we know that two means or two proportions are unequal, then why we hypothesize that they are equal and essentially testing them? As we normally write equality for H_0 and unequality for H_1, that is,

$$H_0 : \mu_1 = \mu_2$$

$$H_1 : \mu_1 \neq \mu_2,$$

the unequality in H_1 says that we are unsure of the direction of the difference. That is, the difference between μ_1 and μ_2 may be positive or negative since

$$\mu_1 \neq \mu_2 \rightarrow \mu_1 - \mu_2 \neq 0.$$

This is the idea behind the two-tailed or two-sided test. For the one-tailed or one-sided test, we might have a situation like:

$$H_0 : \mu_1 = \mu_2 \text{ or } H_0 : \mu_1 = \mu_2$$

$$H_1 : \mu_1 > \mu_2 \quad H_1 : \mu_1 < \mu_2.$$

Example 5.29

Let us go back to Example 5.28 of testing a drug for 75% effectiveness hypothesis. In this case, our hypothesis will be H_0 for at least 75% and H_1 for <75%. If we denote the "effectiveness" by "e", then we will have:

$$H_0 : e \geq 75\%$$

$$H_1 : e < 75\%$$

(5.65)

This is the case of the one-tailed test.

The company could have hypothesized the effectiveness of the drug as 75%. In that case, we would have had:

$$H_0 : e = 75$$

$$H_1 : e \neq 75.$$

(5.66)

In the latter case, an alternative hypothesis specifies that the parameter is not true on either side of the value specified by H_0, which is called **two-sided** (or **two-tailed**).

Generally, a one-sided test makes it easier to reject the H_0 when H_1 is true. For a large sample, two-sided, 0.05 level t-test puts a probability of 0.025 on each side. It needs a t statistic of less than -1.96 or >1.96 to reject the null hypothesis of no difference in means. A one-sided test puts all of the probabilities into a single tail. It rejects the hypothesis for values of t less than -1.645. Therefore, a one-sided test is more likely to reject H_0 when the difference is in the expected direction.

This makes one-sided tests very attractive to those whose definition of success is having a statistically significant result.

Decision of acceptance or rejection of a null hypothesis is based on probability rather than certainty. Thus, performing error is possible. When the null hypothesis is rejected while it is really true, a **type I error** has occurred. The probability of type I error is α. That is,

$$\alpha \equiv P(\text{Type I error}) = P(\text{Rejecting } H_0 | H_0 \text{ is true}). \tag{5.67}$$

The α **is referred to** as the **significant level**. Thus, for instance, if $\alpha = 0.05$, it would mean that there is a 5% chance that when the null hypothesis is true, it is rejected.

When the null hypothesis is accepted, while it is not true, again an error has occurred. The error occurred in this case is referred to as **type II error**. In other words, **a type II error occurs when a false null hypothesis is not rejected**.

The probability of type II error is denoted by β. That is,

$$\beta \equiv P(\text{Type II error}) = P(\text{Accepting } H_0 | H_0 \text{ is false}). \tag{5.68}$$

Example 5.30

Consider a person who is accused of a crime and has to show up on a criminal jury in the United States, where a person is assumed to be innocent unless proven guilty. So the jury considers the assumption of innocence that indicates the null hypothesis H_0. Evidence is presented to the jury and the jury decides. Its verdict is whether the evidence shows that the person is innocent, that is, H_0 is true, or the person is guilty of the crime accused of, that is, H_0 is false. Thus, there could be possibility of some type of error or no error, depending upon which of the four possible cases we are dealing with.

  i. The accused is innocent and the jury decided he/she is guilty. Hence, no error has occurred.
  ii. The accused is innocent and the jury decided he/she is guilty. Here, the jury has made a mistake and the error is of type I.
  iii. The accused is guilty and the jury decided he/she is innocent. Hence, no error has occurred.
  iv. The accused is guilty, but the jury decided he/she is innocent. Hence, an error has occurred and it is of type II (Table 5.3).

Hence, considering the probabilities α and β for this example, conditional probabilities could be interpreted as follows:

$$\alpha = P\left(\text{Verdict is "guilty"} | \text{person is "innocent"}\right)$$

$$\beta = P\left(\text{Verdict is "innocent"} | \text{person is "guilty"}\right).$$

In case the assumption would have been "guilty", as in some countries, then the types of errors will reverse, as well as α and β.

Hypothesis testing for this example can be set up as follows:

TABLE 5.3

Example of Types of Hypothesis Testing Error, I and II

	Possibilities	Person Really Is (Condition of H_0)	
		Innocent (H_0 is True)	Guilty (H_0 is False)
Jury's Verdict	Innocent (correct action)	No error (correct action)	**Type II error**
(Possible Action)	(Fail to reject H_0)		
	Guilty (reject H_0)	**Type I error**	No error (correct action)

H_0 : The person is innocent

H_1 : The person is guilty.

Note 5.30

If H_0 is rejected, the conclusion is that H_1 is true. If H_0 is not rejected, the conclusion is that H_0 may be true. Hence, in our example, if the verdict is "not guilty", it would mean that the evidence was not strong enough to reject the null hypothesis, which creates doubt of whether the assumption of innocent is really true. This doubt may be quantified.

Thus, to test a hypothesis,

i. It is assumed that H_0 is true until it is proven not so
ii. H_0 is rejected only if the probability that it is true is very small;
iii. If H_0 is rejected, while it is really true, type I error will occur, with probability α (standard practice is to choose value $\alpha = 0.05$ or $\alpha = 0.01$);
iv. If H_0 is accepted, while it is really incorrect, type II error will occur, with probability β;
v. α and β are not independent; in fact, if the sample size increases, both α and β will decrease since the sampling error decreases;
vi. A small sample size may lead to type II error; that is, H_1 might be true, but because the sample is small, H_0 may be failed to be rejected, although it should have.

Definition 5.16

The probability of finding the observed results when the null hypothesis, H_0, is true is referred to as the *p*-value, or **calculated probability**.

The *p*-value can also be described in terms of rejecting H_0 when it is actually true. The *p*-value is used to determine the statistical significance in a hypothesis test.

Here is how we should interpret the *p*-value:

→ **Low *p*-value (typically, $\leq \alpha$):** Data is unlikely with a true null; that is, sample provides enough evidence that the null hypothesis can be rejected. However, it cannot evaluate which of two competing cases is more likely: (1) The null is true but the sample was unusual or (2) the null is false.

→ **High *p*-value (typically, $> \alpha$):** Data is likely with a true null. That is, it indicates weak evidence against the null hypothesis: one fails to reject it.

Note 5.31

The *p*-values very close to the cutoff point, 0.05, are considered to be **marginal**; that is, it could go either way.

Note 5.32

The *p*-values are *not* the probability of making a mistake by rejecting a true null hypothesis, a type I error.

Note 5.33

The hypothesis test and CI give the same result.

Example 5.31

Suppose we are interested in the population mean, μ. Let:

$$H_0 : \mu = 100$$

$$H_1 : \mu > 100$$

Also, suppose that the *p*-value is 0.001, which is significant, and we reject the null hypothesis. This means that μ should be more than 100. If we construct a CI for μ, this interval should not contain 100. It may be something like (105, 110). Hence, both hypothesis test and CI give us the same result in different forms.

Note 5.34

Graphically, the *p*-value is the area in the tail of a probability distribution. It is the area to the right of the test statistic if the alternative hypothesis is greater (upper-tailed test), and it is the area to the left of the test statistic if the alternative hypothesis is lesser (lower-tailed test). If the alternative hypothesis is not equal (the two-tailed test), then the *p*-value, if the area is less than the negative value of the test statistic, and the area is more than the test statistic. For a two-sided alternative, the *p*-value is the probability of deviating only in the direction of the alternative away from the null hypothesis value (Figure 5.6).

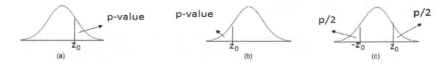

FIGURE 5.6   *p*-values with Z-test statistic. (a) Upper-tailed test, (b) lower-tailed test, and (c) two-tailed test.

Let Z_0 be the test statistic. Following shows the calculation of *p*-value for a given hypothesis.

H_0	H_1	Name of the Test	*p*-value
$\mu = \mu_0$	$\mu > \mu_0$	Upper-tailed	$P(Z > z_0)$
$\mu = \mu_0$	$\mu < \mu_0$	Lower-tailed	$P(Z > z_0)$
$\mu = \mu_0$	$\mu \neq \mu_0$	Two-tailed	$P(Z > z_0) + P(Z < -z_0)$ or $2P(Z > z_0)$

Example 5.32 (Two-Sided Test)

Suppose an institution of higher learning, say a university A, claims that its graduates are as high quality as (=) of a well-known university B, for having high-quality student products. In order to determine the claim is true, the university A conducts a two-tailed test. The null hypothesis states that the university A's graduates are equivalent to the university B's graduates over a period of 10 years, while the alternative hypothesis states that this is not the case (≠).

The university A uses the *p*-value as 0.05. If the test concludes that the *p*-value is 0.03, which is <0.05, then university A needs to accept the fact that there is strong evidence against its claim and, hence, its claim is rejected. That is, the graduates of university A are not as high quality as of the university B.

Example 5.33 (One-Sided Test)

Consider a delivery career that claims its delivery times are 3 days or less on average. However, we might think it's more than that. So, we conduct a hypothesis test. The null hypothesis, H_0, in this test would mean that the delivery time of at most 3 days (≤3) is incorrect. The alternative hypothesis, H_1, in this case, is that the mean time is >3 days (> 3). We randomly sample some delivery times and run the data through the hypothesis test. The *p*-value turns out to be 0.005, which is much <0.05. In real terms, there is a probability of 0.005 that we will mistakenly reject the career's claim that its delivery time is ≤3 days. Since typically we are willing to reject the null hypothesis when this probability is <0.05, we conclude that the mail career's claim is wrong; their delivery times are in fact more than 3 days on average. Of course, we could be wrong by having sampled an unusually high number of late deliveries just by chance.

Example 5.34 (One-Sided Test)

A university is experiencing problems with its first mathematics course that all entering students must enroll in. The mathematics department decides to implement a plan consisting of mandatory attending a tutorial laboratory by students in a special tutorial program on a course section with 25 students who have randomly registered in. The department chair hypothesizes that with all tutorial activities, the class average on the final examination score will increase to be >70 out of 100. We want to test the chair's claim with a significant level of 0.05.

Answer

Suppose the final examination for the section was recorded as in the following table:

67	86	70	79	56
83	63	69	77	84
55	48	68	78	93
89	52	88	99	61
64	72	83	40	39

Let X be a random variable representing the random sample of scores of the final examination and μ its mean. Then, we set the hypothesis as:

$$H_0 : \mu = 70\%$$

$$H_1 : \mu > 70\%$$

Using the statistical software Minitab, the data is entered and running descriptive statistics is as follows (Figure 5.7):

Descriptive Statistics: Final Examination Statistics

Variable	N	Mean	SE Mean	StDev
Final examination	25	70.52	3.24	16.21

One-Sample T: Final Examination Descriptive Statistics

N	Mean	StDev	SE Mean	95% Lower Bound for μ
25	70.52	16.21	3.24	64.97

μ: mean of final examination

Test

Null hypothesis	$H_0: \mu = 70$
Alternative hypothesis	$H_1: \mu > 70$

t-Value	p-Value
0.16	0.437

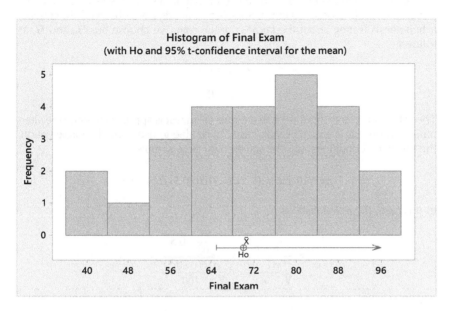

FIGURE 5.7 Histogram for Example 5.34.

Thus, for the sample size of $n = 25$, a t-test was performed. As mentioned before, when n is <30, the t-test is used. The t-value, mean, and standard deviation of the sample are found to be 0.16, 70.52, and 16.21, respectively. Also, the SE has been shown as 3.24. This number is the standard deviation divided by the square root of the sample size. The high value of p, 0.437, shows that the hypothesis mean = 70 is rejected and >70 is accepted.

Note 5.35

There are times that unfair or biased coins are needed to be used. The biasedness is due to the fact that the chance of occurrence of one side is not the same as the chance of occurrence of the other side. So suppose we flip a coin that we have been told is a fair coin. We flip this coin 20 times and tail appears 15 times. Hence, we might think we got a biased coin. To test the biasedness, we hypothesize that the coin is unbiased, that is, fair. This is our null hypothesis, H_0.

Example 5.35 (One-Sided Test)

As a project in a statistics course at a university, a coin is randomly chosen. Two students are to test the fairness of the coin. One student believes the coin is "fair" and the other believes it is "biased toward heads". To test this difference, use $\alpha = 0.05$ and conduct a hypothesis test about the fairness of the coin.

Answer

Let the proportion of number of heads appearing in a set of tosses be denoted by p. So, the coin is tossed 100 times. Suppose 55 heads were observed. To conduct a hypothesis testing about the fairness of the coin, we choose our H_0 and H_1 as follows:

$$H_0 : p = 0.5$$

$$H_1 : p > 0.5.$$

The set hypothesis in this case is that a one-tailed test is appropriate since the alternative hypothesis is a greater than case. To be able to use normal approximation, the following conditions must be satisfied for large sample:

$$np \geq 10 \text{ and } np(1-p) = (100)(0.5)(0.5) \geq 10.$$

In this case, the test statistic is:

$$Z = \frac{\hat{p} - p}{\sqrt{\frac{p(1-p)}{n}}} = \frac{\frac{55}{100} - 0.5}{\sqrt{\frac{0.5(1-0.5)}{100}}} = 1.$$

Thus,

$$\text{the } p\text{-value} = P(Z > 1) = 0.1587.$$

For the p-value greater than α, the null hypothesis will not be rejected at $\alpha = 0.05$. That is, there is no evidence to support the claim of $p > 0.5$. Figure 5.8 shows the p-value (shaded area) for a right-tailed test ($H_1: p > 0.5$).

Note 5.36

In testing a hypothesis to use the p-value and α, the following criteria will be used:

i. The null hypothesis will be rejected if p-value $\leq \alpha$.
ii. The null hypothesis will not be rejected (or fail to be rejected) if p-value $> \alpha$.

Example 5.36 (One-Sided Test)

A civil rights group claims that the percentage of minority teachers in a particular school district is about 20%. The superintendent of the school district selects a random sample of 145 school teachers from this school district in order to test whether the mentioned percentage is <20%. Assume there are 20 minority teachers in the selected sample. Conduct a hypothesis test using the level of significance, $\alpha = 0.05$, to test the claim.

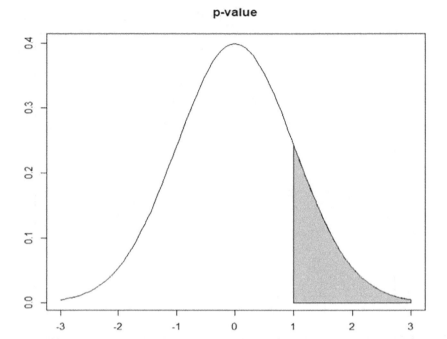

FIGURE 5.8 The *p*-value (shaded area) for a right-tailed test (H_1: $p > 0.5$), Example 5.34.

Answer

For testing hypothesis in this case, we have:

$$H_0 : p = 0.2$$

$$H_1 : p < 0.2.$$

The large sample conditions to satisfy for this example are as follows:

$$np = (145)(0.2) > 10 \text{ and } np(1-p) = (145)(0.2)(0.8) > 10.$$

In this case, the test statistic is:

$$Z = \frac{\hat{p} - p}{\sqrt{\dfrac{p(1-p)}{n}}} = \frac{\dfrac{20}{145} - 0.2}{\sqrt{\dfrac{0.2(1-0.2)}{145}}} = -1.87.$$

Thus,

$$\text{the } p = \text{value} = P(Z < -1.87) = 0.0307.$$

For the *p*-value less than α, the null hypothesis will be rejected at $\alpha = 0.05$. That is, the superintendent of the school district has evidence to support the claim that the percentage of minority teachers in the school district is <20%. Table 5.2 shows the *p*-value (shaded area) for a left-tailed test (H_1: $p < 0.2$) (Figure 5.9).

p-value

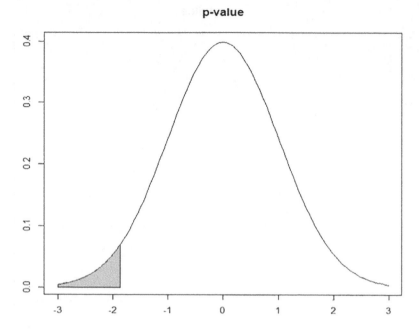

FIGURE 5.9 The *p*-value (shaded area) for a left-tailed test ($H_1: p < 0.2$), Example 5.36.

Example 5.37 (One-Sided Test)

Suppose two students are to test the fairness of a coin in hand. One student believes the coin is "fair" and the other believes it is "biased toward heads". Use $\alpha = 0.05$, choose a random sample by tossing the coin 30 times, and test a hypothesis to indicate whether or not the first student's claim is supported by the results.

Answer

Let *X* be a random variable representing the random sample of size 30 that is taken. Hence, $n = 30$. If the coin were fair, the probability of getting a head would be $p = 0.5$. Thus, $\tilde{p} = 15$ heads. Thus, based on the information available, our hypothesis will be:

$$H_0 : \tilde{p} = 15$$

$$H_1 : \tilde{p} > 15.$$

The set hypothesis in this case is that a one-tailed test is appropriate since the second student believes the coin is biased toward heads.

Of course, the sample distribution in this case is binomial with parameters $n = 30$ and $p = 0.5$. With the sample size being large, we can approximate this pmf by normal pdf with sample mean $\bar{X} = 15$ and sample standard deviation of binomial

$$S = \sqrt{np(1-p)} = \sqrt{(30)(0.5)(0.5)} = 2.7386.$$

To use the standard normal pdf, let

$$Z = \frac{X - \overline{X}}{S} = \frac{X - 15}{2.7386}.$$

Hence, $Z \sim N(0,1)$. To apply the correction for continuity, we subtract 0.5 from X if the observed number of heads is >15, and we add 0.5 to X if the observed number of heads is <15.

Alternatively, since $\sqrt{(0.5)\left(\frac{0.5}{20}\right)} = 0.1118,$

$$z = \frac{\frac{0.5}{20} - 0.5}{0.1118}.$$

Hence, since $P(Z > 1.65) = 0.95 = 1 - \alpha$, H_a should be rejected if $Z > 1.65$. Equivalently, since $X = Z.0.05 + 0.5 + 10$, H_a should be rejected if $X > 14.18$. Yet, equivalently, since $X = Z \cdot 0.1118 + \frac{0.5}{20} + 0.5$, H_a should be rejected if $p > 0.709$.

Thus, $z = \frac{15 - 0.5 - 10}{0.05} = 2.01$ or $z = \frac{0.75 - \frac{0.5}{20} - 0.5}{0.1118} = 2.01$. Not applying the correction for continuity, we would have $z = \frac{15 - 10}{0.05} = 2.24$. This does not change our previous conclusion. In practice, correction for continuity tends not to be made, especially when N is large, but it is still a good idea to do it. Since the computed z-value is >1.65, we reject H_a.

Example 5.38 (One-Sided Test)

Consider the treatment of a type of cancer which is done through taking a medicine with ten levels of doses. History has proved that if patient starts with dose level 2, treatment has no effect. Hence, the treatment starts with level 3. A pharmacology company claims that its researchers have developed a new drug that can start the same treatment with level 2 to reduce the toxicity in the patient.

To test this claim, a statistician is hired to conduct a hypothesis test. With a one-tailed test, the hypothesis of no difference is rejected if and only if the patients taking the drug have toxicity levels significantly lower than those of controls. Outcomes in which patients taking the drug have toxicity levels higher than those of controls are treated as failing to show a difference no matter how much higher they may be.

5.3 COMPARISON OF MEANS AND ANALYSIS OF VARIANCE (ANOVA)

So far, we have discussed how to make inference on a single population. In practice, sometimes one may need to compare two or more populations. Here are some examples: (1) An economist may be interested in comparing the average income of three states to find out which state has significantly higher average income than the other two. (2) An engineer may be interested in comparing the average strength of concrete prepared using different brands of cements. (3) A dean of a college within a university may be interested in comparing the average academic performances of several departments. Such inferences can be made using the mean comparison techniques that will be discussed below.

5.3.1 INFERENCE ABOUT TWO INDEPENDENT POPULATION MEANS

Suppose two independent populations, P_1 and P_2, are under study. We summarize the features for these populations in Table 5.4.

5.3.1.1 Confidence Intervals for the Difference in Population Means

As discussed before (Definition 5.2.3), inferential statistics has both CIs and hypothesis testing. We first consider the CIs for the difference in populations.

Definition 5.17

The **expected value and the variance of the difference in their sample means** are defined, respectively, as:

$$E\left(\overline{X} - \overline{Y}\right) = \mu_1 - \mu_2,$$ (5.69)

and

$$\sigma\left(\overline{X} - \overline{Y}\right) = \sigma_{\overline{X}-\overline{Y}} = \sqrt{\frac{\sigma_1^2}{m} + \frac{\sigma_2^2}{n}}.$$ (5.70)

Note 5.37

From this definition, we see that the difference between the sample means is an unbiased estimator of the difference in the corresponding population means.

We now consider two sizes for sample sizes, namely, <30 and ≥30.

Case 1. Two large sample sizes, $m \geq 30, n \geq 30$

In this case, the $100(1-\alpha)\%$ CI for the difference between the means of the two populations is given by

---

TABLE 5.4

Features for Populations

Populations	Population Means	Population Variances	Random Sample	Sample Sizes	Sample Means	Sample Variances
P_1	μ_1	σ_1^2	$x_1, x_2, \ldots, x_m$	m	$\overline{X}$	S_1
P_2	μ_2	σ_2^2	$y_1, y_2, \ldots, y_n$	n	$\overline{Y}$	S_2

$$\left(\bar{X} - \bar{Y}\right) \pm Z_{\frac{\alpha}{2}} \sqrt{\frac{S_1^2}{m} + \frac{S_2^2}{n}}. \tag{5.71}$$

Example 5.39

An economist was interested in the average family incomes of residents in two cities A and B, with populations P_1 and P_2, respectively. He randomly selected 100 families from city A and 125 families from city B and found the following data (using some notations from Table 5.5).

We want to calculate the 95% CI for the difference in the average incomes of the two cities.

Answer

Here $\alpha = 0.05$, or 5%. Since sample sizes are large and population variances are unknown, using standard normal distribution, we will have $Z_{\frac{\alpha}{2}} = 1.96$. Also, we can use the sample variances instead of the population variances. Thus, the 95% CI for the difference in the means can be calculated as follows:

$$\left(\bar{X} - \bar{Y}\right) \pm Z_{\frac{\alpha}{2}} \sqrt{\frac{S_1^2}{m} + \frac{S_2^2}{n}} = 2,000 - 2,050 \pm 1.96 \sqrt{\frac{10^2}{100} + \frac{12^2}{125}}$$

$$= -50 \pm (1.96)(2.875) = (\$-55.636, \$-44.365).$$

Hence, we are 95% confident that the difference in the average incomes between the two cities A and B is $(\$-55.636, \$-44.365)$.

5.3.1.2   Hypothesis Test for the Difference in Population Means

Let us consider the following hypotheses:

$$H_0 : \mu_1 - \mu_2 = 0$$

$$H_1 : \mu_1 - \mu_2 > 0 \text{ upper-tailed test}$$

$$\mu_1 - \mu_2 < 0 \text{ upper-tailed test}$$

$$\mu_1 - \mu_2 \neq 0 \text{ two-tailed test.}$$

TABLE 5.5
Features for Populations, Example 5.39

Populations	Sample Sizes	Sample Means	Sample Standard Deviations
P_1	$m = 100$	$\bar{X} = \$2,000$	$S_1 = \$10$
P_2	$n = 125$	$\bar{Y} = \$2,050$	$S_2 = \$12$

Test statistic

$$Z = \frac{(\bar{X} - \bar{Y}) - (\mu_1 - \mu_2)}{\sqrt{\dfrac{S_1^2}{m} + \dfrac{S_2^2}{n}}}.$$

Rejection criteria:

Rejection of the hypothesis is determined by the following criteria, based on the alternative hypothesis.

- If the alternative is $\mu_1 - \mu_2 > 0$, we reject H_0 if $Z \geq Z_\alpha$.
- If the alternative is $\mu_1 - \mu_2 < 0$, we reject H_0 if $Z \leq -Z_\alpha$.
- If the alternative is $\mu_1 - \mu_2 \neq 0$, we reject H_0 if $|Z| \geq Z_{\frac{\alpha}{2}}$.

Example 5.40

Let us return to Example 5.39 and test whether the average income of city B is higher than the average income of city A, using the level of significance 0.05.

Answer

Here, in this example, the null and alternative hypotheses are as follows:

$$H_0 : \mu_1 - \mu_2 = 0$$

$$H_1 : \mu_1 - \mu_2 < 0.$$

Test statistic

$$Z = \frac{(\bar{X} - \bar{Y}) - (\mu_1 - \mu_2)}{\sqrt{\dfrac{S_1^2}{m} + \dfrac{S_2^2}{n}}} = \frac{(2,000 - 2,050) - (0)}{\sqrt{\dfrac{10^2}{100} + \dfrac{12^2}{125}}} = -34.08.$$

Note 5.38

The critical value in this case is: $-Z_{0.05} = -1.645$.

Therefore, $Z < -Z_\alpha$. This means that we reject the null hypothesis at the level of significance of 0.05. That is, we have evidence to support that the average income of city B is higher than the average income of city A.

Case 2. Both populations are normally distributed (Two-sample t)

In this case, the $100(1 - \alpha)\%$ CI for the difference in the means of the two populations is given by

$$(\bar{X} - \bar{Y}) \pm t_{\frac{\alpha}{2}, v} \sqrt{\dfrac{S_1^2}{m} + \dfrac{S_2^2}{2}}, \tag{5.72}$$

where $t_{\frac{\alpha}{2}, v}$ represents the critical value of the t-distribution with the level of significance $\frac{\alpha}{2}$ and the degrees of freedom v. Furthermore, v is given by the following:

$$v = \frac{\left(\dfrac{S_1^2}{m} + \dfrac{S_2^2}{n}\right)^2}{\dfrac{\left(\dfrac{S_1^2}{m}\right)^2}{m-1} + \dfrac{\left(\dfrac{S_2^2}{n}\right)^2}{n-1}} \cdot \tag{5.73}$$

Example 5.41

An automobile engineer is interested in comparing the average amount of carbon dioxide emits per each mile driven of two brands of cars. Therefore, he randomly selects ten cars from each brand and measures the amount of carbon dioxide (CO_2) emitted by each car and finds the following statistics (Table 5.6).

Assuming that the amount of CO_2 emitted is normally distributed for both brands of cars, we want to calculate the 95% CI for the difference in true means.

Answer

First, the degree of freedom is calculated as follows:

$$v = \frac{\left(\dfrac{S_1^2}{m} + \dfrac{S_2^2}{n}\right)^2}{\dfrac{\left(\dfrac{S_1^2}{m}\right)^2}{m-1} + \dfrac{\left(\dfrac{S_2^2}{n}\right)^2}{n-1}} = v = \frac{\left(\dfrac{0.3^2}{10} + \dfrac{0.1^2}{10}\right)^2}{\dfrac{\left(\dfrac{p \cdot 3^2}{10}\right)^2}{10-1} + \dfrac{\left(\dfrac{S_2^2}{n}\right)^2}{10-1}} = 10.97.$$

$$v = \frac{\left(\dfrac{s_1^2}{m} + \dfrac{s_2^2}{n}\right)^2}{\dfrac{\left(s_1^2/m\right)^2}{m-1} + \dfrac{\left(s_2^2/n\right)^2}{n-1}} = \frac{\left(\dfrac{0.3^2}{10} + \dfrac{0.1^2}{10}\right)^2}{\dfrac{\left(0.3^2/10\right)^2}{10-1} + \dfrac{\left(0.1^2/10\right)^2}{10-1}} = 10.97.$$

So, we take the integer value of the d.f., $v = 10$

Using the t-table, the critical value $t_{\frac{\alpha}{2}, v} = t_{0.025,10} = 2.228$, the 95% CI is

$$\left(\bar{X} - \bar{Y}\right) \pm t_{\frac{\alpha}{2}, v}\sqrt{\dfrac{S_1^2}{m} + \dfrac{S_2^2}{n}} = (0.75 - 0.76) \pm 2.228\sqrt{\dfrac{0.3^2}{10} + \dfrac{0.1^2}{10}}$$

$$= (-0.2328, 0.2128).$$

TABLE 5.6
Features for Populations, Example 5.41

Car Brand	Sample Size	Sample Average Amount of Emitted CO_2 (in Pounds)	Sample Standard Deviation of Emitted CO_2 (in Pounds)
1	10	0.75	0.3
2	10	0.76	0.1

Hence, we are 95% confident that the difference in the average amount of CO_2 emitted by both car brands is between -0.2328 and $0.2128\,g$.

   Hypothesis test

$$H_0 : \mu_1 - \mu_2 = 0$$

$$H_1 : \mu_1 - \mu_2 > 0 \text{ upper-tailed test}$$

$$\mu_1 - \mu_2 < 0 \text{ upper-tailed test}$$

$$\mu_1 - \mu_2 \neq 0 \text{ two-tailed test.}$$

Test statistic

$$T = \frac{\left(\bar{X} - \bar{Y}\right) - \left(\mu_1 - \mu_2\right)}{\sqrt{\dfrac{S_1^2}{m} + \dfrac{S_2^2}{n}}}.$$

Critical value, $t_{\alpha,v}$, where

$$v = \frac{\left(\dfrac{s_1^2}{m} + \dfrac{s_2^2}{n}\right)}{\dfrac{\left(\dfrac{S_1^2}{m}\right)^2}{m-1} + \dfrac{\left(\dfrac{S_2^2}{n}\right)^2}{n-1}}.$$

Rejection criteria:

   Rejection of the hypothesis is determined by the following criteria, based on the alternative hypothesis.

- If the alternative is $\mu_1 - \mu_2 > 0$, we reject H_0 if $T \geq t_{\alpha,v}$.
- If the alternative is $\mu_1 - \mu_2 < 0$, we reject H_0 if $T \leq -T_{\alpha,v}$.
- If the alternative is $\mu_1 - \mu_2 \neq 0$, we reject H_0 if $|T| \leq t_{\frac{\alpha}{2},v}$.

Example 5.42

Consider Example 5.41 about the average amount of carbon dioxide. Test whether the average amount of carbon dioxide emitted by the first brand is higher than that of the second brand using the level of significance 0.05.

Answer

Here, in this example, the null and alternative hypotheses are as follows:

$$H_0 : \mu_1 - \mu_2 = 0$$

$$H_1 : \mu_1 - \mu_2 > 0.$$

Test statistic

$$T = \frac{\left(\bar{X} - \bar{Y}\right) - \left(\mu_1 - \mu_2\right)}{\sqrt{\dfrac{S_1^2}{m} + \dfrac{S_2^2}{n}}} = \frac{(0.75 - 0.76) - (0)}{\sqrt{\dfrac{0.3^2}{10} + \dfrac{0.1^2}{10}}} = -0.1.$$

The critical value, in this case, is $t_{\alpha,v} = t_{0.05,10} = 1.812$.
Hypothesis test

$$H_0 : \mu_1 - \mu_2 = 0$$

$$H_1 : \mu_1 - \mu_2 > 0 \text{ upper-tailed test}$$

$$\mu_1 - \mu_2 < 0 \text{ upper-tailed test}$$

$$\mu_1 - \mu_2 \neq 0 \text{ two-tailed test.}$$

Test statistic

$$T = \frac{(\bar{X} - \bar{Y}) - (\mu_1 - \mu_2)}{\sqrt{\dfrac{S_1^2}{m} + \dfrac{S_2^2}{n}}}.$$

Critical value, $t_{\alpha,v}$, where

$$v = \frac{\left(\dfrac{S_1^2}{m} + \dfrac{S_2^2}{n}\right)^2}{\dfrac{\left(\dfrac{S_1^2}{m}\right)^2}{m-1} + \dfrac{\left(\dfrac{S_2^2}{n}\right)^2}{n-1}}.$$

Note 5.39

$T < t_{\alpha,v}$. This means we fail to reject the null hypothesis at the level of significance 0.05; that is, we do not have evidence to support that the average amount of carbon dioxide emitted by the first brand is higher than that emitted by the second one.

Example 5.43

Refer to the Example 5.40 again. Test whether the average income of a city B is higher than the average income of city A using the level of significance of 0.05.

Case 3 Both populations are normally distributed and population variances are equal (pooled t)

Usually population parameters are unknown. However, there might be an evidence that both populations have the same variance, that is, $\sqrt{(3/4)^2} = (3/2)^2$.

$100(1 - \alpha)\%$ CI for the difference in means of two populations is given by

$$(\bar{X} - \bar{Y}) \pm t_{\frac{\alpha}{2},m+n-2} S_p \sqrt{\frac{1}{m} + \frac{1}{n}}, \tag{5.74}$$

where

$$S_p = \sqrt{\frac{(m-1)^2 S_1^2 + (n-1)S_2^2}{m+n-2}}. \tag{5.75}$$

Hypothesis test
  Test statistic

$$T = \frac{(\bar{X} - \bar{Y}) - (\mu_1 - \mu_2)}{S_p\sqrt{\dfrac{1}{m} + \dfrac{1}{n}}}, \qquad (5.76)$$

where S_p is given in (5.75).
  Rejection criteria:
  Rejection of the hypothesis is determined by the following criteria, based on the alternative hypothesis.

- If the alternative is $\mu_1 - \mu_2 > 0$, we reject H_0 if $T \geq t_{\alpha, m+n-2}$.
- If the alternative is $\mu_1 - \mu_2 < 0$, we reject H_0 if $T \leq -T_{\alpha, m+n-2}$.
- If the alternative is $\mu_1 - \mu_2 \neq 0$, we reject H_0 if $|T| \geq t_{\frac{\alpha}{2}, m+n-2}$.

5.3.2 CONFIDENCE INTERVAL FOR THE DIFFERENCE IN MEANS OF TWO POPULATIONS WITH PAIRED DATA

In this section, we consider the inference between the two dependent samples, which come as a pair of data. To construct the difference between two means, we use the paired t-confidence intervals.

Let $x_1, x_2, \ldots, x_m$ and $y_1, y_2, \ldots, y_n$ denote the first and second sample data from populations P_1 and P_2, respectively. Let $d_i = y_i - x_i, i = 1, 2, \ldots, n$ be the difference between ith observations between the second samples and the first samples.

The $100(1 - \alpha)\%$ CI for the difference in the means of two populations is given by

$$\bar{d} \pm t_{\frac{\alpha}{2}, n-1} \frac{S_D}{\sqrt{n}}, \qquad (5.77)$$

where S_D is the standard deviation of the difference between the samples.

Example 5.44

The latest of a particular brand of a car model is said to emit more carbon dioxide than usual. Engineers of this company designed a system and installed it on ten cars. They measured the amount of carbon dioxide emitted in each mile before and after the installation of the unit. A data set measured in grams is shown in Table 5.7.

TABLE 5.7

Collected Measured Data Set, Example 5.44

Before, x_i	0.76	0.74	0.70	0.76	0.73	0.72	0.74	0.77	0.79	0.73
After, y_i	0.70	0.75	0.69	0.76	0.70	0.71	0.78	0.75	0.77	0.74

Using this data set, we want to construct a 99% CI for the difference in the mean carbon dioxide amounts emitted before and after the installation of the unit.

Answer

We first calculate the difference in the given sample points and list them below:

$y_i - x_i$	−0.06	0.01	−0.01	0	−0.03	−0.01	0.04	−0.02	−0.02	0.01

The average value of the differences is $\bar{d} = -0.009$. From (5.77), the 99% CI for the average difference is obtained as:

$$\bar{d} \pm t_{\frac{\alpha}{2},n-1}\frac{S_D}{\sqrt{n}} = -0.009 \pm 2.262\left(\frac{0.026854}{\sqrt{10}}\right) = (-0.02821, 0.010209).$$

Using R program, we obtain the following:

```
Before< c(0.76,0.74,0.70,0.76,0.73,0.72,0.74,0.77,0.79,0.73)
After<-c(0.70,0.75,0.69,0.76,0.70,0.71,0.78,0.75,0.77,0.74)
t.test(After, Before, paired=TRUE, conf.level=0.95)
```

Output
```
data:   After and Before
t = -1.0598, df = 9, p-value = 0.3168
alternative hypothesis: true difference in means is not equal
to 0
95 percent confidence interval:
 -0.02820985  0.01020985
sample estimates:
mean of the differences
                 -0.009
```

According to the above output, the CI for the difference in the means is (−0.02820985, 0.01020985), which is very close to the prior interval.

5.3.3 ANALYSIS OF VARIANCE (ANOVA)

In everyday life, at times we may need to compare more than one population to make decisions, for instance, comparing several brands of smart phones for the average price or the average capacity. Companies offering these products need to have good advertising tool to persuade customers to buy one product versus another. Thus, marketing strategists may want to find ways to present their products showing better features than their competitors. This example may be extended to other cases. For instance, suppose three medical treatment options are available to use on patients with similar disease. Reviewing the test results, the treatment that takes the least amount of time, on the average, to cure the disease may be the best option among all three. Thus, in order to make a confident and reliable decision, documents are needed to support the choice of treatment. In such cases, one needs to compare population averages in order to make a decision. The statistical approach for such procedures is referred to as the **analysis of variance or ANOVA**. In other words, ANOVA is a statistical technique that compares the means of different populations.

In comparing two population means that was discussed before, when the number of populations is large, the two-population-mean comparison can be extended to several populations by taking two populations at a time. ANOVA is a method to reduce the error accompanied by comparisons of population means. This is because ANOVA compares the means of populations by identifying the sources of variation of a dependent or response variable, which is in numerical form. Depending on the number of factors associated, there are several types of ANOVA, such as one-way ANOVA and multifactor ANOVA. Here, our attention is only on the one-way ANOVA.

Note 5.40

ANOVA is available for both parametric (score data) and nonparametric (ranking/ ordering) data. In this section, we discuss the parametric ANOVA, and in Chapter 6, the nonparametric version will be discussed.

5.3.3.1 ANOVA Implementation Steps

All types of ANOVA follow the basic principles outlined below. However, as the number of groups and the interaction effects increase, the sources of variation will become more complex.

Step 1.
> The mean is calculated for each group. Using the example of education and sports teams from above, the mean education level is calculated for each sport team.

Step 2.
> The overall mean is then calculated for all groups combined.

Step 3.
> Within each group, the total deviation of each individual's score from the group mean is calculated. This tells us whether the individuals in the group tend to have similar scores or whether there are many variations between the different people in the same group. This concept is referred to as the **within-group variation** (sum of squares).

Step 4.
> Next, each group's mean deviation from the overall mean is calculated. This is called the **between-group variation** (sum of squares).

Step 5.
> The ANOVA table is prepared by considering all the sources of variations (sum of squares), degrees of freedom, and the mean square values to calculate the test statistic, F, that is, the ratio of the **between-group variation** to the **within-group variation**. If there is significantly greater **between-group variation** than the **within-group variation** (i.e., when the F statistic

is larger), then it is likely that the difference between the groups is statistically significant. Statistical software calculates the p-value to determine whether it is significant or not.

5.3.3.2 One-Way ANOVA

When comparing means across different groups that involve only one categorical variable (independent) and one numerical variable (response), we use one-way ANOVA. For instance, suppose a farmer is interested in studying the type of fertilizer that makes the plant to grow more. In this case, he is interested only in the type of fertilizer. Hence, this involves only one factor. Similarly, an educator may be interested in how the teaching method (face-to-face, online, and hybrid) impacts the students' understanding. In this case, it also involves only one factor, which is the teaching method.

The following definitions are now necessary.

Definition 5.18

An independent categorical variable that impacts the dependent numerical variable is referred to as a **factor** or **treatment**. The different values that a factor can assume are referred to as the **levels of a factor**.

Definition 5.19 ANOVA Model

For $i = 1, 2, \ldots, c$ and $j = 1, 2, \ldots, n_i$, where n is the sum of observations $n_1, n_2, \ldots, n_c$, where c is the number of levels, that is, $n = n_1 + n_2 + \cdots + n_c$. Let us denote the jth observation of the ith treatment by $Y_{i,j}$. Let us also denote the common effect of the experiment by α_j. We further denote the effect of the ith treatment by α_i. Finally, we denote the error of the jth observation of the ith treatment by $e_{i,j}$. Then, the one-way ANOVA model, which describes the relationship between the response and the treatment, is expressed as follows:

$$Y_{i,j} = \mu + \alpha_i + e_{i,j}. \tag{5.78}$$

We now summarize our notations for a single-factor case with c number of levels in Table 5.8.

The overall mean of the observations is calculated as follows:

$$\bar{y} = \frac{y_{1,1} + y_{1,2} + \ldots + y_{c,nc}}{n_1 + n_2 + \ldots + n_c} = \frac{\sum_{i=1}^{c} \sum_{j=1}^{n_i} y_{i,j}}{\sum_{k=1}^{c} n_k} = \frac{\sum_{i=1}^{c} \sum_{j=1}^{n_i} y_{i,j}}{n}, \tag{5.79}$$

where n is the total number of observations.

The total sum of squares (SST) is the total sum of the squares of the difference between each observation and the grand (overall) mean. It is calculated as follows:

TABLE 5.8

Collected Measured Data Set

	Levels of a Factor			
	1	2	i	c
	$y_{1,1}$	$y_{2,1}$	$y_{i,1}$	$y_{c,1}$
	$y_{1,2}$	$y_{2,2}$	$y_{i,2}$	$y_{c,2}$
	...	...	...	...
	$y_{1,n1}$	$y_{2,n2}$	$y_{i,ni}$	$y_{c,nc}$
Means	$\bar{y}_1$	$\bar{y}_2$	$\bar{y}_i$	$\bar{y}_c$
Sum of Squares	$\sum_{j=1}^{n_i}\left(y_{1j}-\bar{y}_1\right)^2$	$\sum_{j=1}^{n_i}\left(y_{2j}-\bar{y}_1\right)^2$	$\sum_{j=1}^{n_i}\left(y_{2j}-\bar{y}_i\right)^2$	$\sum_{j=1}^{n_i}\left(y_{c,i}-\bar{y}_c\right)^2$

$$\text{SST} = \sum_{i=1}^{c}\sum_{j=1}^{n_i}\left(y_{i,j}-\bar{y}\right)^2. \tag{5.80}$$

The treatment sum of squares (SSTr) is the total sum of the squares of the difference between the group mean and the grand mean. It is calculated as follows:

$$\text{SSTr} = \sum_{i=1}^{c}\sum_{j=1}^{n_i}\left(y_{i,j}-\bar{y}\right)^2 = \sum_{j=1}^{ni}n_j\left(\bar{y}_i-\bar{y}\right)^2. \tag{5.81}$$

The error sum of squares (SSE) is the total sum of the squares of the difference between each observation and the group means. It is calculated as follows:

$$\text{SSE} = \sum_{i=1}^{c}\sum_{j=1}^{n_i}\left(y_{i,j}-\bar{y}_i\right)^2. \tag{5.82}$$

The degrees of freedom (d.f.) of the analysis in a one-way ANOVA is distributed as follows:

$$\begin{cases} \{\text{Total d.f.} = \text{Treatment d.f.} + \text{Error d.f.} \\ \left(n_1+n_2+...+n_c\right)-1 = (c-1)+(n-c) \\ (n-1) = (c-1)+(n-c). \end{cases} \tag{5.83}$$

The mean squares are calculated as the ratio of sum of squares to the corresponding degrees of freedom. Hence, **the treatment mean squares (MSTr)** is

$$\text{MSTr} = \frac{\text{SSTr}}{\text{d.f.}} = \frac{\sum_{i=1}^{c}\sum_{j=1}^{n_i}\left(\bar{y}_i-\bar{y}\right)^2}{c-1}, \tag{5.84}$$

and the error mean squares (MSE) is:

$$MSE = \frac{SSE}{d.f.} = \frac{\sum_{i=1}^{c}\sum_{j=1}^{n_i}(y_{i,j} - \bar{y}_i)^2}{n-c}. \tag{5.85}$$

The **F-ratio** is the ratio between MSTr and MSE that follows an F-distribution with $(c-1)$ and $(n-1)$ degrees of freedom. It is calculated as follows (Figure 5.10):

$$\frac{MSTr}{MSE} = \frac{\dfrac{\sum_{i=1}^{c}\sum_{j=1}^{n_i}(\bar{y}_i - \bar{y})^2}{c-1}}{\dfrac{\sum_{i=1}^{c}\sum_{j=1}^{n_i}(y_{ij} - \bar{y}_i)^2}{n-c}} \approx F_\alpha(c-1, n-c). \tag{5.86}$$

We will now discuss the **complete ANOVA process**. We start with:

Research Question

The aim of this analysis is to compare the population means $\mu_1, \mu_2, \ldots, \mu_c$ for each c population. Hence, the hypotheses are as follows:

$$H_0 : \mu_1 = \mu_2 \ldots = \mu_c$$

H_1 : At least one of the means different.

Assumptions

In ANOVA, the following assumptions are made about the data:

i. Observations are independent.
ii. Response variable is the interval or ratio level (i.e., continuous).
iii. Independent variable has two or more categories (levels).
iv. Population of the response variable is normally distributed.
v. Populations have equal variances (homogeneous).

ANOVA Table

Table 5.9 for ANOVA shows the test statistics and decision criteria in order to make the decision with the hypotheses.

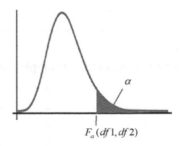

FIGURE 5.10 Rejection region of F distribution with d.f.1 and d.f.2.

TABLE 5.9

ANOVA Table

Source of Variation	Sum of Squares	Degrees of Freedom	Mean Squares	F-Ratio
Treatment	SSTr	$(c-1)$	$MSTr = \dfrac{SSTr}{(c-1)}$	$\dfrac{MSTr}{MSE}$
Error	SSE	$(n-c)$	$MSE = \dfrac{SSE}{(n-c)}$	
Total	SST	$(n-1)$		

Conclusion

If the test statistic F-ratio is higher than the critical value, $F_\alpha(\text{d.f.1}, \text{d.f.2})$, then the null hypothesis is rejected. Otherwise, the null hypothesis has failed to be rejected.

Example 5.45

A farmer was interested in identifying the best amount of fertilizer of a certain brand that he needs to use for the vegetable plants. He randomly selected 15 plants of a certain vegetable of same size and planted in 30 pots, which have similar environmental parameters such as soil, amount of water, and sunlight. He randomly selected five plants and assigned one pound of fertilizer, and he randomly selected another five plants and assigned two pounds of fertilizer. No fertilizer was applied to the remaining five plants. After a month, he measured the growth of each plant (in inches) and the measurements are summarized in Table 5.10.

Using this data, conduct a one-way ANOVA to check whether the average growth of the plant due to different amounts of fertilizer is the same, using the level of significance as 0.05.

Answer

In this case, the factor is the fertilizer and the levels are amounts of fertilizer 0, 1, and 2 pounds. Thus, $c = 3, n_1 = n_2 = n_3 = 5$, and $n_1 + n_2 + n_3 = 15$. Now, from Table 5.9, We have the following Table 5.11:

The sum of the squares is calculated as follows:

$$y = \frac{1+0+2+2+0+3+0+1+0+1+3+4+6+3+4}{15} = \frac{30}{15} = 2$$

TABLE 5.10

Summary of Measurements, Example 5.45

Fertilizer		Growth				
$0 \xrightarrow{\text{pound}}$	1	0	2	2	0
$1 \xrightarrow{\text{pounds}}$	3	0	1	0	1
$2 \xrightarrow{\text{pounds}}$	3	4	6	3	4

TABLE 5.11
Summary of Measurements, Example 5.45

Fertilizer	Growth of Plants (in inch) Due to Fertilizer					Means
	Plants No. 1	2	3	4	5	
0 —pound→	1	0	2	2	0	$\bar{y}_1 = \dfrac{5}{5} = 1$
1 —pounds→	3	0	1	0	1	$\bar{y}_2 = \dfrac{5}{5} = 1$
2 —pounds→	3	4	6	3	4	$\bar{y}_3 = \dfrac{20}{5} = 1$

$$SST = \sum_{i=1}^{3}\sum_{j=1}^{n_i}\left(y_{i,j} - \bar{y}\right)^2 = (1-2)^2 + (0-2)^2 + (2-2)^2 + (2-2)^2 + (0-2)^2$$

$$+ (3-2)^2 + (0-2)^2 + (1-2)^2 + (0-2)^2 + (1-2)^2$$

$$+ (3-2)^2 + (4-2)^2 + (6-2)^2 + (3-2)^2 + (4-2)^2$$

$$= 1+4+0+0+4+1+4+1+4+1+1+4+16+1+3 = 46.$$

$$SSTr = \sum_{i=1}^{c}\sum_{j=1}^{n_i}\left(\bar{y}_i - \bar{y}\right)^2 = \sum_{j=1}^{n_i} n_j \left(\bar{y}_i - \bar{y}\right)^2 = 5(1-2)^2 + 5(1-2)^2 + 5(4-2)^2 = 30.$$

$$SSE = \sum_{i=1}^{3}\sum_{j=1}^{n_i}\left(y_{i,j} - \bar{y}_i\right)^2 = (1-1)^2 + (0-1)^2 + (2-1)^2 + (2-1)^2 + (0-1)^2$$

$$+ (3-1)^2 + (0-1)^2 + (1-1)^2 + (0-1)^2 + (1-1)^2$$

$$+ (3-4)^2 + (4-4)^2 + (6-4)^2 + (3-4)^2 + (4-4)^2$$

$$= 0+1+1+1+1+4+1+0+1+0+1+0+4+1+0 = 16.$$

Let μ_1, μ_2, and μ_3 be the average growth of three fertilizer levels in inches. Then, our hypotheses are as follows:

$$H_0 : \mu_1 = \mu_2 = \mu_3$$

$$H_1 : \text{At least one of the means is different}$$

From Table 5.9, ANOVA table, we have (Table 5.12):
    Rejection criteria:

Test statistic = 11.25
Critical value = $F_\alpha(\text{d.f.1}, \text{d.f.2}) = F_\alpha(v_1, v_2) = F_{0.05}(2,12) = 3.89$, where d.f. is denoted by v.

TABLE 5.12

ANOVA Table, Example 5.45

Source of Variation	Sum of Squares	Degrees of Freedom	Mean Squares	F-Ratio
Treatment	SSTr = 30	$(c-1) = 3 - 1 = 2$	$MSTr = \dfrac{SSTr}{(c-1)} = \dfrac{30}{2} = 15$	$\dfrac{MSTR}{MSE} = 11.25$
Error	SSE = 16	$(n-c) = 15 - 3 = 12$	$MSE = \dfrac{SSE}{(n-1)} = \dfrac{16}{12} = 1.33$	
Total	SST = 46	$(n-1) = 15 - 1 = 14$		

Thus, the test statistic is in the rejection region, as 11.25 > 3.89.

Conclusion

Since F-ratio is higher than the critical value, $F_\alpha(\text{d.f.1,d.f.2}) = 3.89$, the null hypothesis is rejected as the level of significance of 0.05. This means that we have evidence to support that at least one of the means is different from the others. Therefore, the average growth of the plant due to different amounts of fertilizer is not the same.

Using R-Program

```
Amount<-c("0lb","0lb","0lb","0lb","0lb","1lb","1lb","1lb","1lb",
"1lb","2lb",  "2lb",  "2lb","2lb","2lb")
Growth<-c(1,0,2,2,0,3,0,1,0,1,3,4,6,3,4)
fit<-aov(Growth~Amount)
summary(fit)
```

Output

```
          Df Sum Sq Mean Sq F value  Pr(>F)
Amount     2     30  15.000   11.25 0.00177 **
Residuals 12     16   1.333
```

According to the above output, the p-value is 0.00177, which compares with the level of significance of 0.05. As the p-value is smaller than the α level, we do reject the null hypothesis at the level of significance of 0.05.

Example 5.46

A researcher was interested in studying about the summer gas price in a particular state. She selected three main cities. In each city, she randomly selected several gas stations to record the gas prices per gallon. The following is the data she gathered:

City A	$2.50	$2.60	$2.65	$2.70	$2.60
City B	$2.70	$2.55	$2.60	$2.55	
City C	$2.70	$2.46	$2.50	$2.70	

$$\bar{y} = \frac{1+0+2+2+0+3+0+1+0+1+3+4+6+3+4}{15} = \frac{30}{15} = 2.$$

Using this information and $\alpha = 0.05$, conduct an ANOVA test to check whether the average gas prices per gallon of three cities in the summer are the same.

Answer

Let μ_1, μ_2, and μ_3 be the average growth of three fertilizer levels in inches. Then, our hypotheses are as follows:

$$H_0 : \mu_1 = \mu_2 = \mu_3$$

$$H_1 : \text{At least one of the means is different.}$$

Let μ_1, μ_2, and μ_3 be the average prices per gallon of gas in summer in three cities. Then, our hypotheses are as follows:

$$H_0 : \mu_1 = \mu_2 = \mu_3$$

$$H_1 : \text{At least one of the means is different.}$$

The sum of squares are calculated as:

$$\bar{y}_1 = \frac{2.50 + 2.60 + 2.65 + 2.70 + 2.60}{5} = \$2.61,$$

$$\bar{y}_2 = \frac{2.70 + 2.55 + 2.60 + 2.55}{5} = \$2.60, \text{ and}$$

$$\bar{y}_3 = \frac{2.70 + 2.46 + 2.50 + 2.70}{5} = \$2.59.$$

$$\bar{y} = \frac{2.50 + 2.60 + 2.65 + 2.70 + 2.60 + 2.70 + 2.55 + 2.60 + 2.55 + 2.70 + 2.46 + 2.50 + 2.70}{13} = 2.60.$$

$$SST = \sum_{i=1}^{3} \sum_{j=1}^{n_i} (y_{i,j} - \bar{y})^2$$

$$= (2.50 - 2.60)^2 + (2.60 - 2.60)^2 + (2.65 - 2.60)^2 + (2.70 - 2.60)^2 + (2.60 - 2.60)^2$$

$$+ (2.70 - 2.60)^2 + (2.55 - 2.60)^2 + (2.60 - 2.60)^2 + (2.55 - 2.60)^2$$

$$+ (2.70 - 2.60)^2 + (2.46 - 2.60)^2 + (2.50 - 2.60)^2 + (2.70 - 2.60)^2$$

$$= 0.0871.$$

$$SSTr = \sum_{i=1}^{c} \sum_{j=1}^{n_i} (\bar{y}_i - \bar{y})^2 = \sum_{j=1}^{n_i} n_j (\bar{y}_i - \bar{y})^2 = 5(2.61 - 2.60)^2 + 4(2.60 - 2.60)^2 + 4(2.59 - 2.60)^2$$

$$= 0.0009.$$

$$SSE = \sum_{i=1}^{3} \sum_{j=1}^{n_i} \left(y_{i,j} - \bar{y}_i \right)^2$$

$$= (2.50 - 2.61)^2 + (2.60 - 2.61)^2 + (2.65 - 2.61)^2 + (2.70 - 2.61)^2$$

$$+ (2.60 - 2.61)^2 + (2.70 - 2.60)^2 + (2.55 - 2.60)^2$$

$$+ (2.60 - 2.60)^2 + (2.55 - 2.60)^2 + (2.70 - 2.59)^2$$

$$+ (2.46 - 2.59)^2 + (2.50 - 2.59)^2 + (2.70 - 2.59)^2$$

$$= 0.0862.$$

From Table 5.9, ANOVA table, we have (Table 5.13):
Rejection criteria:

Test statistic = 0.052
Critical value = 4.10

Thus, the test statistic is not in the rejection region, as $0.052 < 4.10$.

Conclusion
We do not reject the null hypothesis at the level of significance 0.05. This concludes that there is no evidence to believe that there is significant difference among the gas prices in the three cities during the summer.
Using R-Program

```
City<-c("A","A","A","A","A","B","B","B","B","C","C","C","C")
Price<-c(2.50,2.60,2.65,2.70,2.60, 2.70, 2.55, 2.60,
2.55,2.70,2.46, 2.50, 2.70)
fit<-aov(Price~City)
summary(fit)
```

Output

```
          Df   Sum Sq   Mean Sq  F value  Pr(>F)
City       2  0.00089  0.000446    0.052    0.95
Residuals 10  0.08620  0.008620
```

TABLE 5.13
ANOVA Table, Example 5.45

Source of Variation	Sum of Squares	Degrees of Freedom	Mean Squares	F-Ratio
Treatment	SSTr = 0.0009	$(c-1) = 3-1 = 2$	$MSTr = \dfrac{0.0009}{2} = 0.00045$	$\dfrac{MSTr}{MSE} = 0.052$
Error	SSE = 0.0862	$(n-c) = 13-3 = 10$	$MSE = \dfrac{0.0862}{10} = 0.00862$	
Total	SST = 0.0871	$(n-1) = 13-1 = 12$		

According to the above output, the p-value is 0.95, which compares with the level of significance of 0.05. As the p-value is higher than the α level, we do not reject the null hypothesis.

EXERCISES

5.1. An engineer is going to compare the average densities of two types of wood. In order to test this, he randomly selected six samples of the first wood, and the average density and the standard deviation are 22.5 and 0.16, respectively. He randomly selected five samples of the second type of wood and found the average density and the standard deviation as 21.9 and 0.24, respectively. Assuming the populations are normally distributed, answer the following questions:
   (i)   Construct a 95% CI for the difference in population means and interpret the interval.
   (ii) Use hypothesis test to confirm the above findings in part (i).

5.2. Amount of a certain acid in women's blood was measured using nine women after taking a special drug. The following data shows that the average amount of acid in the blood differs before and after taking the drug, using the level of significance of 0.05.

Before	14	17	15	17	13	18	15	12	14
After	12	16	13	15	14	10	12	12	11

5.3. A marketing company compares the average amount of satisfaction viewers of the two channels get by watching these TC channels. Using two random samples of people who watch these two channels, a survey was made to rate the two channels based on a five-point scale, with five being the highest satisfaction and one being the lowest. Calculate a 95% CI for the difference in the average satisfaction of watching these two TV channels. Interpret your result.

Channel 1	Channel 2
Sample size = 120	Sample size = 159
Average = 3.51	Average = 3.24
StDev = 0.51	StDev = 0.52

5.4. The department of transportation of a major city examines the average time to commute from three neighboring suburbs to the city. The commute times in minutes were collected using three random samples, which are given in the following table:

Suburb A	Suburb B	Suburb C
10.5	12.4	9.8
11.0	11.5	10.1
12.6	13.5	10.5
10.7	12.4	9.7
11.2	13.8	10.0
	11.5	11.0
	11.9	

Test whether the average commute time from each suburb to the main city is the same or different using the level of significance of 0.05.

5.5. A novel airline is going to compare the average air fare between two destinations by considering three popular airlines. A random sample of air fares was collected during a particular month, which is given in the following table.

Airline A	Airline B	Airline C
$101	$151	$101
$108	$149	$109
$98	$160	$198
$107	$112	$186
$111	$126	$160

Based on the collected data, compare the average air fare among the three airlines and state whether they are the same or different. Use $\alpha = 0.05$.

6 Nonparametric Statistics

6.1 WHY NONPARAMETRIC STATISTICS?

It is believed that the idea about nonparametric is as nearly old as statistics, as early works in this theory go back to fifteenth and sixteenth centuries. Arbuthnot, a mathematician, investigated whether the number of male births is higher than the number of female births, during the period of 1629–1710. This is considered as the first sign test. Among statisticians, it is believed that the first work in nonparametric statistics belongs to the paper written by Hotelling (1895–1973) and Pabst (1936), who discussed about the rank correlation. The origin of the term "nonparametric" goes back to 1942 in a paper by Wolfowitz (1910–1981). Introduction of nonparametric in the literature moved the theory of statistics beyond the parametric setting.

In inferential statistics, we considered some types of distribution regarding the underlining data. Statistical procedures based on distribution regarding the population are referred to as the **parametric methods**. For example, when calculating the confidence intervals or when conducting hypothesis test, we assume that the data comes from a normal distribution (or from t-distribution) and we use parameters such as the mean and the standard deviation. However, not all distributions contain parameters as normal distribution does. Hence, in those cases, we cannot use parametric methods; instead, we turn to use statistical procedures that are not based on distributional assumptions. Such methods are referred to as the **distribution free-techniques** or **nonparametric techniques**.

In parametric methods, we usually make assumptions that the population is normally distributed and is symmetric about the mean. In such cases, all calculations are made based on these critical assumptions. But what if the normality feature is not present? Naturally, the use of mean as a measure of central tendency will not be available any longer. Thus, the nonparametric statistics steps in and help us in such cases letting us use the median instead and do all necessary calculations based on the median. Thus, if we desire to analyze data using parametric tests and the assumptions are not satisfied, nonparametric statistical procedures can be used instead.

Here are some advantages and disadvantages of nonparametric methods:

1. Clearly due to the fewer number of assumptions about the data, nonparametric methods can be applied to a larger variety of data compared to the parametric methods. When considering nominal and ranked data, for instance, it is very difficult to use parametric methods to analyze such data. This is also the case when there are outliers in the data.
2. Calculations in the nonparametric procedures are easy to perform compared to those in the parametric methods.

3. It is well known that the parametric techniques are more efficient than the nonparametric techniques. Hence, nonparametric methods are applied when assumptions of parametric procedures fail.
4. Unfortunately, most calculators do not support nonparametric procedures. Hence, calculations for nonparametric procedures need to be done manually or by computer programming.
5. In nonparametric statistics, the actual values of the data are not necessarily of interest. It is only the rank that matters, as the result. Thus, the exact values of the data are not calculated, which are opposite to the parametric approach.

6.2   CHI-SQUARE TESTS

Chi-square tests are considered as appropriate for nominal-level and categorical-level measurements such as major of students, brand of a soda, students' classification (freshman, sophomore, junior, and senior), and gender of students. Sometimes it is necessary to investigate the proportions or relative frequencies, for example, the brand of soda that is preferred by young men or preference of engineering or biology by students as major. Thus, unlike in the parametric case, here in nonparametric case, there is no variable to be measured; rather, the number of objects (frequency) or proportion of objects is determined. Furthermore, we do not make assumptions about the population that follows a certain parametric distribution.

The chi-square test is applied for nonparametric tests mainly with the following three cases:

1. Goodness-of-fit,
2. Testing of independence,
3. Testing for homogeneity.

We now explain each of these three items.

6.2.1   GOODNESS-OF-FIT

The Chi-Square goodness-of-fit is used to compare the frequency (with m number of categories) of a single variable with its theoretical distribution. As the name "goodness-of-fit" implies, this test validates how well the observed frequencies follow the theoretical (expected) frequencies of m number of categories of the variable of interest. This is tested using the Chi-square goodness-of-fit as follows:

Hypotheses : H_o : Population distribution and the theoretical distribution are the same.

H_a : Population distribution and the theoretical distribution are not the same.

(6.1)

$$Test\ statistic : \chi^2 = \sum_{i=1}^{m} \frac{(O_i - E_i)^2}{E_i},$$

(6.2)

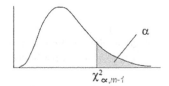

FIGURE 6.1 Critical value of chi-square distribution.

where O_i represents the observed value of ith category of the variable, and E_i represents the expected value of ith category of the variable.

Rejection criteria : Reject the null hypothesis if $\chi^2 > \chi^2_{\alpha,m-1}$ at α level of significance,

$$(6.3)$$

where $\chi^2_{\alpha,m-1}$ represents the critical value, which is given in hi-Square table with level of significance α and degrees of freedom $m-1$.

Note 6.1

The Chi-Square goodness-of-fit is appropriate if the expected frequency of each category of the variable is higher than 5.

As an illustration, Figure 6.1 shows the critical value $\chi^2_{\alpha,m-1}$ and the shaded region, the area of the right side of the critical value. This is also called the **rejection region**. If the calculated test statistics χ^2 falls on the rejection region, that is, if $\chi^2 > \chi^2_{\alpha,m-1}$, the null hypothesis is rejected.

Example 6.1

An administrator of a college assumes that each of their five majors has about equal number of students. The majors are Biology, Chemistry, Mathematics, History, and Social Sciences. Using a random sample of 2,500 students, the administrator finds the following information (Table 6.1).

We want to test whether there is enough evidence to support administrator's claim, using the level of significance $\alpha = 0.05$.

TABLE 6.1

Observed Distribution for Example 6.1

Major	Biology	Chemistry	Mathematics	History	Social Sciences
Number of students	550	420	500	525	505

Answer

We consider "major" as a category. Hence, there are five categories in this case, that is, $m = 5$. The hypotheses can, now, be considered as follows:

H_o: Each major has about the same number of students.
H_a: Each major does not have about the same number of students.

Using, the test statistic (6.2), we will have:

$$\chi^2 = \sum_{i=1}^{5} \frac{(O_i - E_i)^2}{E_i}$$

$$= \frac{(550 - 500)^2}{500} + \frac{(420 - 500)^2}{500} + \frac{(500 - 500)^2}{500} + \frac{(525 - 500)^2}{500} + \frac{(505 - 500)^2}{500}$$

$$= 5 + 12.8 + 0 + 1.25 + 0.05$$

$$= 19.1.$$

The critical value can be found from the Chi-Ssquare table, which is $\chi^2_{0.05, 5-1} = \chi^2_{0.05, 4} = 9.488$. This value is smaller than the test statistic of 19.1. Therefore, we reject the null hypothesis at the level of significance of 0.05. In other words, the administrator's claim is not accepted.

Using TI83/84 calculator:
We can calculate the test statistics using TI83 or TI84 calculator, as follows:

1. Select STAT button and select EDIT.
2. Enter Observed values into L_1 and Expected values into L_2.
3. Click STAT and then select TESTS.
4. Choose X²GOF Test; select L_1 as the Observed and L_2 as the Expected. Enter 4 as the *df* and then select CALCULATE.
5. At this point, you should see the test statistic value 19.1 and the *p*-value, *df*, and contribution from each category, etc.
6. Then, compare the *p*-value with α level of 0.05, as explained below to conclude:

If the *p*-value is less than or equal to α level, then the null hypothesis is rejected at the significance level, α.

Using R:
We can calculate the test statistics using R as follows:

```
Chisq.test(c(550,420,500,525,505))
```

The following output should appear:

```
Chi-squared test for given probabilities
data: c(550, 420, 500, 525, 505)
X-squared = 19.1, df = 4, p-value = 0.0007512
```

As it can be seen from the output, the p-value is 0.0007512, which is less than the α level of 0.05. Therefore, the null hypothesis is rejected at the level of significance of 0.05.

6.2.2 TEST OF INDEPENDENCE

Chi-square test can be used to test whether there is a relationship between two categorical variables in the population. Here, each individual sample point is classified into one of the above two random variables, which makes a frequency distribution of dimension two. These are called the **contingency tables**.

The following is the procedure to use **chi-square test of independence**:

$$Hypotheses: H_o : \text{Two variables (or levels of variables) are independent.}$$
$$(6.4)$$
$$H_a : \text{Two variables (or levels of variables) are dependent.}$$

Test statistic: The same as (6.2).
Assuming the null hypothesis, H_o, the expected frequencies, E_{ij}, of the cell number (i, j) can be calculated by

$$E_{ij} = \frac{(\text{Total of row number } i)(\text{Total of row number } j)}{\text{Total number of observations}}. \qquad (6.5)$$

Rejection criteria: The same as (6.3).

The degrees of freedom, denoted by df, is:

$$df = (\text{number of rows} - 1)(\text{number of columns} - 1) = (r-1)(c-1). \qquad (6.6)$$

Example 6.2 Calculation of Expected Value

Let us consider a two-way contingency table with two variables A (with n levels) and B (with m levels) as follows: The column and row totals in each corresponding cell are represented by $C_1, C_2, ... C_m$ and $R_1, R_2, ... R_n$, respectively. Consider a general cell (i, j).

		Variable B						
		1	2	...	j	...	m	Total
	1	$a_1 b_1$						R_1
	2							R_2
Variable A	$\vdots$			...		...		$\vdots$
	i							R_i
	$\vdots$			...		...		$\vdots$
	n			...		...		R_n
	Total	C_1	C_2	...	C_j	...	C_m	N

Let E_{ij} be the expected value of the row number i and the column number j. If the two variables A and B are independent, then

$$P(a_i b_j) = P_{ij} = P(a_i)P(b_j) = \frac{R_i}{N} \frac{C_j}{N}.$$

Hence, the expected value of the cell (i, j) can be found as

$$E_{ij} = NP_{ij} = \frac{R_i \times C_j}{N} = \frac{(\text{Total of row number } i)(\text{Total of row number } j)}{N}.$$

Example 6.3

Let us refer to Example 6.1. Suppose the administrator of the college wants to test whether the levels of college education and the major are independent of each other or not. To test the independence or dependence of the college education and the major, the college completes the contingency table, which is given in Table 6.2.

Answer

We test the above hypothesis using a Chi-Square test of independence at the level of significance of 0.05. Hence, the hypotheses are as follows:

H_o: Levels of college education and major are independent.
H_a: Levels of college education and major are dependent.

Assuming the null hypothesis, H_o, the expected frequencies, E_{ij}, of the cell number (i, j) can be calculated using (6.5). Thus, the expected frequencies can be calculated using (6.3) as follows:

$$E_{11} = \frac{(800)(550)}{2,500}, E_{12} = \frac{(800)(420)}{2,500}, E_{13} = \frac{(800)(500)}{2,500}, E_{14} = \frac{(800)(525)}{2,500}, E_{15} = \frac{(800)(505)}{2,500},$$

$$E_{21} = \frac{(725)(550)}{2,500}, E_{22} = \frac{(725)(420)}{2,500}, E_{23} = \frac{(725)(500)}{2,500}, E_{24} = \frac{(725)(525)}{2,500}, E_{25} = \frac{(725)(505)}{2,500},$$

$$E_{31} = \frac{(540)(550)}{2,500}, E_{32} = \frac{(540)(420)}{2,500}, E_{33} = \frac{(540)(500)}{2,500}, E_{34} = \frac{(540)(525)}{2,500}, E_{35} = \frac{(540)(505)}{2,500},$$

$$E_{41} = \frac{(435)(550)}{2,500}, E_{42} = \frac{(435)(420)}{2,500}, E_{43} = \frac{(435)(500)}{2,500}, E_{44} = \frac{(435)(525)}{2,500}, E_{45} = \frac{(435)(505)}{2,500}.$$

Instead of calculating E_{41} using the previous equation, one can use the following method as well.

$$E_{41} = 550 - (E_{11} + E_{21} + E_{31}), E_{42} = 420 - (E_{12} + E_{22} + E_{32}), E_{43} = 500 - (E_{13} + E_{23} + E_{33}),$$

$$E_{44} = 525 - (E_{14} + E_{24} + E_{34}), E_{45} = 505 - (E_{15} + E_{25} + E_{35}).$$

TABLE 6.2
Level of Education and Major (Example 6.3)

Level\Major	Biology	Chemistry	Mathematics	History	Social Sciences	Total
Freshman	175	125	200	150	150	800
Sophomore	175	125	150	150	125	725
Junior	125	90	75	125	125	540
Senior	75	80	75	100	105	435
Total	550	420	500	525	505	2,500

The expected frequencies are shown in Table 6.3.

Now, we have both observed values, O_{ij}, and expected values, E_{ij}, to complete the problem. Here, the test statistic (6.2) can be revised for two variables as:

$$\chi^2 = \sum_{i=1}^{m}\sum_{j=1}^{n}\frac{\left(O_{ij} - E_{ij}\right)^2}{E_{ij}}. \tag{6.7}$$

Thus,

$$\chi^2 = \sum_{i=1}^{m}\sum_{j=1}^{n}\frac{\left(O_{ij} - E_{ij}\right)^2}{E_{ij}}$$

$$= \frac{(175-176)^2}{176} + \frac{(125-134.4)^2}{134.4} + \frac{(200-160)^2}{160} + \frac{(150-168)^2}{168} + \frac{(150-161.6)^2}{161.6} + \cdots$$

$$= 43.21.$$

In addition, the critical value $\chi^2_{\alpha,d.f} = \chi^2_{0.05,12} = 21.03$ at α level of significance. Therefore, the test statistic is greater than the critical value. This suggests that the null hypothesis should be rejected at the significance level of $\alpha = 0.05$. This means that there is no evidence to support the claim that levels of college education and major are dependent.

TABLE 6.3
Expected (Theoretical) Values (Example 6.3)

Level\Major	Biology	Chemistry	Mathematics	History	Social Sciences	Total
Freshman	176.0	134.4	160.0	168.0	161.6	800
Sophomore	159.5	121.8	145.0	152.3	146.5	725
Junior	118.8	90.7	108.0	113.4	109.1	540
Senior	95.7	73.1	87.0	91.4	87.9	435
Total	550	420	500	525	505	2,500

Using TI 83/84 calculator:

1. Enter the contingency table of data as a matrix. To do this, select 2nd and "X⁻¹".
2. Now select "[A]", change the dimension of the matrix from 1×1 to "4×5", and hit enter.
3. Now enter all the data in Table 6.3 to the matrix A.
4. Select "STAT" and then "χ^2-test" to perform the chi-square test.
5. Select matrix "A" as Observed and leave the Expected as "B", then hit "Calculate".
6. Now you will have the output with test statistics, *df*, and the *p-value*.
7. Consider the *p*-value and level of significance to conclude.

Using R:

First of all, the frequency table should be created with two rows.

```
Fresh<-c(175,125,200,150,150)
sophomore<-c(175,125,150,150,125)
junior<-c(125,90,75,125,125)
senior<-c(75,80,75, 100,105)
LevelMajor<-data.frame(rbind(fresh, sophomore, junior, senior))
chisq.test(LevelMajor)
```

Output:

```
Pearson's Chi-squared test
data:  LevelMajor
X-squared = 43.231, df = 12, p-value = 2.063e-05
```

As it can be seen in the output, the *p*-value is 2.063e-05, which is very close to zero and less than α level of 0.05. Therefore, the null hypothesis is rejected at the level of significance $\alpha = 0.05$. This means that there is no evidence to believe that both the levels of college education and the major are independent.

6.2.3 TEST OF HOMOGENEITY

This particular test is mainly applicable to a single variable, which is of categorical nature. This variable represents two or more populations. Our interest is to determine whether each population has the same frequency distribution of the variable or a different one.

Suppose k populations $p_1, p_2, \ldots, p_k$ are under consideration. Our interest is to test whether these proportions are the same across all the k populations. To perform this test, we can construct the hypotheses of the test of homogeneity as follows:

$$Hypotheses: H_0 : p_1 = p_2 = \cdots p_k$$
$$H_a : p_i \neq p_j, \text{ for some } i, j. \tag{6.8}$$

Test statistic: The same as (6.2), where the null hypothesis, H_o, and the expected frequencies, E_{ij}, of the cell number (i, j) can be calculated using (6.3).

Rejection criteria: The null hypothesis is rejected if (6.3) is true.

In order to understand how to use chi-square test, consider the following example.

Example 6.4

Consider a manufacturer that produces electric bulbs. Assume that the factory has five production lines. The quality controller of the manufacturer randomly collects samples of the produced bulbs from each line as detailed in Table 6.4. We are to conduct a chi-square test of hypothesis using the level of $\alpha = 0.05$ to test whether the proportion of defective bulbs produced by each production line is different.

Answer

In this example, the hypotheses can be written as follows:

H_o: Proportions of defective bulbs for all production lines are the same, that is, $p_1 = p_2 = p_3 = p_4 = p_5$ (here p_i is the proportion of defective bulbs in line number i).

H_a: Proportions of defective bulbs for all production lines are not the same, that is, $i, j = 1,2,3,4,5, p_i \neq p_j$.

As explained before, the expected frequencies, E_{ij}, of the cell number (i, j) can be calculated using (6.5) as given in Table 6.5.
From Table 6.5, the test statistic can be calculated as follows:

$$\chi^2 = \sum_{i=1}^{m} \sum_{j=1}^{n} \frac{\left(O_{ij} - E_{ij}\right)^2}{E_{ij}} + \frac{(44 - 44.09)^2}{44.09} + \frac{(58 - 57 - 31)^2}{57.31} + \cdots + \frac{(8 - 7.09)^2}{7.09} = 1.64.$$

TABLE 6.4
Observed Values for Example 6.4

Production Line	1	2	3	4	5
Number of non-defectives	44	58	64	43	52
Number of defectives	6	7	6	8	8

TABLE 6.5
Expected Values for Example 6.4

Production Line	1	2	3	4	5	Total
# Non-defectives	44.09	57.31	61.72	44.97	52.91	261
# Defectives	5.91	7.69	8.28	6.03	7.09	35
Total	49	65	70	45	60	296

In addition, the

$$df = (\text{number of rows} - 1)(\text{number of columns} - 1) = (2 - 1)(5 - 1) = 4$$

and using α level of significance of 0.05, then the critical value, $\chi^2_{\alpha, d \cdot f} = \chi^2_{0.05,4} = 9.49$. Therefore, the test statistic is less than the critical value. This suggests that we do not reject the null hypothesis at the level of significance $\alpha = 0.05$. That is, we have evidence to support the claim that the proportions of defective bulbs across all the production lines are not the same.

Using R:

```
defective<-c(44, 58, 64, 43, 52)
nondefctive<-c(6, 7, 6, 8, 8)
items<-data.frame(rbind(defective,nondefctive))
chisq.test(items)
```

Output:

```
Pearson's Chi-squared test
data:  items
X-squared = 1.6419, df = 4, p-value = 0.8012
```

Note 6.2

The above p-value is 0.8012, which is higher than $\alpha = 0.05$. Therefore, we fail to reject the null hypothesis at the level of significance $\alpha = 0.05$.

6.3   SINGLE-SAMPLE NONPARAMETRIC STATISTIC

In statistics, sometimes we need to make inference based on a single sample. In other words, there may be situation where we have a single sample to analyze in order to make inference about the population parameters. In nonparametric statistics, we use the single-sample sign test procedure to make inference about the population parameters.

6.3.1   SINGLE-SAMPLE SIGN TEST

In inferential statistics, inferences about population parameters are based on the collected sample statistics. Usually in parametric approach, when conducting a hypothesis test about the population mean, μ, two cases are considered. If it is possible to collect a sample over 25 observations, it is considered as a large sample. Therefore, large sample inference can be applied for those cases. Similarly, if the sample size is less than 25 observations, small sample inference can be applied. For this case, it is required to have a sample from a normally distributed population.

If the above two criteria are not met, we have to conduct a hypothesis test using the nonparametric approach. As explained before, in nonparametric approach, we consider the population median, denoted by $\tilde{\mu}$, instead of population mean, μ. Hence, let $\tilde{\mu}_0$ be the hypothesized value of the population median. Here, we consider the following two cases based on the sample sizes (Table 6.6):

TABLE 6.6
Single-Sample Hypothesis Test Using Sign Test

	Two-Tailed Test	Lower-Tailed Test	Upper-Tailed Test
Hypotheses	$H_o : \tilde{\mu} = \tilde{\mu}_0$	$H_o : \tilde{\mu} = \tilde{\mu}_0$	$H_o : \tilde{\mu} = \tilde{\mu}_0$
	$H_a : \tilde{\mu} \neq \tilde{\mu}_0$	$H_a : \tilde{\mu} < \tilde{\mu}_0$	$H_a : \tilde{\mu} > \tilde{\mu}_0$
Test statistic	K is the minimum value of the number of minus or plus signs	K is the number of positive signs	K is the number of minus signs
Rejection criteria	Reject the null hypothesis if $p-value \leq \alpha$ level. Here, $p\text{-}value$ is calculated using binomial distribution with $n = sample\ size$ and $p = 0.5$.		

Case I: Small sample ($n < 25$)
Case II: Large sample ($n \geq 25$)
Test statistics:

$$Z_0 = \frac{(2K+1-n)}{\sqrt{n}},\tag{6.9}$$

where K is calculated as explained in Case I.

Conclusion: Reject the null hypothesis if the test statistic $K \leq$ critical value. Here, the critical value is calculated using the standard normal distribution. Another approach is to compute the p-value. With the p-value, we can reject the null hypothesis if the p-value < level of significance α.

Steps in Single-Sample Sign Test

1. Arrange the observations from the lowest to the highest.
2. Assign the minus sign (−) to the observations whose values are less than $\tilde{\mu}_0$ and the plus sign (+) to the observations whose values are greater than $\tilde{\mu}_0$. If there are any observations, whose values are equal to $\tilde{\mu}_0$, just discard it.
3. Calculate the test statistic (K or Z_0 as explained in Case I and Case II).
4. Compare the critical value using the appropriate table for the given α level and the sample sizes.
5. Reject the null hypothesis if test statistic $K \leq$ critical value, which is given by the sign test table.

Example 6.5

A sociologist collected a data set that contains monthly salary of 15 families in a particular city as follows:

$1,900	$2,100	$1,750	$1,800	$2,050
$2,300	$2,500	$1,250	$2,800	$2,200
$1,950	$2,100	$2,000	$1,800	$1,750

He claims that the median monthly salary of a family in that city is not less than $2,000. Using the level of significance, $\alpha = 0.05$, determine if his collected data supports his claim.

Answer

Let $\tilde{\mu}$ denote the median income of a family of the city and $\tilde{\mu}_0 = \$2,000$. Then, our aim is to conduct the following hypothesis test:

$$H_o : \tilde{\mu} = \$2,000.$$

$$H_a : \tilde{\mu} < \$2,000.$$

Let us consider Table 6.7:

There are seven minus signs and seven plus signs in the table. Hence, $K = 7$. According to the sign test table, the critical value is 3. According to the rejection criteria, the null hypothesis is rejected if test statistic $K \le$ critical value. Here, test statistic $K = 7$, which is less than the critical value of 3. Therefore, the null hypothesis is not rejected or failed to reject the null hypothesis $\alpha = 0.05$. In other words, this means that there is no statistical evidence to support that the population median is less than $2,000.

TABLE 6.7
Monthly Family Incomes

Income	Sign
$1,250	−
$1,750	−
$1,750	−
$1,800	−
$1,800	−
$1,900	−
$1,950	−
$2,000	
$2,050	+
$2,100	+
$2,100	+
$2,200	+
$2,300	+
$2,500	+
$2,800	+

Using R:

```
library(BSDA)
income<-c(1250, 1750, 1750, 1800, 1800, 1900, 1950, 2000, 2000,
2100, 2100, 2200, 2300, 2500, 2800)
SIGN.test(income, md = 2000, alternative= "less",conf.level=
0.95)
```

Output:

```
One-sample Sign-Test
data: income
s = 6, p-value = 0.5
alternative hypothesis: true median is less than 2000
```

According to the output, the p-value is 0.5, which is higher than the α level of 0.05. Therefore, the null hypothesis cannot be rejected at the level of significance of 0.05. That is, the sociologist's claim is valid.

Example 6.6

An owner of a store believes that the median number of water bottle packs he sells each day is more than 10 packs. Using his previous records, he randomly selects 30-day sells as one is displayed below:

11	8	20	15	6	8	16	15	10	9
7	14	17	7	1	16	10	9	18	7
12	6	18	9	5	11	7	4	17	5

Using the level of significance, $\alpha = 0.05$, determine if his collected data supports the storeowner's claim.

Answer

Let $\tilde{\mu}$ denote the median number of water bottle packs the storeowner sells per day. Hence, we need to conduct the following hypothesis test about $\tilde{\mu}$.

$$H_o : \tilde{\mu} = 10$$

$$H_a : \tilde{\mu} > 10.$$

The given data is organized in Table 6.8.

Thus, K, in this case, is the number of negative signs, which is 15. Sample size being 30 is considered to be large. Hence, we use the test statistics as (6.9). Thus,

$$Z_0 = \frac{(2K+1-n)}{\sqrt{n}} = \frac{(2 \times 15 + 1 - 30)}{\sqrt{30}} = 0.1826.$$

Also,

$$p - \text{value} = P(Z > 0.1826) = 0.4276,$$

TABLE 6.8

Monthly Sells of Water Bottle Packs

Number of Packs	Sign	Number of Packs	Sign
1	–	10	
4	–	10	
5	–	11	+
5	–	11	+
6	–	12	+
6	–	14	+
7	–	15	+
7	–	15	+
7	–	16	+
7	–	16	+
8	–	17	+
8	–	17	+
9	–	18	+
9	–	18	+
9	–	20	+

which is greater than the significance level of $\alpha = 0.05$. Therefore, at $\alpha = 0.05$, the storeowner's claim cannot be rejected.

Using R:

```
library(BSDA)
bottles<-c( 11, 8, 20, 15, 6, 8, 16, 15, 10, 9,
7, 14, 17, 7, 1, 16, 10, 9, 18, 7,
12, 6 ,18, 9, 5, 11, 7, 4, 17,5)
SIGN.test(bottles, md = 10, alternative= "greater",conf.level=
0.95)
```

Output:

```
One-sample Sign-Test
data:  bottles
s = 13, p-value = 0.7142
alternative hypothesis: true median is greater than 10
95 percent confidence interval:
```

According to the output, the p-value is greater than 0.05. Therefore, at the level of significance of 0.05, the null hypothesis cannot be rejected.

6.3 TWO-SAMPLE INFERENCE

When conducting two-sample inference under parametric approach, we consider two-sample t-test and paired t-test depending on the dependence or independence of the two samples. When the parametric assumptions are not met, we have to look for nonparametric methods. Therefore, similar to the parametric approach, we consider the two cases of independent and dependent sample inferences separately.

6.3.1 INDEPENDENT TWO-SAMPLE INFERENCE USING MANN–WHITNEY TEST

When inferring about two independent samples, Mann–Whitney test is used and this is considered as the nonparametric version of the two-sample t-test. We use this nonparametric test when testing a hypothesis of equality of medians in two independent samples. In other words, we can test whether the two populations have the same distributions (same mediansor different distributions. There are some different names that are frequently referred to this test such as **Mann–Whitney U-test**, **Mann–Whitney–Wilcoxon test**, **Wilcoxon rank-sum test**, and **Wilcoxon–Mann–Whitney test**.

Test of hypotheses regarding this tes

$$H_o : \text{The two populations are equal.}$$

$$H_a : \text{The two populations are not equal.}$$

(6.10)

Mann–Whitney test is usually performed as a two-sided test as shown in (6.10). One-sided hypothesis is conducted if we are interested in identifying a positive or negative shift in one population with respect to another population. For small sample sizes, the following steps are used to conduct the Mann–Whitney test.

Steps in Mann–Whitney Test

Let n_1 and n_2 be the sizes of the two samples. The following steps should be taken into account:

1. Combine both samples and arrange the data from the smallest to the largest.
2. Rank each data point as $1, 2, \ldots, n_1 + n_2$. If there are ties (identical observations), then replace the rank of each observation by the average of ranks of the identical observations. For instance, if the sixth observation and the seventh observation are identical (same value), then assign the rank of 6.5 (average of 6 and 7) instead of their previous ranks of 6 and 7.
3. Let R_1 and R_2 be the ranks of the observations from the first and the second samples. Calculate the following two statistics, U_1 and U_2. Let's label the smaller value of them as U, that is,

$$U = \min(U_1, U_2),$$

$$U_1 = n_1 n_2 + \frac{1}{2} n_1 (n_1 + 1) - R_1,$$

$$U_2 = n_1 n_2 + \frac{1}{2} n_2 (n_2 + 1) - R_2.$$

4. Now, compare the critical value that can be found from the table for Mann–Whitneytest, based on the significance level and the sample sizes with the value of U. Reject the null hypothesis if $U \leq$ critical value.

Example 6.7

A study was conducted to examine whether the accuracy of throwing the basketball to the basket depends on the handedness of the player. To test this, a group of 17 players were randomly selected and each player was given 20 chances to shoot the ball from the free throw line. Table 6.9 shows the number of successful shots they made.

Using $\alpha = 0.05$, test whether the given data supports the claim that handedness of the player affects the accuracy when shooting the ball.

Answer

If the handedness does not impact the accuracy, the accuracy of the left handers and the right handers should be the same. Therefore, the distribution of scores for both left and right handers should be the same. Hence, the related hypotheses can be defined as follows:

H_o: Distributions of the scores for both right and left handers are the same.
H_a: Distributions of the scores for right handers are higher than those for left handers.

Now, in Table 6.10, we arrange the data set to find the rank for each data point. Since,

$$U_1 = n_1 n_2 + \frac{1}{2} n_1 (n_1 + 1) - R_1,$$

$$U_2 = n_1 n_2 + \frac{1}{2} n_2 (n_2 + 1) - R_2,$$

we have:

$$U_1 = 10 \times 7 + \frac{1}{2} \times 10 \times (10 + 1) - 83 = 42,$$

$$U_2 = 10 \times 7 + \frac{1}{2} \times 7 \times (7 + 1) - 70 = 28.$$

Hence, $U = 28$. Thus, according to the table for Mann–Whitney test, the critical value is 17. Since $U \leq$ critical value is not satisfied, at the level of significance, $\alpha = 0.05$, the null hypothesis will not be rejected. That is, there is no evidence to support the claim that the handedness of the basketball player affects the accuracy of shooting the ball at the level of significance of 0.05.

TABLE 6.9
Left Handers' vs Right Handers' Accuracy When Shooting the Ball

Right handers	15	16	18	14	15	11	15	17	10	12
Left handers	17	12	13	16	18	14	16			

TABLE 6.10
Organized Ranks of Left Handers vs Right Handers

Right-Handed Data	Right-Handed Data Ranks	Left-Handed Data	Left-Handed Data Ranks
10	1		
11	2		
12	3.5	12	3.5
		13	5
14	6.5	14	6.5
15	9		
15	9		
15	9		
16	12	16	12
		16	12
17	14.5	17	14.5
18	16.5	18	16.5
	$R_1 = 83$		$R_2 = 70$

Using R:

```
Rhand<-c(15,16,18,14,15,11,15,17,10,12)
Lhand<-c(17,12,13,16,18, 14,16)
wilcox.test( Rhand,Lhand, alternative = "greater",conf. level =
0.95)
```

Output:

```
Wilcoxon rank sum test with continuity correction
data:  Rhand and Lhand
W = 28, p-value = 0.7696
alternative hypothesis: true location shift is greater than 0
```

As it can be seen in the output, the p-value is 0.7696, which is greater than the α level of 0.05. Therefore, the null hypothesis at the level of significance of 0.05 cannot be rejected.

Example 6.8

An administrator of a particular university wants to test whether the median GPAs (grade point averages) for two different majors, A and B, in the college are the same. To conduct this test, he randomly selects two samples of students and records GPAs of each major as shown in Table 6.11.

Using the level of significance of 0.05, test whether the population median GPAs for two majors differ.

TABLE 6.11

GPAs of Majors A and B

A 2.70 3.15 1.77 2.45 1.80 2.92 2.91 2.98 2.70 1.65 4.00 2.68 2.89 2.85 2.25
B 2.06 1.29 1.76 1.20 1.54 1.23 1.64 1.47 2.07 1.38 1.43 2.01

Answer

Let us define the related hypotheses as follows:

H_o : Distributions of the PGAs for both majors are *the* same.

H_a : Distributions of the PGAs for both majors are different.

Now, we arrange the data set to find the ranks for each data point as displayed in Table 6.12.
  Since

$$U_1 = n_1 n_2 + \frac{1}{2} n_1 (n_1 + 1) - R_1,$$

$$U_2 = n_1 n_2 + \frac{1}{2} n_2 (n_2 + 1) - R_2,$$

TABLE 6.12

Ranks of GPA of Major A and Major B

Major A Data	Major A Data Rank	Major B Data	Major B Data Rank
1.65	9	1.20	1
1.77	11	1.23	2
1.80	12	1.29	3
2.25	16	1.38	4
2.45	17	1.43	5
2.68	18	1.47	6
2.70	19	1.54	7
2.70	20	1.64	8
2.85	21	1.76	10
2.89	22	2.01	13
2.91	23	2.06	14
2.92	24	2.07	15
2.98	25		
3.15	26		
4.00	27		
	$R_1 = 290$		$R_2 = 88$

we have:

$$U_1 = 15 \times 12 + \frac{1}{2}15(15+1) - 290 = 10,$$

$$U_2 = 15 \times 12 + \frac{1}{2}12(12+1) - 88 = 170.$$

Also, $U = \min(U_1, U_2) = \min(10, 170) = 10$. Thus, $U = 10$. Thus, according to the table for Mann–Whitney test, the critical value is 49. Therefore, it satisfies $U \leq Critical\ Value$. Hence, at the level of $\alpha = 0.05$, the null hypothesis is rejected.

Using R:

```
Major_A<-c(2.70,3.15,1.77,2.45,1.80,2.92,2.91,2.98,2.70,1.65,
        4.00,2.68,2.89,2.85,2.25)
Major_B<-c(2.06,1.29,1.76,1.20,1.54,1.23,1.64,1.47,2.07,1.38,
        1.43,2.01)
wilcox.test(Major_B, Major_A, alternative="two.sided",
conf.level = 0.95)
```

Output:

```
Wilcoxon rank sum test with continuity correction
data: Major_A and Major_B
W = 10, p-value = 0.0001045
alternative hypothesis: true location shift is not equal to 0
```

In the output, the p-value is smaller than the level of significance of 0.05. Hence, the null hypothesis is rejected. This means that we favor the alternative hypothesis of two different distributions of GPAs in two majors. This favors the claim that the GPAs of both majors are different.

At times, we come across two data sets in which one set depends on the other. For instance, consider a study to examine the impact of a particular medical treatment. In this case, we can assign the treatment to a certain number of randomly selected subjects and take measurements before and after the application of the treatment. Then, the second set of measurements will depend on the first set of measurements. This kind of dependence has also considered as a paired sample case. In a situation like this, the Wilcoxon signed-rank test is used to make an inference regarding the two samples.

Let $x_1, x_2, \ldots, x_n$ be the first data set and $y_1, y_2, \cdots, y_n$ be the second data set, which depends on the first. Let $d_i = y_i - x_i$; $i = 1, 2, \ldots, n$ and $\tilde{\mu}_D$ be the difference in the two population medians (i.e., $\tilde{\mu}_D = \tilde{\mu}_Y - \tilde{\mu}_X$). To make inference between $\tilde{\mu}_X$ and $\tilde{\mu}_Y$, the following hypotheses should be tested. So, we will arrange the values of d_i. Let R_- denote the sum of the ranks of the negative values of d_i and R_+ denote the sum of the ranks of the positive values of d_i. We now take the following steps:

Steps in Single-Sample Sign Test

1. For each pair, subtract the second value from the first (i.e., $d_i = y_i - x_i$; $i = 1, 2, \ldots, n$).
2. Find the absolute values of the difference, $|d_i|$, and discard the zeros.

TABLE 6.13

Paired Sample Hypothesis Test (R_α Values Are from the Wilcoxon Signed-Rank Table)

	Two-Tailed Test	Lower-Tailed Test	Upper-Tailed Test
Hypotheses	$H_o : \tilde{\mu}_D = 0$	$H_o : \tilde{\mu}_D = 0$	$H_o : \tilde{\mu}_D = 0$
	$H_a : \tilde{\mu}_D \neq 0$	$H_a : \tilde{\mu}_D < 0$	$H_a : \tilde{\mu}_D > 0$
Test statistic	$R = \min\{R_+, \mid R_- \mid\}$	$R = R_+$	$R = \mid R_- \mid$
Rejection criteria	Reject H_o if $R < R_{\frac{\alpha}{2}}$	Reject H_o if $R < R_\alpha$	Reject H_o if $R < R_\alpha$

3. Rank absolute values of the differences from the smallest to the largest. If two values have the same absolute values, assign the average value of their ranks to each (as explained before).
4. Assign the minus sign (–) to the observations whose d_i values are negative, and assign the plus sign (+) to the observations whose d_i values are positive.
5. Take the sum of the ranks of both positives and negatives. Let R_+ and R_- represent the two values, respectively (Table 6.13).

Case I: Small sample hypothesis test
Case II: Large sample hypothesis test

Test statistic,

$$Z_0 = \frac{R - \dfrac{n(n+1)}{4}}{\sqrt{\dfrac{n(n+1)(2n+1)}{24}}},$$ (6.11)

where R is the value calculated in Case I.

Conclusion: Reject the null hypothesis by comparing Z_0 with the critical value calculated using the standard normal distribution. In another way, we reject H_o if p-value < level of significance α or test statistic $K \leq$ critical value.

Example 6.9

A pharmaceutical company believes that its new drug lowers the blood pressure. The company wants to examine this claim and asks a researcher to conduct a research. The researcher randomly selects 10 people for this study, and he measures the participants' blood pressures before and after taking the drug. The collected data is given in Table 6.14. Let X and Y be the variables representing the blood pressures before and after taking the drug, respectively.
At $\alpha = 0.05$, does the data support the belief of the company?

TABLE 6.14
Blood Pressure before and after Taking the Drug

X (Blood Pressure before Taking the Drug)	Y (Blood Pressure after Taking the Drug)	$d = Y - X$	Rank of $\mid d \mid$	Signed Ranks
150	148	−2	5	−5
145	148	3	7.5	+7.5
160	159	−1	2	−2
155	145	−5	5	−9
149	150	1	2	+2
150	148	−2	5	−5
151	150	−1	2	−2
145	135	−10	10	−10
163	160	−3	7.5	−7.5
157	155	−2	5	−5

Answer

Let $\tilde{\mu}_D$ denote the median of the difference in blood pressures after and before taking the drug. We choose the hypotheses as follows:

$$H_o : \tilde{\mu}_D = 0$$

$$H_a : \tilde{\mu}_D < 0$$

Here, $R_+ = 7.5 + 2 = 9.5$ and $R_- = -5 - 3 - 9 - 5 - 2 - 10 - 7.5 - 5 = -45.5$. Hence, the test statistics is $R = R_+ = 9.5$. From the Wilcoxon signed-rank table, R_α value for $n = 10$ and $\alpha = 0.05$ is 10. Therefore, at $\alpha = 0.05$, the null hypothesis is rejected since $R < R_\alpha$. This means that we have no evidence to support the company's claim that at the level of confidence, the new drugs lower the blood pressure.

Using R

```
before<-c(150,145,160,155,149,150,151,145,163,157)
after<-c(148,148,159,145,150,148,150,135,160,155)
wilcox.test(after, before, paired=TRUE, alternative = "less",
conf.level=0.95)
```

Output:

```
Wilcoxon signed rank test with continuity correction
data: after and before
V = 9.5, p-value = 0.03629
alternative hypothesis: true location shift is less than 0
```

As it can be seen in the output, the p-value is 0.03629, which is less than the α level of 0.05. Therefore, the null hypothesis is rejected at the level of significance of 0.05.

Example 6.10

A researcher wants to compare the gas prices in two consecutive months in a particular city. He randomly selects 10 gas stations in that city and records the average gas prices in two consecutive months as given in Table 6.15.

Does the data in Table 6.15 show that at $\alpha = 0.05$, the gas prices in two months are different?

Answer

Let $\tilde{\mu}_D$ denote the median of the difference in gas prices as calculated in Table 6.13. Here, the hypotheses can be written as follows:

$$H_o : \tilde{\mu}_D = 0$$

$$H_a : \tilde{\mu}_D \neq 0$$

Thus, $R_+ = 39$ and $R_- = -16$. Then, the test statistics $R = \min\{R_+, |R_-|\} = 16$. From the Wilcoxon signed-rank table, R_α value for $n = 10$ and $\alpha = 0.05$ is 10. Therefore, the null hypothesis at $\alpha = 0.05$ is not rejected since $R < R_\alpha$.

Using R:

```
Month_1<-c(2.75,2.70,3.00,2.85,2.90,2.70,2.85,3.05,2.98,2.90)
Month_2<-c(2.86,2.73,2.90,2.84,3.05,2.65,2.87,3.17,2.90,3.10)
wilcox.test(before, after, paired=TRUE, alternative = "two.
sided")
```

Output:

```
Wilcoxon signed rank test with continuity correction
data: before and after
V = 45.5, p-value = 0.07258
alternative hypothesis: true location shift is not equal to 0
```

TABLE 6.15
Gas Prices

Month # 1 (X)	Month # 2 (Y)	$d = Y - X$	Rank of $|d|$	Signed Ranks
2.75	2.86	0.11	0.11	7
2.70	2.73	0.03	0.03	3
3.00	2.90	−0.10	0.10	−6
2.85	2.84	−0.01	0.01	−1
2.90	3.05	0.15	0.15	9
2.70	2.65	−0.05	0.05	−4
2.85	2.87	0.02	0.02	2
3.05	3.17	0.12	0.12	8
2.98	2.90	−0.08	0.08	−5
2.90	3.10	0.20	0.20	10

Since the *p*-value is greater than the level of significance of 0.05, the null hypothesis is not rejected.

6.3.2   DEPENDENT TWO-SAMPLE INFERENCE USING WILCOXON SIGNED-RANK TEST

In this section, we consider the inference between two dependent samples. To conduct inference regarding the two population medians, we use the Wilcoxon signed-ranked test. This test is the counterpart of the parametric approach of the paired *t*-test. As discussed in Section 6.3.1, let $x_1, x_2, \ldots x_n$ and $y_1, y_2, \ldots, y_n$ denote the first and second data sets, respectively. Let $d_i = y_i - x_i$; $i = 1, 2, \ldots, n$ and $\tilde{\mu}_D$ denote the difference between the two population medians, that is, $\tilde{\mu}_D = \tilde{\mu}_Y - \tilde{\mu}_X$.

From Table 6.14, the hypotheses are as follows:

$$H_o : \tilde{\mu}_D = 0, \quad H_a : \tilde{\mu}_D > 0 \quad \text{(Right-tailed test)}$$

$$H_o : \tilde{\mu}_D = 0, \quad H_a : \tilde{\mu}_D < 0 \quad \text{(Left-tailed test)}$$

$$H_o : \tilde{\mu}_D = 0, \quad H_a : \tilde{\mu}_D \neq 0 \quad \text{(Two-tailed test)}.$$

Case I
Small sample hypothesis test

- For right-tailed test, the test statistic is $R = |R_-|$, where R_- is the sum of the negative signed ranks.
- For left-tailed test, the test statistic is $R = |R_+|$, where R_+ is the sum of the positive signed ranks.
- For the two-tailed test, the test statistic is $R = \min(|R_-|, R_+)$.

The R_-, R_+ and the rejection criteria are as discussed in Section 6.3.1.

Case II
Large sample hypothesis test: The test statistic is calculated as follows:
Here, T is the test statistic defined in (6.11).

Example 6.11

Consider the mathematics majors graduated from mathematics department in six colleges C1, C2, C3, C4, C5, and C6 in a state for the fall of 2017 and 2018 as listed below:

College #	Fall 2017	Fall 2018
C1	20	16
C2	8	6
C3	6	7
C4	12	20
C5	10	4
C6	25	15

Using the level of significance, $\alpha = 0.05$, does this data provide evidence to support that the median of number of mathematics graduates has decreased in fall 2017 compared with that in fall 2018?

Solution

In this case, the hypothesis is as follows (Table 6.16):

$$H_o : \tilde{\mu}_D = 0, \quad H_a : \tilde{\mu}_D < 0$$

This is a left-tailed test case. Hence, the test statistic is $R = R_+ = 1 + 5 = 6$. Hence, from the table of Wilcoxon signed-rank test (Table 6.D in Appendix), the critical value is $R_\alpha = 2$. This means that $R > R_\alpha$. Thus, at the level of significance $\alpha = 0.05$, the null hypothesis will not be rejected. Therefore, there is no support for the alternative hypothesis of the median number of mathematics graduates in fall 2018 to be lower than that in fall 2017.

Using R:

```
library(MASS)
Fall_2017<-c(20,8,6,12,10,25)
Fall_2018<-c(16,6,7,20,4,15)
wilcox.test( Fall_2018, Fall_2017,paired=TRUE, alternative =
"less", conf.level = 0.95)
```

Output:

```
data: Fall_2018 and Fall_2017
V = 6, p-value = 0.2188
alternative hypothesis: true location shift is less than 0
```

According to the output given by R, the p-value is 0.2188, which is greater than 0.05. Therefore, there is insufficient evidence to support the alternative hypothesis.

TABLE 6.16

Mathematics Majors Graduated in Fall 2017 and Fall 2018

College #	Fall 2017	Fall 2018	Difference, Fall 2018–Fall 2017	$\lvert d \rvert$	Ranks $\lvert d \rvert$	Signed Rank
C1	20	16	−4	4	3	−3
C2	8	6	−2	2	2	−2
C3	6	7	1	1	1	1
C4	12	20	8	8	5	5
C5	10	4	−6	6	4	−4
C6	25	15	−10	10	6	−6

Example 6.12

A researcher conducts a titration using two different indicators. Her aim is to understand whether the median time to complete the titration, using two indicators A and B, is different with $\alpha = 0.05$. In order to find out, she conducts an experiment seven times and records the data as listed below:

	Time	
Experiment	Using Indicator A	Using Indicator B
1	2.25	2.30
2	2.15	2.17
3	2.05	2.06
4	2.35	2.43
5	2.40	2.34
6	2.29	2.26
7	2.22	2.29

Is the median time to complete the titration, using two indicators A and B, different with $\alpha = 0.05$?

Answer

We calculate the numbers listed in Table 6.17.
We conduct the following hypothesis:

$$H_o : \tilde{\mu}_D = 0, \quad H_a : \tilde{\mu}_D \neq 0$$

According to the above data, $R_+ = 4 + 2 + 1 + 6 + 7 = 20$ and $R_- = -5 - 3 = -8$. Therefore, for two-tailed test, the test statistic is $R = \min(|R_-|, R_+) = \min(|-8|, 20) = 20$. Hence, the table value for this test using $n = 7$ and $\alpha = 0.05$ is $R_{\alpha/2} = 2$. Using this information, we have $R > R_{\alpha/2}$.

TABLE 6.17
Time to Complete the Titrations

Experiment	Time Using Indicator A	Time Using Indicator B	Time Difference (B − A)	$|d|$	Ranks $|d|$	Signed Rank
1	2.25	2.30	0.05	0.05	4	4
2	2.15	2.17	0.02	0.02	2	2
3	2.05	2.06	0.01	0.01	1	1
4	2.35	2.43	0.08	0.08	6	6
5	2.40	2.34	−0.06	0.06	5	−5
6	2.29	2.26	−0.03	0.03	3	−3
7	2.22	2.29	0.07	0.07	7	7

Therefore, using $\alpha = 0.05$, the null hypothesis will not be rejected. In other words, there is enough evidence to support the claim that the median time to complete the titration using indicators A and B is different.

Using R:

```
library(MASS)
Indicator_A<-c(2.25,2.15,2.05,2.35,2.40,2.29,2.22)
Indicator_B<-c(2.30,2.17,2.06,2.43,2.34,2.26,2.29)
wilcox.test(Indicator_B, Indicator_A, paired=TRUE, alternative =
"two.sided",conf.level = 0.99)
```

Output:

```
data: Indicator_B and Indicator_A
V = 20, p-value = 0.375
alternative hypothesis: true location shift is not equal to 0
```

According to the output given by R, the p-value is 0.375, which is greater than 0.05. Therefore, we have insufficient evidence to support the alternative hypothesis.

6.4   INFERENCE USING MORE THAN TWO SAMPLES

So far, we have discussed both single-sample and two-sample inferences using non-parametric approach. In practice, there are situations where there are more than two populations involved in inference problems. Therefore, we should consider the case of nonparametric inference based on more than one sample. For instance, suppose an instructor wants to compare the performances of students in three sections of elementary statistics classes. If this situation violates the parametric assumptions, the prior nonparametric procedures cannot be applied. When considering more than two samples, they may be independent or dependent. We will discuss both of these cases in the subsections that follow.

6.4.1   INDEPENDENT SAMPLE INFERENCE USING THE KRUSKAL–WALLIS TEST

When analyzing more than two samples in nonparametric statistics, we adhere to Kruskal–Wallis test. This is an extension of the Mann–Whitney test that was discussed earlier. In other words, Kruskal–Wallis test is considered as the nonparametric version of parametric procedure of one-way ANOVA.

With this nonparametric procedure, we will test the following hypothesis:

H_0 : All groups have the same median.

H_a : At least one group has a different median.

Test statistic is:

$$H = \left[\frac{12}{N(N+1)} \sum_{i=1}^{C} \frac{T_i^2}{n_i} \right] - 3(N+1), \tag{6.12}$$

where N is the total in all the groups, T_i; $i = 1, 2, \ldots, C$ represents the total rank of each group i, and n_i represents the total observations in each group i.

Steps in Kruskal–Wallis Test

1. Rank all of the scores in all the groups without considering which group they belong to. If there are ties, take the average of them as the rank.
2. Calculate T_i values for each group, and compute the test statistic H.
3. Find the critical value from the chi-square table, with $df = C - 1$ degrees of freedom.
4. Reject the null hypothesis if the test statistic $K \leq$ critical value.

Example 6.13

Does the mode of teaching influence the student's academic success?

An instructor wants to check whether his mode of teaching has an impact on student's success in introductory statistics course. He randomly selects 35 students from his three sections of the introductory statistics classes that are taught face-to-face, hybrid, and online. He records the final grades for each student in each section as the measurement of student's academic success. Does this data show that the mode of teaching impact the student's performance, using the level of confidence of 0.01?

Solution

Hypothesis: H_o : Grades of all teaching methods have the same median value.

H_a : At least one median is different.

Test statistic:
$$H = \left[\frac{12}{N(N+1)} \sum_{i=1}^{C} \frac{T_i^2}{n_i} \right] - 3(N+1)$$

$$H = \left[\frac{12}{35(35+1)} \left(\frac{313^2}{15} + \frac{183^2}{10} + \frac{134^2}{10} \right) \right] - 3(35+1)$$

$$= 3.20$$

Critical value $= \chi^2_{0.01,C-1} = \chi^2_{0.01,2} = 9.21$ $d.f = C - 1 = 3 - 1 = 2$.

Conclusion: Based on the critical value and the test statistics, test statistic $K \leq$ critical value. Therefore, we do not reject the null hypothesis at the level of significance of 0.01; that is, we do not have evidence to claim that at least one of the teaching methods has a different median than the others (Tables 6.18 and 6.19). This suggests that the mode of teaching does not have a significant impact on student's performance.

TABLE 6.18

Teaching methods vs students' performances

No	Face-to-Face	Hybrid	Online
1	85	73	73
2	79	72	62
3	80	50	60
4	95	91	61
5	65	78	78
6	70	79	79
7	75	85	55
8	91	60	80
9	88	73	81
10	69	95	70
11	72		
12	95		
13	68		
14	84		
15	77		

TABLE 6.19

Ranked Data for Teaching Methods vs Students' Performances

No	Face-to-Face	Rank	Hybrid	Rank	Online	Rank
1	65	7	50	1	55	2
2	68	8	60	3.5	60	3.5
3	69	9	72	13	61	5
4	70	10.5	73	15	62	6
5	72	12	73	15	70	10.5
6	75	17	78	19.5	73	15
7	77	18	79	22	78	19.5
8	79	22	85	28.5	79	22
9	80	24.5	91	31.5	80	24.5
10	84	27	95	34	81	26
11	85	28.5				
12	88	30				
13	91	31.5				
14	95	34				
15	95	34				
	T_i	313		183		134

Using R:

```
F2F<-c(85,79,80,95,65,70,75,91,88,69,72,95,68,84,77)
Hybrid<-c(37,72,50,91,78,79,85,60,73,95)
Online<-c(73,62,60,61,78,79,55,80,81,70)
dataTeaching<-list(g1=F2F,g2=Hybrid,g3=Online)
kruskal.test(dataTeaching, conf.level=0.99)
```

Output:

```
Kruskal-Wallis rank sum test
data: dataTeaching
Kruskal-Wallis chi-squared = 3.1236, df = 2, p-value = 0.2098
```

According to the above output, the p-value is 0.2098, which is higher than the α level of 0.01. Therefore, we fail to reject the null hypothesis at the level of significance of 0.01.

Example 6.14

The research and development unit of a company, who manufactures electric light bulbs, wants to compare the median lifetime of its light bulbs with the median lifetimes of light bulbs at similar two other competitors' companies. The following data is available for three selected types of light bulbs (Table 6.20).

Consider the following collected data set. Let brand 1, brand 2, and brand 3 be the three types of selected types of light bulbs. Use the level of confidence of 0.01.

Solution

Before the analysis, let's arrange the above data based on their ranks (Table 6.21).

Hypothesis: H_o : Median Lifetime of each brand is the same.

H_a : At least one brand has a different median lifetime than other brands.

Test statistic: $$H = \left[\frac{12}{N(N+1)} \sum_{i=1}^{C} \frac{T_i^2}{n_i} \right] - 3(N+1)$$

$$H = \left[\frac{12}{23(23+1)} \left(\frac{181.5^2}{10} + \frac{30^2}{7} + \frac{64.54^2}{6} \right) \right] - 3(23+1)$$

$$= 17.50$$

Critical value $= \chi^2_{0.01,C-1} = \chi^2_{0.01,2} = 9.21 \, d.f = C - 1 = 3 - 1 = 2.$

Conclusion: Here, the test statistic K (17.50)$\leq$critical value (9.21). Therefore, we reject the null hypothesis at the level of significance of 0.01; that is, we have enough evidence to support the alternative hypothesis. This suggests that at least one of the brands has median lifetimes.

TABLE 6.20

Lifetimes (in 1,000 hours) of Three Types of Light Bulbs

No	Brand 1	Brand 2	Brand 3
1	10.1	8.7	9.0
2	11.5	8.5	9.1
3	10.5	9.0	9.4
4	9.5	8.3	8.9
5	10.2	8.6	9.5
6	11.5	8.4	10.0
7	10.3	8.9	
8	9.7		
9	9.9		
10	11.2		
Totals T_i	104.4	60.4	55.9

TABLE 6.21

Ranked Data of the Lifetimes

No	Brand 1	Rank	Brand 2	Rank	Brand 3	Rank
1	9.5	12.5	8.3	1	8.9	6.5
2	9.7	14	8.4	2	9	8.5
3	9.9	15	8.5	3	9.1	10
4	10.1	17	8.6	4	9.4	11
5	10.2	18	8.7	5	9.5	12.5
6	10.3	19	8.9	6.5	10	16
7	10.5	20	9	8.5		
8	11.2	21				
9	11.5	22				
10	11.5	23				
T_i		181.5		30		64.54

Using R:

```
Brand1<-c(9.5,9.7,9.9,10.1,10.2,10.3,10.5,11.2,11.5,11.5)
Brand2<-c(8.3,8.4,8.5,8.6,8.7,8.9,9)
Brand3<-c(8.9,9,9.1,9.4,9.5,10)
dataBulbs<-list(g1= Brand1, g2=Brand2,g3= Brand3)
kruskal.test(dataBulbs, conf.level=0.99)
```

Output:

```
Kruskal-Wallis rank sum test
data: dataBulbs
Kruskal-Wallis chi-squared = 17.517, df = 2, p-value = 0.0001572
```

According to the above output, the p-value is 0.0001572, which is less than the α level of 0.01. Therefore, we reject the null hypothesis at the level of significance of 0.01.

EXERCISES

6.1. Consider a car seller who sells four brands (A, B, C, and D) of cars. He says that he does not give any priority to any brand of cars and considers all brands equally important. In a given day, the number of cars available with brands A, B, C, and D is 50, 40, 45, and 60, respectively. Using significance $\alpha = 0.05$, test to find out if there is sufficient evidence to conclude that the car seller is right with his claim.

6.2. A survey was conducted to find out the number of TV-watching time (per week) by people of different age groups. The following table summarizes the finding:

Age Group	# Hours (Per Week)
< 5 years	15
5 years – 15 years	17
15 years – 35 years	20
35 years – 55 years	25
55 years – 75 years	30
> 75 years	33

Using the available data and $\alpha = 0.05$, find out if the TV-watching time is evenly distributed across all age groups.

6.3. A school administrator wants to find out school kid's means of transportation to school. She wants to know whether all transportation means are equally distributed or not. She has the following data available:

Means of Transportation	Walking	Bicycle	Car
Number of students	200	175	215

Test whether the transportation means are evenly distributed at the level of significance of 0.1.

6.4. A mayor of a particular city says that he is happy with the city because his city has an equal proportion of higher-income, middle-income, and lower-income families. A survey provided the following details about this city using a random sample of 500 families. Does this collected data support the claim of the mayor? Use the α level of 0.1.

Income level	Lower	Middle	Higher
# families	160	165	175

6.5. A manufacture of a certain candy brand says that in each of the bag of candies, there is an equal number of candies of red, green, yellow, and white colors. A randomly selected bag consists of 15, 25, 13, and 16 red, green, yellow, and white color candies, respectively. Is there a truth of what candy manufacture says as far as the distribution of colors of candies is concerned? Use the $\alpha = 0.05$ to test the above.

6.6. There is a claim that the gender and the smoking are independent. A sociologist attempts to test the above claim using the following collected data.

	Gender	
Smoking	Male	Female
Yes	150	100
No	75	60

6.7. A car manufacture says his buyers that when manufacturing cars, the type of the car (Sport Utility Vehicle, SUV or non-SUV) and the color of the car (white, black, blue, and red) are independent. Randomly collected sample from the latest records indicates the following information. Use the level of significance of 0.05 to test the car manufacture's claim.

	Color		
Car Type	White	Black	Blue
SUV	100	110	105
Non-SUV	95	115	105

6.8. An instructor is teaching three different classes. After the final examinations are over, he is curious about the students' grades distributions. Based on the following summary of grades, the inductor can claim that grades and the classes are independent, at the level of significance of 0.05.

		Grades			
		A	B	C	F
Class #	1	10	15	20	5
	2	8	20	18	4
	3	15	15	17	3

6.9. Gas price depends on the location of the gas station in a big city. Following data shows the gas price recorded from 15 locations of the above city.

$2.75	$3.00	$2.80	$2.90	$2.85
$3.05	$2.95	$2.98	$3.05	$2.88
$2.78	$3.04	$2.92	$2.96	$2.89

Does the above data show that the average gas price of this city is over $2.95? Conduct a hypothesis test to check this using the level of significance of 0.01.

6.10. The compressive strength of a particular concrete should be at least 30,000 pound-force psi. An engineer randomly collected 12 samples of this concrete in order to find their compressive strengths. Following are his measurements of compressive strengths in psi units.

28,900	31,100	29,500	30,090	30,195	29,875
29,995	28,710	31,560	29,180	32,295	31,965

Based on the collected sample of data, the engineer has evidence to support that the average compressive strength of concrete is at least 30,000 psi. Use the level of significance of 0.05.

6.11. New car manufacturer claims that the average gas mileage for their latest model is more than 35mpg. A research team collected 15 observations randomly to test the above manufacture's claim. They selected the level of significance as 0.05. Does the following data support the manufacture's claim?

34.7	36.5	34.9	36.4	35.8	35.1
35.7	34.5	36.9	35.4	35.0	35.5
36.0	35.3	33.9			

6.12. An administrator of a particular school district wanted to study about student's performance for mathematics SAT examination. He found that the distribution of this score is very skewed and conducted a nonparametric study with the collected random sample of data. His hypotheses were as follows.

$$H_o : \text{Median SAT score is 600.}$$

$$H_a : \text{Median SAT score is above 600.}$$

After running the test, the p-value for the test was 0.03. Interpret this value, and state your conclusion using $\alpha = 0.05$.

6.13. A tire manufacturing company rejects tires if the median diameter of a tire is not equal to 10 inches. Suppose a quality controller randomly selects a sample of 30 tires to test whether the median diameter is different from 10 inches. Does the following data help the quality controller to confirm his prior assumption using the level of significance $\alpha = 0.01$?

10.01	10.02	9.09	10.03	10.05	10.01	10.05	10.03	10.06	9.08
9.95	10.04	9.89	10.13	10.25	10.41	9.45	9.93	10.16	9.98
10.02	10.03	9.39	9.93	10.05	9.89	9.05	10.13	9.76	9.38

6.14. Community of Summer Gate is complaining about purity of the water in its neighborhood lake. The officers in homeowners' association ask a researcher to investigate the purity of the water of the lake. The researcher randomly selects 12 samples of water from random location of the lake and measures their pH values. The measurements are given below.

   9.1 8.8 6.7 7.6 8.7 6.9 7.9 8.2 8.5 8.6 8.6 8.9

   Do the above pH values indicate that the pH value of the water is above 8.5? Use $\alpha = 0.05$ to test this.

6.15. A farmer wants to compare the effect of two fertilizers on the growth of a particular crop. He randomly selects two samples of plants of the same size and plants them in an area where both samples have homogeneous environments other than the different fertilizers. After two months, he measured the lengths (in cm) of all the plants in both samples as shown in the following table.

| Fertilizer A | 5.5 | 5.4 | 5.3 | 5.0 | 5.2 | 5.3 | 5.1 | 5.5 | 5.4 | 5.3 |
| Fertilizer B | 5.6 | 5.8 | 5.5 | 5.8 | 5.6 | 5.8 | 5.0 | 5.9 | 5.4 | 5.7 |

   Using the above data, the farmer finds out that the average growth of the crop using the fertilizer B is higher than that of the crop using the fertilizer A using $\alpha = 0.05$.

6.16. A car seller sells two brands (A & B) of cars that were manufactured in last year. With his experience, he thinks that the median gas mileage of brand A is higher than that of brand B. He randomly selects two samples of cars from both brands and measures the distance the car can run using one unit of gas. The measurements are given below.

| A | 34.7 | 36.5 | 34.9 | 36.4 | 35.8 | 35.1 | 35.7 | 34.5 | 36.9 | 35.4 |
| B | 32.5 | 33.5 | 32.5 | 33.4 | 32.8 | 34.1 | 33.7 | 34.6 | | |

   Does the above data provide evidence to support the seller's prior assumption using $\alpha = 0.05$?

6.17. A manufacturer introduces a formula to enhance the performance of the car engine. Before he advertises this new product, the manufacturer assigns his research and development (R&D) team to investigate this product. He assumes that this novel formula increases the gas mileage of cars. The R&D team selects a random sample of 10 cars (different brands) and records the gas mileage due to one unit of gas before and after using the novel formula. Does the following measurement suggest that the new formula improves the gas mileage using $\alpha = 0.05$?

| Before | 34.7 | 38.5 | 30.9 | 39.4 | 37.8 | 15.1 | 26.7 | 24.5 | 26.9 | 25.4 |
| After | 35.0 | 39.2 | 31.5 | 39.4 | 38.7 | 15.0 | 25.7 | 24.6 | 26.9 | 26.7 |

6.18. Domestic violence has become a major issue in a particular city. One of the NGOs introduces an educational program to educate people about the domestic violence. They randomly select 8 men who involved in domestic violence during the last year. These men are educated for a certain number of days and their behaviors are observed for another year. Following table shows the collected information about their involved domestic violence before and after the participation into the program.

Domestic violence (before)	3	4	2	3	4	5	1	3
Domestic violence (after)	2	2	2	1	0	3	0	2

The NGO wants to understand the effectiveness of their educational program on domestic violence. Does the above information indicate that the program has been able to reduce the domestic violence at α level of 0.1?

6.19. After taking the orange harvest, a farmer wants to know whether all of his orange fields produce oranges with same average weights or not. As a means of testing this, he collects random samples from each orange field and measures their weights (in grams) as displayed below. Do the following weights indicate that all of the fields produce oranges with same average weights or not? Use the level of significance of 0.01 to test this.

Field A	139.5	141.1	135.6	137.9	139.5
	138.7	137.6	138.9	136.9	139.0
Field B	140.5	142.1	144.5	145.3	144.4
	145.3	147.8	143.7		
Field C	130.5	132.1	134.4	135.1	134.0
	135.5	137.8	133.3	136.2	129.6
	136.9	137.9			

7 Stochastic Processes

7.1 INTRODUCTION

One of the applications of probability, and, in fact, a generalization of random vector, which is vastly used in biology, cell biology, and birth-and-death (B-D) processes, is the concept of stochastic processes. Historically, after Kolmogorov started the modern probability theory, as we know it today, in 1933, it was interrupted by World War II, after a decade. However, in 1940s, stochastic or random processes, as applications of probability in potential theory, gained attention.

However, stochastic processes were known decades before through the theory of queues originated by the Danish engineer, Agner Krarup Erlang (1878–1929) (Figure 7.1), beginning of the twentieth century (1917); see Haghighi and Mischev (2014).

Even Albert Einstein in his paper published in 1905 on physical observation of Brownian motion (random motion of particles suspended in a fluid or a gas resulting from their collision with the fast-moving molecules in the fluid) using ideas from the kinetic theory of gases.

FIGURE 7.1   Agner K. Erlang 1878–1929.

But now martingale (a sequence of random variables with certain conditions) was developed by an American mathematician Joseph Doob (1910–2004), who detailed in his book *Stochastic Processes* published in 1953. The idea of martingale was derived from random walk, Brownian motion, branching, and Poisson processes.

After the brief historic comments regarding how the stochastic process started, we move to formal definition of stochastic process. However, before that, we remind the readers from Chapter 2 that $(\Omega, \mathbb{S}, P)$ is a probability space, where P is a probability measure on a σ-field $\mathbb{S}$.

Definition 7.1

Let T be a subset of non-negative real numbers, $[0,\infty)$. A **stochastic (or random) process** is a sequence (or a family) of random variables, say $X_t, t \in T$, indexed by T, defined on a probability space $(\Omega, \mathbb{S}, P)$. The set T is referred to as the **index set** or the **parameter space**. If T is the set of non-negative integers, $\mathbb{N}_0$, the stochastic process is referred to as the **discrete-time,** denoted by $\{X(t), t \in T\}$ and if it is an interval $[0,\infty)$, it is referred to as the **continuous-time,** denoted by $\{X(t), t \in T\}$, stochastic process. If the probabilistic rules do not change with time, the process is called the **stationary process**.

Definition 7.2

For a stochastic process $\{X_t, t \in T\}$ or $\{X(t), t \in T\}$, the set of all possible values of random variable X_t or $X(t)$, that is, the value of a stochastic process at any epoch of time, denoted by S, is referred to as the **state space** of the process in discrete case and the **phase space** in continuous case, state space in both cases, if there is no confusion, which is a mathematical space like a subset of Euclidean n-dimensional space. This space reflects the different values the stochastic process can take. The state at which the process starts at, denoted by X_0 or $X(0)$, is referred to as the **initial state**.

Note 7.1

It should be noted that if T is finite such as $T = \{1, 2, \ldots, n\}$, we can find probability density function (pdf) or probability mass function (pmf) of the process through the joint distribution of random vector $\{X_{t_1}, X_{t_2}, \ldots, X_{t_n}\}$. However, for an infinite case of T, finding pdf or pmf is somewhat problematic.

Note 7.2

When the index set, say T, consists of only one element, that is, it is a singleton, $T = \{1\}$, the process $\{X_t, t \in T\}$ becomes a single random variable X_1. On the other hand, when T is a finite set such as $T = \{1, 2, \ldots, n\}$, the process will become a random vector, that is, $\{X_{t_1}, X_{t_2}, \ldots, X_{t_n}\}$.

Example 7.1

Consider a random sample $\{X_1, X_2, ..., X_n\}$ of size n from a population with an unknown distribution function. The function F_n (x) referred to as the **empirical distribution function** is defined as

$$F_n(x) = \frac{\text{Number of sample points } X_j \leq x}{n} = \frac{1}{n}\sum_{j=1}^{n} I_x(X_j), \quad -\infty < x < \infty, \qquad (7.1)$$

where I_x is the indicator function. Since $F_n(x)$ is a random variable for each x, the sequence $\{F_n(x), \ -\infty < x < \infty\}$ is a stochastic process.

Example 7.2

As an example of a discrete index set, consider the net number count of delivery and returns of a particular item in an online store at time t within 24 hours of operation. Hence, the process of delivery and returning is a stochastic process with the index set T as the set $\{0,1,2,...,24\}$ and with the state space as $S = \{0,\pm1,\pm2,...\}$.

Example 7.3

As an example of a continuous index set, let us consider arrival of customers at a bank, which is assumed to be random in nature. The number of customers arrived during the interval $[0,t]$ is a stochastic process in which the index set is $T = [0,\infty)$ and the state space $S = \{0,1,2,...\}$. This is also an example of a **pure birth process** that we will consider later.

Example 7.4

Consider the mathematics department at a university with 30 faculty and staff that has 5 computers for the staff. The IT unit of the university has assigned a technician for computers in this department, who is capable of repairing one failed computer on the day of failure. Due to the age of these computers, there is a backlog of requests for service and, hence, a failed computer must get in a waiting line (queue). The average rate of failure for these computers during the past year has been 20% at any working day, and failure of a computer is independent of failure of other computers in the office. In such a situation, the process of quantity of the backlogs at time t will have the set of states as $S = \{0,1,2,3,4\}$. This is because there is a chance for each computer to fail on a day. Hence, while the technician is preparing one, a maximum of four computers are waiting to be repaired.

7.2 RANDOM WALK

Suppose you are standing at 0. Flip a coin. If the coin comes up heads, move to the right by one step. If it comes up tails, move to the left by one step. Flip the coin again. If it comes up heads, move a step to the right, and if it comes up tails, move a step to the left. Repeat and continue this process. *Must you come back to zero?* This is the question Pólya (1887–1985) laid on the table and then answered. He essentially expressed the idea of random walk in a very simple way. Perhaps it is because George Pólya is considered as the Father of Problem Solving in Mathematics Education, although he is a mathematician who made important contributions to probability theory, number theory, the theory of functions, and the calculus of variations.

Another way random walk can be illustrated is as follows: A drunken man is to walk on a road that runs east and west, that is, a fifty percent probability in either of the two directions. Being intoxicated, he is likely to. Take a step forward (east) as backward (west). He decides to start walking. From each new position, he is again as likely to go forward as backward. Each of his steps is of the same length, but of random direction—forward or backward. We will ask the students to investigate and show the position of the walker or the **state of the random walk after *n* steps**.

This example shows that a random walk is a process consisting of a sequence of discrete steps of fixed length. For example, the random thermal perturbations in a liquid form a random walk process, known as **Brownian motion**. The collisions of molecules in a gas also form a random walk that is responsible for diffusion.

Yet, another example of a random walk is transport molecules that play a crucial role in cell viability. Among others, linear motors transport cargos along rope-like structures from one location of the cell to another in a probabilistic fashion. So, each step of the motor, either forward or backward, bridges a fixed distance and requires several biochemical transformations, which are modeled as internal states of the motor. While moving along the rope, the motor can also detach, and the walk is interrupted. Many complex processes take place in living cells. Transport of cargos across the cytoskeleton is fundamental to cell viability and activity. To move cargos between the different cell parts, cells employ **molecular motors**. The motors are responsible for a huge variety of tasks, ranging from cell division to DNA replication and chemical transport. A special class of such motors operate by transporting cargos along the so-called cellular microtubules, namely, rope-like structures that connect, for instance, the cell nucleus and outer membrane. One particular example for such motors is kinesin V, common in eukaryotic cells. Due to the periodic molecular structure of the microtubules, the steps of kinesin have all the same length equal to 8 nm. Under normal conditions present in living cells, this motor performs a random walk in one dimension with a drift on the microtubule possibly stopped by detachment from the tubule.

Now, we formulize a random walk mathematically. As a simple application of a stochastic process, we consider a sequence of independent and identically distributed (iid) random variables. Each random variable may have possible values 1 with probability $\frac{1}{2}$ and -1 with the same probability. These probabilities may be more general as p and $1 - p$, respectively. In some cases, 0 could be a possible value. Let S_n be a partial sum of $X_1, X_2, \ldots$, that is, with $n = 1, 2, \ldots$.

Definition 7.3

Let $X_1, X_2, \ldots$, be a sequence of iid random variables. Also, let $S_n = X_1 + X_2 + \cdots + X_n$. Then, the sequence of sums, $\{S_n, n \geq 1\}$, which is the integer-time stochastic process, is referred to as a **simple random walk** or **one-dimensional random walk**, based on $\{X_j, j \geq 1\}$. The word "simple" will be dropped when higher dimensions are considered.

The term was introduced by Carl Pearson in 1905, although its discrete version was originated by Albert Einstein and Marian Smoluchowski (a Polish physics, 1872–1917) in the same time (Figure 7.2).

As it can be seen, S_n is the difference between the positive and negative occurrences of events in the first n trials. That is, the sum of the **increments** is given below:

$$X_k - X_{k-1}, k = 1, 2, \ldots, n, n = 1, 2, \ldots, \tag{7.2}$$

(a)

(b)

FIGURE 7.2 (a) Marian Ritter von Smolan, Smoluchowski (1872–1917). (b) Albert Einstein (1879–1955).

Note 7.3

The increments $X_k - X_{k-1}, k = 1, 2, \ldots, n$, are independent of $\{X_n\}, n = 1, 2, \ldots$. Also, each increment $X_k - X_{k-1}, k = 1, 2, \ldots, n$, has a Bernoulli pmf with the probability of success as p. That is,

$$P(X_k - X_{k-1} = 1) = p \text{ and } P(X_k - X_{k-1} = -1) = 1 - p,$$

$$k = 1, 2, \ldots, n, \ n = 1, 2, \ldots, X_0 = 0.$$

(7.3)

Definition 7.4

Refer to the Definition 7.3. Let $\tau > 0$ be a positive real number. A term in $\{S_n, n \geq 1\}$ for which $S_n \geq \tau$ is referred to as the **threshold** at τ.

Note 7.4

Two questions may arise. We leave the answers as exercises. (1) What is the probability that the sequence $\{S_n, n \geq 1\}$ contains a term for which a threshold at T is crossed? (2) What is the distribution of the smallest n for which $S_n \geq \tau$?

Example 7.5

For $p \leq \dfrac{1}{2}$ and $j > 0$, find the probability that the sequence $S_1, S_2, \ldots$ reaches or exceeds j.

Answer

We leave it as an exercise to prove that the answer to this question is as follows:

$$P\left\{ \bigcup_{n=1}^{\infty} (S_n \geq j) \right\} = \left(\frac{p}{1-p} \right)^j.$$

(7.4)

Relation (7.4) is referred to as the **probability that the random walk crosses a threshold** at j.

Example 7.6

Imagine a walker who starts at point 0, walking randomly one step (with the same length throughout the process) forward (+1) and one step backward (−1). That is, staying put is not allowed. Suppose he stops after ten steps. During the process of walking, if he walked four steps forward, he will be at position $S_{10} = -2$.

Note 7.5

If we allow him to stay put during the walking process, we can assign 0 to values of the random variables.

Example 7.7

In this example, we will show how we may construct a simple random walk process. We start with the probability space $(\Omega, \mathbb{S}, P)$ and the stochastic process $\{X_k, k = 1,2,...\}$. So, we suppose that $X_1, X_2,...$ are iid random variables, each with two values +1 and −1, with probabilities p and $1 - p$, respectively. Let $S_n = X_1 + X_2 + ... + X_n$. Now if there are j positive steps out of the n total steps, then $S_n = 2j - n$ (why?). Hence,

$$P(S_n = 2j - n) = \frac{n!}{j!(n-j)!} p^j (1-p)^{n-j}. \tag{7.5}$$

So, the probability of position of the walker having four forward steps, with the probability of each such a step as $\frac{2}{3}$, in a total of ten steps, is:

$$P(S_{10} = -2) = \frac{10!}{(4!)(6!)}\left(\frac{2}{3}\right)^4\left(\frac{1}{3}\right)^6 = 0.0569.$$

Note 7.6

If $X = \{X_1, X_2,..., X_n\}$, then

$$\lim_{n \to \infty} \frac{S_n}{n} = E(X) = \overline{X}. \tag{7.6}$$

Hence, the position of the walker depends upon the value of $\overline{X}$ as it is negative, 0, or positive. If $\overline{X} < 0$, then S_n will tend to drift downward, and if $\overline{X} > 0$, then S_n will tend to drift upward. Of course, position for $\overline{X} > 0$ can be obtained from the $\overline{X} < 0$ by the use of $\{-S_n, n \geq 1\}$.

We now offer some examples of random walk involving mathematical formulizations.

Example 7.8 Gambler's Ruin Problem

Two gamblers, designated by G_1 and G_2 (Gambler 1 and Gambler 2, respectively), play a simple gambling game. On each round (trial or betting) of the game, a coin is tossed, and if $H =$ "Heads" results, then G_1 wins $1 from G_2. On the other hand, if $T =$ "Tails" results from the coin tossing, then G_2 wins $1 from G_1. The probabilities associated with "Heads" and "Tails" on the coin are unknown to the players so that both players feel free to continue the rounds. Game continues until one of the players loses all his/her money and is thereby "ruined".

Note 7.7

For the game to be "fair", the coin used in the model should have $P(H) = P(T) = \dfrac{1}{2}$.

On the other hand, if we wished to model a more general series of success/failure trials, where levels of skill or other features are considered, we could use a biased coin. For example, a game where G_2 is three times as skilled as G_1 might use a model with $P(H) = \dfrac{1}{4}$ and $P(T) = \dfrac{3}{4}$.

For this example, suppose G_1 has \$3 and G_2 has \$1 before the game begins and the coin is biased. We also suppose that, unknown to the players, the coin is biased with $P(H) = \dfrac{1}{3}$ and $P(T) = \dfrac{2}{3}$. We can study the game by keeping track of the money that G_1 has at the end of each round of the game. Over the course of the game, G_1 will have a fortune ranging from \$0 to \$4.

Figure 7.3 illustrates the possible dollar amounts of G_1's fortune as well as the probabilities for his moving from one fortune value to the next during the process.

In Figure 7.3, the values 0, 1, 2, 3, and 4 are the so-called states of the process, and these represent possible dollar amounts (fortunes) that G_1 might have during the course of the trials. Arrows between states of the form $x \rightarrow (p_{x,y}) \rightarrow y$ indicate that $p_{x,y}$ is the probability that G_1 moves from \$x to \$y (given that he/she has \$x) on the next round of the game. For example, $2 \leftarrow \left(p_{3,2} = \dfrac{2}{3} \right) \leftarrow 3$ means that if G_1 has \$3 (state 3), then the probability that he/she loses a round and moves to \$2 (state 2) is 2/3. The arrowhead beneath state 3 contains the initial probability of beginning at state 3 and this probability is 1 since G_1 begins with \$3.

Let us consider state 0. Here, G_1 has \$0 and the game has stopped. Thus, we write $p_{0,0} = 1$ since there are no subsequent rounds. Similarly, we indicate game's end at state 4 (G_1 has all \$4) by $p_{4,4} = 1$. If the game ends at state 0 (G_1 loses all his money), we say that the process has been **absorbed** into state 0 since state 0 cannot be exited once entered. Similarly, state 4 is an **absorbing state**. A state that is not absorbing is called a **transient state**.

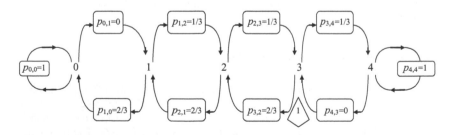

FIGURE 7.3 Transition probability diagram for Example 7.8.

Note 7.8

For each state, the sum of the probabilities leaving that state is 1.

The transition probability diagram (Figure 7.3) illustrates an example of a **random walk**. In this process, we "walk" back and forth (or right and left) from state to state and the direction of a step (transition) is determined in a random fashion by the coin toss.

In Example 7.7, we had five states 0, 1, 2, 3, and 4. Let us now generalize the number of states and assume that the state space (the set of all states) has $N+1$ elements, that is, $S = \{0,1,2,...,N\}$. Recall also that states in the Gambler's Problem represent the amount of money (fortune) of one gambler. The gambler bets \$1 per round and wins or loses. He is ruined if he reaches state 0. The probability is assumed to be $p > 0$ and that of losing is $q > 0$ with $p + q = 1$. In terms of transition probabilities, this assumption means that $p_{k,k+1} = p$ and $p_{k,k-1} = q$, for $k = 1,2,...,N-1$. We further assume that $p_{0,0} = p_{N,N} = 1$; that is, $p_{0,0}$ and $p_{N,N}$ are the so-called absorbing boundaries. All other transition probabilities are assumed to be zero.

Note 7.9

Transition of a state to itself is not allowed, that is, $p_{k,k} = 0$ for all k except $k = 0$ and $k = N$.

The transition matrix for this random walk is as follows:

$$
P = \begin{bmatrix}
1 & q & 0 & 0 & 0 & 0 & \cdots & 0 & 0 \\
0 & 0 & q & 0 & 0 & 0 & \cdots & 0 & 0 \\
0 & p & 0 & 0 & 0 & 0 & \cdots & 0 & 0 \\
0 & 0 & p & 0 & 0 & 0 & \cdots & 0 & 0 \\
0 & \ddots & \ddots & \ddots & \ddots & \ddots & \cdots & 0 & 0 \\
\vdots & \vdots & \ddots & \ddots & \ddots & \ddots & \cdots & \vdots & \vdots \\
0 & 0 & 0 & 0 & 0 & 0 & \cdots & q & 0 \\
0 & 0 & 0 & 0 & 0 & 0 & \cdots & 0 & 0 \\
0 & 0 & 0 & 0 & 0 & 0 & \cdots & p & 1
\end{bmatrix}
$$

The states 0 and N are absorbing states, and all other states are transient. Hence, this is an example of a random walk with absorbing boundaries.

For example, if walker steps forward with a probability of 0.75, then the probability that he/she is at location 4 in five trials is:

$$p_4(5) = \binom{5}{4}(0.75)^4(0.25)^{5-4} = 0.3955.$$

7.3   POINT PROCESS

As another case of a stochastic process, we briefly discuss the point process.

Definition 7.5

A set of random points $\{T_1, T_2, ..., T_r\}$ such that $T_1 < T_2 < ... < T_r$, $\lim_{r \to \infty} T_r = \infty$, in some space, often a subset of multidimensional Euclidean space, say $\mathbb{R}^r$, for some r, is a stochastic process referred to as a **point process**. In $\mathbb{R}^1$, each T_j is a random variable. The numbers $T_r, r = 1, 2, ...$, are referred to as the **event times** or **epochs**.

Note 7.10

> According to the definition above, a point process is a strictly increasing sequence of real numbers, which does not have a finite limit point.

Note 7.11

> As examples of point process, we can mention the arrival time epochs of customers at a service station and in biology by the moments of impulses in nerve fibers.

Note 7.12

> We note that $T_r, r = 1, 2, ...$, do not necessarily have to be points of time. They could be points other than times like location of potholes in a road. However, since the number of potholes is finite, in the real world, to satisfy the conditions stated in Definition 7.4, we need to consider finite sample from a point process. The arrival times are of less interest than the number arrived in an interval of time.

Definition 7.6

Let $N(t), t > 0$ represent the number of events occurred in an interval of time $(0, t], t > 0$, at epochs $T_k, k = 1, 2, ...$, that is, $N(t) = \max\{k, T_k \leq t\}$. The sequence $\{N(t), t \geq 0\}$, which is a nondecreasing function of t and is a random process, is referred to as the **counting process** belonging to the point process $\{T_1, T_2, ..., T_r, T_1 \geq 0\}$. It is referred to as the **simple counting process** if one event can occur at a time. It is also referred to as the **random point process** if the time epochs $T_k, k = 1, 2, ...$ are random variables, with $P(\lim k \to \infty T_k) = 1$. It is further referred to as the **marked point process**.

Note 7.13

The increment $N(t+s) - N(t)$ will contain the number of events that occur in the interval $(t, t+s]$. That is why $\{N(t), t \geq 0\}$ is called the **counting process**, that is, counting the number of events that occurred by time t.

Note 7.14

We can summarize a counting process $\{N(t), t \geq 0\}$ as follows:

1. For each t, the values of $N(t)$ are nonnegative integers.
2. The function $N(t)$ is a nondecreasing function with respect to t.
3. The function $N(t)$ is a right continuous function.

The third property is merely a convention, and it interprets as follows: Let two events occur at times 5 and 8, that is, $N(5) = 1$, $N(8) = 1$ and

$$N(t) = \begin{cases} 1, & t \in (5,8) \\ 0, & t<5. \end{cases}$$

Definition 7.7

Let us consider arrival epochs of tasks to a system. We denote the first n inter-arrival times by $X_i, i = 1, 2, \ldots, n$, that is, $X_i - X_{i-1}, i = 1, 2, \ldots, n$, is the time elapsed between the $(i-1)^{th}$ arrival and the $i^{th}, i = 1, 2, \ldots, n$, arrival. We denote the nth arrival by S_n. Hence, for positive iid random variables, $X_1, X_2, \ldots, S_n = X_1 + X_2 + \cdots + X_n$. That is, each arrival epoch is the sum of $X_1, X_2, \ldots, X_n$. Then, not only is the sequence $\{S_n, n \geq 1\}$ a special case of a random walk, but also it is a sequence of epochs of **reoccurrences**, or **recurrent** or more commonly referred to as the **renewal counting process**, denoted by $\{N(t), n > 0\}$.

Note 7.15

Because the process probabilistically starts over at each arrival epoch, the process is referred to as the **renewal process**. In this case, the process $\{S_n, n \geq 1\}$ should cross the threshold $T > 0$.

Note 7.16

We leave it as an exercise to show that the arrival epochs and the counting random variables are related as follows:

$$\{S_n \leq t\} = \{N(t) \geq n\} \text{ and } \{S_n > t\} = \{N(t) < n\}. \tag{7.7}$$

Definition 7.8

A counting process $N(t)$ is a **Poisson process** with parameter (or rate) λ if the following conditions are true:

1. $N(0) = 0$.
2. It has independent increments, that is, if $(s_1, t_1) \cap (s_2, t_2) = \varnothing$, then the increment $N(t_1) - N(s_1)$ and $N(t_2) - N(s_2)$ are independent.
3. The number of events in any interval of length t is the Poisson pmf with parameter λt, that is,

$$P\big(N(t+s) - N(s) = j\big) = \frac{(\lambda t)^j}{j!} e^{-\lambda t}, \quad j = 0, 1, 2, \dots, \tag{7.8}$$

$$E[N(t+s) - N(s)] = \lambda t, \tag{7.9}$$

and

$$\lim_{\Delta \to 0} \big\{ P\big(N(\Delta) = 1\big) \big\} = \lambda \Delta e^{-\lambda \Delta} \approx \lambda \Delta, \tag{7.10}$$

$$\lim_{\Delta \to 0} \big\{ P(N(\Delta) \geq 2) \big\} = o\big(\Delta^2\big) \ll \lambda \Delta. \tag{7.11}$$

Note 7.17

Last properties state that in an interval with a very small length, events occur singly with probability proportional to the length of the interval. This is why λ is referred to as the **rate**.

Note 7.18

It is interesting to investigate the existence of the Poisson process. Proof is beyond the level of this book. It is a very sophisticated proof.

Example 7.9

Consider a biased coin with high probability for a tail and low probability for a head to occur. We toss this coin n times with n to probability of success (head) as $p = \dfrac{\lambda}{n} < \dfrac{1}{2}$, where λ in a real number $<n$. Denote the number of heads in a period of time $[0,t]$ by $N_n(t)$. We leave it as an exercise to show that as n approaches infinity, the number of heads occurred in $(s, t]$ is a binomial with number of trials equal to $n(t-s) \pm 2$ with $p = \dfrac{\lambda}{n}$, which converges to a Poisson pmf with parameter $t-s$.

Example 7.10

In this example, we show a property of a simple queue, which we will discuss later, as an application of a Poisson process. We consider arrival of tasks to a system to be a Poisson process with parameter λ. We denote the inter-arrival times by $T_1, T_2, \ldots$, that is, T_n is the time elapsed between the $(n-1)$st and nth arrivals, or the time between consecutive arrivals of tasks. We want to find the cumulative density function (cdf) of the inter-arrival times.

Answer

From the assumption for the arrivals to be Poisson, we conclude that pmf of arriving tasks follows a Poisson pmf with parameter λ. Hence, if the first arrival time is $>t$, there will not be any arrival task. In other words,

$$P(T_1 > t) = P[N(t) = 0] = e^{-\lambda t}. \tag{7.12}$$

Thus, T_1 is an exponential random variable with parameter λ. On the other hand,

$$P(T_2 > t|T_1 = s) = P(\text{No arrival in } (s, s+t]|T_1 = s$$

$$= P(N(t) = 0) = e^{-\lambda t}, s, t > 0. \tag{7.13}$$

This is because the tasks in $(s, s+t]$ are not affected by what happens in $[0, s]$. Hence, T_2 is independent of T_1 and also is exponential with parameter λ. We leave it as an exercise by using mathematical induction, to show that T_3 is independent of T_1 and T_2, and so on.

Theorem 7.1

Let $T_1, T_2, \ldots$ be iid exponential random variables with parameter λ. We also let S_n denote the sum of the first n of these random variables, $T_1, T_2, \ldots$, that is, $S_n = T_1 + T_2 + \cdots + T_n$, which presents the waiting time for the nth arrival. We further assume that $N(t)$ denotes the arrival making the n largest such that $S_n \leq t$. Then, the following statements are true:

1. $E(S_n) = \dfrac{n}{\lambda}$. $\tag{7.14}$
2. The pdf of S_n is $\Gamma(n, \lambda)$, the two-parameter gamma pdf, that is,

$$fS_n = \frac{\lambda^n t^{n-1}}{(n-1)!} e^{-\lambda t}. \tag{7.15}$$

Proof:
1. Since $T_1, T_2, \ldots, T_n$ are independent and the expected value of each T_j, $j = 1, 2, \ldots, n$, is $\dfrac{1}{\lambda}$, $E(S_n) = \dfrac{n}{\lambda}$.

2. From (7.6), we have:

$$P(S_n > t) = P(N(t) < n) = \sum_{j=0}^{n-1} \frac{(\lambda t)^j}{j!} e^{-\lambda t}. \qquad (7.16)$$

Differentiating (7.16) with respect to t, we obtain:

$$-fs_n = \sum_{j=0}^{n-1} \frac{1}{j!} \left[-(\lambda t)^j \lambda e^{-\lambda t} + j(\lambda t)^{j-1} \lambda e^{-\lambda t} \right]$$

$$= \lambda e^{-\lambda t} \sum_{j=0}^{n-1} \left[-\frac{(\lambda t)^j}{j!} + \frac{j(\lambda t)^{j-1}}{(j-1)!} \right]$$

$$= -\lambda \frac{(\lambda t)^{n-1}}{(n-1)!} e^{-\lambda t}.$$

Hence, the (7.15) is true.

Example 7.11

Suppose occurrences of events in a process are distributed according to a Poisson pmf with parameter.

    i. Find the probability that the eighth event occurs 2 or more time units after the seventh event.
    ii. Find the expected time of occurrence of the eighth event.
    iii. Find the probability that the eighth event occurs after the time epoch 10.
    iv. Find the probability that two events occur in the time interval [1,3] and three events in [2,4].

Answer

    i. The question we have to answer in this item is finding

$$P\big(\text{The eighth event occurs later than time } 10\big).$$

The answer is yet $e^{-\lambda t}$. This is because one can restart the Poisson process at any event.

    ii. From (7.12), the answer is $\frac{8}{\lambda}$.

    iii. We need to find $P\big(\text{The eighth event occurs later than time } 10\big)$. This is equivalent to finding:

$$P(S_8 > 10) = P\{N(10)\} = P\{N(10) < 8\}.$$

We have two choices as how to find this probability. That is,

---

TABLE 7.1

Finding the Values of a Stochastic Process, Example 7.11(iii)

λ	0.5	1	3	5
$P(S_8 > 10)$	0.8666	0.2202	5.2337e-07	3.4640e-14

---

a. $P(S_8 > 10) = \int_{10}^{\infty} \frac{(\lambda t)^7}{7!} \lambda e^{-\lambda t} dt,$ (7.17)

or

b. $P(N(10) < 8) = \sum_{i=0}^{7} \frac{(10\lambda)^i}{7!} \lambda e^{-10\lambda}.$ (7.18)

Using (a) with the aid of two-line codes of MATLAB:

```
L=5; %L=lambda
fun=@(t) (L/(factorial(7))).*((L.*t).^7).*(exp(-L.*t))
p=integral(fun,10,inf)
```

we obtain the values for different values of λ as in Table 7.1.

iv. To find the probability that two events occur in the time interval [1,3] and three events in [2,4], symbolically, is P(Two events in [1,3] and three events in [2,4]). To find this probability, we use the conditional probability as follows:

P(2 events in [1,3] and 3 events in [2,4])

$$= \sum_{i=0}^{2} P\left(2 \text{ events in } [1,3] \text{ and } 3 \text{ events in } [2,4] \mid j \text{ events in } [2,3]\right)$$

$$.P\left(j \text{ events in } [2,3]\right)$$

$$= \sum_{i=0}^{2} P\left(2 - j \text{ events in } [1,2] \text{ and } 3 - j \text{ events in } [3,4]\right) \quad (7.19)$$

$$.P\left(j \text{ events in } [2,3]\right)$$

$$= \sum_{i=0}^{2} \frac{\lambda^{2-j}}{(2-j)!} e^{-\lambda} \frac{\lambda^{3-j}}{(3-j)!} e^{-\lambda} \frac{\lambda^j}{j!} e^{-\lambda}.$$

$$= \frac{\lambda^3 (\lambda^2 + 6\lambda + 6)}{12} e^{-3\lambda}$$

Again, for different values of λ, we can find the values of this probability as in Table 7.2.

TABLE 7.2

Finding the Values of a Stochastic Process, Example 7.11(iv)

λ	0.5	1	3	5
P(2 events in [1,3] and 3 events in [2,4])	0.0215	0.0539	0.0092	1.9438e-04

Definition 7.9

A discrete stochastic process $\{X_n, n = 0,1,\ldots\}$ on a countable state space $S = \{X_0, X_1,\ldots\}$ is called the **discrete Markov process** or a **Markov chain;** if for $i, j \in S$ and $n \geq 0$, the following conditional probability, called the **Markov property**, or **forgetfulness** property, holds:

$$P\{X_{n+1} = j | X_0, X_1, \ldots, X_n\} = P\{X_{n+1} = j | X_n\}. \tag{7.20}$$

The Markov property (7.20) states that at any epoch n, with state X_n, the next state X_{n+1} is conditionally independent of the past states $X_0, X_1, \ldots, X_n$. That is, the next state depends upon the present state only, regardless of the history.

Denoted by $P_{j,i}$, (7.20) would mean the conditional probability of being in state j given that the transition took place from state i, that is,

$$P_{ji} = P\{X_{n+1} = j | X_n = i\}, \tag{7.21}$$

with

$$\sum_{j \in s} p_{ji} = 1, i \in S. \tag{7.22}$$

In other words, **given the current state of the process being i, transition to the next state, j, is independent of the history of the process prior to the current state.**

This idea was originated by the Russian Mathematician Andrey (Andrei), Andreyevich Markov (Markoff): June 14, 1856, to July 20, 1922 (Figure 7.4). See: https://en.wikipedia.org/wiki/Andrey_Markov#/media/File:AAMarkov.jpg.

It can be seen from (7.21) that the conditional probability $P_{j,i}$ is the probability of **transition from state i to state j in *one* step.**

FIGURE 7.4   Andrei Markov.

That is, transition probabilities do not depend upon the time parameter n. Hence, the Markov chain is referred to as **time homogenous**. It is nonhomogeneous if the process $\{X_n, n = 0, 1, \ldots\}$ is time dependent.

Definition 7.10

For a homogenous Markov chain, the matrix **P**, whose elements are $P_{j,i}$, is called the **transition matrix**. In general, **P** looks like the following matrix:

$$
P = \begin{array}{c} i/j \\ 0 \\ 1 \\ 2 \\ 3 \\ \vdots \end{array}
\begin{array}{c} p_{ij} \\ \left(\begin{array}{ccccc}
0 & 1 & 2 & 3 & \cdots \\
p_{0,0} & p_{1,0} & p_{2,0} & p_{3,0} & \cdots \\
p_{0,1} & p_{1,1} & p_{2,1} & p_{3,1} & \cdots \\
p_{0,2} & p_{1,2} & p_{2,2} & p_{3,2} & \cdots \\
p_{0,3} & p_{1,3} & p_{2,3} & p_{3,3} & \cdots \\
& \cdots & & \vdots & \\
& & \ddots & \ddots &
\end{array} \right)
\end{array}
\qquad (7.23)
$$

Note 7.19

We leave it as exercises to show that:

i. The exponential distribution function is the only continuous probability distribution having the forgetfulness property, and conversely, a continuous probability distribution function having such a property is the exponential cdf.
ii. The geometric mass function is the only discrete pmf having the forgetful property.

We now generalize (7.21) as follows.

Definition 7.11

The **n-steps transition probability from state i to state j**, denoted by $p_{ji}^{(n)}$, that is, transition from state i to state j in n steps, is defined as follows:

$$p_{ji}^{(n)} = P\{X_{k+n} = j | X_k = i\}, i, j \in S, n = 0, 1, \ldots, \tag{7.24}$$

where $p_{ji}^{(0)}$ is defined as follows:

$$p_{ji}^{(0)} = \begin{cases} 1, & j=1, \\ 0, & j \neq i. \end{cases} \tag{7.25}$$

In other words, X_k represents the state of the process after k steps at i; if it moves n more steps, its state after $k + n$ steps is j.

Example 7.12

Consider a machine that is constantly working unless it stops due to failure. Instances $1,2,\ldots,n$ are the epochs of failures that will be recorded. We denote the failure states of the machine at time instance k by X_k with state space $S = \{0,1,2,\ldots,n\}$. Hence, if the machine fails at epoch i, its next failure will be at epoch $j, j > i$. If it fails at epoch n, it is indicated that it should either be discarded or be replaced by a new machine and so the next state will be the state 0. Hence, transition probabilities are as follows:

   i. p_{ii}: The machine stopped working and the next move will be stay put because it is not repaired yet.

   ii. p_{nn}: The machine stopped working and it is not replaceable.

   iii. $p_{0n} = 1 - p_{nn}$: The machine stopped working and it reached its limit. Hence, it is being replaced by a new machine.

Thus, we have constructed a Markov chain with transition matrix, denoted by **P**, as follows:

$$p_{i,j}$$

$$P = \begin{array}{c} i/j \\ 0 \\ 1 \\ 2 \\ 3 \\ \vdots \\ n \end{array} \begin{pmatrix} 0 & 1 & 2 & 3 & \cdots & n \\ p_{0,0} & p_{1,0} & p_{2,0} & p_{3,0} & \cdots & p_{n,0} \\ 0 & p_{1,1} & p_{2,1} & p_{3,1} & \cdots & p_{n,1} \\ 0 & 0 & p_{2,2} & p_{3,2} & \cdots & p_{n,2} \\ 0 & 0 & 0 & p_{3,3} & \cdots & p_{n,3} \\ \vdots & \vdots & \vdots & \vdots & \cdots & \vdots \\ p_{0,n} & 0 & 0 & \cdots & p_{n-1,n} & p_{n,n} \end{pmatrix}$$

Example 7.13

Let us go back to Example 7.10. Suppose we choose the usual working period of the machine as number of years. So, X_k, in this case, denotes the number of years that the machine is continuously working. This is an example of what is referred to as the **success runs**. Thus, we consider a sequence of independent events with success or failure at each of its occurrences. The events in this case are working of the machine in different periods of times with "success" as "working" and "failure" as the "breakdown" of the machine. Thus, consider the working of the machine in years. Then, X_k denotes the number of years the machine is working after its last failure and before its kth failure again. Hence, if the number of years the machine continued working is i, it means that the next event occurs with the probability of success as $p_i, 0 < p_i < 1$, and stopping to work with probability $1 - p$. If there is no time limit, the transition matrix for this Markov chain will be as follows:

$$P = \begin{pmatrix} 1-p_0 & p_0 & 0 & \cdots & & \\ 1-p_1 & 0 & p_1 & 0 & \cdots & \\ 1-p_2 & 0 & 0 & p_2 & \cdots & \\ \cdots & \cdots & \cdots & \cdots & \cdots & \cdots \\ p_n & \cdots & \cdots & 0 & 0 & p_k \\ \cdots & \cdots & \cdots & \cdots & \cdots & \cdots \end{pmatrix}. \tag{7.26}$$

Example 7.14

In a special case of Example 7.11, let us consider tossing an unbiased coin repeatedly with the same probability of occurrence of a head (success) for each toss as p and for tail as $q = 1 - p$. That is, we have a sequence of independent Bernoulli trials with the probability of success p and of failure $q = 1 - p$. This is referred to as the **Bernoulli process**. Let X_n denote the number of successes in n trials, $n \geq 1$. Thus, X_n has a binomial distribution with parameters n and p. Suppose i heads appeared in n trials, that is, $X_n = i$. Then, the result of the next trial is the state X_{n+1}, which is:

$$X_{n+1} = \begin{cases} j = i+1, & \text{A head will appear with probability } p, \\ j = i, & \text{A tail } \textit{will} \text{ appear with probability } 1 - p, \end{cases} \tag{7.27}$$

regardless of values of $X_1, X_2, \ldots, X_{n-1}$. Hence, X_n is a Markov chain with transition probabilities

$$\begin{cases} p_{i+1,i} = p, \\ p_{i,i} = 1 - p, \\ p_{ji} = 0, & \text{otherwise.} \end{cases} \tag{7.28}$$

In such a case, X_n is a **binomial Markov chain**, which is a special case of a random walk.

Example 7.15

The **failure–repair (FR) model** is one of the standard examples of a Markov chain. It is as follows. Suppose an office has a computer for its daily routine work. The administrator staff turns on the computer as she comes to work each day. Turning on the computer is referred to as a **trial**. The computer fails to turn on (work) with probability q. The probability that it turns on and functions normally is p, $p + q = 1$. Each time the computer fails to work, a repairperson (an IT personnel) is called. The computer cannot be used until it is completely repaired. We want to find the probability that the computer is in working condition at a specific number of trials.

Answer

It is clear that the process is a random walk. Let {0, 1} denote the state space of this Markov chain, that is, 0 for failure and 1 for working condition. Thus, if we define the random variable $X_n, n = 0, 1, 2, \ldots$, as the state of the computer after n trials, then for $n = 0, 1, 2, \ldots$, we have:

$$X_n = \begin{cases} 1, & \text{The machine is up after trial } n, \\ 0, & \text{The machine is down after trial } n. \end{cases}$$

We assume that the process with working condition is state 1. Since there are only two states 0 and 1 for this process, we define the probability of being in any of the state after n step by $p_k^{(n)}, k = 0, 1$, and p_k, if $n = 1$. We also denote the transition probability of transiting from state i to state j in n step by $p_{ji}^{(n)}$ and p_{ji} if $n = 1$. Hence, for $k = 0, 1, 2, \ldots$, inductively, we will have:

For $k = 0$,

$$p_1^{(0)} = 1 \quad \text{and} \quad p_0^{(0)} = 0, \quad \text{by assumption.}$$

For $k = 1$,

$$p_1 = p_{10} p_0^{(0)} + p_{11} p_1^{(0)} = (1)(0) + (p)(1) = p,$$

and

$$p_0 = p_{10} p_1^{(0)} = (q)(1) = q.$$

For $k = 2$,

$$p_1^{(2)} = p_{10} p_0^{(1)} + p_{11} p_1^{(1)} = (1)(q) + (p)(p) = q + p^2 = 1 - q + q^2,$$

and

$$p_0^{(2)} = p_{10} p_1^{(1)} = (p)(q) = q - q^2.$$

Following the same pattern, we can find the other probabilities as follows:

For $k = n$,

$$p_1^{(n)} = 1 + (-q) + (-q)^2 + \cdots + (-q)^n = \sum_{k=0}^{n} (-q)^k = \frac{1 - (-q)^{n+1}}{1 + q},$$

and

$$P_0^{(n)} = 1 - \frac{1-(-q)^{n+1}}{1+q} = \frac{q+(-q)^{n+1}}{1+q}.$$

The transition matrix $\mathbf{P}^{(n)}$ with elements $p_{ji}^{(n)}$, $i = 0,1$; $j = 0,1$; $n = 0,1,2,...$, is as follows:

$$\mathbf{P}^{(n)} = \begin{array}{c} n/k \\ 0 \\ 1 \\ 2 \\ 3 \\ 4 \\ \vdots \end{array} \begin{pmatrix} 0 & 1 \\ 0 & 1 \\ q & 1-q \\ q-q^2 & 1-q+q^2 \\ q-q^2+q^3 & 1-q+q^2-q^3 \\ q-q^2+q^3-q^4 & 1-q+q^2-q^3+q^4 \\ \vdots & \vdots \end{pmatrix}$$

For instance, if the machine breaks down and 5% of the times is used, then the probability that the machine is "in working condition" after five times turned on is 95.2%, because

$$p_1^{(5)} = \frac{1-(-0.5)^6}{1+0.5} = 0.952.$$

Example 7.16

Traditionally, **Rat in a Maze** is a standard example used for a Markov chain. It can also be used as a puzzle or a game that a **player** may use for fun. However, we use a **robot** as a generic name. So, here is the problem: Suppose we have a maze and we assume the following:

   i. The maze has five cells, say Cell 1, Cell 2, ..., Cell 5.
  ii. The maze has no entrance.
 iii. To start, the rat is put in a cell at random, unless otherwise stated.
 iv. Each cell has one or more exit ways to the neighboring cells.
  v. There is a way to exit out the maze referred to as the **Out Window**, denoted by O.
 vi. The connections of cells are as follows:
    Cell 1 to Cell 2 or Cell 4
    Cell 2 to Cell 1, Cell 3 or Cell 4
    Cell 3 to Cell 2 or Cell 5
    Cell 4 to Cell 1 or Cell 2 or Cell 5
    Cell 5 Cell 3 or Cell 4 to Cell O,
    which are illustrated in Figure 7.5.
 vii. After a move (transition) has occurred, the robot cannot remember in what cell it was before (the Markov property).
viii. Transitions from one cell to another are independent of each other.
 ix. Choices to move out of a cell from each pathway out to neighboring cells have the same probabilities.

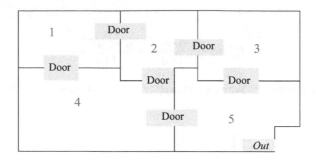

FIGURE 7.5   Graph of a two-dimensional maze with five cells, Example 7.17.

    x. No transition is possible from the exit window O, other than exiting the maze.
    xi. Staying put or moving from a cell to itself in the next move is not allowed.

We want to find the average number (mean) of moves for the robot to exit the maze starting from any cell.

Answer

To begin with, let us denote the transition probabilities from Cell i to Cell j, including the "exit out window", by p_{ji}, such that, for the exit out window, O, we have:

$$p_{ji} = \begin{cases} 1, & i=O, j=O, \\ \dfrac{1}{3}, & i=5, j=O, \\ 0, & \text{otherwise}, \end{cases}$$

that is, p_{OO}, symbolically, means the probability of permanent exit. Since no transition is possible from state O, this state is referred to as the **absorbing state**. In other words, as soon as the chain enters this state, it will remain there forever.
    Also, staying put not being allowed, that is,

$$p_{ii} = 0, i, j = 1, 2, 3, 4, 5.$$

Let us denote by X_n, the cell that robot will visit immediately after the nth move. Thus, the process $\{X_n\}$ is a Markov chain with state space $S = \{O, 1, 2, 3, 4, 5\}$. According to the structure of the maze and the assumptions associated with it, we will develop Table 7.3, the transition probability matrix **P**, and Table 7.4, as follows.

Note 7.20

In Table 7.3, the sum of each row is 1, as expected.

TABLE 7.3

Transition Probabilities, $p_{j,i}$, Example 7.16

Cell i $j/p_{j,i}$	0	1	2	3	4	5
1	$p_{01}=0$	$p_{11}=0$	$p_{21}=\frac{1}{2}$	$p_{31}=0$	$p_{41}=\frac{1}{2}$	$p_{51}=0$
2	$p_{02}=0$	$p_{12}=\frac{1}{3}$	$p_{22}=0$	$p_{32}=\frac{1}{3}$	$p_{42}=\frac{1}{3}$	$p_{52}=0$
3	$p_{03}=0$	$p_{13}=0$	$p_{23}=\frac{1}{2}$	$p_{33}=0$	$p_{43}=0$	$p_{53}=\frac{1}{2}$
4	$p_{04}=0$	$p_{14}=\frac{1}{3}$	$p_{24}=\frac{1}{3}$	$p_{34}=0$	$p_{44}=0$	$p_{54}=\frac{1}{3}$
5	$p_{05}=\frac{1}{3}$	$p_{15}=0$	$p_{25}=0$	$p_{35}=\frac{1}{3}$	$p_{45}=\frac{1}{2}$	$p_{55}=0$

TABLE 7.4

The Cell Number That the Robot Will Visit Immediately after the nth Move, Example 7.16

Cell Numbers i j	States	0	1	2	3	4	5
1	X_1			2 with $p_{31}=\frac{1}{2}$		4 with $p_{41}=\frac{1}{2}$	
2	X_2		1 with $p_{12}=\frac{1}{3}$		3 with $p_{53}=\frac{1}{3}$	4 with $p_{42}=\frac{1}{3}$	
3	X_3			2 with $p_{23}=\frac{1}{2}$			5 with $p_{53}=\frac{1}{2}$
4	X_4		1 with $p_{14}=\frac{1}{3}$	2 with $p_{24}=\frac{1}{3}$			5 with $p_{54}=\frac{1}{2}$
5	X_5	0 with $p_{05}=\frac{1}{3}$			3 with $p_{35}=\frac{1}{3}$	4 with $p_{45}=\frac{1}{3}$	

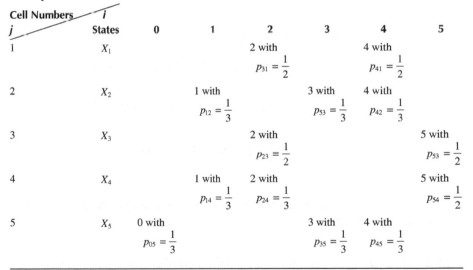

$$p_{ij}$$

$$
P = \begin{array}{c}
i/j \\ 1 \\ 2 \\ 3 \\ 4 \\ 5
\end{array}
\left(
\begin{array}{cccccc}
O & 1 & 2 & 3 & 4 & 5 \\
0 & 0 & 1/2 & 0 & 1/2 & 0 \\
0 & 1/3 & 0 & 1/3 & 1/3 & 0 \\
0 & 0 & 1/2 & 0 & 0 & 1/2 \\
0 & 1/3 & 1/3 & 0 & 0 & 1/3 \\
1/3 & 0 & 0 & 1/3 & 1/3 & 0
\end{array}
\right).
$$

So far, we have set up the problem and now are ready to answer the question. So, let us denote the starting location of the robot in Cell s by $X_s, s = 1, 2, 3, 4, 5$. For instance, if at the start it is standing in Cell 1, then $X_s = 1$, and if it starts from Cell 3, $X_s = 3$. Now, let us start by denoting the minimum number of moves for the robot to exit, if it begins from cell number $s, s = 1, 2, 3, 4, 5$, by $\alpha_{Os}, s = 1, 2, 3, 4, 5$, defined by

$$\alpha_{Os} = \min\{m \geq 0 \text{ such that } X_m = O | X_s = s, s = 1, 2, 3, 4, 5\}. \qquad (7.29)$$

We now have to calculate the values of $E(\alpha_{Os})$, $s = 1, 2, 3, 4, 5$. So, let us start with $s = 1$. It, then, moves to the possible neighboring cell. In this case, we have to find $E(\alpha_{O1})$. From Table 7.4, we see that the robot cannot directly exit from Cell 1; it rather has only two choices, Cell 2 and Cell 4, and then other neighboring cells before it can exit. However, we know that from Cell 1, it can move to Cell 2 with one move and probability 1/2. Thus, $E(\alpha_{O1} | X_s = 2) = 1 + E(\alpha_{O2})$. Since from Cell 1 it could also go to Cell 4 with 1 move and with probability 1/2, we will have $E(\alpha_{O1} | X_s = 4) = 1 + E(\alpha_{O4})$. Here is the role of the Markov property. That is, once the robot is in Cell 2, for example, it will not remember how he got there, that is, in which cell it was before or even before that. Thus, the chance of remaining number of moves to exit from the cell it is in now would be the same as if it were in Cell 1. Based on this argument, we will have the following:

$$E(\alpha_{O1} | X_s = 2) = P\{X_s = O | X_s = 2\}$$

$$= +E(\alpha_{O1} | X_s = 4) = P\{X_s = O | X_s = 4\}$$

$$= \left(\frac{1}{2}\right)(1 + E(\alpha_{O2})) + \left(\frac{1}{2}\right)(1 + E(\alpha_{O4}))$$

$$= 1 + \left(\frac{1}{2}\right)E(\alpha_{O3}) + \left(\frac{1}{2}\right)E(\alpha_{O4})$$

In a similar fashion, we can find the rest of the expected values, except the last one, $E(\mu_{O5})$ Thus, we will have:

$$E(\mu_{O2}) = E(\mu_{O2} | X_2 = 1) P\{X_2 = 1 | X_O = 0\} + E(\mu_{O2} | X_2 = 3) P\{X_2 = 3 | X_O = 0\}$$

$$+ E(\mu_{O2} | X_2 = 4) P\{X_2 = 4 | X_O = 0\}$$

$$= \left(\frac{1}{3}\right)(1 + E(\alpha_{O1})) + \left(\frac{1}{3}\right)(1 + E(\alpha_{O3})) + \left(\frac{1}{3}\right)(1 + E(\alpha_{O4}))$$

$$= 1 + \left(\frac{1}{3}\right)(E(\alpha_{O1})) + \left(\frac{1}{3}\right)(E(\alpha_{O3})) + \left(\frac{1}{3}\right)(E(\alpha_{O4})).$$

$$E(\alpha_{03}) = E(\alpha_{03}|X_3 = 2)P\{X_3 = 2|X_O = 0\} + E(\alpha_{03})$$

$$= E(\alpha_{03}|X_3 = 5)P\{X_3 = 5|X_O = 0\}$$

$$= \left(\frac{1}{2}\right)(1 + E(\alpha_{02})) + \left(\frac{1}{2}\right)(1 + E(\alpha_{05}))$$

$$= 1 + \left(\frac{1}{2}\right)(E(\alpha_{02})) + \left(\frac{1}{2}\right)(E(\alpha_{05})).$$

$$E(\alpha_{04}) = E(\alpha_{04}|X_4 = 1)P\{X_4 = 1|X_O = 0\}$$

$$= E(\alpha_{04}|X_4 = 2)P\{X_4 = 2|X_O = 0\} + E(\alpha_{04}|X_4 = 5)P\{X_4 = 5|X_O = 0\}$$

$$= \left(\frac{1}{3}\right)(1 + E(\alpha_{01})) + \left(\frac{1}{3}\right)(1 + E(\alpha_{02})) + \left(\frac{1}{3}\right)(1 + E(\alpha_{05}))$$

$$= 1 + \left(\frac{1}{3}\right)(E(\alpha_{01})) + \left(\frac{1}{3}\right)(E(\alpha_{02})) + \left(\frac{1}{3}\right)(E(\alpha_{05})).$$

$$E(\alpha_{05}) = E(\alpha_{05}|X_5 = O)P\{X_5 = O|X_O = 0\}$$

$$= +E(\alpha_{05}|X_5 = 3)P\{X_5 = 3|X_O = 0\} + E(\alpha_{05}|X_5 = 4)P\{X_5 = 4|X_O = 0\}$$

$$= \left(\frac{1}{3}\right)(1) + \left(\frac{1}{3}\right)(1 + E(\alpha_{03})) + \left(\frac{1}{3}\right)(1 + E(\alpha_{04}))$$

$$= 1 + \left(\frac{1}{3}\right)(E(\alpha_{03})) + \left(\frac{1}{3}\right)(E(\alpha_{04})).$$

Therefore, we have five equations with five unknowns $E(\mu_{o_1}), E(\alpha_{o_2}), E(\alpha_{o_3}),$ $E(\alpha_{o_4}),$ and $E(\alpha_{o_5})$ as follows:

$$\begin{vmatrix} E(\alpha_{01}) - \left(\frac{1}{2}\right)E(\alpha_{02}) - \left(\frac{1}{2}\right)E(\alpha_{04}) = 1, \\[2mm] E(\alpha_{02}) - \left(\frac{1}{3}\right)E(\alpha_{01}) - \left(\frac{1}{3}\right)E(\alpha_{03}) - \left(\frac{1}{3}\right)E(\alpha_{04}) = 1, \\[2mm] E(\alpha_{03}) - \left(\frac{1}{2}\right)E(\alpha_{02}) - \left(\frac{1}{2}\right)E(\alpha_{05}) = 1, \\[2mm] E(\alpha_{04}) - \left(\frac{1}{3}\right)E(\alpha_{01}) - \left(\frac{1}{3}\right)E(\alpha_{02}) - \left(\frac{1}{3}\right)E(\alpha_{05}) = 1, \\[2mm] E(\alpha_{05}) - \left(\frac{1}{3}\right)E(\alpha_{03}) - \left(\frac{1}{3}\right)E(\alpha_{04}) = 1. \end{vmatrix}$$

Thus, using MATLAB, we can solve this system of five linear equations with five unknowns and obtain:

$$\begin{cases} E(\alpha_{O1}) = 20.1818 \\ E(\alpha_{O2}) = 19.7273 \\ E(\alpha_{O3}) = 17.3636 \\ E(\alpha_{O4}) = 18.6364 \\ E(\alpha_{O5}) = 13.0000. \end{cases}$$

What these values say are that if the robot starts in Cell 1, it will exit after about 20 moves, on the average. It is the same for others; that is, if the robot starts in Cell 2, it will exit after about 20 moves, on the average, after about 17 moves if it starts in Cell 3, after about 19 moves if it starts in Cell 4, and after about 13 moves if it starts in Cell 5, which is the least number of moves.

We can now extend Definition 7.7 for continuous random variables. In fact, we will have similar results as in the discrete case with different terminologies.

Definition 7.12

A stochastic process with continuous time is denoted by $\{T(t), t \geq 0\}$ and is referred to as a **Markov process**. The **transition probability** from state i to state j in a time interval of length j, will be denoted by $p_{ji}(t)$, is defined as

$$p_{ji}(t) = P\{T(t+s) = j | T(t) = i\}.$$

Similar definitions and formulae for Markov chain follow accordingly by replacing n by t. For instance, the continuous case of (7.18) is:

$$P = P\{T > t + s | T > t\} = P\{T > s\}, \forall s, t \geq 0. \tag{7.30}$$

Transition probabilities may be represented by matrix, referred to as the **transition matrix**, denoted by $\mathbf{P}(t) = (p_{yx(t)})$, where $p_{yx}(t)$ is the general element.

Note 7.21

The transition probabilities are nonnegative and sum to 1.

It is assumed that at least one transition from state x to another state in $[0, t]$, for all $t \geq 0$, occurs. For this reason, the transition matrix $\mathbf{P}(t)$ is called a **stochastic matrix**. The transition probabilities satisfy what is referred to the **Chapman–Kolmogorov equations.**

$$p_{yx}(t+s) = \sum_{k=0}^{\infty} p_{kx}(t)p_{yk}(s) \tag{7.31}$$

or in matrix form

$$\mathbf{P}(t+s) = \mathbf{P}(t)\mathbf{P}(s), \ \forall s, t \in [0,\infty). \tag{7.32}$$

Note 7.22

Similar equations can be derived for discrete-time Markov chains.

Definition 7.13

A Markov process that is independent of time is referred to as the **stationary** or **time-homogeneous** or simply **homogeneous Markov process**.

7.4   CLASSIFICATION OF STATES OF A MARKOV CHAIN/PROCESS

States of a Markov chain (similarly of a Markov process) are given names for reasons so that the terms explain the case of those states.

Definition 7.14

For a Markov chain, a state j is called **accessible** from state i, denoted by $i \rightarrow j$, if there is a positive probability, say $p_{ji}^{(n)} > 0$, for some $n \geq 0$. That is, there is a possibility of reaching the state j from state i in some number, n, of steps. On the other hand, if a state j is **not accessible** from state i, then $p_{ji}^{(n)} = 0$, for all $n \geq 0$. In this case, if the chain started from the state i, it will never visit the state j.

Definition 7.15

If a state j is accessible from a state i $(i \rightarrow j)$ and the state i is accessible from the state j $(j \rightarrow i)$, then we say the states i and j **communicate**, denoted by $i \leftrightarrow j$. In this case, $p_{ji}^{(n)} > 0$ and $p_{ji}^{(n)} > 0$.

We leave as an exercise to prove that $i \leftrightarrow j$ is an equivalence relation, that is,

   i. $i \rightarrow j$,
   ii. $i \rightarrow j$ implies $j \rightarrow i$,
   iii. $i \rightarrow k$ and $k \rightarrow j$ together imply that $i \rightarrow j$.

Definition 7.16

The accessibility relation, defined above, divides states into **classes**. All states within each class communicate to each other. However, no pair of states in different classes will communicate with each other.

Definition 7.17

A Markov chain is called **irreducible** if it contains only one class. Otherwise, the chain is **reducible**.

Example 7.17

Suppose we have a chain with three states 1, 2, and 3, with its transition matrix as follows:

$$
P = \begin{array}{c} i/j \\ 1 \\ 2 \\ 3 \end{array} \begin{pmatrix} 1 & 2 & 3 \\ 1/2 & 1/2 & 0 \\ 1/2 & 1/4 & 1/4 \\ 0 & 1/3 & 2/2 \end{pmatrix}
$$

Is this chain irreducible?

Answer

Consider, for instance, $3 \leftrightarrow 2$ and $2 \leftrightarrow 1$. Thus, the chain has only one class, and hence, it is irreducible.

Example 7.18

Suppose we have a chain with four states 1, 2, 3, and 4, with its transition matrix as follows:

$$
P = \begin{array}{c} i/j \\ 1 \\ 2 \\ 3 \\ 4 \end{array} \begin{pmatrix} 1 & 2 & 3 & 4 \\ 1/2 & 1/2 & 0 & 0 \\ 1/2 & 1/2 & 0 & 0 \\ 0 & 0 & 1/4 & 3/4 \\ 0 & 0 & 0 & 1 \end{pmatrix}
$$

Is this chain irreducible?

Answer

Looking at state transitions, we see that only states 1 and 2 commute. That is, there are three classes as {1,2},{3}, and {4}. Thus, the chain is reducible.

Definition 7.18

In a Markov chain, a state i from the state space S is referred to as **recurrent** if it will be revisited. That is, if

$$P(i \text{ be revisited}|X_0 = i) = 1.$$

The state i is called **transient** if

$$P(i \text{ be revisited}|X_0 = i) < 1.$$

Note 7.23

From this definition, it can be seen that in a Markov chain starting from any recurrent state, it can be revisited infinitely many times or will not be revisited at all.

Example 7.19

Let us return to the Example 7.18. It is clear that state 4 is recurrent: Since if the process starts with state 4, it can be revisited infinitely many times with probability 1, that is, $P(4 \text{ be revisited}|X_0 = 4) = 1$. State 3 is transient since $P(3 \text{ be revisited}|X_0 = 3) = 1/4 < 1$. State 1 is recurrent. This is because in order to move out of 2 to go to state 1, we have to stay in state 2. But the probability of staying at state 2 is $(1/2)^n, n \to \infty$, that is, 0. Thus, we will exit state 2 and revisit state 1. With a similar argument, we can see that state 2 is also recurrent.

Note 7.24

We leave the proof of the following properties as exercises.

i. A state is recurrent if and only if $\sum_{n=1}^{\infty} P_{ii}^n = \infty$.
ii. If a state is recurrent and $i \to j$, then also $j \to i$.
iii. If a closed subset S only has finitely many states, then there must be at least one recurrent state among them. In particular, any finite Markov chain must contain at least one recurrent state.
iv. In any class, either all states are recurrent or all states are transient. In particular, if the Markov chain is irreducible, then either all states are recurrent or all states are transient.
v. Any recurrent class is a closed subset of states.

Example 7.20

Suppose we have a Markov chain with four states 1, 2, 3, and 4, all transient, with its transition matrix as follows:

$$
P = \begin{array}{c} i/k \\ 1 \\ 2 \\ 3 \\ 4 \end{array}
\begin{pmatrix}
 & 1 & 2 & 3 & 4 \\
0 & 0 & 1/2 & 1/2 \\
1 & 0 & 0 & 0 \\
0 & 1 & 0 & 0 \\
0 & 1 & 0 & 0
\end{pmatrix}
$$

Is every state recurrent?

Answer

We leave it as an exercise to show by inspection that it is indeed accessible, and hence, every state is recurrent.

Definition 7.19

A state i, for which $p_{ii} = 1$, that is, $p_{ji} = 0, j \neq i$, is referred to as an **absorbing state** (or an **absorbing barrier**). A nonabsorbing state is referred to as a **transient state**.

7.5   MARTINGALES

In 1934, the French mathematician Paul Pierre Lévy (1886–1971) initiated the discrete stochastic process describing waging. The idea was extended by Joseph Leo Doob to other areas of stochastic processes, including the decomposition theory and martingale. Later, in 1939, Jean Ville defined the concept of martingale for continuous stochastic process.

Definition 7.20

Let $X_1, X_2, \ldots, X_n$ be a sequence of random variables with finite expected values, that is, $E\left(|Y_n|\right) < \infty$. Let also $\{X_n\}$ be a discrete-time stochastic process such that:

$$E\left(X_{n+1} | X_1, X_2, \ldots, X_n\right) = X_n. \tag{7.33}$$

Such a process is referred to as the **discrete-time martingale**.

Example 7.21

In a gambling game, if the games are fair, a gambler's capital is a martingale. Here is why. Let the game consist of tossing a fair coin repeatedly. The game rule is that a gambler wins \$1 if the coin comes up heads and loses \$1 if it comes up tails. Suppose that the leftover capital of a gambler after n losses is X_n. Also, suppose that the gambler's conditional expected capital after the next trial, given the history, equals the gambler's current capital. Thus, the sequence is a martingale.

Definition of martingale stated in the Example 7.21 may be generalized as follows.

Definition 7.21

Let us consider two sequences of random variables $X_0, X_1, X_2, \ldots$ and $Y_0, Y_1, Y_2, \ldots$, Choose a random sample from each of the sequences, say $X_0, X_1, X_2, \ldots, X_n$ and $Y_0, Y_1, Y_2, \ldots, Y_n$, respectively. The sequence $Y_0, Y_1, Y_2, \ldots$, with finite expected value, that is, if for all $n \geq 0$, $E\left(|Y_n|\right) < \infty$. is called a **martingale with respect to** $X_0, X_1, X_2, \ldots$, if for all $n \geq 0$, the following holds:

$$E\left(Y_{n+1} | X_0, X_1, X_2, \ldots, X_n\right) = Y_n. \tag{7.34}$$

Note 7.25

If $X_n = Y_n$, then this definition reduces to the definition of basic martingale.

Note 7.26

A similar definition may be stated for continuous-time martingale.

Definition 7.22

Let $\{X_n\}, n = 1, 2, \ldots$, be a sequence of random variables. The random variable ξ is referred to as a **stopping time** with respect to $\{X_n\}, n = 1, 2, \ldots$, if for each k, the occurrence or nonoccurrence of the event $\xi = k$ depends only on the values of $X_1, X_2, \ldots, X_k$.

Note 7.27

What the definition states is that at any particular time k, one can look at the sequence so far and tell if it is time to stop. For instance, referring to the gambler's example, the time at which a gambler leaves the gambling table, which might be a function of his previous winnings, is the stopping time since he might leave only when he goes broke, but he cannot choose to go or stay based on the outcome of games that have not been played yet.

Note 7.28

Some authors define the stopping time by requiring only that the occurrence or nonoccurrence of the event $\xi = k$ is probabilistically independent of $X_{k+1}, X_{k+2}, ...,$ but not that it is completely determined by the history of the process up to instant k. This is a weaker condition than we stated in the above, but it is strong enough to serve in some of the proofs in which stopping times are used.

The martingale may be defined differently. We define it through an example. Suppose a company just established a business. Traditionally, for such businesses, it is expected not to gain much for a while before business is picking up. Clearly, the asset invested is subject to move probabilistically based on the changes in the market and fix costs. Now let us suppose the asset of the company at epoch k is denoted by a stochastic process $\{X_k\}$, which is subject to the market knowledge at epoch k. This factor is referred to as a **filtration** and is denoted by $\{\mathcal{F}_k\}$. In other words, we are defining a stochastic process $\{X_k\}$, meaning "what we have at instant k", and $\{\mathcal{F}_k\}$, meaning "what we know at instant k". Thus, for a while, expected gain after each transaction would be 0. Hence, estimated "what the company will have at instant $k + 1$", based on what it knows at time k, will be X_k. In other words, the stochastic process defined is a **martingale** if $E(X_{k+1}|\mathcal{F}_k) = X_k$.

We now state these definitions below.

Definition 7.23

Let $\mathcal{F}$ be a σ-field. A sequence $\{\mathcal{F}_k, k \geq 0\}$ of subfields of $\mathcal{F}$ is called a **filtration** in the σ-field if

$$\mathcal{F}_k \subset \mathcal{F}_{k+1}, \quad \forall k \geq 0. \tag{7.35}$$

Definition 7.24

Let $\{X_k\}, k = 1, 2, ...,$ be a sequence of random variables. Then, $\{X_k\}, k = 1, 2, ...,$ is referred to as **adapted to the filtration** if X_n is $\mathcal{F}_n$-measurable for all $k \geq 0$. In other words, the events $\{X_k\}, k = 1, 2, ...$ are in $\mathcal{F}_k$, that is, $\{X_k\} \in \mathcal{F}_k$.

We can now state the general definition of martingale.

Definition 7.25

Let $\{\mathcal{F}_k, k \geq 0\}$ be a filtration in $\mathcal{F}$. Also, let $\{Y_k\}, k = 0, 1, 2, ...,$ be a sequence of random variables adapted to that filtration. Then, $\{Y_k\}, k = 0, 1, 2, ...,$ is a **martingale**, with respect to $\{\mathcal{F}_k, k \geq 0\}$, if:

$$E(|Y_k|) < \infty \tag{7.36}$$

and

$$E\left(Y_{k+1}\middle|\mathcal{F}_k\right) = Y_k.\tag{7.37}$$

Example 7.22

Let us refer to the Gambler's Ruin example we saw earlier. So, a gambler wins or loses \$1 in each round of betting, with equal chances and independently of the past events. He starts betting with the determination that he will stop playing games when either he won x dollars or he lost y dollars. We want to answer the following questions:

  i. What is the probability that he will be winning when he stops playing further?
 ii. What is the expected number of his betting rounds before he will stop playing further?

Answer

To answer the questions, we model the gambling with simple symmetric random walk. So, let ξ_j, $j = 1,2,...$, be iid random variables with common distribution as follows:

$$P\left(\xi_j = +1\right) = \frac{1}{2}, P\left(\xi_j = -1\right) = \frac{1}{2}, \quad j = 1,2,...\tag{7.38}$$

Let, also, $\mathcal{F}_n = \left(\xi_j, 0 \le j \le n\right), n \ge 0$ be their filtration. We define

$$S_0 = 0, \quad S_n = \sum_{j=1}^{n} \xi_j, n \ge 1.\tag{7.39}$$

We also define the following stopping times

$$T_L = \inf\left\{n > 0 : S_n = -y\right\}, T_R = \inf\left\{n > 0 : S_n = x\right\}, T = \min\left\{T_L, T_R\right\}.\tag{7.40}$$

Note 7.29

$$\begin{cases} \{\text{The gamber wins } x \text{ dollars}\} = \{T = T_R\}, \\ \{\text{The gamber loses } y \text{ dollars}\} = \{T = T_L\}. \end{cases}$$

i. Using stopping time property, we have:

$$E(S_T) = E(S_0) = 0.\tag{7.41}$$

Hence,

$$-yP(T = T_L) + xP(T = T_R) = 0.$$

Since

$$P(T = T_L) + P(T = T_R) = 1,$$

we have:

$$P(T = T_L) = \frac{x}{x+y}, P(T = T_R) = \frac{y}{x+y}.$$

ii. We first prove that $M_n = S_n^2 - n$ is a martingale:

$$E\left(M_{n+1}|\mathcal{F}_n\right) = E\left(S_{n+1}^2|\mathcal{F}_n\right) - (n+1)$$

$$E\left(S_n^2 + 2S_n\xi_{n+1} + 1|\mathcal{F}_n - (n+1)\right) = \cdots = M_n.$$

Now applying the stopping time property, we will have:

$$0 = E(M_T) = E\left(S_T^2 - T\right) = P(T = T_L)y^2 + P(T + T_R)x^2 - E(T).$$

Thus, from part (i), we have:

$$E(T) = xy.$$

Example 7.23

Suppose we have U number of urns and B number of balls. We randomly place the balls in urns, one at a time. All balls have the same chance to be selected from urns. That is, the probability of a ball to be selected from any urn is $1/B$. Also, if a ball is selected randomly from an urn, placing it in another urn has the same probability as any other. In other words, selection of an urn has a probability of $1/U$. Now, we select a ball at random. Let the number of balls in the first urn at time k be denoted by X_k. For $k \geq 0$, let $\mathcal{F}_k = \sigma\left(X_i, 1 \leq i \leq k\right)$ be the filtration generated by the process $k \to X_k$. We want to calculate $E\left(X_{k+1}|\mathcal{F}_k\right)$.

Answer

$$E\left(X_{k+1}|\mathcal{F}_k\right) = (X_k + 1)\frac{B - X_k}{B}\frac{1}{U} + (X_k - 1)\frac{X_k}{B}\frac{U-1}{U}$$

$$+ X_k\left(\frac{B - X_k}{B}U - 1K + \frac{X_k}{B}\frac{1}{U}\right)$$

$$= X_k\frac{B-1}{B} + \frac{1}{U}.$$

For an example, if there are $B = 10, U = 3$, and $X_3 = 4$, then

$$E\left(X_4|\mathcal{F}_3\right) = (4)\frac{10-1}{10} + 13 = \frac{59}{15} \approx 4.$$

7.6   QUEUEING PROCESSES

The word "queue" was first appeared in the work of Danish Mathematician Agner Krarup Erlang (1878–1929) to mean "waiting line", introduced by Erlamg, see Erlang (1917). The theory of queues generates a process, which is another example of a stochastic process, and it is referred to as the **queueing process**. It presents features of processes related to mass servicing when random fluctuation is developed, such as telephone traffic that Erlang started with, stock processes in economics, machine repair, car traffic, patient treatment in hospitals, and computer job processing.

A queueing process consists of several key processes. They are **arrival**, **service**, and **departure** processes. Each of these processes has its own features. Below, we list some of the features for (i) arrival, (ii) service, and (iii) the system as a whole.

i. **Arrivals** may arrive **singly** or by **bulk (group, or batch)**, with **reneging (abandonment)**, **depending on time** or **independent of time**. There might be a limitation on the batch size; it may have a minimum and a maximum number of arrivals. The arrival source may be **finite** or **infinite**.

ii. **Service** may be conducted **singly** or by **batch**, service with **vacation**. Service process may have an order referred to as the **queue discipline**. This feature could be **first-come–first-served (FCFS)** or **first-in–first-out (FIFO)**, **random service**, **priority service**, **last-come–first-served (LCFS)**, **batch (bulk) service**, etc. There might be more than one server; that is, the queue is with a **single server** or **multiple servers**. In case of multiple servers, the server may be set in **parallel**, **series**, or **mixed**.

iii. **The queueing system** may consist of a **buffer** or a waiting room with finite or infinite capacity to hold the arrivals to wait for their services to begin. The system may allow **feedback** to the original or intermediator queues for further service. The system may also allow **splitting** leaving the service station. Within the system, each process may have a distribution associated with it. For instance, arrivals may arrive according to a Poisson distribution. On the other hand, a service may be given according to an exponential distribution. Feedback process may be based on a Bernoulli or geometric distributions, and splitting may be based on the Bernoulli distribution.

In relation to the purpose of this chapter and the goal of this book, from here to the end of this chapter, we will offer some models of queueing theory that describe applications of probability, statistics, and stochastic processes. Some of these models will be presented for the purpose of basic theories and some as applications in the engineering, economics, and sciences.

7.6.1   THE SIMPLEST QUEUEING MODEL, M/M/1

The queueing model with Poisson arrival, exponential service time, and a single server is now known as **the simplest type of a queueing model**. Almost all types of features and properties one can think of have been found about this model, although some of the time-dependent cases have not been tackled due to their complexities.

Since this model is simple, it is a good example for more complicated ones in the theory.

So, here is the model. The model consists of arrival process, service process, and one server. Tasks (or customers) arrive according to a Poisson distribution with parameter λ, where λ is a constant. If the service station is empty, the task's service starts immediately by the single server being available in the station. If there are already tasks in the system, they form a single queue and enter the service station one at a time, in order of arrivals. In other words, service will be **FCFS** or **FIFO**.

The service time of tasks is assumed to have an exponential distribution with parameter μ, where μ is a constant. Thus, the system that consists of Markovian arrival process (MAP), Markovian service time process and there is a single server is denoted by *M/M/1*.

Note 7.30

A historical note regarding the *M/M/1* and *M/M/1/N* notations is described as follows. Standard queueing systems are symbolized, generically, by the notation *A/B/C/D* (introduced by Kendall in 1953). The letters represent the arrival type (for instance, deterministic, stochastic; Markovian, non-Markovian, or general) and the service type (such as deterministic or probabilistic; Markovian, Erlangian, or even general). The symbols also indicate the number of servers in parallel and the capacity of the system (i.e., the capacity of the buffers plus the service stations). Hence, *M/M/1/K* means that the inter-arrival times are exponential (Markovian); that is, the arrival process is a Poisson process, and service times distribution is also exponential (Markovian) with a single server and a finite buffer with capacity $N - 1$, with total capacity N. When the capacity of the buffer is infinite, the letter N is dropped.

Now that the *M/M/1* queueing model has been defined, we may ask what are the minimum measures we can find about it?

The answer is that, minimally, we can find out about

i. Distribution of the number of tasks in the system, referred to as the **queue length**.
ii. The average time the server is **busy**.
iii. The average **service times**. To find these measures, we need to formulate the model mathematically and then try to answer the questions.

To mathematically formulate the model, we start by supposing that at a particular time instance, we wish to have j tasks in the system, which is referred to as the **state of the system**. How this would be possible? There are three possibilities, as follows:

i. There are already j tasks in the system. Hence, no **transition** should occur; that is, no arrival and no departure should occur.

ii. There are $j - 1$ tasks in the system at the instance; that is, the system is short of one task and one task should arrive.

iii. There are $j + 1$ tasks in the system at the instance; that is, the system has one task over and one task should complete its service and exit the system.

Note 7.31

Later, we will see that $M/M/1$ queueing model is an example of a stochastic process referred to as the **birth-and-death process** that only allows transitions at any time as stated in (i) through (iii) above and travels forward and backward among the integers in one-step transitions, if not staying in the same place.

We pause here to elaborate transition probabilities from the Chapman–Kolmogorov term mentioned earlier in (7.34). From the system (7.34), we have:

$$\frac{p_{yx}(t+s) - p_{yx}(t)}{s} = \sum_{k \neq x} \frac{p_{kx}(s)}{s} p_{yk}(t) - \frac{1 - p_{xx}(s)}{s} p_{yx}(t). \tag{7.42}$$

Passing (7.41) to the limit as $s \to 0$, we obtain:

$$p'_{yx}(t) = \sum_{k \neq x} r_{kx} p_{yk}(t) - r_x p_{yx}(t), \quad t \geq 0, \tag{7.43}$$

which is referred to as **Kolmogorov's forward equations for the transition probabilities**, where

$$r_x = \lim_{s \to 0} \frac{1 - p_{xx}(s)}{s} \tag{7.44}$$

and

$$r_{yx} = \lim_{s \to 0} \frac{p_{yx}(t)}{s}, \quad y \neq x. \tag{7.45}$$

The quantities r_x and r_{yx}, used in (7.40) and after, are referred to as the **transition rates** of the Markov chain. The term "r_x" is the **unconditional rate** of leaving the state x to make a transition to any other state, and the term "r_{yx}" is the **conditional rate** of making a transition from state x to state y.

Note 7.32

It is to note that, in addition to the property $r_{yx}(t) \geq 0$, it is generally assumed that

$$\sum_{y \neq x} p_{yx}(t) = 1, \quad t \geq 0, \ x = 0, \pm 1, \pm 2, \ldots$$

Thus,

$$\sum_{x, x \neq y} r_{yx}(t) = r_x, \qquad x = 0, \pm 1, \pm 2, \ldots$$

Limits of (7.44) and (7.45) exist if $p_{xx}(0) = 1$, and consequently, $p_{yx}(0) = 0$ if $y \neq x$. Thus,

$$p'_{xx}(0) = \lim_{t \to 0} \frac{dp_{xx}(t)}{dt} = -r_x \qquad (7.46)$$

and

$$p'_{yx}(0) = \lim_{t \to 0} \frac{dp_{yx}(t)}{dt} = -r_{yx}, \qquad y \neq x. \qquad (7.47)$$

If we were to allow s to approach 0, before reaching the limit, from (7.44) and (7.45), we, respectively, obtain the following:

$$p_{xx}(s) = 1 - sr_x + o(s) \qquad (7.48)$$

and

$$p_{yx}(s) = 1 - sr_{yx} + o(s), \qquad y \neq x, \qquad (7.49)$$

where $o(z)$ (read as little-o) is the little-o of z, which is generally defined as follows.

Definition 7.26

We say $f(k) = o(f(k))$, read as little-o of $f(k)$, if for every $C > 0$, there is a K such that $f(k) \leq Cg(k)$, for all $k \geq K$. Symbolically, it is almost the same as stating

$$\lim_{k \to \infty} \frac{f(k)}{g(k)} = 0, \qquad (7.50)$$

except when $g(k)$ is let to be zero infinitely often.

From (7.41), we will have:

$$p_{yx}(t + s) = \sum_{k=0}^{\infty} p_{kx}(t) p_{ky}(s). \qquad (7.51)$$

Thus,

$$p'_{yx}(t) = \sum_{k \neq y} r_{ky} p_{xk}(t) - r_y p_{yx}(t), \qquad t \geq 0. \qquad (7.52)$$

Relation (7.52) is referred to as **Kolmogorov's forward equations for the transition probabilities**, with the assumption that

$$\sum_{k \in \mathbb{Z}} p_k(t) = 1. \tag{7.53}$$

If $\mathbb{Z}$, the set of integers, in (7.53) is replaced by a finite subset of it, it is referred to as the **normalizing condition**.

Note 7.33

From (7.49), we can easily see that

$$p'_x(t) = \sum_{k \neq y} r_{yk} p_k(t) - r_y p_y(t), \quad t \geq 0, \ y \in \mathbb{Z}. \tag{7.54}$$

We now develop the system of differential–difference equations for $M/M/1$ system based on Chapman–Kolmogorov's backward and forward equations, equation (7.43). Such a system, when developed, can be solved using different methods, such as generating function and matrix methods. We will see that once the system is developed, it will be similar for both discrete and continuous time cases. We refer the detail of this development to Montazer-Haghighi (1976), Haghighi and Mishev (2013a, b, 2014, 2016a), and Haghighi et al. (2011a,b).

Note 7.34

We should remind the reader of three properties of the exponential pdf that we have showed in earlier chapter.

  i. For a Poisson arrival process, the inter-arrival times follow an exponential distribution with the same parameter.
 ii. Using the exponential distribution, we can describe the state of the system at time t as the number of tasks in the system.
iii. The exponential distribution is memoryless.

We now start by assuming that the capacity of the system is infinite. We also let the probability that at time t there are n tasks in the system be denoted by $p_n(t)$, $n = 0, 1, 2, \ldots$. The memoryless property of the exponential pdf implies that the system forgets the history. That is, at each epoch of time, only three states are to be concerned, namely, the current state, one state backward, and one state forward. For a queueing system, the three cases were mentioned above under the transitions. Thus, within a small period of time Δt, we have:

$$\begin{cases} p_0(t + \Delta t) = \lambda(1 - \Delta t) p_0(t) + \mu \Delta t p_1(t) + o(\Delta t), \\ p_n(t + \Delta t) = \lambda \Delta t p_{n-1}(t) + [1 - (\lambda + \mu)\Delta](t) p_n(t) + \mu \Delta t p_{n+1}(t) + o(\Delta t), n = 1, 2, \ldots, \end{cases} \tag{7.55}$$

where $o(\Delta t)$ is the "little-o" defined earlier. That is, $\lim_{\Delta t \to 0} = \dfrac{o(\Delta t)}{\Delta t} = 0$. In (7.55), we pass to the limit as $\Delta t \to 0$ and obtain the following **transient** (or **time-dependent**) system of differential–difference equations for the $M/M/1$ queueing system:

$$\begin{cases} p_0'(t) = -\lambda p_0(t) + \mu p_1(t), \\ p_n'(t) = \lambda p_{n-1}(t) - (\lambda + \mu)p_n(t) + \mu p_{n+1}(t), \quad n = 1, 2, \ldots \end{cases} \tag{7.56}$$

For the stationary case, that is, time-independent case, the derivatives on the left hand of (7.56) should be 0. Thus, moving the negative terms to the other side, the **stationary** or **time-independent system of difference equations** are:

$$\begin{cases} \lambda p_0 = \mu p_1, \\ (\lambda + \mu)p_n = \lambda p_{n-1} + \mu p_{n+1}, \quad n = 1, 2, \ldots, \end{cases} \tag{7.57}$$

with the sum of probabilities to be one,

$$\sum_{n=0}^{\infty} p_n = 1, \tag{7.58}$$

which is called **the normalizing equation**.

The system (7.57) along with (7.58) can be solved recursively. To do this, we define

$$\rho = \frac{\lambda}{\mu}, \tag{7.59}$$

which needs to be <1, for the system (7.57) to have a solution. (Why?) Thus, from the first equation of (7.57), we will have:

$$p_1 = \rho p_0. \tag{7.60}$$

Substituting (7.60) in the second equation of (7.57), we find. Continuing this way, we will have:

$$p_n = \rho^n p_0, \quad n = 0, 1, 2, \ldots \tag{7.61}$$

Using (7.58), from (7.61), we will have:

$$p_0 + \rho p_0 + \rho^2 p_0 + \cdots = 1 \tag{7.62}$$

or

$$p_0 = 1 - \rho. \tag{7.63}$$

Thus,

$$p_n = (1-\rho)\rho^n, \quad n = 0,1,2,... \tag{7.64}$$

There is another method of solving the system (7.57), that is, the generating function method. We will use this method here, as well.

We discussed the probability generating function in Chapter 6. Hence, we define the generating function for p_n by $G(z), |z| < 1$ as:

$$G(z) \sum_{n=0}^{\infty} p_n Z^n, \quad |z| < 1. \tag{7.65}$$

We leave it as an exercise that after applying the probability generating function (7.63) on the system (7.57) and performing some algebra, the following yields:

$$G(z) = \sum_{n=0}^{\infty} (1-\rho)\rho^n z^n = \frac{1-\rho}{1-\rho^z}. \tag{7.66}$$

Now, writing the power series (7.64) by its McLaurin expansion, we obtain the same values for p_0 and p_n as in (5.63) and (5.64), respectively.

From the generating function (7.66), we can obtain all moments, by taking derivatives repeatedly. Thus, the first derivative, for instance, yields the average number of tasks in the system, that is, in the waiting line and in the service. Thus, denoting the number of tasks in the system by N, the mean by L_s, and variance by $\text{Var}_s(N)$, we will have:

$$L_s = E(N) = \frac{\rho}{1-\rho} = \frac{\lambda}{\mu-\lambda} \text{ and } V_s = \text{Var}_s(N) = \frac{\rho}{(1-\rho)^2}. \tag{7.67}$$

We could also find the average and the variance of the number of the tasks waiting in line, denoted by L_q and V_q, respectively, as:

$$L_s = L_q + \rho. \tag{7.68}$$

Hence,

$$L_q = L_s - \rho = \frac{\lambda^2}{\mu(\mu-\lambda)} \text{ and } V_q = \frac{\rho^2(1+\rho-\rho^2)}{(1-\rho)^2}. \tag{7.69}$$

Let us now try to find the mean time spent in the system, denoted by W_s. To do that, we use a very well-known relation for this purpose, called **Little's formula**. It gives the relation between the mean number of tasks in the system and the mean time spent in the system. The **Little's formula** states:

$$L_s = \lambda W_s, \tag{7.70}$$

from which the mean number of tasks in the system is:

$$W_s = \frac{L_s}{\lambda} = \frac{1}{\mu - \lambda}. \tag{7.71}$$

The service distribution is exponential with parameter μ, and the mean service time is $1/\mu$. Thus, denoting by W_q and V_q, the mean and the variance of waiting time in the waiting line, we, respectively, have:

$$W_s = W_q + \frac{1}{\mu} \text{ and } W_q = W_s - \frac{1}{\mu} = \frac{\lambda}{\mu(\mu - \lambda)}. \tag{7.72}$$

Denoted by $VW_s = \mathrm{Var}(W_s)$ and $VW_q = \mathrm{Var}(W_q)$, the variance of length of the time a task spends in the system and the number of tasks waiting in line, respectively, are:

$$VW_s = \mathrm{Var}(W_s(N)) = \frac{1}{(1-\rho)^2 \mu^2} \text{ and } VW_q = \frac{\rho(2-\rho)}{(1-\rho)^2 \mu^2}. \tag{7.73}$$

So far, we have performed our entire analysis assuming infinite size buffer. Now, suppose we look at the case with finite buffer. So, we assume the buffer capacity to be N. In this case, the notation $M/M/1$ will change to $M/M/1/N$. The system and the solution will also be affected. Hence, the distribution of the number of tasks in the system, the mean number of tasks in the system, and the mean number of tasks lost due to the finiteness of the buffer, are, respectively, as follows:

$$p_{N_n} = \begin{cases} \frac{1-\rho}{1-\rho^{N+1}} \rho^n & \text{if } \rho \neq 1, n = 0,1,2,N, \\ \frac{1}{N+1} & \text{if } \rho = 1, n = 0,1,2,N. \end{cases} \tag{7.74}$$

$$L_{N_s} = \begin{cases} \frac{\rho}{1-\rho} - \frac{(N+1)\rho^{N+1}}{1-\rho^{N+1}}, & \rho \neq 1, \\ \frac{N}{2}, & \rho = 1. \end{cases} \tag{7.75}$$

$$L_{N_{\mathrm{loss}}} = \begin{cases} \frac{\lambda(1-\rho)\rho^N}{1-\rho^{N+1}} & \text{if } \rho \neq 1, \\ \frac{\lambda}{N+1} & \text{if } \rho = 1. \end{cases} \tag{7.76}$$

Note 7.35

We moved from infinite buffer to finite buffer. So, if we go back from finite to infinite, we should be able to get the same results back. Thus, letting $N \to \infty$, we leave it as an exercise to show that relations (7.74) and (7.75) should yield the same results for the infinite-capacity case.

Example 7.24

Suppose we have a single-server queue with Poisson arrivals, exponential service times, and infinite buffer capacity. Let us suppose that the mean of the inter-arrival times is 15 minutes and the mean of the service times is 10 minutes. We want to find the following:

i. The mean waiting time in the waiting line,
ii. The mean number of tasks in the waiting line,
iii. The mean waiting time in the system,
iv. The mean number of tasks in the system,
v. The proportion of times the server is not serving (i.e., it is idle).

Answer

The system described in this example is an *M/M/1* queue with $\lambda = \dfrac{1}{15}$ and $\mu = \dfrac{1}{10}$.
Hence, $\rho = \dfrac{\lambda}{\mu} = \dfrac{10}{15}$. Thus, we have the following answers to the questions:

i. The mean waiting time in the waiting line is:

$$L_q = \frac{\rho^2}{1-\rho} = \frac{\left(\dfrac{10}{15}\right)^2}{1-\dfrac{10}{15}} = \frac{0.44}{0.33} = 1.33 \text{ tasks.}$$

ii. The mean number of tasks in the waiting line is:

$$W_q = \frac{L_q}{\lambda} = \frac{1.33}{\dfrac{1}{15}} = \frac{1.33}{0.07} = 19 \text{ minutes.}$$

iii. The mean waiting time in the system is:

$$W_s = W_q + \frac{1}{\mu} = 19 + 10 = 29 \text{ minutes.}$$

iv. The mean number of tasks in the system is:

$$L = \lambda W_s = \frac{1}{15}(29) = 1.93 \text{ tasks.}$$

v. The proportion of times the server is not serving (i.e., the **idle time**) is:

$$\text{The idle time is} = 1 - \rho = 1 - \frac{10}{15} = 0.33.$$

Example 7.25

Suppose we have a computer in an office that works on one job at a time. Let the arrival distribution of jobs to this computer be Poisson and the execution time distribution be exponential. Let us also suppose that arrival of jobs has a rate of 20 per hour, that is, $\lambda = 20$, and execution of jobs has a rate of 40 jobs per hour, that is, $\mu = 40$. We want to find

    i. The mean number of jobs in the computer,
    ii. The variance of the number of jobs in the system,
    iii. The mean number of jobs in the waiting line,
    iv. The variance of the number of jobs in the waiting line,
    v. The mean waiting time in the system,
    vi. The mean waiting time in the waiting line,
    vii. The variance of the waiting time in the system,
    viii. The variance of the waiting time in the waiting line.
        In case the computer capacity is limited to four jobs at a time, we want to find
    ix. The mean number of jobs in the system,
    x. The mean number of jobs lost due to the finiteness of the capacity.

Answer

In this case, we have queueing systems *M/M/1* and *M/M/1/4*. Also, from the given information, we have $\rho = \dfrac{\lambda}{\mu} = \dfrac{1}{2}$, which, of course, is <1, as required. Thus, for the case of *M/M/1*, we have:

i. $L_s = \dfrac{\rho}{1-\rho} = \dfrac{\frac{1}{2}}{\frac{1}{2}} = 1.$

ii. $V_s = \dfrac{\rho}{(1-\rho)^2} = \dfrac{\frac{1}{2}}{\left(1-\frac{1}{2}\right)^2} = 2.$

iii. $L_q = L_s - \rho = 1 - \dfrac{1}{2} = \dfrac{1}{2}.$

iv. $V_q = \dfrac{\rho^2\left(1+\rho-\rho^2\right)}{(1-\rho)^2} = \dfrac{\left(\frac{1}{2}\right)^2\left(1+\frac{1}{2}-\left(\frac{1}{2}\right)^2\right)}{\left(1-\frac{1}{2}\right)^2} = 1.25.$

v. $W_s = \dfrac{L_s}{\lambda} = \dfrac{1}{20}.$

vi. $W_q = W_s - \dfrac{1}{\mu} = \dfrac{1}{20} - \dfrac{1}{40} = \dfrac{1}{40}.$

vii. $VW_s = \dfrac{1}{(1-\rho)^2 \mu^2} = \dfrac{1}{\left(1-\dfrac{1}{2}\right)^2 (40)^2} = \dfrac{1}{400}.$

viii. $VW_q = \dfrac{\rho(2-\rho)}{(1-\rho)^2 \mu^2} = \dfrac{\dfrac{1}{2}\left(2-\dfrac{1}{2}\right)}{\left(1-\dfrac{1}{2}\right)^2 (40)^2} = \dfrac{3}{1{,}600}.$

For the finite case with $N = 4$, since $\rho = \dfrac{1}{2} < 1$, we have:

ix. $L_{Ns} = \dfrac{\rho}{1-\rho} - \dfrac{(N+1)\rho^{N+1}}{1-\rho^{N+1}} = \dfrac{\dfrac{1}{2}}{1-\dfrac{1}{2}} - \dfrac{(4+1)\left(\dfrac{1}{2}\right)^{4+1}}{1-\left(\dfrac{1}{2}\right)^{4+1}} = 1 - \dfrac{\dfrac{5}{32}}{1-\dfrac{1}{32}} = \dfrac{26}{31}.$

x. $L_{N_{\text{loss}}} = \dfrac{\lambda(1-\rho)\rho^N}{1-\rho^{N+1}} = \dfrac{20\left(1-\dfrac{1}{2}\right)\left(\dfrac{1}{2}\right)^4}{1-\left(\dfrac{1}{2}\right)^{4+1}} = \dfrac{\dfrac{10}{16}}{\dfrac{31}{32}} = \dfrac{20}{31}.$

7.6.2 AN M/M/1 QUEUEING SYSTEM WITH DELAYED FEEDBACK

A simple queueing system may include some features that would make it more challenging, yet interesting. Different models may be analyzed by different methods; each may be quite challenging and thus involves quite mathematical theories. The model we are to discuss is a part of two connected models. We first tackle the busy period, and then with some additional features, we will discuss the queue length. Hence, we start this model with the following additional features to the standard M/M/1 (Haghighi and Mishev, 2016a):

i. **Feedback**,
ii. **Delay** (delays occur in many different instances such as computer breakage, heavy traffic, telephone signal outage, and power outage during a surgery. For instance in a case like machine breakdown, a task needing service has to wait until the machine is repaired before the task resumes its service. The same time of delay may occur on the way to return for receiving further service),
iii. **Splitting**.

Thus, the model we are to start an M/M/1 queuing system with the aforementioned three features is illustrated in Figure 7.6. The model can be viewed as a **tandem (series) queueing system**. These two research papers are authored by Haghighi

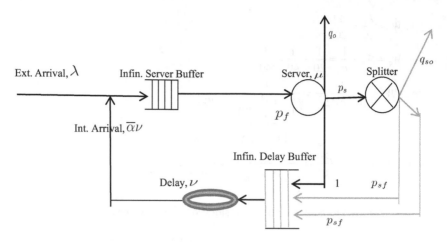

FIGURE 7.6   Single processor with infinite server buffer, splitting and delay-general batch feedback.

and Mishev. This was published in the *International Journal of Mathematics in Operational Research* that appeared in Vol. 8, No. 2, 2016, titled *Busy Period of a Single-Server Poisson Queueing System with Splitting and Batch Delayed-Feedback.*

Note 7.36

Historically, Takács (1963a,b) introduced the study of queueing systems with instantaneous Bernoulli feedback. Montazer-Haghighi (1976), one of his doctoral students, extended the systems in his dissertation, to include a multi-station services in which he also briefly discussed the delayed-feedback single-server queueing system as a particular case of tandem queueing models without offering a solution. The idea started to develop. Nakamura (1971) seems the first who conducted a detailed discussion of the delayed-feedback queueing system followed by Hannibalsson and Disney (1977) who considered queue length with exponential service times. Later, Disney and Kiessler (1987) considered queues with delayed feedback as a traffic process. Kleinrock and Gail (1996) discussed an *M/M/*1 with random delayed feedback. Haghighi et al. (2011a,b) considered a similar queue, but with an additional feature, a splitting device set that a serviced task might choose to go through after the service station. A detailed discussion of this concept appeared in Haghighi and Mishev (2013a,b). Haghighi et al. (2008) discussed a delayed-service model that was expanded in Haghighi and Mishev (2014) as an *M/G/*1 queueing system with processing time consisting of two independent parts, the delay time and the service time (Figure 7.7).

Here is the model Haghighi and Mishev (2016b) considered as an expansion of their earlier work, Haghighi et al. (2011a,b). This model consists of two types of arrivals: **external** (coming from outside) and **internal** (the **feedback** and **splitting** tasks).

FIGURE 7.7 (a) Lajos Takács (August 21, 1924, to December 4, 2015) https://www.pvamu. edu/sites/mathematics/journal/aam/2015/vol-10-issue 2. (b) Aliakbar Montazer Haghighi (September 29, 1940) https://www.pvamu.edu/bcas/departments/mathematics/faculty-and-staff/amhaghighi/. (c) Dimitar P. Mishev (Michev) https://www.pvamu.edu/bcas/departments/mathematics/faculty-and-staff/dimichev/.

The processing time for this system consisted of two independent parts, the **service time** and the **delay time**. So, we describe the model as follows:

i. External tasks arrive singly from an infinite source according to a **Poisson process** with parameter λ.

ii. There is an infinite-capacity **buffer** set before the **service station**.

iii. A **single server** is placed in the service station.

iv. When a task arrives and finds the server busy, it will join the waiting line.

v. The service time is an **exponentially distributed** random variable with parameter μ.

vi. Service is on the basis of FCFS.

vii. After a task leaves the service station, one of the following three events may occur:

    a. It may leave the system forever with probability $q_0, 0 \le q_0 \le 1$.

    b. It may **feedback** the service station for further service with probability $q_f, 0 \le q_f \le 1$.

    c. It may go to a unit called the **splitter** with probability $q_s, 0 \le q_s \le 1$. The splitting is immediate; that is, the time required to split is negligible, with $q_o + q_f + q_s = 1$.

viii. The splitter unit receives a task and splits it into two subtasks:

    a. One returns to the service station with probability 1

    b. The other

        1. Either leaves the system forever with probability $q_{so}, 0 \le q_{so} \le 1$, or

        2. Feedbacks to the service station with probability $p_{sf}, 0 \le p_{sf} \le 1$; $q_{s0} + p_{sf} = 1$.

ix. The feedback to the service station is not immediate. There is another infinite buffer, referred to as the **delay station**, between the server and the first buffer such that a returnee task must go through it to get back to the service station again.

x. A **mover** moves the tasks entered the delay station to the waiting line buffer.

xi. There is a processing procedure in the delay station. Tasks in the delay-station group have varying random sizes between two natural numbers minimum k and maximum $K, 1 \le k \le K$.

xii. The group (batch) sizes may be represented by random variables, denoted by X, with probability distribution function and values between b and B, inclusive, as:

$$P\{X = x\} = \alpha_x, k \le x \le K, \tag{7.77}$$

with mean random size of a group denoted by $\bar{\alpha}$.

xiii. Both the mean and the variance of X are positive and finite.

xiv. The delay times have exponential distribution with parameter ν.

xv. Based on the property of the exponential distribution mentioned earlier, internal arriving tasks arrive from the delay station by groups that follow a Poisson distribution with parameter δ batches per unit time.

xvi. The average internal arriving task rate is $\bar{\alpha}\nu$ tasks per unit time.

xvii. The total rate into the buffer is the sum of the two rates of external and internal arrivals, that is, $\lambda + \bar{\alpha}v$.

xviii. The returnee groups open up at the service buffer and will be served one at a time according to their order of arrivals into the delayed station.

xix. The return of a task is an event independent of any other event and, in particular, independent of the number of its returns.

Since we have already discussed the queue length and losses to a system, for a queueing model, we will now analyze the busy period of the server in its service station.

Analysis

Let us denote the expected attendance rates to the service station, the traffic intensity (or the load) for the service station, and the same for the delay station by $\lambda_1, p_1, \lambda_2$, and p_2, respectively. Then,

$$\begin{cases} \lambda_1 = \lambda + \bar{\alpha}v, \\ \lambda_2 = \lambda_1 \left[p_f + p_s \left(1 + p_{sf} \right) \right], \end{cases} \tag{7.78}$$

or

$$\begin{cases} \lambda_1 = \lambda + \bar{\alpha}v, \\ \lambda_2 = (\lambda + \bar{\alpha}v) \left[p_f + p_s \left(1 + p_{sf} \right) \right]. \end{cases} \tag{7.79}$$

From (7.78), we have:

$$p_i \equiv \frac{\lambda_i}{\mu_i}, i = 1, 2, \tag{7.80}$$

where $\mu_1 = \mu$ and $\mu_2 = \bar{\alpha}v$ with each load restricted to be less than one. Also, the intensity of the system, or total load, denoted by ρ_{sys}, is:

$$\rho_{sys} \equiv \frac{\lambda_1 + \lambda_2}{\mu + \bar{\alpha}v} = \frac{(\lambda + \bar{\alpha}v) \left[1 + p_f + p_s \left(1 + p_{sf} \right) \right]}{\mu + \bar{\alpha}v}. \tag{7.81}$$

Of course, (7.81) is restricted to be less than one.

Since there are two types of arrival, singly with Poisson and in bulk, also with Poisson, but with a different parameter, the distribution of the combined arrival is not immediately known. Because of the case, the service station may be looked at as a $G/M/1$ with two arrival types: external (tasks arrive singly) of rates λ and internal (tasks arrive in bulks) of rates v and service (in batches) of rate μ. It is well known that $G/M/1$ and $M/G/1$ are dual queueing systems. However, the $G/M/1$ queue may be analyzed in a variety of ways like using the method of supplementary variables. Our approach is based on an appropriately imbedded Markov chain at the arrival epoch of a task. Hence, based on the assumptions of the model, the mean arrival rate is $1/(\lambda + \bar{\alpha}v)$, the service times are iid random variables exponentially distributed with mean service times $1/\mu$ regardless of the source of arrivals, and the traffic load ρ_1 is given in (7.80).

Moving toward the goal, we briefly pass through the number of tasks in the service station and leave the detail as an exercise.

Considering just the service station, it is an $M/M/1$ queueing system. We list some of the well-known properties that we need. Suppose that the service station starts with i, $i \geq 0$, tasks. Let the number of tasks, at time t, $t \geq 0$, at the service station, including the one being served be a random variable, denoted by $\xi(t)$. We also denote the probability of having m tasks in the service station at time t by $\phi_m(t)$, that is,

$$\phi_m(t) = P\{\xi(t) = m\}. \tag{7.82}$$

Going to the steady-state case, we let

$$\phi_m = \lim_{t \to \infty} \phi_m(t) = \lim_{t \to \infty} P\{\xi(t) = m | \xi(0) = i\}. \tag{7.83}$$

The transient probability distribution of the number of tasks in the service station at time t for the $M/M/1$ queue is:

$$\phi_m(t) = e^{-(\lambda_1 + \mu)t} \left[\rho_1^{\frac{m-i}{2}} I_{m-i}\left(2\mu\sqrt{\rho_1 t}\right) + \rho_1^{\frac{m-i-1}{2}} I_{m+i+1}\left(2\mu\sqrt{\rho_1 t}\right) \right.$$

$$\left. + \left(1 - \rho_1\right)\rho_1^m \sum_{j=m+i+2}^{\infty} \rho_1^{-\frac{i}{2}} I_j\left(2\mu\sqrt{\rho_1 t}\right) \right], \tag{7.84}$$

where I_r is the modified Bessel function of the first kind given by:

$$I_r(x) = \sum_{m=0}^{\infty} \frac{1}{m!\,\Gamma(m+r+1)} \left(\frac{x}{2}\right)^{2m+r} \tag{7.85}$$

and $\Gamma(u)$ is the gamma function given by:

$$\Gamma(u) = \int_0^{\infty} t^{u-1} e^{-t} dt, \tag{7.86}$$

with $\Gamma(n) = (n-1)!$, when n is a positive integer and i is the initial number of tasks in the station. Thus, the steady-state queue length distribution at the service station, including the one in service, will be geometric as follows:

$$\phi_m = \rho_1\left(1 - \rho_1\right)^m, \quad m = 0, 1, 2, \dots \tag{7.87}$$

Therefore, the transient and steady-state distributions of the number of tasks in the service station are obtained through (7.84) and (7.87), respectively. Thus, denoting by ξ, the steady-state queue length of the service station, the mean and the variance of the queue length are known, respectively, as

$$E(\xi) = \frac{\rho_1}{1-\rho_1}, \quad \text{and} \quad \mathrm{Var}(\xi) = \frac{\rho_1}{1-\rho_1}. \qquad (7.88)$$

We are now ready to address the busy period of server. We first define the following three terms.

Definition 7.27

i. The **server's busy period** is the period starting from one state of the system to the immediate next state. That is, let us set the time $t = 0$ when there are i tasks in the service station, and service starts when a new task arrives. The time period until the service of the $(i+1)\,st$ ends and the number in the station returns to i is a busy period. We denote the **length of a busy period** by B.
ii. By an **idle period**, it is meant the period during which there is no task to be served; that is, the station is empty.
iii. Combination of the two periods, (i) server's busy period and (ii) idle period, is referred to as a **busy cycle**.

Note 7.37

If, during a busy period, a total of l tasks join the service station (internally and externally), where l is a nonnegative integer, then the server's busy period will go through the state transitions: $i+1 \rightarrow i+1+l \rightarrow l \rightarrow i, l = 0,1,2,\dots.$

Note 7.38

During a server's busy period, when an individual task's service is completed, another task will take its place. That is, tasks are born and die during this period. Hence, a busy period contains a renewal process.

We now assume that the service station is empty at time $t = 0$, that is, $i = 0$ in (7.84), and service starts when the first task arrives. It is well known that the distribution of B can be determined through the adjusted transition system of differential–difference equations. Thus, the pdf of a busy period, B, denoted by $f(t)$, will be:

$$f(t) \equiv \phi_0'(t), \qquad (7.89)$$

where $\phi_0(t)$ is the probability of the service station being empty at time t. The function $\phi_0(t)$ can be interpreted as the probability of $Y < t$. Hence,

$$f(t) = \frac{1}{t\sqrt{\rho_1}} e^{-(\lambda_1+\mu)t} I_1\left(2t\sqrt{\lambda_1\mu}\right), \quad t > 0, \qquad (7.90)$$

where I_1 is the modified Bessel function of the first kind given by (7.85). Thus, the distribution of the length of a busy period may be obtained by integrating (7.90) over the interval (0, t]. The mean and variance of a busy period, respectively, will be:

$$E\{B\} = \frac{1}{\mu - \lambda_1} \qquad (7.91)$$

and

$$\mathrm{Var}\{B\} = \frac{1 + \rho_1}{\mu^2 - \left(1 - \rho_1\right)^3}. \qquad (7.92)$$

It is clear, from (7.91), that if $\mu \leq \lambda_1$, B can be infinity with some positive probability.

Denoting the length of an idle period by Δ, since the inter-arrival times to the service station are exponentially distributed with parameter λ_1, the pdf of Δ, denoted by $\psi_\Delta(t)$, is

$$\psi_\Delta(t) = \lambda_1 e^{-\lambda_1 t}, \quad t > 0. \qquad (7.93)$$

Therefore, (7.90) and (7.93) provide the distribution of the busy and idle periods, respectively.

Example 7.26

Suppose that the distribution of the random variable representing the batch sizes, α_x, is a uniform distribution:

$$P\{X = x\} = \alpha_x = \begin{cases} \dfrac{x - k}{K - k}, & k \leq x \leq K, \\ 0, & \text{otherwise.} \end{cases} \qquad (7.94)$$

From (7.94), we have:

$$\bar{\alpha} = \frac{k + K}{2}. \qquad (7.95)$$

The model under consideration may be thought of as a discrete-time single-server tandem queueing system with two queues each having an infinite-space buffer. Two types of tasks are arriving in the system: single-task external arrivals and batches from the second queue. The first queue feeds the service station, some of the departures from the service station feed the second queue, and there is no external arrival to the second queue. This tandem queueing system has multiple exit windows.

To calculate the density and the distribution of a busy period, we partition the time interval into subintervals. As an illustration, we choose a finite time interval of length 2 units, that is, $[0, T] = [0, 2]$. The left endpoint of this interval is a singularity of the pdf defined in (7.90). Problems that will occur due to this singularity like biasness may be addressed using the method offered by Kim and Witt (2013) in the estimation of Little's law parameters when the interval under consideration is finite and the initial busy period is not zero. However, we choose a more basic method as follows.

We choose subintervals with equal lengths, denoted by h, and in this case, $h = 0.0001$. Thus, for the interval $[0,T] = [0,2]$, the number of subintervals will be $n = \dfrac{T}{h} = \dfrac{2}{0.0001}$, with the beginning and end partition points as $t_0 = 0$ and $t_n = 2$.

We also choose the following arbitrary data satisfying conditions imposed previously: $p_f = 0.1$, $q_0 = 0.3$, $p_s = 0.6$, $q_{so} = 0.1$, $p_{sf} = 0.9$, $k = 2$, and $K = 5$.

Note 7.39

The exact values of the mean and variance of a busy period are given by (7.91) and (7.92), respectively. However, we can find their approximated values using discretized values of $f(t)$. This way, we verify our approximation of $f(t)$ using a partitioning of the time interval. Table 7.5 shows these values.

7.6.2.1 Number of Busy Periods

In general, the elapsed time from time $t = 0$ to the epoch of observing $i+1$ tasks in the service station is, sometimes, referred to as the **age of the busy period at $t = 0$** (this is actually the initial idle period). Using this terminology,

a. We denote by ψ_0 the age of the busy period at $t = 0$, which from (7.93) is $\psi_0(0) = \lambda_1$, and by X_1 the remaining time of the first period, that is, the duration of time from arrival of the $(i+1)$st task until the epoch of the station returns to containing j tasks for the first time thereafter.

b. Keeping the age in mind, we denote by $X_1, X_2, \ldots$, the random variables representing the length of successive busy periods, after the age before beginning of the first one.

We assume that the random variables $X_2, X_3, \ldots$, are independent and identically distributed with cdf as $F(t)$ and are independent of X_1, where

$$F_1(t) = P\{X_1 \le t\} \tag{7.96}$$

TABLE 7.5

Expected Value and Variance of Length of a Busy Period for Three Sets of Data Points

Data Set	λ	μ	v	$E(B)$ Using Formula	$E(B)$ Using Approximation	Var(B) Using Formula	Var(B) Using Approximation
1	15	180	30	0.0167	0.0167	0.0001	0.0014
2	10	60	10	0.0667	0.0637	0.0005	0.0233
3	3	30	3	0.6667	0.1818	0.0084	0.1059

---

and

$$F(t) = P\{X_n \le t, n = 2, 3, \ldots\}. \tag{7.97}$$

Note 7.40

Derivative of $F(t)$, given in (7.97), is the same as $f(t) \equiv \phi_0^1(t)$, given in (7.90), for $t = 2, 3, \ldots$. Generally, $F_1(t) \ne F(t)$ except when $\psi_0 = 0$. Thus, clearly, we have:

$$F_1(t) = P\{X - \psi_0 \le t | X > \psi_0\} = \frac{P\{\psi_0 < X < t + \psi_0\}}{P\{X > \psi_0\}}$$

$$\frac{F(t + \psi_0) - F(\psi_0)}{1 - F(\psi_0)}, \quad t \ge 0. \tag{7.98}$$

Suppose we initially choose a busy period with age distribution as $\alpha(\psi_0)$ with its pdf as $d\alpha(\psi_0)$. Then, (7.98) can be rewritten as:

$$F_1(t) = \int_0^\infty \frac{F(t + \psi_0) - F(\psi_0)}{1 - F(\psi_0)} d\alpha(\psi_0). \tag{7.99}$$

Concentrating on just the busy periods of the service station and ignoring possible idle periods, it is clear that successive busy periods of the server constitute renewal cycles. Hence, we can easily answer the following two questions:

(Q1) How old is the current busy period?
(Q2) On average, how many busy periods will be observed?

To answer these questions, first note that when we ignore the idle periods, the epoch of ending each busy period can be considered as an arrival epoch. Then, we will have a sequence of arrivals, as each busy period ends, that is, $\{X_n\}$. In other words, ending busy periods form a renewal or recurrent process.

Now, let the partial sums of sequence $\{X_n\}$ of the busy periods be denoted by

$$S_n = \begin{cases} 0, & n = 0, \\ \sum_{i=1}^{n} X_i, & n = 1, 2, \ldots \end{cases} \tag{7.100}$$

That is, S_n counts the number of busy periods. Let us denote the distribution of S_n by $F_n(t)$, that is,

$$F_n(t) = P\{S_n \le t\}, \quad t \ge 0, \ n \ge 0. \tag{7.101}$$

Note 7.41

$F_n(t)$ denotes the n-fold convolution of F with itself.

Note 7.42

$F_0(t) = 0$ for $t \geq 0$ and $F_1(t)$ has already been defined above through (7.98) and (7.99). Thus, we have:

$$F_{n+1}(t) = \int_0^t F_n(t-s) \, dF(s), \quad n = 1, 2, \dots, \tag{7.102}$$

Now, let us define the random variable $N(t)$ to count the number of busy periods within the interval $(0, t]$. Then, the values of $N(t)$ in the sequence $\{N(t), t > 0\}$ are positive integers. The sequence $\{N(t)\}$ is increasing; that is, $N(s) - N(t)$ if $s < t$. Note that if $s < t$, then the difference $N(t) - N(s)$ is the number of busy periods that occurred during the interval $[s, t]$. Accordingly, $S_n = T$ will be the time the nth busy period has concluded. Counting from this point, the ith subsequent busy period will occur at $S_{n+i} - S_n = X_{n+1} + \dots + X_{n+i}$. Thus, given $S_n = T$, $\{N(T+t) - N(T), 0 \leq t < \infty\}$ is a renewal counting or recurrent process.

There is an extensive study on the behavior of the ratio $N(t)/t$, which is a random variable for each value of t and is the time-averaged renewal rate over the interval $(0, t]$, that is, the number of busy periods per unit time. It is well known that (e.g., Täcklind (1944))

$$\lim_{t \to \infty} \frac{N(t)}{t} = \frac{1}{E(X)}, \tag{7.103}$$

where $E(X)$ is given in (7.98). Hence, in our case, we will have:

$$\lim_{t \to \infty} \frac{N(t)}{t} = \mu - \lambda_1. \tag{7.104}$$

Note 7.43

The process $\{S_n, n = 0, 1, 2, \dots, S_0 = 0\}$ can be considered as a random walk in the sense that when the system has completed its $n-1$ busy periods, that is, the system is in state S_{n-1}, then X_n units of time later, it would be in state S_n. Thus, the following equivalence holds:

$$\{N(t) \geq n\} \equiv \{S_n \leq t\}. \tag{7.105}$$

Relation (7.105) implies that

$$P\{N(t) \geq n\} = P\{S_n \leq t\}. \tag{7.106}$$

Therefore, from (7.101) and (7.106), we have:

$$P\{N(t) = n\} = P\{S_n(t) \le t\} - P\{S_{n+1}(t) \le t\}$$
$$= F_n(t) - F_{n+1}(t), \quad n = 0, 1, 2, \ldots \tag{7.107}$$

If we denote the expected number of busy periods in the time interval $(0,t)$ by $H(t)$, then from (7.107), we will have:

$$H(t) \equiv E\{N(t)\} = \sum_{n=1}^{\infty} n[F_n(t) - F_{n+1}(t)]. \tag{7.108}$$

Since

$$n[F_n(t) - F_{n+1}(t)] = F_1(t) - F_2(t) + 2[F_2(t) - F_3(t)] + 2[F_3(t) - F_4(t)] + \cdots,$$

from (7.108), we will have:

$$H(t) \equiv E\{N(t)\} = \sum_{n=1}^{\infty} F_n(t). \tag{7.109}$$

Thus, if we assume that $X_1 = \eta, 0 \le \eta \le t$, then we will have **Takács's renewal equation** (Takács 1958)

$$H(t) = F(t) + \int_0^t H(t - \eta) dF(\eta). \tag{7.110}$$

It can also be shown that (Prabhu 2007, p. 111)

$$E\{N(t)^2\} = E\{N(t)\} + 2\int_0^t H(t - \eta) dH(\eta). \tag{7.111}$$

From (7.98) and (7.110), we can find the average duration time that the server will be busy within any time interval $(0, t]$. If the duration of a busy period instead of (7.98) had an exponential distribution with parameter β, that is, $F(t) = 1 - e^{\beta t}, \ t \ge 0$, then from (7.97), we would have had:

$$F_1(t) = \frac{\left(1 - e^{-\beta(t + \psi_0)}\right) - \left(1 - e^{\beta \psi_0}\right)}{1 - e^{\beta \psi_0}} = F(t), \tag{7.112}$$

and from (7.101), we would have:

$$F_n(t) = 1 - \sum_{i=0}^{n-1} e^{-\beta t} \frac{(\beta t)^i}{i!}, \quad n = 1, 2, \ldots \tag{7.113}$$

Hence, in this hypothetical case,

$$P\{N(t) = n\} = F_n(t) - F_{n+1}(t) = e^{-\beta t}\frac{(\beta t)^n}{n!}, \quad n = 0,1,2,\dots, \quad (7.114)$$

which is a Poisson distribution with parameter β. Therefore, the average number of busy periods in this case would have been

$$E\{N(t)\} = \beta. \quad (7.115)$$

Thus, (7.108) and (7.110), in general case, and (7.114) and (7.115), for the special case, answer the questions (Q1) and (Q2), respectively.

7.6.3 A MAP SINGLE-SERVER SERVICE QUEUEING SYSTEM

This research paper is also authored by Haghighi and Mishev (2016b). It was published in the *International Journal of Mathematics in Operational Research* that appeared in Vol. 9, No. 1, 2016, under the title *Stepwise Explicit Solution for the Joint Distribution of Queue Length of a MAP Single-server Service Queueing System with Splitting and Varying Batch Size Delayed-Feedback*. It is an extension of the previous model and also an extension of the one discussed in Haghighi et al. (2011b).

In this model, we will discuss some additional features involved in the real-life situation development of a queue. Here the features, including the new ones, that describe the model are as follows:

i. Two Poisson arrivals as a **MAP queueing network**.
ii. Arrivals from two sources: singly from outside with Poisson process with λ, and from inside, the delay station to the service station, by **batch** (size varies between two natural numbers, minimum k and maximum, $K, 1 \le k \le K$).
iii. An infinite-capacity buffer is set before the service station.
iv. There is a single-server service station.
v. The service time distribution is exponential with parameter μ.
vi. Based on the property of the exponential distribution mentioned earlier, internal arriving tasks arrive from the delay station by groups that follow a Poisson distribution with parameter δ batches per unit time.
vii. The returnee groups open up at the service buffer and will be served one at a time according to their order of arrivals into the delay station.
viii. When an arrival task finds the server busy, it will join the **waiting line**.
ix. Service is on basis of FCFS.
x. The feedback to the service station is not immediate, and it will be through the delay station.
xi. The return of a task is an event independent of any other event and, in particular, independent of the number of its returns.
xii. There is another infinite buffer, referred to as the **delay station**, set between the server and the first buffer such that a returnee task must go through it to get back to the service station again.

xiii. The delay times have exponential distribution with parameter ν.

xiv. After a task leaves the service station, one of the following three events may occur:

    a. It may leave the system forever with probability q_0, $0 \le q_0 \le 1$.

    b. It may **feedback** the service station for further service with probability q_f, $0 \le q_f \le 1$.

    c. It may go to a unit called the **splitter** with probability q_s, $0 \le q_s \le 1$; the splitting is immediate; that is, the time required to split is negligible, with $q_0 + q_f + q_s = 1$.

xv. Splitter unit has an infinite buffer.

xvi. The splitter unit receives a task and splits it into two subtasks, and then either of the following two events may occur:

    a. One returns to the service station, through the delay station, with probability 1.

    b. The other either leaves the system forever with probability q_{so}, $0 \le q_{so} \le 1$, or feedbacks to the service station with probability p_{sf}, $0 \le p_{sf} \le 1$, $q_{so} + p_{sf} = 1$.

xvii. Tasks in the delay-station group with varying random sizes.

xviii. The group (batch) sizes are represented by random variables, denoted by X, with probability distribution function and values between b and B, inclusive, as:

$$P\{X = x\} = \alpha_x, k \le x \le K, \tag{7.116}$$

with mean random size of a group denoted by $\bar{\alpha}$.

xix. Both the mean and the variance of X are positive and finite.

xx. The average internal arriving task rate is $\bar{\alpha}\nu$ tasks per unit time.

xxi. The total rate into the service station buffer is the sum of the two rates of external and internal arrivals, that is, $\lambda + \bar{\alpha}\nu$.

xxii. A **mover** moves the tasks entered the delay station to the service station buffer.

xxiii. For the analysis purpose, **truncation, augmentation, tridiagonalization, and supplementary variables** methods on infinite block matrices with infinite block matrix elements will be applied.

xxiv. Duality properties of G/M/1 and M/G/1 is another method used for analysis.

xxv. Algorithm as how the parameters must be chosen will be given.

xxvi. Stepwise algorithm to compute the joint pdf of queue length will be given.

xxvii. Due to approximation, error analysis will be performed.

We are looking for some characteristics of the model like average queue length in both stations.

7.6.3.1 Analysis of the Model

To understand the process of splitting, we offer an example. In the United States, the creation of the Health Protection Agency (HPA), the Department of Health (DH), and the Home Office (HO) agreed that there should be a review of port health.

A joint DH, HO, and HPA Steering Group is established to oversee the work, and a joint HPA and HO Project Group was asked to undertake the review.

Let us consider the service station as the port of inspection of regulations of both DH and HO. Suppose a person subject to immigration control (an arrival) attends the service station. The inspection of all documents may lead to release the person with no problem. However, it may be necessary to reinspect the documents and so the person is sent back to the service station. On the other hand, it might be necessary for the documents to be reviewed by the HPA and HO Joint Project. At this review point, the individual with all documents is sent back to the service station for further inspection where the documents will be reviewed. This may lead to releasing the documents with no problem or sending them (as a new individual) to the service station for yet another inspection. The rest of the model follows as stated above.

We will develop an algorithm to find not only the mentioned characteristics, but also the joint distribution of the number of tasks in the system, that is, in both stations.

Let the random variables $\xi_1(t)$ and $\xi_2(t)$ denote the number of tasks, at time t, $t \geq 0$, at the service station, including the one being served, and at the delay station, including those being processed (i.e., the batch ready to be moved to the service station), respectively. Let us also denote the joint probability of these random variables by $\psi_{m,n}(t)$, that is,

$$\psi_{m,n}(t) = P\{\xi_1(t) = m, \xi_2(t) = n\}. \tag{7.117}$$

Thus, $\{\xi_1(t) = m, \xi_2(t) = n\}$ is an irreducible continuous-time Markov chain.

We consider the stationary process. Hence, we define the state of the system as (m, n), where m and n denote the number of tasks in the service station and the delay station, respectively. Thus,

$$\psi_{m,n}(t) = \lim_{t \to \infty} \psi_{m,n}(t). \tag{7.118}$$

Of course, tasks in service and being processed are included in m and n, respectively, and since splitting is immediate, only those on their way to the delay station are to be considered as part of the delay station. That is, the state of the system is the event $\{\xi_1 = m, \xi_2 = n\}$. Thus,

$$\psi_{m,n} = \lim_{t \to \infty} \psi_{m,n}(t) = \lim_{t \to \infty} P\{\xi_1(t) = m, \xi_2(t) = n | \xi_1(0) = m_0, \xi_2(0) = n_0\}; \tag{7.119}$$

that is, the steady-state probability of having m tasks in the service station and n tasks in the delay station, given the initial states in each station, exists and is independent of the initial state.

Let the expected attendance rates to the service station and the delay station, respectively, be denoted by λ_1 and λ_2. Then, from the description of the model, a graphical presentation in Figure 7.6, it can be seen that

$$\begin{cases} \lambda_1 = \lambda + \bar{\alpha}v, \\ \lambda_2 = \lambda + \left[p_f + p_s \left(1 + p_{sf} \right) \right] \end{cases} \tag{7.120}$$

Then, we may write the traffic intensities in the system, for the service station, ρ_1, the delay station, ρ_2, and the system, ρ_{sys}. We note that in our case, the traffic intensity of the system is the ratio of the total arrivals into the system that is, the external plus internal, denoted by λ_1) to the total processing rate that is, the service and the moving rates, denoted by $\mu + \bar{\alpha}v$). Thus, we have:

$$\rho_1 = \frac{\lambda_1}{\mu}, \rho_2 = \frac{\lambda_2}{\bar{\alpha}v}, \text{ and } \rho_{sys} = \frac{\lambda_1}{\mu + \bar{\alpha}v}, \tag{7.121}$$

each restricted to be less than one.

Among our goals is finding the steady-state joint probabilities of the number of tasks in both stations, denoted by $\psi_{m,n}$ for $m, n \geq 0$. We first analyze the service station.

7.6.3.2 Service Station

Realistically, in many manufacturing processes, tasks (jobs or orders) arrive from various sources, such as vendors, shifts, and assembly lines, to a common processing area. In such cases, the arrival process can no longer be assumed to form a renewal process. Hence, *MAP* seems to be a natural choice. *MAP* was originally introduced by Neuts (1989), and queueing systems with it have been extensively analyzed through the matrix analytic method in the literature. Thus, it is a fairly general arrival process. By appropriately choosing the parameters, the underlying assumption of *MAP* can be made as a renewal process.

Since all feedback items have to go through the delay station with some probability and a feedback event is assumed to be independent of any other event in the system, we can consider the service station as an *MAP/M/*1 with two arrival sources with rates λ and $\bar{\alpha}v$, from the external and internal sources, respectively, and the service rate μ.

It is well known that *G/M/*1 may be considered to be the dual of the *M/G/*1 queue. The *G/M/*1 queue may also be analyzed in a variety of ways like an alternative approach using the **method of supplementary variables**.

The service discipline is assumed to be FCFS. Hence, based on the assumptions mentioned above, the mean arrival rate is $1/(\lambda + \bar{\alpha}v)$, and the service times are independent and identically exponentially distributed with the mean service times as $1/\mu$ regardless of the source of arrivals and traffic load ρ_1 given in (7.121).

Note 7.44

Although the arrivals at the delay station are dependent upon departures from the service station, as long as the service station is not empty, non-existing departures from the service station arrive to the delay station with assigned probability, and thus, the dependence between the service station and the delay station virtually discontinues. Therefore, at the arrival to the service station, the two Poisson streams are independent, and hence, the mean arrival would be as mentioned. In the case either the service station or the delay station is empty, then we will have the state (0, 0), or the delay station will continue its processing without new arrival, which does not constitute any dependence.

7.6.3.3 Number of Tasks in the Service Station

Based on the discussion above, recognizing the service station as an $MAP/M/1$ queueing system with the mean inter-arrival and service times as $1/(\lambda + \bar{a}v)$ and $1/\mu$, respectively, many measures will be known and are available. We will list some of those below. Let us assume that the service station starts with i, $i \geq 0$, tasks in its buffer and being served. Denoted by $\phi_m(t)$, the probability of having m tasks in the service station at time t is,

$$\phi_m(t) = P\{\xi_1(t) = m\}. \tag{7.122}$$

For the steady-state case, we let

$$\phi_m \equiv \lim_{t \to \infty} \phi_m(t) = \lim_{t \to \infty} P\{\xi_1(t) = m | \xi_1(0) = i\}. \tag{7.123}$$

As one of the known properties of the $MAP/M/1$ queue, the transient probability distribution of the number of tasks in the service station at time t is:

$$\phi_m(t) = e^{-(\lambda_1 + \mu)t} \left[\rho_1^{\frac{m-i}{2}} I_{m-i}\left(2\mu\sqrt{\rho_1 t}\right) + \rho_1^{\frac{m-i-1}{2}} I_{m+i+1}\left(2\mu\sqrt{\rho_1 t}\right) \right.$$

$$\left. + (1-\rho_1)\rho_1^m \sum_{j=m+i+2}^{\infty} \rho_1^{\frac{1}{2}} I_j\left(2\mu\sqrt{\rho_1 t}\right) \right], \tag{7.124}$$

where ρ_1 is given in (7.121), and I_r is the modified Bessel function of the first kind given by

$$I_r(x) = \sum_{m=0}^{\infty} \frac{1}{m!\Gamma(m+r+1)} \left(\frac{x}{2}\right)^{2m+r}, \tag{7.125}$$

where $\Gamma_{(u)}$ is the gamma function given by

$$\Gamma(u) = \int_0^{\infty} t^{u-1} e^{-t} dt, \tag{7.126}$$

with $\Gamma(n) = (n-1)!$, when n is a positive integer and i is the initial number of tasks in the station.

The steady-state queue length distribution at the service station, including the one being serviced, is geometric as follows:

$$\phi_m = \rho_1(1-\rho_1)^m, \quad m = 0, 1, 2, \ldots \tag{7.127}$$

Thus, the transient and steady-state distributions of number of tasks in the service station are given by (7.124) and (7.127), respectively. Hence, denoting by ξ_1 the steady-state queue length of the service station, the mean and variance of the queue length are known, respectively, as

$$E(\xi_1) = \frac{\rho_1}{1-\rho_1}, \text{ and } \text{Var}(\xi_1) = \frac{\rho_1}{1-\rho_1}, \tag{7.128}$$

where ρ_1 is given in (7.121).

7.6.3.4 Stepwise Explicit Joint Distribution of the Number of Tasks in the System: General Case When Batch Sizes Vary between a Minimum k and a Maximum K

For the case of $k = K = 1$, the joint distribution function of the system's queue length, using generation function equation, is given in Haghighi et al. (2011a,b) by the following theorem (note that since the batch sizes are the same and equal to 1, from (7.116), we have $\bar{\alpha} = 1$).

Theorem 7.2

Let $a = q_0\mu$, $b = (p_f + p_s q_{so})\mu$, $c = p_s p_{sf}\mu$ with $a+b+c = \mu$, and $d = q_o - p_s p_{sf}$. If $d > 0, \dfrac{\lambda_1}{\mu} < 1$, and $\dfrac{\lambda_2}{v} < 1$, then for $k = K = 1$, the joint distribution function of the number of tasks in each station exists and can be obtained by the coefficient of Maclaurin expansion of solution of the functional equation for the generating function of the distribution function

$$A(w,z)G_1(w,z) = B_1(w,z)G_1(0,z) + B_2(w,z)G_1(w,0), \tag{7.129}$$

where A, B_1, B_2, and $G_1(w,z)$ are defined below:

$$A(w,z) = cz^3 + bz^2 + \left[a + \lambda w^2 - (\lambda + \mu + v)w\right]z + vw^2, \tag{7.130}$$

$$B_1(w,z) = cz^3 + bz^2 + (a - \mu w)z, \tag{7.131}$$

$$B_2(w,z) = vw^2 - vwz, \tag{7.132}$$

and

$$G_1(w,z) = \sum_{m=0}^{\infty}\sum_{n=0}^{\infty} \psi_{m,n} w^m z^n, |z| < 1, |w| < 1 \tag{7.133}$$

is the generating function of the joint distribution function of the number of tasks in each station. The index of G indicates the value of k, which in this case is 1.

Now, for the general case, when $1 \le k \le K$ and $K \ge 2$, the matrix for the system of balance equations for the queue length is as follows (see Figure 7.6):

$$
\hat{Q} = \begin{bmatrix}
\hat{\mathbf{B}}_1 & \hat{\mathbf{A}}_0 & 0 & 0 & 0 & \cdots & 0 & 0 & 0 & 0 & 0 & 0 & \cdots \\
\hat{\mathbf{B}}_2 & \hat{\mathbf{A}}_1 & \hat{\mathbf{A}}_0 & 0 & 0 & \cdots & 0 & 0 & 0 & 0 & 0 & 0 & \cdots \\
\hat{\mathbf{B}}_3 & \hat{\mathbf{B}}_2 & \hat{\mathbf{A}}_1 & \hat{\mathbf{A}}_0 & 0 & \cdots & 0 & 0 & 0 & 0 & 0 & 0 & \cdots \\
\hat{\mathbf{B}}_4 & \hat{\mathbf{B}}_2 & \hat{\mathbf{B}}_2 & \hat{\mathbf{A}}_1 & \hat{\mathbf{A}}_0 & \cdots & 0 & 0 & 0 & 0 & 0 & 0 & \cdots \\
0 & \ddots & \ddots & \ddots & \ddots & \ddots & \cdots & \cdots & \cdots & 0 & 0 & \cdots \\
\vdots & \vdots & \ddots & \ddots & \ddots & \ddots & \ddots & \ddots & \vdots & 0 & 0 & \cdots \\
\hat{\mathbf{B}}_K & \hat{\mathbf{B}}_{K-1} & \hat{\mathbf{B}}_{K-2} & \cdots & 0 & \hat{\mathbf{B}}_2 & \hat{\mathbf{A}}_2 & \hat{\mathbf{A}}_0 & 0 & \cdots & 0 & 0 & \cdots \\
\hat{\mathbf{B}}_{K+1} & \hat{\mathbf{B}}_K & \hat{\mathbf{B}}_{K-1} & \cdots & 0 & 0 & \hat{\mathbf{B}}_2 & \hat{\mathbf{A}}_1 & \hat{\mathbf{A}}_0 & 0 & 0 & 0 & \cdots \\
0 & \hat{\mathbf{B}}_{K+1} & \hat{\mathbf{B}}_K & \cdots & 0 & 0 & 0 & \hat{\mathbf{B}}_2 & \hat{\mathbf{A}}_1 & \hat{\mathbf{A}}_0 & 0 & 0 & \cdots \\
0 & 0 & \hat{\mathbf{B}}_{K+1} & \cdots & 0 & 0 & 0 & 0 & \hat{\mathbf{B}}_2 & \hat{\mathbf{A}}_1 & \hat{\mathbf{A}}_0 & 0 & \cdots \\
\vdots & \vdots & \vdots & \vdots & \vdots & \vdots & \vdots & \vdots & \vdots & \vdots & \vdots & \vdots & \vdots
\end{bmatrix}
$$

$$(7.134)$$

where block matrices $\hat{\mathbf{A}}_0, \hat{\mathbf{A}}_1, \hat{\mathbf{B}}_1, \hat{\mathbf{B}}_2, \hat{\mathbf{B}}_3, \dots, \hat{\mathbf{B}}_K$, and $\hat{\mathbf{B}}_{K+1}$ are square infinite-size matrices with the elements of each described, respectively, through (7.135)–(7.142), below (note that rather than using $\bar{\alpha}$, we write all possible discrete values of k):

$$
a_0(i,j) = \begin{cases}
q_0\mu & i = j & j = 1,2,3,\dots, \\
\left(p_f + p_s q_{so}\right)\mu & i = j+1, & j = 1,2,3,\dots, \\
p_s p_s f\mu & i = j+2, & j = 1,2,3,\dots, \\
0 & \text{otherwise}
\end{cases}
$$

$$(7.135)$$

$$
a_1(i,j) = \begin{cases}
-(\lambda+\mu), & i = j = 1, \\
-(\lambda+\mu+v), & i = j = 2, \\
-(\lambda+\mu+2v), & i = j = 3, \\
\cdots \quad \cdots & \cdots, \\
-[\lambda+\mu+(K-1)v], & i = j = K \\
-(\lambda+\mu+Kv), & i = j = K+1, K+2, K+3, \dots, \\
0, & \text{otherwise.}
\end{cases}
$$

$$(7.136)$$

$$\hat{b}_1(i,j) = \begin{cases} -\lambda, & i = j = 1, \\ -(\lambda + v), & i = j = 2, \\ -(\lambda + 2v), & i = j = 3, \\ \cdots & \cdots, \\ -[\lambda + (K-1)v], & i = j = K \\ -(\lambda + Kv), & i = j = K+1, K+2, K+3, \ldots, \\ 0, & \text{otherwise.} \end{cases} \tag{7.137}$$

$$\hat{b}_2(i,j) = \begin{cases} \lambda, & i = j, i = 1,2,3,\ldots, \\ v, & i = 1, \ j = 2, \\ 0, & \text{otherwise.} \end{cases} \tag{7.138}$$

$$\hat{b}_3(i,j) = \begin{cases} 2v, & i = 1, j = 3, \\ 0, & \text{otherwise.} \end{cases} \tag{7.139}$$

$$\hat{b}_4(i,j) = \begin{cases} 3v, & i = 1, j = 3, \\ 0, & \text{otherwise.} \end{cases} \tag{7.140}$$

$$\vdots \qquad\qquad \vdots$$

$$\hat{b}_K(i,j) = \begin{cases} (K-1)v, & i = 1, j = K, \\ 0, & \text{otherwise.} \end{cases} \tag{7.141}$$

$$\hat{b}_{K+1}(i,j) = \begin{cases} Kv, & i = j - K, j = K+1, K+2,\ldots, \\ 0, & \text{otherwise.} \end{cases} \tag{7.142}$$

7.6.3.5 An Illustrative Example

The system matrix (7.134) for the special case $K = 3$, $k = 1,2,3$, is as follows (note that if $k = 2$ or 3, or both, then the system needs to be adjusted accordingly):

i. $\lambda\Psi_{0,0} = q_o\mu\Psi_{1,0}$, $\quad m = 0; n = 0; k = 1,2,3,$

ii. $(\lambda + v)\Psi_{0,1} = q_o\mu\Psi_{1,1} + (p_f + p_s q_{so})\mu\Psi_{1,0}$, $\quad m = 0; n = 1; k = 1,2,3,$

iii. $(\lambda + v)\Psi_{0,n} = q_o\mu\Psi_{1,n} + (p_f + p_s q_{so})\mu\Psi_{1,n-1} = p_s P_{sf}\mu\Psi_{1,n-2}$, $\quad m = 0; n \geq 2;$
$k = 1,2,3,$

iv. $(\lambda + \mu)\Psi_{1,0} = \lambda\Psi_{0,0} + q_o\mu\Psi_{1,0} + v\Psi_{0,1}$, $\quad m = 0; n = 0; k = 1,2,3,$

v. $(\lambda + \mu + v)\Psi_{1,1} = \lambda\Psi_{0,1} + q_o\mu\Psi_{1,1} + (p_f + p_s q_{so})\mu\Psi_{2,0}$, $\quad m = 1; n = 1;$
$k = 1,2,3,$

vi. $(\lambda + \mu + \upsilon)\Psi_{1,n} = \lambda\Psi_{0,n} + q_o\mu\Psi_{2,n} + (p_f + p_s q_{so})\mu\Psi_{2,n-1} + p_s p_{sf}\mu\Psi_{2,n-2},$
$\quad m = 1; n \geq 2; k = 1, 2, 3,$

vii. $(\lambda + \mu)\Psi_{2,0} = \lambda\Psi_{1,0} + q_o\mu\Psi_{3,0} + \upsilon\Psi_{1,1} + \upsilon\Psi_{0,2}, \quad m = 2; n = 0; k = 1, 2, 3,$

viii. $(\lambda + \mu + \upsilon)\Psi_{2,1} = \lambda\Psi_{1,1} + q_o\mu\Psi_{3,1} + (p_f + p_s q_{so})\mu\Psi_{3,0}, \quad m = 2; n = 1;$
$\quad k = 1, 2, 3,$

ix. $(\lambda + \mu + \upsilon)\Psi_{2,n} = \lambda\Psi_{1,n} + q_o\mu\Psi_{3,n} + (p_f + p_s q_{so})\mu\Psi_{3,n-1} + p_s p_{sf}\mu\Psi_{3,n-2},$
$\quad m = 2; n \geq 2; k = 1, 2, 3,$

x. $(\lambda + \mu)\Psi_{m,0} = \lambda\Psi_{m-1,0} + q_o\mu\Psi_{m+1,0} + \upsilon(\Psi_{m-1,1} + \Psi_{m-2,2} + \Psi_{m-3,3}), \quad m \geq 3;$
$\quad n = 0; k = 1, 2, 3,$

xi. $(\lambda + \mu + v)\Psi_{m,1} = \lambda\Psi_{m+1,1} + q_o\mu\Psi_{m+1,1} + (p_f + p_s q_{so})\mu\Psi_{m+1,0} + v\Psi_{m-3,4m},$
$\quad m \geq 3; n = 1; k = 1, 2, 3,$

xii. $(\lambda + \mu + v)\psi_{m,2} = \lambda\psi_{m-1,2} + q_o\mu\psi_{m+1,2} + (p_f + p_s q_{so})\mu\psi_{m+1,1} +$
$\quad p_s p_{sf}\mu\psi_{m+1,0} + v\psi_{m-3,5m}, \quad m \geq 3; n = 2; k = 1, 2, 3,$

xiii. $(\lambda + \mu + v)\Psi_{m,n} = \lambda\Psi_{m-1,n} + q_o\mu\Psi_{m+1,n} + (p_f + p_s q_{so})\mu\Psi_{m+1,n-1} +$
$\quad p_s p_{sf}\mu\Psi_{m+1,n-2} + v\Psi_{m-3,n+3} \quad m \geq 3; n = 3; k = 1, 2, 3,$

xiv. $\displaystyle\sum_{m=0}^{\infty}\sum_{n=0}^{\infty}\Psi_{m,n} = 1.$ $\hspace{4cm}$ (7.143)

The system (7.143) may be summarized in matrix form as follows:

$$\hat{Q} = \begin{bmatrix}
\hat{B}_1 & \hat{A}_0 & 0 & 0 & 0 & \cdots & 0 & 0 & 0 & 0 & \cdots \\
\hat{B}_2 & \hat{A}_1 & \hat{A}_0 & 0 & 0 & \cdots & 0 & 0 & 0 & 0 & \cdots \\
\hat{B}_3 & \hat{B}_2 & \hat{A}_1 & \hat{A}_0 & 0 & \cdots & 0 & 0 & 0 & 0 & \cdots \\
\hat{B}_4 & \hat{B}_3 & \hat{B}_2 & \hat{A}_1 & \hat{A}_0 & \cdots & 0 & 0 & 0 & 0 & \cdots \\
0 & \hat{B}_4 & \hat{B}_3 & \hat{B}_2 & \hat{A}_1 & \hat{A}_0 & \ddots & \cdots & \cdots & \cdots & \cdots \\
0 & 0 & \ddots & \ddots & \ddots & \ddots & \ddots & \cdots & \cdots & \cdots & \cdots \\
\vdots & \vdots & 0 & \ddots & \ddots & \ddots & \ddots & \cdots & \cdots & \cdots & \cdots \\
0 & 0 & \cdots & \cdots & \hat{B}_4 & \hat{B}_3 & \hat{B}_2 & \hat{A}_1 & \hat{A}_0 & \vdots & \cdots \\
0 & 0 & \cdots & 0 & 0 & \hat{B}_4 & \hat{B}_3 & \hat{B}_2 & \hat{A}_1 & \hat{A}_0 & \cdots \\
0 & 0 & \cdots & 0 & 0 & 0 & \hat{B}_4 & \hat{B}_3 & \hat{B}_2 & \hat{A}_1 & \cdots \\
\vdots & \vdots & \vdots & \vdots & \vdots & \vdots & \vdots & \vdots & \vdots & \vdots & \cdots
\end{bmatrix}$$ (7.144)

where block matrices $\hat{A}_0, \hat{A}_1, \hat{B}_1, \hat{B}_2, \hat{B}_3, \ldots,$ and $\hat{B}_4$ are the square infinite-size matrices and elements of each described, respectively, below:

$$\hat{a}_0(i, j) = \begin{cases} q_o\mu, & i = j, & j = 1, 2, 3, \ldots, \\ (p_f + p_s q_{so})\mu, & i = j+1, & j = 1, 2, 3, \ldots, \\ p_s p_{sf}\mu, & i = j+2, & j = 1, 2, 3, \ldots, \\ 0, & \text{otherwise.} \end{cases}$$ (7.145)

$$\hat{a}_1(i,j) = \begin{cases} -(\lambda+\mu), & i=j=1, \\ -(\lambda+\mu+v), & i=j=2, \\ -(\lambda+\mu+2v), & i=j=3, \\ -(\lambda+\mu+3v), & i=j=4,5,\dots, \\ 0, & \text{otherwise.} \end{cases} \tag{7.146}$$

$$\hat{b}_1(i,j) = \begin{cases} -\lambda, & i=j=1, \\ -(\lambda+v), & i=j=2, \\ -(\lambda+2v), & i=j=3, \\ -(\lambda+3v), & i=j=4,5,\dots, \\ 0, & \text{otherwise.} \end{cases} \tag{7.147}$$

$$\hat{b}_2(i,j) = \begin{cases} \lambda, & i=j, i=1,2,3,\dots, \\ v, & i=1, \ j=2, \\ 0, & \text{otherwise.} \end{cases} \tag{7.148}$$

$$\hat{b}_3(i,j) = \begin{cases} 2v, & i=1, j=3 \\ 0, & \text{otherwise.} \end{cases} \tag{7.149}$$

$$\hat{b}_4(i,j) = \begin{cases} 3v, & i=j-3, j=4,5,\dots, \\ 0, & \text{otherwise.} \end{cases} \tag{7.150}$$

The normalizing equation for this system is:

$$\sum_{m=0}^{\infty}\sum_{n=0}^{\infty}\Psi_{m,n} = 1. \tag{7.151}$$

The literature shows that an exact solution for the system represented by (7.131) and in the special case given by (7.143) is forbidden. Thus, an algorithmic solution would be the second best. The method we will use can be found in Haghighi et al. (2011a,b). What is extra in this case is that since the system is an infinite one and the matrix and its blocks are all of infinite sizes, we will have to truncate the process for computation purpose and manually create infinity. To guarantee the theoretical convergence, we will use the *augmentation method* that already exists in the literature; see Gibson and Seneta (1987), for instance. To be able to use some of the existing properties of Markov processes, another new idea has been introduced in this paper, that is, tridiagonalization of our matrix **Q** by combining blocks within the matrix.

We now start the algorithm. Let $\mathbf{X} = \langle \mathbf{X}_0, \mathbf{X}_1, \mathbf{X}_2, \dots, \mathbf{X}_n, \dots \rangle^T$ be an infinite-size column vector, where each of $\mathbf{X}_0, \mathbf{X}_1, \mathbf{X}_2, \dots$, is a column vector, in its own right, as follows:

$$\mathbf{X}_0 = \langle x_{0,0}, x_{0,1}, x_{0,2}, \dots \rangle^T,$$

$$\mathbf{X}_1 = \langle x_{1,0}, x_{1,1}, x_{1,2}, \dots \rangle^T,$$

$$\mathbf{X}_1 = \langle x_{1,0}, x_{1,1}, x_{1,2}, \dots \rangle^T.$$

$$\vdots \quad \vdots$$

(7.152)

Further, let matrix Q^T be a $K + 2$ diagonal block matrix as follows:

$$\mathbf{Q}^T =
\begin{bmatrix}
\mathbf{B}_1^T & \mathbf{B}_2^T & \mathbf{B}_3^T & \cdots & \mathbf{B}_K^T & \mathbf{B}_{K+1}^T & 0 & 0 & 0 & \cdots & \cdots \\
\mathbf{A}_0^T & \mathbf{A}_1^T & \mathbf{B}_2^T & \mathbf{B}_3^T & \cdots & \mathbf{B}_K^T & \mathbf{B}_{K+1}^T & 0 & 0 & \cdots & \ddots \\
0 & \mathbf{A}_0^T & \mathbf{A}_1^T & \mathbf{B}_2^T & \mathbf{B}_3^T & \cdots & \mathbf{B}_K^T & \mathbf{B}_{K+1}^T & 0 & \cdots & \ddots \\
0 & 0 & \mathbf{A}_0^T & \mathbf{A}_1^T & \mathbf{B}_2^T & \mathbf{B}_3^T & \cdots & \mathbf{B}_K^T & \mathbf{B}_{K+1}^T & 0 & \ddots \\
0 & 0 & 0 & \mathbf{A}_0^T & \mathbf{A}_1^T & \mathbf{B}_2^T & \mathbf{B}_3^T & \cdots & \mathbf{B}_K^T & \mathbf{B}_{K+1}^T & \ddots \\
\ddots & \ddots & \ddots & \ddots & \ddots & \ddots & \ddots & \ddots & \ddots & \ddots & \ddots \\
\vdots & \vdots & \vdots & \vdots & \vdots & \vdots & \vdots & \vdots & \ddots & \ddots & \ddots
\end{bmatrix}$$

(7.153)

where block matrices $\mathbf{A}_0, \mathbf{A}_1, \mathbf{B}_1, \mathbf{B}_2, \dots, \mathbf{B}_K$, and $\mathbf{B}_{K+1}$ are the square infinite-size matrices, whose elements are described by (7.154)–(7.161), respectively, with its transpose as

$$\mathbf{Q} =
\begin{bmatrix}
\mathbf{B}_1 & \mathbf{A}_0 & 0 & 0 & 0 & \cdots & 0 & 0 & 0 & 0 & \cdots & \cdots \\
\mathbf{B}_2 & \mathbf{A}_1 & \mathbf{A}_0 & 0 & 0 & \cdots & 0 & 0 & 0 & 0 & \cdots & \cdots \\
\mathbf{B}_3 & \mathbf{B}_2 & \mathbf{A}_1 & \mathbf{A}_0 & 0 & \cdots & 0 & 0 & 0 & 0 & \cdots & \cdots \\
\mathbf{B}_4 & \mathbf{B}_3 & \mathbf{B}_2 & \mathbf{A}_1 & \mathbf{A}_0 & \cdots & 0 & 0 & 0 & 0 & \cdots & \cdots \\
0 & \ddots & \ddots & \ddots & \ddots & \ddots & \ddots & \cdots & \cdots & \cdots & \cdots & \cdots \\
\vdots & \vdots & \ddots & \ddots & \ddots & \ddots & \ddots & \ddots & \vdots & \cdots & \cdots & \cdots \\
\mathbf{B}_K & \mathbf{B}_{K-1} & \mathbf{B}_{K-2} & \cdots & 0 & \mathbf{B}_3 & \mathbf{B}_2 & \mathbf{A}_1 & \mathbf{A}_0 & 0 & 0 & \cdots \\
\mathbf{B}_{K+1} & \mathbf{B}_K & \mathbf{B}_{K-1} & \cdots & 0 & 0 & \mathbf{B}_3 & \mathbf{B}_2 & \mathbf{A}_1 & \mathbf{A}_0 & 0 & \cdots \\
0 & \mathbf{B}_{K+1} & \mathbf{B}_K & \cdots & 0 & 0 & 0 & \mathbf{B}_3 & \mathbf{B}_2 & \mathbf{A}_1 & \mathbf{A}_0 & \cdots \\
\vdots & \vdots & \vdots & \vdots & \vdots & \vdots & \vdots & \vdots & \vdots & \vdots & \vdots & \vdots
\end{bmatrix},$$

(7.154)

where the block matrices $\mathbf{A}_0, \mathbf{A}_1, \mathbf{B}_1, \mathbf{B}_2, \ldots, \mathbf{B}_K$, and $\mathbf{B}_{K+1}$ are the square infinite-size matrices, where the elements of each are described, respectively, below:

$$a_0(i,j) = \begin{cases} q_0\mu, & i=j, & j=1,2,3,\ldots, \\ (p_f + p_s q_{so})\mu, & i=j+1, & j=1,2,3,\ldots, \\ p_s p_{sf}\mu, & i=j+2, & j=1,2,3,\ldots, \\ 0, & & \text{otherwise.} \end{cases} \tag{7.155}$$

$$a_1(i,j) = \begin{cases} 1+(\lambda+\mu), & i=j=1, \\ 1+(\lambda+\mu+v), & i=j=2, \\ 1+(\lambda+\mu+2v), & i=j=3, \\ \vdots & \\ 1+[\lambda+\mu+(K-1)v], & i=j=K, \\ 1+[\lambda+\mu+Kv], & i=j=K+1,K+2,K+3,\ldots, \\ 0, & \text{otherwise.} \end{cases} \tag{7.156}$$

$$b_1(i,j) = \begin{cases} 1+\lambda, & i=j=1, \\ 1+\lambda+v, & i=j=2, \\ 1+(\lambda+2v), & i=j=3, \\ \vdots & \\ 1+[\lambda+(K-1)v], & i=j=K, \\ 1+\lambda+Kv, & i=j=K+1,K+2,K+3,\ldots, \\ 0, & \text{otherwise.} \end{cases} \tag{7.157}$$

$$b_2(i,j) = \begin{cases} -\lambda, & i=j, i=1,2,\ldots, \\ -v, & i=1, \ j=2, \\ 0, & \text{otherwise.} \end{cases} \tag{7.158}$$

$$b_3(i,j) = \begin{cases} -2v, & i=1, j=3, \\ 0, & \text{otherwise.} \end{cases} \tag{7.159}$$

$$b_K(i,j) = \begin{cases} -(K-1)v, & i=1, j=K, \\ 0, & \text{otherwise.} \end{cases} \tag{7.160}$$

$$b_{K+1}(i,j) = \begin{cases} -Kv, & i=j-K, j=K+1,K+2,K+3,\ldots, \\ 0, & \text{otherwise.} \end{cases} \tag{7.161}$$

Thus, we may summarize the system (7.134) in a matrix equation form below:

$$(\mathbf{I} - \mathbf{Q})\mathbf{X} = 0, \quad \sum_{i=0}^{\infty} X_i = 1, \tag{7.162}$$

where $\mathbf{Q}$ is given in (7.156) and the vector $\mathbf{X}$ is defined by (7.151).

At this point, we want to use the **truncation** method along with combining block matrices and transposes to obtain the truncated block matrix (7.162) below. We start with matrix $\mathbf{Q}$ defined by $\mathbf{Q}$ (7.154) that is an infinite block matrix with each block as an infinite matrix. We choose τ rows and τ columns from each block starting from the northwest corner of matrix Q. Hence, each block matrix is now finite of size $\tau \times \tau$. This ends the truncation part, and we have created a new block matrix (7.162) as below:

$$_\tau \mathbf{Q}^\mathbf{T} = \begin{bmatrix}
_\tau\mathbf{B}_1^T & _\tau\mathbf{B}_2^T & _\tau\mathbf{B}_3^T & \cdots & _\tau\mathbf{B}_K^T & _\tau\mathbf{B}_{K+1}^T & 0 & 0 & 0 & \cdots & \cdots \\
_\tau\mathbf{A}_0^T & _\tau\mathbf{A}_1^T & _\tau\mathbf{B}_2^T & _\tau\mathbf{B}_3^T & \cdots & _\tau\mathbf{B}_K^T & _\tau\mathbf{B}_{K+1}^T & 0 & 0 & \cdots & \ddots \\
0 & _\tau\mathbf{A}_0^T & _\tau\mathbf{A}_1^T & _\tau\mathbf{B}_2^T & _\tau\mathbf{B}_3^T & \cdots & _\tau\mathbf{B}_K^T & _\tau\mathbf{B}_{K+1}^T & 0 & \cdots & \ddots \\
0 & 0 & _\tau\mathbf{A}_0^T & _\tau\mathbf{A}_1^T & _\tau\mathbf{B}_2^T & _\tau\mathbf{B}_3^T & \cdots & _\tau\mathbf{B}_K^T & _\tau\mathbf{B}_{K+1}^T & 0 & \ddots \\
0 & 0 & 0 & _\tau\mathbf{A}_0^T & _\tau\mathbf{A}_1^T & _\tau\mathbf{B}_2^T & _\tau\mathbf{B}_3^T & \cdots & _\tau\mathbf{B}_K^T & _\tau\mathbf{B}_{K+1}^T & \ddots \\
\ddots & \ddots & \ddots & \ddots & \ddots & \ddots & \ddots & \ddots & \ddots & \ddots & \ddots \\
\vdots & \vdots & \vdots & \vdots & \vdots & \vdots & \vdots & \vdots & \ddots & \ddots & \ddots
\end{bmatrix},$$

$$\tag{7.163}$$

where $_\tau\mathbf{A}_0, {}_\tau\mathbf{A}_1, {}_\tau\mathbf{B}_1, {}_\tau\mathbf{B}_2, {}_\tau\mathbf{B}_3, \ldots, {}_\tau\mathbf{B}_K$, and $_\tau\mathbf{B}_{K+1}$ are $\tau \times \tau$ matrices located at northwest corners of $\mathbf{A}_0, \mathbf{A}_1, \mathbf{B}_1, \mathbf{B}_2, \mathbf{B}_3, \ldots, \mathbf{B}_K$, and $\mathbf{B}_{K+1}$, respectively.

Truncation alone does not do complete what the trick. So, we combine blocks as appropriate to make the matrix $\mathbf{Q}$ a tridiagonal block matrix. This is a novel way of creating a tridiagonal matrix (7.164), with elements as in (7.165) through (7.168) below, after truncation:

$$_\tau\mathbf{Q} = \begin{bmatrix}
\mathbf{B}_{1c} & \mathbf{A}_{0c} & 0 & 0 & 0 & 0 & \cdots \\
\mathbf{B}_{2c} & \mathbf{A}_{1c} & \mathbf{A}_{0c} & 0 & 0 & 0 & \cdots \\
0 & \mathbf{B}_{2c} & \mathbf{A}_{1c} & \mathbf{A}_{0c} & 0 & 0 & \cdots \\
0 & 0 & \mathbf{B}_{2c} & \mathbf{A}_{1c} & \mathbf{A}_{0c} & 0 & \cdots \\
0 & 0 & 0 & \mathbf{B}_{2c} & \mathbf{A}_{1c} & \mathbf{A}_{0c} & \cdots \\
0 & 0 & \ddots & \ddots & \ddots & \ddots & \ddots \\
\vdots & \vdots & \vdots & \vdots & \vdots & \vdots & \cdots
\end{bmatrix}, \tag{7.164}$$

where $\mathbf{B}_{1c}, \mathbf{B}_{2c}, \mathbf{A}_{0c}$, and $\mathbf{A}_{1c}$ are $K_T \times K_T$ matrices as follows:

$$\mathbf{B}_{1c} = \begin{bmatrix} {}_\tau\mathbf{B}_1 & {}_\tau\mathbf{A}_0 & 0 & 0 & 0 & 0 \\ {}_\tau\mathbf{B}_2 & {}_\tau\mathbf{A}_1 & {}_\tau\mathbf{A}_0 & 0 & 0 & 0 \\ {}_\tau\mathbf{B}_3 & {}_\tau\mathbf{B}_2 & {}_\tau\mathbf{A}_1 & {}_\tau\mathbf{A}_0 & 0 & 0 \\ \cdots & \cdots & \cdots & \cdots & \cdots & \vdots \\ {}_\tau\mathbf{B}_{K-1} & {}_\tau\mathbf{B}_{K-2} & {}_\tau\mathbf{B}_{K-3} & \cdots & {}_\tau\mathbf{A}_1 & {}_\tau\mathbf{A}_0 \\ {}_\tau\mathbf{B}_K & {}_\tau\mathbf{B}_{K-1} & {}_\tau\mathbf{B}_{K-2} & \cdots & {}_\tau\mathbf{A}_2 & {}_\tau\mathbf{A}_1 \end{bmatrix}, \quad (7.165)$$

$$\mathbf{B}_{2c} = \begin{bmatrix} {}_\tau\mathbf{B}_{K+1} & {}_\tau\mathbf{B}_K & {}_\tau\mathbf{B}_{K-1} & \cdots & {}_\tau\mathbf{B}_3 & {}_\tau\mathbf{B}_2 \\ 0 & {}_\tau\mathbf{B}_{K+1} & {}_\tau\mathbf{B}_K & \cdots & {}_\tau\mathbf{B}_4 & {}_\tau\mathbf{B}_3 \\ 0 & 0 & {}_\tau\mathbf{B}_{K+1} & \cdots & {}_\tau\mathbf{B}_5 & {}_\tau\mathbf{B}_4 \\ \vdots & \vdots & \vdots & \ddots & \ddots & \ddots \\ \vdots & \vdots & \vdots & \ddots & {}_\tau\mathbf{B}_{K+1} & {}_\tau\mathbf{B}_K \\ 0 & 0 & 0 & 0 & 0 & {}_\tau\mathbf{B}_{K+1} \end{bmatrix}, \quad (7.166)$$

$$\mathbf{A}_{0c} = \begin{bmatrix} 0 & 0 & 0 & \cdots & 0 & 0 \\ 0 & 0 & 0 & \cdots & 0 & 0 \\ 0 & 0 & 0 & \cdots & 0 & 0 \\ \vdots & \vdots & \vdots & \vdots & 0 & \vdots \\ 0 & 0 & \vdots & \ddots & 0 & 0 \\ {}_\tau\mathbf{A}_0 & 0 & 0 & \cdots & 0 & 0 \end{bmatrix}, \quad (7.167)$$

$$\mathbf{A}_{1c} = \begin{bmatrix} {}_\tau\mathbf{A}_1 & {}_\tau\mathbf{A}_0 & 0 & 0 & 0 & 0 \\ {}_\tau\mathbf{B}_2 & {}_\tau\mathbf{A}_1 & {}_\tau\mathbf{A}_0 & 0 & 0 & 0 \\ {}_\tau\mathbf{B}_3 & {}_\tau\mathbf{B}_2 & {}_\tau\mathbf{A}_1 & {}_\tau\mathbf{A}_0 & 0 & 0 \\ \cdots & \cdots & \cdots & \cdots & \cdots & \vdots \\ {}_\tau\mathbf{B}_{K-1} & {}_\tau\mathbf{B}_{K-2} & {}_\tau\mathbf{B}_{K-3} & \cdots & {}_\tau\mathbf{A}_1 & {}_\tau\mathbf{A}_0 \\ {}_\tau\mathbf{B}_K & {}_\tau\mathbf{B}_{K-1} & {}_\tau\mathbf{B}_{K-2} & \cdots & {}_\tau\mathbf{B}_2 & {}_\tau\mathbf{A}_1 \end{bmatrix}. \quad (7.168)$$

The transpose of (7.164) is:

$$\mathbf{{}_\tau Q}^T = \begin{bmatrix} \mathbf{B}_{1c}^T & \mathbf{B}_{2c}^T & 0 & 0 & 0 & 0 & \cdots \\ \mathbf{A}_{0c}^T & \mathbf{A}_{1c}^T & \mathbf{B}_{2c}^T & 0 & 0 & 0 & \cdots \\ 0 & \mathbf{A}_{0c}^T & \mathbf{A}_{1c}^T & \mathbf{B}_{2c}^T & 0 & 0 & \cdots \\ 0 & 0 & \ddots & \mathbf{A}_{1c}^T & \mathbf{B}_{2c}^T & 0 & \cdots \\ 0 & 0 & \ddots & \ddots & \ddots & \ddots & \cdots \\ 0 & 0 & \ddots & \ddots & \ddots & \ddots & \cdots \\ \vdots & \vdots & \vdots & \vdots & \vdots & \vdots & \cdots \end{bmatrix}. \quad (7.169)$$

As it can be seen, the sum of rows of $_\tau Q^T$ is not 1. To make it so, we use the first column augmentation (see Bean and Latouche (2010)). Thus, we would modify the individual blocks of $_\tau Q^T$, denoting it by $_\tau \tilde{Q}$, as follows: Let

$$\tilde{A}_0 = A_{0e^1}^T \tilde{A}_1 = A_{1c}^T, \tilde{B}_1 = B_{1c}^T + \left(I - B_{1c}^T - B_{2c}^T\right) \cdot 1 \cdot e_1^T,$$

$$\tilde{B}_2 = B_{2c}^T, D = \left(I - A_{0c}^T - A_{1c}^T - B_{2c}^T\right) \cdot 1 \cdot e_1^T, \tilde{B}_0 = A_{0c}^T + \tilde{D},$$

(7.170)

where e_1 is a column vector with the first element 1 and all others as 0, and 1 is a column vector with all elements 1 each. Therefore,

$$_\tau \tilde{Q} = \begin{bmatrix}
\tilde{B}_1 & \tilde{B}_2 & 0 & 0 & 0 & 0 & 0 \\
\tilde{B}_0 & \tilde{A}_1 & \tilde{B}_2 & 0 & 0 & 0 & 0 \\
\tilde{D} & \tilde{A}_0 & \tilde{A}_1 & \tilde{B}_2 & 0 & 0 & 0 \\
\tilde{D} & 0 & \tilde{A}_0 & \tilde{A}_1 & \tilde{B}_2 & 0 & 0 \\
\tilde{D} & 0 & 0 & \tilde{A}_0 & \tilde{A}_1 & \tilde{B}_2 & \vdots \\
\tilde{D} & 0 & \ddots & \ddots & \ddots & \ddots & \ddots \\
\vdots & \vdots & \vdots & \vdots & \vdots & \vdots & \vdots
\end{bmatrix}.$$

(7.171)

As τ approaches infinity, the stationary distribution $_\tau X$ of $_\tau \tilde{Q}$ converges to that of Q. Having constructed the truncated stochastic matrix $_\tau \tilde{Q}$ (in the sense that the sum of the rows equals 1), the next move is to determine the steps to solve the system (7.134).

At this point, we are to find a matrix G from the minimal nonnegative solution of the nonlinear matrix equation $G = \sum_{i=0}^{\infty} G^i A_i$ (see Neuts (1989) and Ramaswami (1988b)). Then, using the matrix G along with Ramaswami formula (see Ramaswami (1988a)), recursively compute the components of the vector $X = \langle X_0, X_1, X_2, ... \rangle^T$. To do that, let R_j be an approximation for G and apply the following algorithm:

Step 1. Write matrices $\tilde{A}_0, \tilde{A}_1, \tilde{B}_1$, and $\tilde{B}_2$, as defined above in (7.169).
Step 2. Let $C = (I - \tilde{A}_1)^{-1}$, where I is an $K_T \times K_T$ identity matrix.
Step 3. Choose $R_0 = 0$, where 0 is a $K_T \times K_T$ zero matrix. Assume that $R_0^0 = 1$. Then, find R_{m+1} for each $m = 0, 1, 2, ...,$, under the condition of Step 4 below, as follows:

$$R_{m+1} = \left(\tilde{B}_2 + R_m^2 \cdot \tilde{A}_0\right) \cdot C \quad m = 0, 1, 2, ...,$$

(7.172)

Step 4.
    Step 4.1. Choose a desired given $\in$.
    Step 4.2. For successive values of m, as 0, 1, 2, ..., find the absolute value of difference between the two matrices found for m and $m+1$.

Step 4.3. Find the largest element of this difference matrix (with i rows and j columns). Check to see if this element is less than $\in$, that is,

$$\max_{i,j} \left| \left(\mathbf{R}_{m+1} - \mathbf{R}_m \right) \right| < \in, \; m = 0, 1, 2, \ldots, \tag{7.173}$$

If (7.173) is true, choose the last $\mathbf{R}_{m+1}$ and write $\mathbf{G} = \mathbf{R}_{m+1}$. Move on to the next step if all eigenvalues of $\mathbf{G}$ are within the unit circle.

Step 5. Compute the following sum of the matrices, denoted by $\bar{\mathbf{B}}_1$:

$$\bar{\mathbf{B}}_1 = \tilde{\mathbf{B}}_1 + \mathbf{G} \cdot \tilde{\mathbf{B}}_0 + \mathbf{G}^2 \cdot (\mathbf{I} - \mathbf{G})^{-1} \cdot \tilde{\mathbf{D}}$$

Step 6. Solve equation $\mathbf{M}\bar{\mathbf{X}}_0 = 0$, where $\mathbf{M} = \bar{\mathbf{B}}_1 - \mathbf{I}$ and $\bar{\mathbf{X}}_0 \langle 1, \tilde{\mathbf{X}}_0 \rangle$, with $\tilde{\mathbf{X}}_0 = \langle \mathbf{X}_1, \mathbf{X}_2, \ldots, \mathbf{X}_n, \ldots \rangle$ as follows:

Step 6.1. Delete the first row of $\mathbf{M}$ and call the remaining matrix as $\mathbf{M}_1$.

Step 6.2. Delete the first column of $\mathbf{M}_1$ and call the remaining matrix as $\mathbf{M}_2$.

Step 6.3. Choose the first column of $\mathbf{M}_2$, multiply it by (-1), and call this matrix as $\mathbf{M}_3$.

Step 6.4. Write $\tilde{\mathbf{X}}_0 = \mathbf{M}_2^{-1} \cdot \mathbf{M}_3$.

Step 7. Write $\tilde{\mathbf{X}}_0 = \mathbf{M}_2^{-1} \cdot \mathbf{M}_3$. Find $\mathbf{X}_n$, $n = 1, 2, 3, \ldots$, as a matrix product as follows:

$$\mathbf{X}_1 = \mathbf{X}_0 \cdot \mathbf{G}, \ldots, \mathbf{X}_n = \mathbf{X}_{n-1} \cdot \mathbf{G}, \ldots,$$

Step 8. Compute the sum, denoted by S, as $S = \mathbf{X}_0 \cdot (\mathbf{I} - \mathbf{G})^{-1} \cdot e$, where e is a $K_T \times 1$ column vector with all elements as 1, or $S = \sum_{i=0}^{\infty} \bar{\mathbf{X}}_i^T$.

Step 9. From Step 7 and Step 8, write

$$\Psi_i = \frac{\mathbf{X}_i}{S}, \; i = 0, 1, 2, \ldots,$$

which is the vector of the unknown joint probabilities as follows:

$$\Psi_i = \left\langle \Psi_{iK}^*, \Psi_{iK+1}^*, \Psi_{iK+2}^*, \ldots, \Psi_{iK+K-1}^* \right\rangle^T, \; i = 0, 1, 2, \ldots,$$

where T denotes the transpose of the vector, and

$$\Psi_{iK}^* = \left\langle \psi_{iK,0}, \psi_{iK,1}, \psi_{iK,2}, \ldots, \psi_{iK,\tau} \right\rangle^T,$$

$$\Psi_{iK+1}^* = \left\langle \psi_{iK+1,0}, \psi_{iK+1,1}, \psi_{iK+1,2}, \ldots, \psi_{iK+1,\tau} \right\rangle^T,$$

$$\Psi_{iK+2}^* = \left\langle \psi_{iK+2,0}, \psi_{iK+2,1}, \psi_{iK+2,2}, \ldots, \psi_{iK+2,\tau} \right\rangle^T,$$

$$\vdots \quad \vdots \qquad\qquad \vdots$$

$$\Psi_{iK+K-i-1}^* = \left\langle \psi_{iK+K-1,0}, \psi_{iK+K-1,1}, \psi_{iK+K-1,2}, \ldots, \psi_{iK+K-1,\tau} \right\rangle^T.$$

We have now algorithmically obtained the joint distribution of the queue length of the system as in Step 9.

Example 7.27

This is a numerical example for the model we just discussed. To show how the steps work, we offer a numerical example with the following data: $\lambda = 5$, $\mu = 25$ and $\nu = 7$, $K = 3$, that is, $k = 1, 2,$ or 3. For distribution of the batch size, represented by a random variable X, we choose a discrete uniform distribution, that is,

$$P(X = 1) = P(X = 2) = P(X = 3) = \frac{1}{3}.$$

Hence, $\bar{\alpha} = (1 + 2 + 3)/3 = 2$. In the computation, we denote the maximum value of x by X, and let it equal to 3. We also choose the following probabilities:

$$p_f = 0.4, p_s = 0.2, q_0 = 1 - (p_f + p_s), p_{sf} = 0.3, \text{ and } q_{so} = 1 - p_{sf}.$$

With the chosen data, we will have $\rho_1 = 0.7600, \rho_2 = 0.8957,$ and $\rho_{sys} = 0.4872$. Hence, the stability conditions are met.

For Step 3 and Step 4, since m has to go to infinity, we choose 50 for the infinity as computation shows it is sufficient. For the infinite sum in Step 8, we take the infinity as 10τ. It should be noted that, in the programming of the numerical example, a computation of the infinite matrices is done through approximating finite square $\tau \times \tau$ matrices. However, we will use the **first-column augmentation method** so that the convergence of the method for the infinite case will be checked through Step 4. We recall that to obtain the distribution of the queue length of the system to be very close to the real values, we have to choose τ large enough so that the error is less than a preassigned value, say $\epsilon = 10^{-15}$. Hence, we have taken $\tau = 700$, and the error analysis shows that we have met this condition (see Tables 7.6 and 7.7).

For tabulation presentation of probabilities, we choose only 18 rows and 12 columns of the distribution matrix to fit in a page. Results are recorded in Table 7.8. However, for a three-dimensional graphic presentation of the probabilities, as in Figure 7.8, we have chosen 60 rows of the probability distribution matrix.

TABLE 7.6

Maximum Differences between Two Consecutive Probabilities in the Distribution of the Queue Size

$\tau = 60, 100, 300, 500, 600, 700$	Max Difference = Error
pdf with $\tau = 100$ - pdf with $\tau = 60$	2.775557561562891e-17
pdf with $\tau = 300$ - pdf with $\tau = 100$	1.249000902703301e-15
pdf with $\tau = 500$ - pdf with $\tau = 300$	1.110223024625157e-16
pdf with $\tau = 600$ - pdf with $\tau = 500$	3.330669073875470e-16
pdf with $\tau = 700$ - pdf with $\tau = 600$	1.665334536937735e-16

TABLE 7.7
Distribution of the Number of Tasks in the System (Both Stations), the First 18 Rows and 12 Columns of the Probability Distribution Matrix

	$\psi_{i,0}$ through $\psi_{i,11}$, $i = 0,1,2,\dots,17$											
ψ_{0j} through ψ_{17j} $j = 0,1,\dots,11$	0	1	2	3	4	5	6	7	8	9	10	11
0	0.1772	0.1468	0.0595	0.0184	0.0064	0.0022	0.0008	0.0003	0.0001	0.0000	0.0000	0.0000
1	0.0886	0.0565	0.0235	0.0078	0.0027	0.0009	0.0003	0.0001	0.0000	0.0000	0.0000	0.0000
2	0.0745	0.0351	0.0149	0.0050	0.0017	0.0006	0.0002	0.0001	0.0000	0.0000	0.0000	0.0000
3	0.0563	0.0258	0.0105	0.0035	0.0012	0.0004	0.0001	0.0000	0.0000	0.0000	0.0000	0.0000
4	0.0355	0.0164	0.0067	0.0022	0.0008	0.0003	0.0001	0.0000	0.0000	0.0000	0.0000	0.0000
5	0.0232	0.0106	0.0043	0.0014	0.0005	0.0002	0.0001	0.0000	0.0000	0.0000	0.0000	0.0000
6	0.0152	0.0069	0.0028	0.0009	0.0003	0.0001	0.0000	0.0000	0.0000	0.0000	0.0000	0.0000
7	0.0099	0.0045	0.0018	0.0006	0.0002	0.0001	0.0000	0.0000	0.0000	0.0000	0.0000	0.0000
8	0.0064	0.0029	0.0012	0.0004	0.0001	0.0000	0.0000	0.0000	0.0000	0.0000	0.0000	0.0000
9	0.0042	0.0019	0.0008	0.0003	0.0001	0.0000	0.0000	0.0000	0.0000	0.0000	0.0000	0.0000
10	0.0027	0.0012	0.0005	0.0002	0.0001	0.0000	0.0000	0.0000	0.0000	0.0000	0.0000	0.0000
11	0.0018	0.0008	0.0003	0.0001	0.0000	0.0000	0.0000	0.0000	0.0000	0.0000	0.0000	0.0000
12	0.0011	0.0005	0.0002	0.0001	0.0000	0.0000	0.0000	0.0000	0.0000	0.0000	0.0000	0.0000
13	0.0007	0.0003	0.0001	0.0000	0.0000	0.0000	0.0000	0.0000	0.0000	0.0000	0.0000	0.0000
14	0.0005	0.0002	0.0001	0.0000	0.0000	0.0000	0.0000	0.0000	0.0000	0.0000	0.0000	0.0000
15	0.0003	0.0001	0.0001	0.0000	0.0000	0.0000	0.0000	0.0000	0.0000	0.0000	0.0000	0.0000
16	0.0001	0.0001	0.0000	0.0000	0.0000	0.0000	0.0000	0.0000	0.0000	0.0000	0.0000	0.0000
17	0.0001	0.0000	0.0000	0.0000	0.0000	0.0000	0.0000	0.0000	0.0000	0.0000	0.0000	0.0000

TABLE 7.8

Differences between Two Consecutive pdf of the Queue with $\tau = 700$ and $\tau = 600$. Numbers Shown Are to Be Multiplied by 10^{-18}.

1	2	3	4	5	6	7	8	9	10
0.16653345369773	0.16653345369773	0.05551115123126	0.02081668171172	0.0006071532165919	0.001301042605983	0.0006505021303491	0.0021684043434497	0.000081315162936	0.000027105054312
0.11102302462516	0.07632783294980	0.0277755575615629	0.0104083408585861	0.00216840434971	0.001192622389734	0.0002710505431 21	0.0016263630325873	0.0000338811317890	0.00001694060658945
0.11102302462516	0.04163363423443	0.01387778780781 4	0.00086736173799 88	0.0017347234759 77	0.000758941520740	0.0002439 45488809	0.0000813 15162936	0.0000027105054312	0.0000067762635 78
0.07632783294980	0.0346944695195 36	0.008673617379884	0.001734723475977	0.001084202172486	0.000542101086243	0.000135525271561	0.0000677762635780	0.0000203287907 34	0.0000059292 30631
0.04163363423443	0.01387778780814	0.006938893903907	0.002168404344971	0.000758941520740	0.000216840434497	0.0000813 15162936	0.0000271050 54312	0.0000101643 95367	0.0000042351 64736
0.017347234759768	0.01040834085861	0.003469446951954	0.001084202172486	0.000325250651746	0.000135525271561	0.0000542102 10108624	0.0000169406 0658945	0.0000059292 30631	0.0000025410 98842
0.01040834085861	0.0060715321659 19	0.002602085213965	0.000867361737988	0.000108420217249	0.000108420217249	0.0000338813 17890	0.0000101643 95367	0.0000042351 64736	0.0000010587 91184
0.008673617379884	0.0034694469 51954	0.001734723475977	0.000542101086243	0.000108420217249	0.0000677762635780	0.0000237169 22523	0.0000084703 29473	0.0000025410 98842	0.0000010587 91184

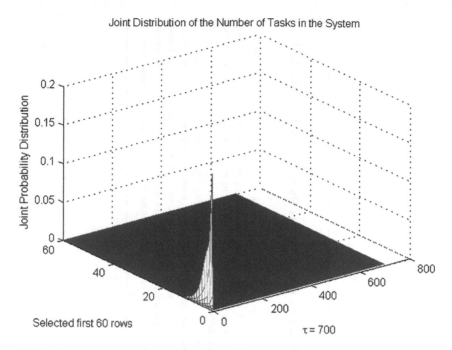

FIGURE 7.8 Graph of the joint probability distribution of the number of tasks in the system.

With $\tau = 700$, for the queue length of the system, mean, second moment, variance, and standard deviation are, respectively, 4.5691, 27.0521, 6.1752, and 2.485.

To validate that the probabilities found form a pdf, we compute and graph the cumulative probabilities as in Figure 7.9. To present this two-dimensional graph of cdf, in order to make sure we are adding all elements of the matrix in some fashion, we choose the following method: Take the first element of the matrix, that is, $\psi_{0,0}$, then add all the rest of elements of the first square box, then add the elements to make the next square box, and so on to the end of the matrix. For instance, take $\psi_{0,0}$ and call it S_1, then add $\psi_{0,0}, \psi_{0,1}, \psi_{1,0}$, and $\psi_{1,1}$ to S_1 and call it S_2. Then, add $\psi_{0,2}, \psi_{2,0}$, and $\psi_{2,2}$ to S_2 and call it S_3. Continue this method to cover all elements of the matrix and graph the $S_1, S_2,...$, to obtain the graph in Figure 7.9.

However, for the three-dimensional cumulative probabilities, we take each row and add its elements one by one. The result is shown in Figure 7.10.

We take the matrix of block matrices of size ten times as much as the size of a block matrix. The difference between elements of two consecutive matrices of probabilities (that we call it error) starts as a large number. However, as the truncation line increases, the number reduces to and soon reaches zero. Figure 7.11 shows a three-dimensional graph of the differences for a set of τ's at 60, 100, 300, 500, 600, and 700. Absolute values of differences of probability matrices for $\tau = 700$ and $\tau = 600$ are listed in Table 7.7. Absolute values of maximum differences of all elements between each two of these values, in the order written, are listed in Table 7.8. Fluctuation of errors for large values of τ such as 600 and 700 is negligible since it occurs at the 16th decimal places. Perhaps, running the

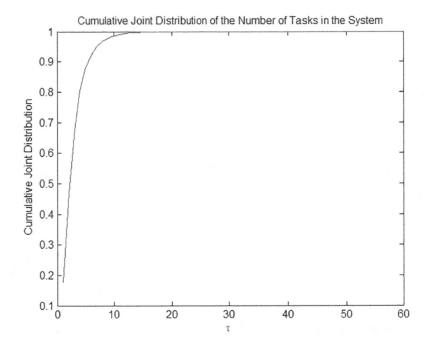

FIGURE 7.9 Graph of the cumulative joint probability distribution of the number of tasks in the system.

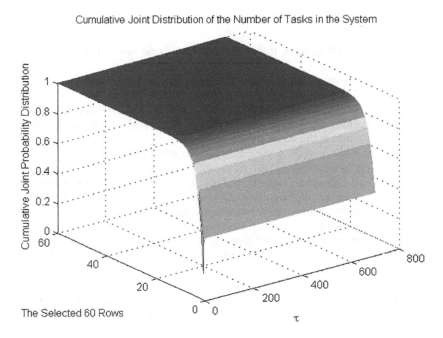

FIGURE 7.10 Three-dimensional graph of the cumulative joint probability distribution of the number of tasks in the system.

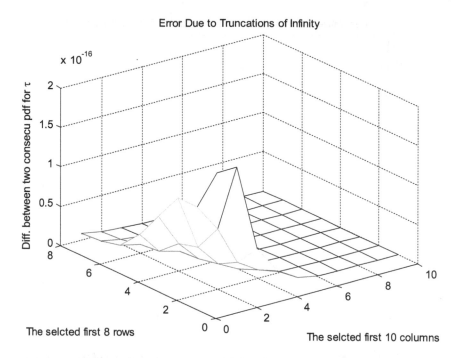

FIGURE 7.11 Three-dimensional graph of differences between two consecutive probabilities of the queue length with $\tau = 700$ and $\tau = 600$.

program with much larger τ, that is, several folds large, we will see a complete stability of the values.

The model we have described is a delayed queueing system with two types of arrival: external single tasks and internal batch tasks, which may be interpreted as a two-station tandem queue with the possibility of immediate splitting after exiting the service station, single-server exponential service, and an exponential batch moving distribution at the delay station.

To minimize the length of the paper, a discussion of the busy period of the service station is referred. It will soon appear in the *International Journal of Mathematics in Operations Research*. Thus, in this paper, we have considered only the queue length and its associated joint distribution. A study of the busy period distribution of the system remains an open problem. With this in mind, here we will complete, as before, the consideration of the model illustrated in Figure 7.6 in the first paper on this topic by Haghighi et al. (2011b).

In offering the stepwise joint distribution, we have also used here two ideas:

1. Duality of two systems $G/M/1$ and $M/G/1$.
2. The independence of two sources of arrivals: external and internal. The external arrivals are generated by the departures from the service station.

With these two ideas, we are able to reduce our model to a $MAP/M/1$ and use some of the existing properties of $MAP/M/1$.

Note 7.45

In the splitting mode, we assumed that the original task returns to the service station with probability 1. However, considering it with probability less than 1 remains to be addressed.

In summary, in this model, we were able to present an algorithmic way of finding the joint distribution of the number of tasks in the system. This includes dealing with infinite block matrices using truncation and augmentation methods and approximating the solution within an arbitrarily small interval of error. Due to the approximations involved, we are restricted to choosing the values of some of the parameters. The numerical example demonstrates how all this works through approximation of solution of the model illustrated in Figure 7.6. We have performed an error analysis to show how the error approaches zero as the matrix sizes approach infinity. We have also offered, to our knowledge, for the first time, an explicit joint distribution of a tandem queue.

7.6.4 MULTI-SERVER QUEUEING MODEL, *M/M/c*

Now, let us consider the same stationary infinite-capacity *M/M/1* queue we discussed in the previous section with the difference that now a number of servers are $c > 1$ identical servers in parallel. Thus, in formulation, the only difference would be in the value of ρ, which in this case would be

$$\rho = \frac{\lambda}{c\mu}. \tag{7.174}$$

The following relation is well known. For instance, see Haghighi and Mishev (2014). We leave it as exercise to prove it for the system *M/M/c*:

$$L_q - \frac{P_0 \left(\dfrac{\lambda}{\mu} \right)^c \rho}{c!(1-\rho)^2}, \tag{7.175}$$

where P_0 denotes the probability that there are no tasks in the system. Accordingly, we will have the mean waiting time in the waiting line as:

$$W_q = \frac{L_q}{\lambda}. \tag{7.176}$$

7.6.4.1 A Stationary Multi-Server Queueing System with Balking and Reneging

We now extend the multi-server queueing system, as discussed above, with two other features regarding arrivals. For a queueing system, it is possible that due to the long lengthy waiting line, a task looks at it and decides not to join. This feature is referred to as the **balking**. On the other hand, it may join the queue anyway, but after staying for a while, it decides to leave the system. This feature is referred to as the **reneging** or **abandonment**. These features appeared in the literature about seven decades ago and have evolved quite a bit. Among contributors are Montazer-Haghighi (1976), Montazer-Haghighi et al. (1986), and Haghighi and Mishev (2006).

Including these two features, here is how the system looks like and how it is analyzed. The system is described as follows:

   i. Tasks arrive from an infinite source to a system with infinite-capacity buffer according to a Poisson distribution of parameter λ.
   ii. There are c servers set in parallel.
   iii. An arriving task may look at the waiting line when all servers are busy and balk with a constant probability β or may decide to join the queue anyway with probability α, $\alpha + \beta = 1$. In other words, the attending rate of tasks is $\alpha\lambda$ and the instantaneous balking rate is $\beta\lambda$.
   iv. If a task balks or reneges and later returns, it will be considered as a new arrival.
   v. After joining the system, when all servers are busy, a task will wait in the buffer for a while and will leave the system before receiving service.
   vi. It is assumed that the length of stay before leaving the system (reneging) is a random variable with an exponential distribution with parameter s.
   vii. It is also assumed that service provided by each server is based on an exponential distribution with parameter μ.
   viii. Finally, services are provided to tasks based on the FCFS.

This system is denoted by an *M/M/c* **with balking and reneging**. Analysis of this system for $c = 1, 2$ is left as an exercise.

Hence, we consider the system with $c \geq 3$. The state when there are k tasks in the system, that is, in the buffer and in the service stations, is referred to as S_k. Based on the assumptions for arrival rate, service rate, balking, and reneging, we have:

$$\text{Average arrival rate} = \begin{cases} \lambda & 0 \leq k \leq c-1, \\ \alpha\lambda, & k \geq c, \end{cases} \tag{7.177}$$

and

$$\text{Average service rate} = \begin{cases} c\mu, & 0 \leq k \leq c-1, \\ c\mu + (k-c)s, & k \geq c. \end{cases} \tag{7.178}$$

The intensity factor in this case is defined as:

$$\rho = \begin{cases} \dfrac{\lambda}{c\mu}, & 0 \leq k \leq c-1, \\ \dfrac{\alpha\lambda}{c\mu + (k-c)s}, & k \geq c. \end{cases} \tag{7.179}$$

Let the **transient-state probability** that there are k tasks in the system at time t be denoted by $P_k(t)$. Then, the **stationary probability** of having k in the system is denoted by P_k, defined by

$$P_k = \lim_{t \to \infty} P_k(t). \tag{7.180}$$

Considering different cases to be in state S_k, the system of differential–difference equations can be written as:

$$\begin{cases} P_0'(t) = -\lambda P_0(t) + \mu P_1(t), \\ P_k'(t) = -(\lambda + k\mu)P_k(t) + \lambda P_{k-1}(t) + (k+1)\mu P_{k+1}(t), & 1 \le k \le c-1, \\ P_c'(t) = -(\alpha\lambda + c\mu)P_c(t) + \lambda P_{c-1}(t) + (c\mu + s)P_{c+1}(t), \\ P_k'(t) = -(\alpha\lambda + c\mu) + (k-c)s)P_k(t) + \alpha\lambda P_{k-1}(t) + [c\mu + (k+1-c)s]P_{k+1}(t), & k > c. \end{cases} \tag{7.181}$$

The time-independent case of the system (7.123), that is, the stationary case, becomes:

$$\begin{cases} 0 = -\lambda P_0 + \mu P_1, \\ 0 = -(\lambda + k\mu)P_k + \lambda P_{k-1}(t) + (k+1)\mu P_{k+1}(t), & 1 \le k \le c-1, \\ 0 = -(\alpha\lambda + c\mu)P_c + \lambda P_{c-1} + (c\mu + s)P_{c+1} \\ 0 = -(\alpha\lambda + c\mu + (k-c)s)P_k + \alpha\lambda P_{k-1} + [c\mu + (k+1-c)s]P_{k+1}, & k > c. \end{cases} \tag{7.182}$$

The system (7.182), indeed, is an extended version of $M/M/1$ case as follows:

$$\begin{cases} \lambda P_0 + \mu P_1, \\ (\lambda + k\mu)P_k = \lambda P_{k-1} + (k+1)\mu P_{k+1}(t), & 1 \le k \le c, \\ (\alpha\lambda + c\mu)p_c = \lambda P_{c-1} + (c\mu + s)P_{c+1}, \\ [\alpha\lambda + c\mu + (k-c)s]P_k + \alpha\lambda P_{k-1} + [c\mu + (k+1-c)s]P_{k+1}, & k > c, \end{cases} \tag{7.183}$$

with the **normalizing equation,**

$$\sum_{k=0}^{\infty} P_k = 1. \tag{7.184}$$

Now, for the solution of the system (7.184), the first two equations yield:

$$P_k = \frac{1}{k!}\left(\frac{\lambda}{\mu}\right)^k P_0, \quad k \le c, \tag{7.185}$$

from which we can compute P_{c-1} and P_c. Then, substituting these values in the third equation of the system (7.183) yields:

$$P_{c+1} = \frac{1}{c!}\left(\frac{\lambda}{\mu}\right)^c \left(\frac{\alpha\lambda}{c\mu+s}\right)P_0. \tag{7.186}$$

Using (7.184) and putting $k = c + 1$ in the last equation of (7.183), we obtain:

$$P_{c+2} = \frac{(\alpha\lambda)^2}{(c\mu+s)(c\mu+2s)}P_c = \frac{1}{c!}\left(\frac{\lambda}{\mu}\right)^c \frac{(\alpha\lambda)^2}{(c\mu+2s)}P_0. \tag{7.187}$$

Finally, we will have a general case for $k > c$ as

$$P_k = \frac{1}{c!}\left(\frac{\lambda}{\mu}\right)^c \frac{(\alpha\lambda)^{k-c}}{\prod_{j=1}^{k-1}(c\mu+js)}P_0, \tag{7.188}$$

with P_0 calculated from (7.186) as:

$$P_0 = \left[\sum_{j=0}^{c}\frac{1}{j!}\left(\frac{\lambda}{\mu}\right)^j + \frac{1}{c!}\left(\frac{\lambda}{\mu}\right)^c \sum_{j=1}^{\infty}\frac{(\alpha\lambda)^j}{\prod_{r=1}^{j}(c\mu+rs)}\right]^{-1}. \tag{7.189}$$

As a validity check if $c = 1$, $\alpha = 1$ (no balking) and $s = 0$ (no reneging), from (7.188) and (7.189), we obtain the distribution for the $M/M/1$. Also, if $\alpha = 1$ (no balking) and $s = 0$ (no reneging), we will have $M/M/c$ system.

7.6.4.2 Case $s = 0$ (No Reneging)

For the case when there is balking only, that is, $\dfrac{1}{s} \to \infty$ or $s \to 0$, we will have:

$$\rho = \frac{\alpha\lambda}{c\mu}, \tag{7.190}$$

and

$$P_k = \begin{cases} \dfrac{(c\rho)^k}{k!}P_0, & 0 \le k \le c, \\[3mm] \dfrac{1}{c!}\left(\dfrac{c}{\alpha}\right)^c (\alpha\rho)^k P_0, & k > c, \end{cases} \tag{7.191}$$

and

$$P_0 = \left[\sum_{j=0}^{c}\frac{(c\rho)^j}{j!} + \frac{\alpha c^c \rho^{c+1}}{c!(1-\alpha\rho)}\right]^{-1}. \tag{7.192}$$

We leave it as an exercise to show that the average number of arrivals attended, denoted by $\bar{A}$, is given by:

$$\bar{A} = \left[\left(\frac{\lambda}{\mu} \right)^c \frac{\alpha\lambda}{(c-1)!} \frac{(c+1)\mu - \alpha\lambda}{(c\mu - \alpha\lambda)^2} + \sum_{k=1}^{c} \frac{1}{(k-1)!} \left(\frac{\lambda}{\mu} \right)^k \right] P_0. \tag{7.193}$$

Back to the case of balking and reneging. An advantage of considering potential arrivals is to have a measure of the virtual load that the system should be prepared for. The balking and reneging features of the system may cause the system to lose some of the potential arrivals.

Definition 7.28

For a multi-server queueing system with balking and reneging, the balking and reneging tasks, coupled with those who attend and finally receive service, constitute the **potential arrivals**. Also, depending upon the situation of a queueing system, the period during which all servers remain busy is referred to a **busy period**.

Now, let $\bar{L}$ and $\bar{B}$ represent, respectively, the average number loss and the average number of potential arrivals during a unit service time within a busy period. When the number of tasks in the system is more than the number of servers in the system,

$$\text{The probability that all servers are busy} = \sum_{k=c}^{\infty} P_k. \tag{7.194}$$

We earlier assumed that when all servers are busy, the probability of losing a task due to balking is a constant, β. Hence,

$$\text{The probability of losing a task due to balking} = \beta \sum_{k=c}^{\infty} P_k. \tag{7.195}$$

Also, we earlier assumed that during a busy period, the average rate of arrivals due to balking is $\alpha\lambda$. Hence,

$$\text{The average loss due to balking during a busy period} = \beta\alpha\lambda \sum_{k=c}^{\infty} P_k. \tag{7.196}$$

Thus, since the average length of a unit service during a busy period is $\frac{1}{c\mu}$, the average number of potential arrivals during a unit service, $\bar{L}$, is

$$\bar{L} = \frac{1}{c\mu} \left(s + \beta\alpha\lambda \sum_{k=c}^{\infty} P_k \right). \tag{7.197}$$

Similar to what we found earlier, the average number of tasks attended during a unit service within a busy period, $\bar{B}$, is

$$\bar{B} = \frac{1}{c\mu}\left(\sum_{k=0}^{c}\lambda P_k + \sum_{k=c+1}^{\infty}\alpha\lambda P_k\right). \qquad (7.198)$$

If we denote the proportion of loss by l, since $\bar{B} = \bar{A} + \bar{L}$, we have:

$$l = \frac{\bar{L}}{\bar{B}}. \qquad (7.199)$$

7.6.4.3   Case $s = 0$ (No Reneging)

In case of balking alone, (7.198) and (7.199), respectively, become:

$$\bar{L} = \frac{\alpha\beta\lambda}{c\mu - a\lambda}\frac{1}{c!}\left(\frac{\lambda}{\mu}\right)^{c}P_0. \qquad (7.200)$$

$$\bar{B} = \frac{\lambda P_0}{c\mu}\left[\sum_{k=0}^{\infty}\frac{1}{k!}\left(\frac{\lambda}{\mu}\right)^{k} + \frac{\alpha^2}{c!}\left(\frac{\lambda}{\mu}\right)^{c+1}\frac{\mu}{c\mu - \alpha\lambda}\right]. \qquad (7.201)$$

Example 7.28

In this example, we analyze a stationary queueing model, which is an urgent care clinic with several beds and the same number of physicians to serve the arriving patients. The following are the assumptions for this model:

  i. The clinic is open to serve patients with no capacity limit; that is, patients arrive from a single infinite source.
  ii. If all beds are occupied at the arrival of a patient, the patient may wait in a waiting room with unlimited capacity.
  iii. Arriving of patients is a Poisson process. That is, arrivals of patients are iid random variables with Poisson distribution with parameter $\lambda = 6$ per hour.
  v. There are five beds available setting in parallel, each with one physician in charge.
  iv. An arriving patient, who finds all five beds occupied, may balk with probability $\beta = 0.3$ or may stay and join the other patients in the waiting room with probability $\alpha = 0.7$. Hence, the instantaneous balking rate is $\beta\lambda = 1.8$, or the arrival rate when all beds are occupied is $\alpha\lambda = 4.2$.
  vi. After entering the clinic, since the patient feels immediate need for care and she/he needs to be visited by a doctor, she/he will renege, that is, $s = 0$.
  vii. Patients are visited by doctors on a FIFO basis.
  viii. The visiting length of times by physicians are iid random variables having an exponential distribution with parameter $\mu = 3$.

ix. If a patient balks and decides to return later, she/he will be considered as a new patient, independent of his/her previous number of balking.
We are looking for:

a. The distribution of the number of patients in the clinic at any time.
b. The probability of the clinic without patients.
c. The average number of arrivals attended, that is, after the balking has been considered.

Answer

a. Let us suppose that the clinic starts operating at time $t = 0$ with no patient waiting. Let also assume that at $t = 0$, one or more of the beds are occupied. Denoted by P_k, the time-independent (stationary) probability is that there are k patients in the clinic. Thus, from (7.190), we have:

$$\rho = \frac{(0.7)(6)}{(5)(3)} = \frac{7}{25}.$$

Now, from (7.192), we have:

$$P_0 = \left[\sum_{j=0}^{5} \frac{\left((5)\left(\frac{7}{25}\right) \right)^j}{j!} + \frac{\left((0.7)(5^5)\left(\frac{7}{25}\right)^{5+1} \right)}{5!\left(1-(0.7)\left(\frac{7}{25}\right) \right)} \right]^{-1} = 0.2467. \qquad (7.202)$$

Thus,

$$P_n = \begin{cases} \dfrac{\left[(5)\left(\frac{7}{25}\right) \right]^k}{k!} P_0, & 0 \le k \le 5, \\[20pt] \dfrac{1}{5!}\left(\frac{5}{0.7} \right)^5 \left((0.7)\left(\frac{7}{25}\right) \right)^k P_0, & k > 5. \end{cases} \qquad (7.203)$$

To calculate these and other values, we use MATLAB programing. Hence, from (7.202), (7.203), and (7.195), we have:

$$P_1 = 0.3454, \ P_2 = 0.2418, \ P_3 = 0.2418, \ P_4 = 0.0395, \ P_k = 0.0111,$$

$$P_6 = 0.0022, \ P_7 = 0.0004, \ P_8 = 0.0001, \ P_9 = 0.0000, \ P_{10} = 0.0000,$$

with $\sum_{0}^{10} P_k = 1$, and

$$\bar{A} = 3.6176,$$

That is, considering the possibility of balking only, on an average, 3.6 patients enter the clinic per hour to receive doctors' visits.

7.7   BIRTH-AND-DEATH PROCESSES

Continuous-time **B-D processes** have shown to be important tools for queueing model, reliability, and inventory systems. B-D processes are good models to represent flows of radioactive, cosmic, and other particles. In the economical sciences, B-D processes are used for describing the development of number of enterprises in a particular area and manpower fluctuations.

Note 7.46

As a historical note, the name "B-D" comes from applications of biology, where the development in time of the number of individuals in populations of organisms is stochastically modeled. The mathematical discussion of stochastic population growth was pointed out by William Feller (1939). In that paper, among other things, Feller discussed the examples of B-D with constant rates of birth and death. Later, David Kendal (1948) gave the complete solution of the system of equations governing the generalized B-D in which the rates were functions of the time.

Note 7.47

Poisson process (sometimes referred to as a point process) is a classic example of a **pure birth**, that is, a B-D process in which death is not to be considered. This is where we start this section with.

Definition 7.29 Pure Birth Process

Assume that over time, a process counts particles in the fashion of moving upward from a state k to the immediate next higher state, $k+1$, or stay put, but never downward to a lower-valued state. That is, the process allows only arrivals or births or walking forward if there is to be a transition, but no departures, deaths, or moving backward. Such a process is referred to as a **pure birth process**.

What the definition states is that a pure birth process at any time t either remains in its current state or transits to the next higher state. Transitions other than $K \to k+1$ or $k \to k$ at time t (where $k \to k$ indicates that the process remains in state k at time t) are not allowed.

Example 7.29 Pure Birth Processes

The following are two examples of pure birth processes:

  i. The number of visitors (by time t) to the Houston Museum of Art since the museum opened.
 ii. The number of telephone calls to the department of mathematics at Prairie View A&M University on workdays, between 8:00 am and 5:00 pm daily during the fall registration period in 2019.

Definition 7.30 Pure Death Process

Suppose that over time, a process counts particles in the fashion of moving backward from a state k to the immediate next lower state, $k-1$, or stay put, that is, $k \to k$, but never forward to an upper-valued state. In other words, the process allows only departures (deaths) if there is to be a transition, but no arrivals (births). Such a process is referred to as a **pure death process**.

What this definition states is that in a pure death process, at any time t, transitions other than $k \to k$ or $k \to k-1$ at time t are not allowed.

Definition 7.31 Birth-and-Death Process (B-D)

The most general type of **B-D process** is a process that allows transitions of a state to another state of the form $k \to k+1, k \to k$, **or** $k \to k-1$ at any time t. It is assumed that the birth-and-death events in a B-D process are independent of each other.

This definition states that in a B-D process, a task travels forward, stay put, or backward among the integers in one-step transitions. In other words, no B-D process allows transitions of more than one step. That is, transitions such as $k \to k \pm i, i = 2,3,\dots$ are not allowed.

The concept of random walk that we discussed earlier is the most general concept of the B-D process. Random walk, however, is a **semi-Markov process** since the holding time in each state has an arbitrary probability distribution function that does not have the forgetfulness property. However, a B-D process may be called a **continuous-time random walk**. So, if the walker takes only forward steps such as $0 \to 1 \to 2 \to 3,\dots$, the chain would be a pure birth process.

Example 7.30 Birth-and-Death Processes

Here are three examples of B-D processes:

  i. Process of programs awaiting execution on a given computer system (at time t).
  ii. The number of customers in a certain grocery store (at time t).
  iii. The population of a particular insect in an area (at time t).

All B-D processes we will consider herein are assumed to have the following properties:

  i. The Markov property
  ii. Stationary transition probabilities.

Hence, because of these properties, a B-D process becomes a Markov process. The state space for a B-D process is the set of nonnegative integers.

Similar to the random walk and simple queueing processes, let us denote by $\lambda_x \cdot \lambda_x > 0$ the forward transition rate from state x to its neighboring upper state $x+1$ and in general:

$$x \to x+1 \to x+2 \to x+3 \to \dots$$

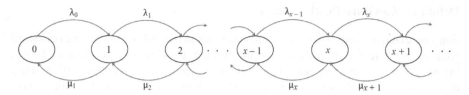

FIGURE 7.12 Transition diagram of a B-D process. Arrows represent the possible transitions between states, and the labels on the arrows are the state transition rates between states.

Staying put in the same position, that is, no move, no arrival, and no departure or no birth or death, the rate is denoted by r_x, $r_x > 0$. On the other hand, backward transitions such as $0 \leftarrow 1 \leftarrow 2 \leftarrow 3 \leftarrow \ldots$ are denoted by μ_x, $\mu_x > 0$.

Note 7.48

From a B-D process, $\lambda_x = 0$ results in a pure death process. In this case, $\mu_x > 0$ $r_x > 0$. If $\mu_x = 0$, it results in a pure birth process. In that case, $\lambda_x > 0$ and $r_x > 0$. Finally, if $r_x = 0$, then it results in a pure B-D process. In that case, $\lambda_x > 0$ and $\mu_x = 0$.

Figure 7.12 is a graphical presentation illustrating the idea of a pure B-D process with birth rate λ_x, $x = 0,1,2,\ldots$, and death rate μ_x, $x = 1,2,\ldots$, which is a standard transition diagram of a pure B-D with infinite population.

7.7.1 Finite Pure Birth

Let us assume that the birth process is a Poisson process. Hence, based on the properties of this process, the times between births will be exponentially distributed.

Clearly, no population can exist with zero original size because in that case, no birth or death can occur. Thus, for an existing population of finite size N, we assume the initial population size to be a number between 1 and N, say n_0. Symbolically,

$$P_n(0) = n_0. \text{ and } P\{n_0 \geq 1\} = 1, n_0 = 1,2,\ldots,N, n = n_0, n_0 + 1,\ldots,N. \quad (7.204)$$

Let us choose Δt sufficiently small so that no more than one birth may occur in it. That is, the probability of more than one birth in Δt is negligible. Now for the purpose of finding the population size n at any time, let us start at time t and look at the population Δt time unit later, that is, at time $t + \Delta t$. Then, for a pure birth case, we have the following two possibilities:

    i. The size at time t is already n. Thus, no birth should occur in the time interval $(t, t + \Delta t)$.

    ii. The size at time t is $n - 1$. Thus, exactly one birth should occur during the time interval $(t, t + \Delta t)$ to make up the size to be n.

Thus, the following probability statements for an occurrence of events during the time interval $[t,t+\Delta t]$, that is, during the period Δt, while the system is not empty, are true:

$$P(\text{no birth}) = 1 - \lambda_x \Delta t + 0(\Delta t), \tag{7.205}$$

$$P(\text{one birth}) = \lambda_x \Delta t + 0(\Delta t), \tag{7.206}$$

$$P(\text{more than one birth}) = 0(\Delta t), \tag{7.207}$$

where $0(\Delta t)$ is the *little* **0** of Δt.

Assuming $0(\Delta t)$ will approach 0 as Δt becomes too small and goes to zero, the $0(\Delta t)$ may be discarded. Hence, from (7.206), the probability of a birth within a time interval $(t,t+\Delta t)$ is $\lambda \Delta t$. Thus, the rate of birth from the entire population will be $n\lambda\Delta t$. Therefore, the probability of no birth within the same interval, that is, the size remaining the same, from (7.205), is $1 - n\lambda\Delta t$. Similarly, we may argue about the transition from size $n-1$ to n that will be $(n-1)\lambda\Delta t$. Thus, if we denote by $P_n(t)$ the probability that the population size is n at time t, then we can write the following:

$$P_n(t+\Delta t) = P_n(t) \cdot P\{\text{no birth occurring in }(t,t+\Delta t)\}.$$

$$= P_{n-1}(t) \cdot P\{\text{exactly one birth occurring in }(t,t+\Delta t)\}. \tag{7.208}$$

$$= P_n(t) \cdot (1 - n\lambda\Delta t) + P_{n-1}(t) \cdot (n-1)\lambda\Delta t.$$

Passing to the limit, (7.208) yields the following:

$$\frac{dP_n(t)}{dt} = (n-1)\lambda P_{n-1}(t) - n\lambda P_n(t), \quad 1 \le n \le N, \tag{7.209}$$

with $P_n(0)$ defined in (7.204).

To solve (7.209), we use the probability generating function method. Thus, after some manipulations, the solution of (7.209) will be:

$$P_n(t) = \binom{n-1}{n_0} e^{n_0 \lambda t} \left(1 - e^{-\lambda t}\right)^{n-n_0}, \tag{7.210}$$

where $P_n(0)$ is defined in (7.204).

Note 7.49

If $n_0 = 1$, then (7.210) is the geometric pmf.

Relation (7.210) can be rewritten and be used easier by choosing the following notations:

$$\xi \equiv e^{\lambda t}, \quad v = n_0, \quad p = n = n_0. \tag{7.211}$$

Thus, (7.210) can be rewritten as:

$$P_n = \binom{v + p - 1}{v}(1 - \xi)^p \xi v, \tag{7.212}$$

which is a pmf of negative binomial distribution. Thus, the mean and variance of the population size, from (7.212), denoted by $L(t)$ and $\sigma^2(t)$ are, respectively, as follows:

$$L(t) = \frac{(n - n_0)e^{\lambda t}}{1 - e^{\lambda t}} \tag{7.213}$$

and

$$\sigma^2(t) = \frac{(n - n_0)e^{\lambda t}}{(1 - e^{\lambda t})^2}, n = n_0, n_0 + 1, \ldots, N, n_0 = 1, 2, \ldots, N. \tag{7.214}$$

We leave it as an exercise to drive the same results, as for the finite pure birth process, the finite pure death process, and the B-D process.

7.7.2 B-D Process

For an infinite B-D process illustrated in Figure 7.12, let us denote by p_{yx} the probability of transition between neighboring states x and y, defined as:

$$p_{yx} = \begin{cases} \lambda_x, & \text{if only one birth occurs, that is, } y = x + 1, \\ \mu_x, & \text{if only one death occurs, that is, } y = x - 1, \\ 0, & \text{otherwise.} \end{cases} \tag{7.215}$$

Note 7.50

From (7.215), it is clear that revising each state is only possible from a neighboring state and may occur infinitely many times until the process is more than one step away from its neighbors.

Note 7.51

Also, from (7.215), transitions from state x to itself would not be possible. This is the case when a birth and a death occur at the same time. For this case, let us assume a positive probability for a case of no birth and no death for a random length of time, like in a random walk process. Then, after the process enters state x, it may stay (sojourns) there for some random length of time. Let us assume that this length of time is exponentially distributed with parameter $(\lambda_x + \mu_x)$. Then, when the process leaves the state x, it either enters state $x+1$ with probability $\lambda_x/(\lambda_x + \mu_x)$ or enters state $x-1$ with probability $\mu_x/(\lambda_x + \mu_x)$. Suppose the process chooses its next state as $x+1$. Then, again, it is assumed that the length of stay of the process in this new state will have an exponential distribution with parameter $(\lambda_{x+1} + \mu_{x+1})$, where λ_{x+1} and μ_{x+1} are the birth and death rates in state $x+1$, respectively. The process will choose the next state, etc.

Now, let $X(t)$ describe the states of the B-D process at time t that are nonnegative integers. In other words, a state of a B-D process is the population size at time t that cannot be negative. Let us also denote by $p_x(t)$ the probability that the process is in state x at time t, that is,

$$p_x(t) = P\{X(t) = x\}. \tag{7.216}$$

We further let $p_{yx}(t)$ be the transition probability of the process moving from state x to state y within the time interval $[0,t]$. In general, this interval could be chosen as $[t, t + \Delta t]$ with the interval length of Δt. Thus, by transition probability $p_{yx}(\Delta t)$, we mean the conditional probability

$$p_{yx}(\Delta t) = P\{X(t + \Delta t) = y \mid X(t) = x\}$$

or

$$p_{yx}(\Delta t) = P\{X(t + \Delta t) = y \mid X(t) = x\} = q_{yx}\Delta t + 0(\Delta t), \tag{7.217}$$

where $0(\Delta t)$ is the little 0 and q_{yx} is the proportionality constant. In other words, when Δt is small, the transition probability from state x to state y in the time interval $[t, t + \Delta t]$ is proportional to the length, Δt, of the time interval, relative to absolute time t, with proportionality constant q_{yx}. For a small Δt, $0(\Delta t)$ is near zero. Hence, dropping the $0(\Delta t)$ part, (7.217) may be rewritten as:

$$p_{yx}(\Delta t) = P\{X(t + \Delta t) = y \mid X(t) + (\Delta t) = y \mid X(t) = x\} \approx q_{yx}\Delta t. \tag{7.218}$$

Because of (7.218), p_{yx} is called the **probability transition rate**.

The transition probabilities, $p_{yx}(t), x, y = 0,1,2,\ldots$, may be found by solving the Kolmogorov forward and backward system of differential–difference equations. Letting $\mathbf{P}(t)$ denote the matrix of transitions with elements $p_{yx}(t)$, that is,

$$\mathbf{P}(t) = \left[p_{yx}(t) \right], \tag{7.219}$$

and $\mathbf{Q}$, representing the transition rate matrix with birth rate $\lambda_x, x = 0,1,2,\ldots$, and death rate $\mu_x, x = 1,2,\ldots$, the uniformly bounded, for the B-D process, we can write:

$$\mathbf{P}'(t) = \mathbf{Q} \cdot \mathbf{P}(t) = \mathbf{P}(t) \cdot \mathbf{Q}, \tag{7.220}$$

where

$$\mathbf{P}(t) = \begin{bmatrix} p_{0,0}(t) & p_{1,0}(t) & \cdots & p_{n,0}(t) & \cdots \\ p_{0,1}(t) & p_{1,1}(t) & \cdots & p_{n,1}(t) & \cdots \\ \vdots & \vdots & \ddots & \vdots & \cdots \\ p_{0,n}(t) & p_{1,n}(t) & \cdots & p_{n,n}(t) & \cdots \\ \vdots & \vdots & \vdots & \vdots & \ddots \end{bmatrix}, \tag{7.221}$$

and

$$\mathbf{Q} = \begin{bmatrix} -\lambda_0 & \lambda_0 & 0 & 0 & 0 & \cdots \\ \mu_1 & -(\lambda_1 + \mu_1) & \lambda_1 & 0 & 0 & \cdots \\ 0 & \mu_2 & -(\lambda_2 + \mu_2) & \lambda_2 & 0 & \cdots \\ 0 & 0 & \mu_3 & -(\lambda_3 + \mu_3) & \lambda_3 & \cdots \\ \vdots & \vdots & \vdots & \vdots & \vdots & \cdots \end{bmatrix}. \tag{7.222}$$

Then, the solution of the Kolmogorov backward equation is well known as

$$\mathbf{P}(t) = e^{\mathbf{Q}t}. \tag{7.223}$$

If we let

$$M = \sup_{i} | \lambda_i + \mu_i | < \infty \tag{7.224}$$

and

$$\mathbf{S} \equiv \frac{1}{M}\mathbf{Q} + \mathbf{I}, \tag{7.225}$$

where $\mathbf{S}$ is a stochastic matrix, then,

$$\mathbf{Q} = M(\mathbf{S} - \mathbf{I}). \tag{7.226}$$

Now from (7.223) and (7.224), we have:

$$\mathbf{P}(t) = e^{\mathbf{Q}t} = \sum_{n=0}^{\infty} \frac{(\mathbf{Q}t)^n}{n!}$$

$$= e^{M(\mathbf{S}-\mathbf{I})t} = e^{M\mathbf{S}t} e^{-Mt}$$

$$= e^{-Mt} \sum_{n=0}^{\infty} \frac{(Mt)^n}{n!}\mathbf{S}^n = e^{Mt} \sum_{n=0}^{\infty} \frac{(Mt)^n}{n!}[s_{yx}]^n, \tag{7.227}$$

where $S_{yx}, x, y = 0, 1, 2, \ldots$, are the elements of matrix $\mathbf{S}$, defined in (7.225).

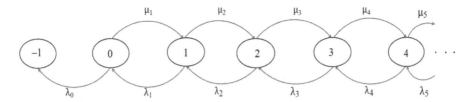

FIGURE 7.13 Transition diagram of the dual process of the B-D process of Figure 7.12. Arrows represent the possible transitions between states, and the labels on the arrows are the state transition rates between states.

Green et al. (2003) dedicated to M. M. Rao, proposed a **dual process** (sometimes referred to as the **inverse process** or **reverse process**) to the B-D, as seen in Figure 7.13.

The purpose of the duality is to facilitate finding the transition probability distributions discretely. Hence, the state "−1" in Figure 7.13 has no physical meaning and/or interpretation. Denoting the transition probability functions of the dual process presented in Figure 7.13 by $p^*_{yx}(t)$, through Theorem 7.3, they showed that the relationship (7.228) holds between a general B-D process and its duality. Thus, if the transient probability functions in either the original B-D process or dual B-D system are known, then the transient probability functions in the other system are known as well.

Theorem 7.3

If $p_{yx}(t)$ and $p^*_{yx}(t)$ are the transient probability functions of the B-D processes corresponding to Figures 7.12 and 7.13, respectively, then, assuming

$$P_{k-1}(t) = 0, \quad \text{for } k > -1,$$

we have:

$$P_{yx}(t) = \sum_{k=x}^{\infty} \left[P^*_{ky}(t) - P^*_{k,y-1}(t) \right] \text{ and } P_{yx}(t) = \sum_{k=0}^{x} \left[P_{ky}(t) - P_{k,y+1}(t) \right], \quad (7.228)$$

for all states $x, y = 0, 1, 2, \ldots$.

Proof:
The proof of this theorem appears as Proposition 2.3 on page 269 of Anderson (1991). It is essentially based on the forward and backward Kolmogorov equations. See also Green et al. (2003). The outline of proof of Theorem 7.3 is as follows:

Consider the finite recurrent B-D chain having transition probabilities diagrammed in Figure 7.14.

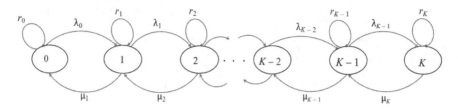

FIGURE 7.14 Transition diagram of a B-D process with staying put allowed. Arrows represent the possible transitions between states, and the labels on the arrows are the state transition rates between states.

For the B-D process represented by its transition diagram in Figure 7.14, assuming that all characters are fractions between 0 and 1, inclusively, we should note the following:

$$r_0 + \lambda_0 = 1,$$

$$r_1 + \lambda_1 + \mu_1 = 1,$$

$$r_2 + \lambda_2 + \mu_2 = 1,$$

$$\ldots$$

$$r_{K-1} + \lambda_{K-1} + \mu_{K-1} = 1,$$

$$r_K + \mu_K = 1.$$

To make sure that the traffic intensities in the dual processes are <1, it is also assumed that:

$$\lambda_0 + \mu_1 \leq 1,$$

$$\lambda_1 + \mu_2 \leq 1,$$

$$\lambda_2 + \mu_3 \leq 1,$$

$$\ldots$$

$$\lambda_{K-1} + \mu_K \leq 1.$$

In case of general B-D process, that is, $r_x \geq 0$, which we described as the random walk in which staying put is allowed, Green et al. (2003) showed that a similar theorem to Theorem 7.3 holds, which is Theorem 7.3 for B-D chains. Although statement of the theorem is on a finite state space, it holds as well for infinite B-D chains.

Figures 7.13 and 7.14 are transition probability diagrams similar to Figures 7.9 and 7.10. Figure 7.11 is the absorbing B-D chain dual to the B-D process represented in Figure 7.15.

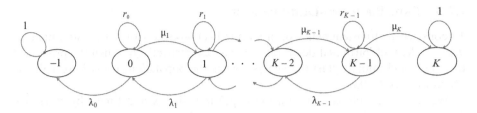

FIGURE 7.15 Transition diagram of the dual process of the B-D process of Figure 7.15. It is an absorbing B-D at states −1 and K. Arrows represent the possible transitions between states, and the labels on the arrows are the state transition rates between states.

In Figure 7.15, it is assumed that all characters are fractions between 0 and 1, inclusively, and that:

$$\lambda_0 + r_0 + \mu_1 = 1,$$

$$\lambda_1 + r_1 + \mu_2 = 1,$$

$$\cdots$$

$$\lambda_{K-1} + r_{K-1} + \mu_K = 1.$$

Theorem 7.4

If $p_{yx}^n(t)$ and $p_{yx}^{*(n)}(t)$ are the n-step transition probabilities of the B-D processes corresponding to Figures 7.14 and 7.15, respectively, then, assuming

$$P_{k,-1}^{*(n)}t = 0, \quad \text{for } k > -1,$$

we have:

$$P_{ji}^{(n)}(t) = \sum_{k=j}^{K}\left[P_{kj}^{*(n)}(t) - P_{k,j-1}^{*n}(t)\right] \text{ and } P_{ji}^{*(n)}(t) = \sum_{k=0}^{i}\left[P_{kj}^{(n)}(t) - P_{k,j+1}^{(n)}(t)\right], \quad (7.229)$$

for $n \geq 0$ and all states $i, j = 0, 1, 2, \ldots, K$.

Proof:
See Green et al. (2003).

Note 7.52

Difficulties facing numerical solution for time-dependent stochastic processes, and in particular, such a system, are well known. However, the transient probability distribution of the population of a finite-state B-D process has been solved using a variety of methods over decades. Obtaining analytic explicit solution is almost impossible.

7.7.3   FINITE BIRTH-AND-DEATH PROCESS

To consider the transient behavior of a finite B-D process with $N+1$ states, namely, $0,1,2,...,N$, with birth and death rates λ_n and μ_n, respectively, when the process is in state n, and initially start the analysis when the population size is i, we will use Mohanty et al. (1993).

We denote the probability that the population size is n at time t by $P_n(t)$. We further denote by $\Psi_n(s)$ the Laplace transform of $P_n(t)$.

7.7.3.1   Analysis

Here, we apply the method offered by Mohanty et al. (1993), using differential–difference equations in an elementary way. We leave it as an exercise to show that $\Psi_n(s)$, $0,1,2,...,N$, satisfies the following set of differential–difference equations:

$$
\left\{
\begin{array}{l}
(\lambda_0 + s)\Psi_0(s) - \mu_1(s) = \delta_{i,0}, \\
(-\lambda_{n-1}\Psi_{n-1}(s)) + (\lambda_n + \mu_n + s)\Psi_n(s) - \mu_{n+1}\Psi_{n+1}(s) = \delta_{i,n}, \quad 1 \le n \le N-1, \quad (7.230) \\
-\lambda_{N-1}\Psi_{N-1}(s) + (\mu_N + s)\Psi_N(s) = \delta_{i,N},
\end{array}
\right.
$$

where $\delta_{i,j}$ is Kronecker's delta function.

To solve the system (7.230), we rewrite it in a matrix form. So, let us denote by $A_{N(s)}$ the determinant of the coefficient matrix of the system (7.230). After some algebraic manipulations, $A_{N(s)}$ may be written as follows:

$$
A_n(s) = C_N(s)
$$

$$
= \begin{bmatrix}
\lambda_0 + s & \sqrt{\lambda_0\mu_1} & & & & \\
\sqrt{\lambda_0\mu_1} & \lambda_2 + \mu_2 + s & \sqrt{\lambda_1\mu_2} & & & \\
& \sqrt{\lambda_1\mu_2} & \lambda_2 + \mu_2 + s & \sqrt{\lambda_2\mu_3} & & \\
& & \sqrt{\lambda_2\mu_3} & \lambda_3 + \mu_3 + s & & \\
& & & & \ddots & \\
& & & & \sqrt{\lambda_{N-1}\mu_{N-1}} + s & \sqrt{\lambda_{N-1}\mu_N} \\
& & & & \sqrt{\lambda_{N-1}\mu_N} & \mu_N + s
\end{bmatrix}
$$

$$
(7.231)
$$

Let $T_k(s)$ and $B_k(s)$ be the determinants of the $k \times k$ matrices formed at the top-left corner and the bottom-right corner of the coefficient matrices, respectively. Set $T_0(s) = B_0(s) = 1$. Thus, using Cramer's rule, the solution of (7.230) is as follows:

$$\Psi_n(s) = \left(\prod_{j=n+1}^{i} \mu_j\right) \frac{T_n(s)B_{N-i}(s)}{A_N(s)}, \quad 0 \le n \le i$$

$$= \left(\prod_{j=i}^{n-1} \lambda_j\right) \frac{T_i(s)B_{N-n}(s)}{A_N(s)}, \quad i+1 \le n \le N,$$

(7.232)

where the first product may be interpreted as

$$\prod_{j=k}^{n} \mu_j = 1,$$

(7.233)

whenever $n < k$.

To be able to invert (7.232), we first express it as a partial fraction. To do so, we note that the right-hand side of (7.232) is a ratio of two polynomials with the degree of numerator less than those with the degree of the denominator. Note that s is a zero of the matrix $C_N(s)$ defined by (7.231) if and only if $-s$ is an eigenvalue of the matrix $E_N(s)$ defined as:

$$E_N = \begin{bmatrix} \lambda_0 & \sqrt{\lambda_0\mu_1} & & & & \\ \sqrt{\lambda_0\mu_1} & \lambda_1 + \mu_1 & \sqrt{\lambda_1\mu_2} & & & \\ & \sqrt{\lambda_1\mu_2} & \lambda_2 + \mu_2 & \sqrt{\lambda_2\mu_3} & & \\ & & \sqrt{\lambda_2\mu_3} & \lambda_3 + \mu_3 & & \\ & & & & \ddots & \\ & & & & \sqrt{\lambda_{N-1}\mu_{N-1}} & \sqrt{\lambda_{N-1}\mu_N} \\ & & & & \sqrt{\lambda_{N-1}\mu_N} & \mu_N \end{bmatrix}$$

(7.234)

Note 7.53

$s = 0$ is an eigenvalue of $E_N(s)$. But each off-diagonal element of $E_N(s)$ is nonzero. Hence, all the eigenvalues are distinct. Hence, s is the only zero of $C_N(s)$.

Note 7.54

Since all the minor elements of $E_N(s)$ are positive, by the S turn sequence property, all other eigenvalues are positive. Therefore, all eigenvalues of the positive semi-definite matrix $E_N(s)$ are real, distinct, and nonnegative. Hence, $A_N(s)$ has exactly $N+1$ distinct zeros, one of which is zero and the rest are the negatives of the eigenvalues of $E_N(s)$.

So, now let us denote the zeros of $A_N(s)$ by z_k, $k = 0,1,2,...,N$ with $z_0 = 0$. Letting the numerator of $\Psi_n(s)$, defined in (7.232), be $G_n(s)$, (3.1.11) may be rewritten as:

$$\Psi_n(s) = \frac{G_n(s)}{A_N(s)}, \quad n = 0,1,2,...,N, \qquad (7.235)$$

and, in turn, it can be expressed in partial fraction as follows:

$$\Psi_n(s) = \frac{\beta_{n,k}}{s - z_k}, \quad n = 0,1,2,...,N, \qquad (2.236)$$

where

$$\beta_{n,k} = \frac{G_n(s)}{\displaystyle\prod_{j=0,j\neq k}^{N} \left(z_k - z_j\right)}, \quad n = 0,1,2,...,N, \qquad (7.237)$$

or

$$\beta_{n,k} = \frac{\displaystyle\prod_{j=0}^{N-n} \left(z_k - s_j\right)}{\displaystyle\prod_{j=0,j\neq k}^{N} \left(z_k - z_j\right)}, \quad n = 0,1,2,...,N, \qquad (2.238)$$

where s_js are the roots of $G_n(s)$. Thus, inverting (7.238), we will have the distribution of the population as:

$$P_n(t) = \beta_{n,0} + \sum_{k=1}^{N} \beta_{n,k} e^{z_k t}, n = 0,1,2,...,N, \qquad (7.239)$$

where $\beta_{n,k}$ are given in (7.236).

For the stationary distribution, denoted by P_n, letting $t \to \infty$, we obtain:

$$P_n = \lim_{t\to\infty} P_n(t) = \beta_{n,0} = \frac{G_n(0)}{\displaystyle\prod_{j=1}^{N} \left(-z_j\right)}. \qquad (7.240)$$

We leave it as an exercise to show that:

$$\beta_{n,0} = \beta_{0,0} \prod_{k=1}^{n} \frac{\lambda_{k-1}}{\mu_k}, \quad 1 \le n \le N. \tag{7.241}$$

Using the normalization equation, we will have:

$$\beta_{0,0} = \left[1 + \sum_{n=1}^{N} \prod_{k=1}^{n} \frac{\lambda_{k-1}}{\mu_k} \right]^{-1}. \tag{7.242}$$

7.7.3.2 Busy Period

To obtain the busy period distribution, we define an m-state busy period to begin with a birth to the system at an instant when the process is in state m to the very next time when the process returns to state $m-1$. We assume that $m=1$ defines the busy period of the process. In this case, the system of differential–difference equations becomes:

$$P'_{m-1}(t) = \mu_m P_m(t),$$

$$P'_m(t) = -\left(\lambda_m + \mu_m\right) P_m(t) + \mu_{m+1} P_{m+1}(t),$$

$$P'_n(t) = -\left(\lambda_n + \mu_n\right) P_n(t) + \lambda_{n-1} P_{n-1}(t) + \mu_{n+1} P_{n+1}(t), \quad m+1 \le n \le N-1,$$

$$P'_N(t) = -\mu_N P_N(t) + \left(\lambda_{N-1} P_{N-1}(t)\right),$$

$$P'_m(t) = -\left(\lambda_m + \mu_m\right) P_m(t) + \mu_{m+1} P_{m+1}(t) \tag{7.243}$$

$$P'_n(t) = -\left(\lambda_n + \mu_n\right) P_N(t) + \lambda_{n-1} P_{n-1}(t) + \mu_{n+1} P_{n+1}(t), \quad m+1 \le n \le N-1,$$

$$P_N^0(t) = -\mu N P_N(t) + \left(\lambda_{N-1} P_{N-1}(t)\right),$$

with $P_m(0) = 1$.

Note 7.55

The first equation of the system (7.243) gives $P'_{m-1}(t)$, which is the pdf of the distribution of the length of a busy period.

To solve the system (7.243), we apply the Laplace transform on the last three equations and obtain the following:

$$\left(\lambda_m + \mu_m + s\right) \Psi_m(s) - \mu_{m+1} \Psi_{m+1}(s) = 1,$$

$$-\lambda_{n-1} \Psi_{n-1}(s) + \left(\lambda_n + \mu_n + s\right) P_n(s) - \mu_{n+1} \Psi_{n+1}(s) = 0, \quad m+1 \le n \le N-1, \tag{7.244}$$

$$-\lambda_{N-1} \Psi_{N-1}(s) + \left(\mu_N + s\right) \Psi_N(s) = 0.$$

Let $D_{N,r}(s), r = m-1,...,N,$, be the determinant of the $r \times r$ matrix formed at the right-bottom corner of the coefficient matrix of the system of equations (7.236). Then, we have:

$$\Psi_m(s) = \frac{D_{N,m+1}(s)}{D_{N,m}(s)}. \tag{7.245}$$

As we argued earlier, we can show that the zeros of $D_{N,m}(s)$ are distinct and negative. Let these be $u_1, u_2, ..., u_{N-m+1}$. Then, the density function of the length of a busy period at time t is given by:

$$P'_{m-1}(t) = \mu_m P_m(t) = \mu_m \sum_{k=1}^{N-m+1} \gamma_k e^{u_k t}, \tag{7.246}$$

where

$$\gamma_k = \frac{\prod\limits_{j=1}^{N-m}(u_k - v_j)}{\prod\limits_{j=1, j \neq k}^{N-m+1}(u_k - u_j)}, \quad k = 1, 2, ..., N-m+1, \tag{7.247}$$

where v_js are the roots of $D_{N,m+1}(s)$.

Note 7.56

Relations (7.246) and (7.247) are similar to (7.239) and (7.238), respectively.

Theorem 7.5

The distribution of a busy period is hyperexponential.

Proof:
We leave it as an exercise to show that γ_ks given by (7.247) are positive. Hence, $\mu_m \frac{\gamma_k}{-\mu_k} > 0$.

Now, $\Psi_m(s)$ given in (7.245) is the same as in (7.235), and thus, it has a partial fraction similar to (7.236). Hence,

$$\Psi_m(s) = \sum_{k=1}^{N-m+1} \frac{D_{N,m+1}(0)}{\prod\limits_{j=0, r \neq k}^{N-m+1}(u_k - u_j)} \frac{1}{s - u_k}. \tag{7.248}$$

Using γ_k, from (7.247), we will have:

$$\sum_{k=1}^{N-m+1} \frac{\gamma_k}{-u_k} = \sum_{k=1}^{N-m+1} \frac{D_{N,m+1}(0)}{\displaystyle\prod_{j=0,j\neq k}^{N-m+1}(u_k-u_j)} \frac{1}{s-u_k}. \tag{7.249}$$

Thus, from (7.245), (7.247), and (7.249), we have the following:

$$\sum_{k=1}^{N-m+1} \frac{\gamma_k}{-u_k} = \Psi_m(0) = \frac{D_{N,m+1}(0)}{D_{N,m}(0)}. \tag{7.250}$$

We leave it as an exercise to show that

$$D_{N,r}(0) = \prod_{k=\gamma}^{N} u_{m+k}. \tag{7.251}$$

Substituting (7.251) in (7.252) completes the proof.

Hence, from Theorem 7.5, we have the mean and the variance of a busy period, respectively, as follows:

$$\sum_{k=1}^{N-m+1} \frac{\mu_m \gamma k}{(u_k)^2} \tag{7.252}$$

and

$$-2\left[\sum_{k=1}^{N-m+1} \frac{\mu_m \gamma k}{(u_k)^3}\right] - \left[\sum_{k=1}^{N-m+1} \frac{\mu_m \gamma k}{(u_k)^2}\right]^2. \tag{7.253}$$

Example 7.31

In this example, we consider a numerical example for the finite B-D model we just discussed. So, we are looking for a numerical distribution of the number of tasks in the system.

Answer

To obtain a numerical distribution, programming language FORTRAN has been used. Here are the items calculated:

  i. The eigenvalues of E_k, (7.239), are computed.
  ii. The zeros of $A_K(s)$ (7.236), denoted by z_k, are the negatives of the eigenvalues of E_k, except that z_0 is set to zero.
  iii. From (7.241), having values of z_k, we can compute $G(z_k)$, and having $\prod_{j=0,j\neq k}(z_k-z_j)$, we can compute $\beta_{n,k}$.

iv. Alternatively, having z_k and s_j, we can compute $\Pi_{j=0, j \neq k}^{K}\left(z_k - z_j\right)$ and

$\Pi_{j=0}^{K-n}\left(z_k - z_j\right)$, and thus, $\beta_{n,k}$. The probabilities are obtained once $\beta_{n,k}$ are
in hand.

v. We will follow the same algorithm for the distribution of a busy period.

The computer method has been checked for various values of N. Some of the probabilities computed by the program along with the negative of the eigenvalues are given in Tables 7.9–7.12. From Tables 7.9, 7.10, and 7.12, we have chosen $N = 7$. Table 7.9 is given as a sample of the numerical values for larger K, that is, $K = 39$.

TABLE 7.9
Transient Probability Distribution of the Number of Tasks in the System with $K = 39$ and $c = 3$

				k or n			
	0	1	2	12	25	38	39
$\lambda = 4, \mu = 1, p = 1, a = 0$							
z_k	0.00000	−0.09446	−0.16227	−2.95394	−9.40526	−13.83590	−13.90509
$p_n(0)$	1.00000	0.00000	0.00000	0.00000	0.00000	0.00000	0.00000
$p_n(1)$	0.07951	0.19954	0.24663	0.00011	0.00000	0.00000	0.00000
$p_n(2)$	0.02901	0.09606	0.15497	0.00534	0.00000	0.00000	0.00000
$p_n(66)$	0.00001	0.00002	0.00004	0.00037	0.00546	0.18341	0.24438
$p_n(200)$	0.00000	0.00000	0.00001	0.00011	0.00445	0.18750	0.25000

TABLE 7.10
Transient Probability Distribution of the Number of Tasks in the System with $K = 7$ and $c = 3$

				k or n				
	0	1	2	3	4	5	6	7
$\lambda = 1, \mu = 4, p = 0.5, a = 5$								
z_k	0.00000	−4.01223	−8.25403	−13.26090	−17.25717	−22.05825	−27.48892	−34.66849
$p_n(0)$	1.00000	0.00000	0.00000	0.00000	0.00000	0.00000	0.00000	0.00000
$p_n(1)$	0.78241	0.19203	0.02357	0.00193	0.00006	0.00000	0.00000	0.00000
$p_n(2)$	0.77892	0.19467	0.02433	0.00203	0.00006	0.00000	0.00000	0.00000
$p_n(3)$	0.77886	0.19471	0.02434	0.00203	0.00006	0.00000	0.00000	0.00000
$p_n(4)$	0.77886	0.19471	0.02434	0.00203	0.00006	0.00000	0.00000	0.00000
$\lambda = 1, \mu = 4, p = 9, a = 5$								
z_k	0.00000	−4.00410	−8.10124	−12.59525	−17.13745	−22.19106	−28.16837	−36.40253
$p_n(0)$	1.00000	0.00000	0.00000	0.00000	0.00000	0.00000	0.00000	0.00000
$p_n(1)$	0.78239	0.19202	0.02356	0.00193	0.00010	0.00000	0.00000	0.00000

(Continued)

TABLE 7.10 (*Continued*)
Transient Probability Distribution of the Number of Tasks in the System with
K = 7 and *c* = 3

					k or *n*			
	0	1	2	3	4	5	6	7
$p_n(2)$	0.77888	0.19466	0.02432	0.00203	0.00010	0.00000	0.00000	0.00000
$p_n(3)$	0.77882	0.19470	0.02434	0.00203	0.00010	0.00000	0.00000	0.00000
$p_n(4)$	0.77882	0.19470	0.02434	0.00203	0.00010	0.00000	0.00000	0.00000
$\lambda = 10, \mu = 9, p = 0.9, a = 8$								
z_k	0.00000	−9.11231	−18.09908	−27.54689	−39.12215	−53.10727	−69.48002	−91.53230
$p_n(0)$	1.00000	0.00000	0.00000	0.00000	0.00000	0.00000	0.00000	0.00000
$p_n(1)$	0.33001	0.36663	0.20366	0.07542	0.01939	0.00406	0.00072	0.00011
$p_n(2)$	0.32997	0.36663	0.20368	0.07544	0.01940	0.00406	0.00072	0.00011
$p_n(3)$	0.32997	0.36663	0.20368	0.07544	0.01940	0.00406	0.00072	0.00011
$\lambda = 10, \mu = 9, p = 0.5, a = 8$								
z_k	0.00000	−9.85973	−20.81287	−28.86287	−39.74234	−51.04649	−61.74538	−79.93031
$p_n(0)$	1.00000	0.00000	0.00000	0.00000	0.00000	0.00000	0.00000	0.00000
$p_n(1)$	0.33404	0.37113	0.20617	0.07636	0.01091	0.00127	0.00012	0.00001
$p_n(2)$	0.33401	0.37113	0.20618	0.07636	0.01091	0.00127	0.00012	0.00001
$p_n(3)$	0.33401	0.37113	0.20618	0.07636	0.01091	0.00127	0.00012	0.00001
$\lambda = 2, \mu = 2, p = 0.8, a = 7$								
z_k	0.00000	−2.18470	−5.08370	−9.00690	−13.34179	−20.21162	−28.49880	−40.07249
$p_n(0)$	1.00000	0.00000	0.00000	0.00000	0.00000	0.00000	0.00000	0.00000
$p_n(1)$	0.42176	0.36568	0.15945	0.04724	0.00544	0.00041	0.00002	0.00000
$p_n(2)$	0.37739	0.37138	0.18298	0.06027	0.00737	0.00059	0.00003	0.00000
$p_n(3)$	0.37251	0.37184	0.18561	0.06179	0.00760	0.00061	0.00004	0.00000
$p_n(4)$	0.37197	0.37189	0.18591	0.06196	0.00763	0.00061	0.00004	0.00000
$p_n(5)$	0.37190	0.37190	0.18594	0.06198	0.00763	0.00061	0.00004	0.00000
$p_n(6)$	0.37190	0.37190	0.18595	0.06198	0.00763	0.00061	0.00004	0.00000
$\lambda = 3, \mu = 1, p = 0.9, a = 1$								
z_k	0.00000	−0.93329	−3.55767	−7.05954	−14.57042	−26.84474	−40.29274	−51.74159
$p_n(0)$	1.00000	0.00000	0.00000	0.00000	0.00000	0.00000	0.00000	0.00000
$p_n(1)$	0.14561	0.25681	0.18890	0.02268	0.02575	0.03599	0.07856	0.24570
$p_n(2)$	0.05129	0.10499	0.08478	0.01097	0.01650	0.04157	0.14770	0.54220
$p_n(17)$	0.00026	0.00079	0.00119	0.00119	0.00801	0.04325	0.19462	0.75069
$p_n(18)$	0.00026	0.00079	0.00119	0.00119	0.00801	0.04325	0.19462	0.75069
$p_n(19)$	0.00026	0.00079	0.00119	0.00119	0.00801	0.04325	0.19462	0.75069
$\lambda = 4, \mu = 1, p = 0.9, a = 1$								
z_k	0.00000	−1.48510	−4.55386	−8.53413	−20.56462	−34.76467	−50.43036	−63.66725
$p_n(0)$	1.00000	0.00000	0.00000	0.00000	0.00000	0.00000	0.00000	0.00000
$p_n(1)$	0.07458	0.16791	0.15176	0.01826	0.02106	0.03305	0.09875	0.43463
$p_n(2)$	0.01554	0.03889	0.03751	0.00484	0.00805	0.02823	0.14417	0.72273
$p_n(11)$	0.00004	0.00015	0.00030	0.00041	0.00365	0.02629	0.15777	0.81138
$p_n(12)$	0.00004	0.00015	0.00030	0.00041	0.00365	0.02629	0.15777	0.81138

TABLE 7.11

Transient Probability Distribution of the Number of Tasks in the System by Randomization Method with $K = 7$, $c = 3$, and m. Being Preasigned according to $m = ft + 4\sqrt{ft} + 5$

k	0	1	2	3	4	5	6	7	$p_k(t)$
$\lambda = 1, \mu = 4, p = 1, a = 0$									
$m = 5$ $p_k(0)$	1.000000	0.000000	0.000000	0.000000	0.000000	0.000000	...	0.00000	1.000000
$m = 32$ $p_k(1)$	0.782350	0.191995	0.023555	0.001924	0.000156	0.000012	...	0.00000	0.999994
$m = 51$ $p_k(2)$	0.778820	0.194639	0.024321	0.002026	0.000169	0.000014	...	0.00000	0.999991
$m = 68$ $p_k(3)$	0.778750	0.194686	0.024336	0.002028	0.000169	0.000014	...	0.00000	0.999984
$m = 85$ $p_k(4)$	0.778748	0.194687	0.024336	0.002028	0.000169	0.000014	...	0.00000	0.999913
$m = 102$ $p_k(5)$	0.778750	0.194688	0.024336	0.002028	0.000169	0.000014	...	0.00000	0.999987

Note: For this table, we have fixed $p_0(0) = 1$.

TABLE 7.12

Execution Time to Obtain the Steady-State Solution for the Present Method for Various Values of N with $c = 3$, $\lambda = 4$, $\mu = 1$, $p = 1$, $\alpha = 0$

N	Execution Time
7	099 seconds
10	2.075 seconds
20	11.24 seconds
50	3 minutes and 52 seconds
75	21 minutes and 53 seconds
100	53 minutes and 58 seconds
125	2 hours and 37 minutes

For the reason given in remark 7.5, we have chosen $p = 1$ and $\alpha = 0$ from Table 7.12. We have also checked our method for various values of the parameters involved, but we have reported examples with the numerical values of the parameters so that we would have $\frac{\lambda}{c\mu} < 1, \frac{\lambda}{c\mu} > 1$, and $\frac{\lambda}{c\mu} = 1$ with various probabilities of balking, as well as various reneging rates. The time units in Tables 7.9–7.12 are cut off at an instance when the steady-state solutions are reached, that is, when convergences have been observed; in these tables, some rows appear to be identical; this is due to round-off. Table 7.12 gives the execution times to obtain the steady-state solution for our method for different values of K.

Note 7.57

1. Computing eigenvalues involves errors to some degree.
2. For analysis of the errors and numerical examples, the reader is referred to Parlett (1980), Wilkinson (1965), and Murphy and O'Donohoe (1975).
3. In our case, to reduce the error, since we know one of the eigenvalues is zero, we have set z_0 and u_0 to zero in the program.
4. The steady-state probabilities obtained from the transient case, using both formulas (7.232) and (7.233) and the direct formula, match exactly.
5. The IMSL performance index in all cases did not exceed 0.06, which according to IMSL (1987) is considered to be "excellent".

Remarks

1. The randomization method, together with an algorithm given in Grassmann (1977a,b), is available to obtain the transient solution, and it works well for non-stiff systems (i.e., systems in which the diagonal elements do not vary considerably in size), but it does not provide exact values of the probabilities. Given the error tolerance, the randomization algorithm determines where the computation should be truncated; that is, the value of m has to be preassigned according to the error tolerance. m is the truncation level in the formula

$$p_k(t) \approx \sum_{n=0}^{m} \pi_k^n (ft)^n \frac{e^{-ft}}{n!}, \qquad (7.254)$$

where $f = \max|a_{ii}|$ in the coefficient matrix of the system of differential–difference equations,

$$\pi_k^n = \sum_{i=0}^{m} \pi_i^{n-1} p_{i,j} \text{ and } p_{i,j} = \frac{a_{i,j}}{f}, i \neq j. \qquad (7.255)$$

For example, for $K = 7, c = 3, \lambda = 1, \mu = 4, p = 1$, and $\alpha = 0$ using the randomization method, Table 7.12 provides the values of probabilities for different truncation levels. In Table 7.12, m is calculated according to Grassmann (1977b) as $m = ft + 4\sqrt{ft} + 5$. By pushing K for larger values, an approximation for the infinite case can be obtained as well.

For specific values of parameters given in Table 7.12, the steady-state solution appears to be reached at $m = 102$. As expected, this shows that to obtain the steady-state solution using randomization, one cannot truncate arbitrarily and m has to be determined for the specific values of λ, μ, p, and α, while for our method, the steady-state distribution is exactly and easily obtained.

For $K = 7, \lambda = 1, \mu = 4, p = 1$, and $\alpha = 0$, randomization took 39.21 seconds to be executed to reach the steady state, while for our method, it took only 0.99 seconds.

2. Our approach in this section yields, with ease, the steady-state solution analytically, while the randomization method does so less easily.
3. Computationally, the method presented in this section gives a reasonably good approximation for the steady state for infinite case; though, it may take longer execution time for larger K (see Table 7.5).
4. In case $p = 1$ and $\alpha = 0$, the method presented in this section works for the finite case perfectly, even for a relatively large K (at least $K = 125$). For $K = 50$, it took only 232 seconds $\ll 4$ minutes to obtain the steady-state solution. This is contrary to the claim by Grassmann (1990) that "Cramer's rule is impractical for all but very small problems". This, perhaps, is one of the small problems since we are dealing with a tridiagonal matrix. Grassmann (1990) also stated that for $K = 50$, the execution time "on a modern supercomputer requires 10^{44} years and this time exceeds the age of the universe by many orders of magnitude" – this, however, seems not to be our experience.
5. For the so-called stiff system, Whitlock mentioned in 1973 that no finite f can be found, and thus, the randomization algorithm is not applicable, while our method works regardless. It is for this reason that we have chosen $p = 1$ and $\alpha = 0$ from Table 3.2.4. However, for some values of the parameters, the diagonal elements became large, making the eigenvalues large and thus causing overflow problem while running the program.

7.8 FUZZY QUEUES/QUASI-B-D

7.8.1 QUASI-B-D

Many important stochastic models involve multidimensional random walks. The two-dimensional case is of particular theoretical and practical importance, often occurring directly or through decomposition of higher-dimensional processes. One such particular case is a two-dimensional Markov processes, referred to as **quasi-birth-and-death (QBD)** processes, whose transitions are skip-free to the left and the right (or up and down), with no restrictions upward or downward, in the two-dimensional lattice. A lattice is a partially ordered set in which any two elements have a least upper bound and a greatest lower bound.

Definition 7.32

Consider a continuous-time Markov process $\{X(t), t \in \mathbb{R}^+\}$, where $\mathbb{R}^+$ denotes the nonnegative real numbers, on the two-dimensional state space $\Omega = \{(i, j) : i \in \mathbb{Z}^+, j \in \{1, 2, ..., S\}\}$, where $\mathbb{Z}^+$ denotes the set of nonnegative integers. The first coordinate, i, is referred to as the **level** and the second coordinate j is referred to as the **phase** of state (i, j). The set $l(i)$ is called the **level** i. Each level may have a finite or infinite number of states, S. Then, the Markov process is called a **QBD process**.

The one-step transitions from each state of the *QBD* process are restricted to states in the same level or in the two adjacent levels, and a **homogeneous** *QBD* process if these transition rates are, additionally, level independent. The infinitesimal generator Q of the Markov process then takes the block tridiagonal form:

$$\mathbf{Q} = \begin{pmatrix} B & A_0 & & \\ A_2 & A_1 & A_0 & \\ & A_2 & A_1 & A_0 \\ & & \ddots & \ddots & \ddots \end{pmatrix}, \tag{7.256}$$

where A_0, A_2 are nonnegative and A_1, B have nonnegative off-diagonal elements and strictly negative diagonals. Each of these matrices has dimension $S \times S$.

7.8.2 A Fuzzy Queueing Model as a QBD

The model we are to discuss in this subsection is based on the basic queue $M/M/1$. However, there are three new features that have been added to:

  i. Fuzzy arrival environment, denoted by *FM*,
  ii. Service fuzzy environment, denoted by *FM*,
  iii. Server's working vacation.

Thus, the model is denoted by **FM/FM/1**, and we refer to it as the **Markovian single-server queue with working vacation in fuzzy environment**. This paper is the work of Kannadasan and Sathiyamoorth (2018). When taking vacation, the server in this model continues to work if needed. With computer technology, this is quite common these days. The term "working vacation" is related to this case, and it is denoted by "WV". There are cases that when a server is on vacation, the service rate may be lowered. Also, cases with multiple vacations have been studied. However, the model we are considering now is as mentioned. For such a model, we will find the busy period, working vacation period, stationary queue length, and waiting time. The numerical example that will follow shows some performance measures for the fuzzy queues. Applications of such a queue can be seen in transportation systems such as bus service, trains, and express elevators, where the service provided is a group that can be served simultaneously, that is, batch servicing in this process.

7.8.2.1 Crisp Model

We first analyze the model in the standard environment, that is, in the absence of the fussiness. We refer to it as the **crisp model**. Thus, we start considering a classic $M/M/1$ queue with arrival rate λ and service rate μ_b. As soon as the queue becomes empty, the server goes on a working vacation. The length of this vacation, denoted by V, is a random variable with exponential pdf with parameter μ_v. When a vacation ends, there are two cases:

 i. There are no tasks in the queue, then another working vacation is taken or
 ii. There are tasks awaiting service, then service starts, and its rate switches to a new rate μ_b when the busy period begins.

Such a system is referred to as ***M/M/1/WV***. For this mode, we assume that inter-arrival times, service times, and working-vacation times are mutually independent. We also assume that the service discipline is FIFO.

Now, consider a classic *M/M/1* queue with arrival rate λ and service rate μ_b. The server is a working vacation of random length at the instants when the queue becomes empty, and the vacation duration V follows an exponential distribution with parameter θ. During a working vacation, an arriving customer is served at a rate of μ_v. When a vacation ends, if still there are no customers in the queue, another working vacation will be taken; otherwise, the service rate switches service from μ_v to μ_b, and a regular busy period starts. Similar to Servi and Finn (2002), this queue is referred to as an ***M/M/1/WV* queue**.

It is assumed that inter-arrival times, service times, and working vacation times are mutually independent. In addition, the service discipline is FIFO. Let $Q_v(t)$ be the number of tasks in the system at time t and let

$$J(t) = \begin{cases} 0, & \text{the system is in a working vacation period at time } t \\ 1, & \text{the system is in a start-up period at time } t. \end{cases}$$

Then, $\{Q_v(t), J(t)\}$ is the *QBD* process with the state space,

$$\Omega = \{0,0\} \cup \{k,j\} : k \geq 1, j = 0,1.$$

7.8.2.2 The Model in Fuzzy Environment

Let us denote by $\bar{\lambda}, \bar{\beta}, \bar{v}$, and $\bar{\theta}$ as the rates of the arrival, service, working vacation service, and working vacation time, respectively, as fuzzy numbers, defined by:

$$\bar{\lambda} = \left\{ w, \mu_{\bar{\lambda}}(w) : w \in S\left(\bar{\lambda}\right) \right\},$$

$$\bar{\beta} = \left\{ w, \mu_{\bar{\beta}}(x) : x \in S\left(\bar{\beta}\right) \right\},$$

$$\bar{v} = \left\{ y, \mu_{\bar{v}}(y) : y \in S\left(\bar{v}\right) \right\},$$

$$\bar{\theta} = \left\{ z, \mu_{\bar{\theta}}(z) : z \in S\left(\bar{\theta}\right) \right\},$$

where $S\left(\bar{\lambda}\right)$, $S\left(\bar{\beta}\right)$, $S\left(\bar{v}\right)$, and $S\left(\bar{\theta}\right)$ are the universal sets of the arrival rate, service rate, working vacation a task is served at a rate, and working vacation time, respectively. They define the function $f(w,x,y,z)$ as the system performance measure related to the fuzzy queuing model under consideration, which depends on the fuzzy membership function $f\left(\overline{\lambda},\overline{\beta},\overline{v},\overline{\theta}\right)$. Applying extension principle of Zadeh (1978), the membership function of the performance measure $f\left(\overline{\lambda},\overline{\beta},\overline{v},\overline{\theta}\right)$ can be defined as:

$$\mu_f\left(\overline{\lambda},\overline{\beta},\overline{v},\overline{\theta}\right) = \sup\left\{\lambda_{\overline{\lambda}}(\omega),\mu_{\overline{\beta}}(x),\mu_{\overline{v}}(y),\mu_{\overline{\theta}}(z)/H = f(\omega,x,y,z)\right\}$$

$$w \in S\left(\overline{\lambda}\right)$$

$$x \in S\left(\overline{\beta}\right) \qquad\qquad\qquad . \quad (7.257)$$

$$y \in S\left(\overline{v}\right)$$

$$z \in S\left(\overline{\theta}\right)$$

Note 7.58

If the α-cuts of $f\left(\overline{\lambda},\overline{\beta},\overline{v},\overline{\theta}\right)$ degenerate to some fixed value, then the system performance will be a crisp number; otherwise, it will be a fuzzy number.

Thus, we have the following measures (we leave the details as exercises):

i. The regular busy period:

$$P_0 = K_0(1-\rho) = \frac{(\mu_b - \lambda)(2\mu_v - N)}{(\mu_b - \lambda)(2\mu_v - N) + \theta \times N}, \qquad (7.258)$$

where $N = \lambda + \theta + \mu_v - \sqrt{(\lambda + \theta + \mu_v)^2 - 4\lambda\mu_b}$.

ii. The working vacation period:

$$P_1 = \frac{(\mu_b - \lambda)(2\mu_v - N) + \theta(2\mu_v - N)}{\mu_b(2\mu_v - N)}. \qquad (7.259)$$

iii. The stationary queue length:

$$E(L) = \left(\frac{\mu_b - \lambda}{\mu_b - \lambda}\right)\left(\frac{(\mu_b - \lambda)}{(\mu_b)} + \frac{(\theta - N)}{\mu_b(2\mu_v - N)}\right)^{-1} \times \left(\frac{\mu_v - N}{2\mu_v}\right)$$

$$+ \left(\frac{N}{2\mu_v}\right)\left(\frac{(\mu_b - \mu v)}{(\mu_b)}\right)\left(\frac{z(2\mu_v - N)}{2\mu - Nz}\right). \qquad (7.260)$$

iv. The waiting time:

$$E(W) = \frac{1}{\mu_b - \lambda} + \left(\frac{2\mu_v(\mu_b - N)}{\lambda(2\mu_v\mu_b - N)}\right)\left(\frac{N}{2\mu_v - N}\right). \qquad (7.261)$$

We now obtain the membership function for the performance measures we found above. Hence, for the system, in terms of this membership function, we have:

$$\mu_{\overline{P}_0}(A) = \sup\left\{\mu_{\overline{\lambda}}(w), \mu_{\overline{\beta}}(x), \mu_{\overline{v}}(y), \mu_{\overline{\theta}}(z)/A\right\}$$

$$w \in S(\overline{\lambda})$$

$$x \in S(\overline{\beta})$$ 　　　　　　　　(7.262)

$$y \in S(\overline{v})$$

$$z \in S(\overline{\theta})$$

where

$$A = \frac{(z-w)(2y-M)}{(x-w)(2y-M)+z[M]}, \text{ where } M = w+z+y-\sqrt{(w+z+y)^2 - 4wy}.$$

$$\mu_{\overline{P}_1}(B) = \sup\left\{\mu_{\overline{\lambda}}(w), \mu_{\overline{\beta}}(x), \mu_{\overline{v}}(y), \mu_{\overline{\theta}}(z)/B\right\}$$

$$w \in S(\overline{\lambda})$$

$$x \in S(\overline{\beta})$$ 　　　　　　　，(7.263)

$$y \in S(\overline{v})$$

$$z \in S(\overline{\theta})$$

where

$$B = \frac{(z-w)(2y-M)+z(2y-M)}{2xyM}.$$

$$\mu_{\overline{EL}}(C) = \sup\left\{\mu_{\overline{\lambda}}(\omega), \mu_{\overline{\beta}}(x), \mu_{\overline{v}}(y), \mu_{\overline{\theta}}(z)/C\right\}$$

$$w \in S(\overline{\lambda})$$

$$x \in S(\overline{\beta})$$ 　　　　　　　，(7.264)

$$y \in S(\overline{v})$$

$$z \in S(\overline{\theta})$$

where

$$C = \left(\frac{x-w}{x-w\mu}\right)\left(\frac{y-w}{y} + \frac{zM}{2xy-M}\right)^{-1} \times \frac{M}{2y} + \left(\frac{M}{2y}\right)\left(\frac{x-y}{x}\right)\left(\frac{u(2y-M)}{2y-uM}\right).$$

$$\mu_{\overline{E(W)}}(D) = \sup\left\{\mu_{\overline{\lambda}}(\omega), \mu_{\overline{\beta}}(x), \mu_{\overline{v}}(y), \mu_{\overline{\theta}}(z)/D\right\}$$

$$w \in S(\overline{\lambda})$$

$$x \in S(\overline{\beta})$$, (7.265)

$$y \in S(\overline{v})$$

$$z \in S(\overline{\theta})$$

where

$$D = \frac{1}{(x-w)} + \left(\frac{(2y(x-y))}{w(2xy-M)}\right)\left(\frac{M}{2y-M}\right).$$

Now, using the fuzzy analysis technique, we can find the membership of $P_0, P_1, E(L)$, and $E(W)$ as a function of the parameter α. Hence, the α-cut approach can be used to develop the membership function of $\overline{P}_0, \overline{P}_1, \overline{E(L)}$, and $\overline{E(W)}$.

7.8.2.3 Performance Measures of Interest

The following performance measures are studied for the model under consideration in fuzzy environment.

i. The regular busy period:

Based on Zadeh's extension principle, $\mu_{P_0}(A)$ is the supremum of minimum over $\mu_{\overline{\lambda}}(w), \mu_{\overline{\beta}}(x), \mu_{\overline{v}}(y), \mu_{\overline{\theta}}(z)$, $A = [A]$ to satisfying $\mu_{P_0}(A) = \alpha$, $0 < \alpha \geq 1$. Thus, we have the following four cases:

i. $\mu_{\overline{\lambda}}(w) = \alpha, \mu_{\overline{\beta}} \geq (x) \geq \alpha, \mu_{\overline{v}}(y) \geq \alpha, \mu_{\overline{\theta}}(z) \geq \alpha.$

ii. $\mu_{\overline{\lambda}}(w) \geq \alpha, \mu_{\overline{\beta}} \geq (x) = \alpha, \mu_{\overline{v}}(y) \geq \alpha, \mu_{\overline{\theta}}(z) \geq \alpha.$

iii. $\mu_{\overline{\lambda}}(w) \geq \alpha, \mu_{\overline{\beta}} \geq (x) \geq \alpha, \mu_{\overline{v}}(y) \geq \alpha, \mu_{\overline{\theta}}(z) \geq \alpha.$

iv. $\mu_{\overline{\lambda}}(w) \geq \alpha \ \mu_{\overline{\beta}} \geq (x) \geq \alpha \ \mu_{\overline{v}}(y) \geq \alpha \ \mu_{\overline{\theta}}(z) = \alpha.$

For Case (i), the lower and upper bounds of α-cuts of $\overline{P}_0$ can be obtained through the corresponding parametric nonlinear programs:

$$\left[\overline{P}_0\right]_\alpha^{L_1} = \min_\Omega\{[A]\} \text{ and } \left[\overline{P}_0\right]_\alpha^{U_1} = \max_\Omega\{[A]\}.$$

Similarly, we can calculate the lower and upper bounds of the α-cuts of $\overline{E(L)}$ for the Cases (ii), (iii), and (iv). By considering the cases, simultaneously the lower and upper bounds of the α-cuts of $\overline{P_0}$ can be written as:

$$\left[\overline{P_0}\right]_\alpha^L = \min_\Omega\{[A]\} \text{ and } \left[\overline{P_0}\right]_\alpha^U = \max_\Omega\{[A]\},$$

such that

$$w_\alpha^L \le w \le w_\alpha^U, x_\alpha^L \le x \le w_\alpha^U, y_\alpha^L \le y \le y_\alpha^U, z_\alpha^L \le z \le z_\alpha^U.$$

If both $\left(\overline{P_0}\right)_\alpha^L$ and $\left(\overline{P_0}\right)_\alpha^U$ are invertible with respect to α, the left and right shape function, $L(A) = \left[\left(\overline{P_0}\right)_\alpha^L\right]^{-1}$ and $R(A) = \left[\left(\overline{P_0}\right)_\alpha^U\right]^{-1}$, can be derived from which the membership function $\mu_{\overline{P_0}}(A)$ can be constructed as

$$\mu_{\overline{P_0}}(A) = \begin{cases} L(A), & (P_0)_{\alpha=0}^L \le A \le (P_0)_{\alpha=0}^U, \\ 1, & (P_0)_{\alpha=1}^L \le A \le (P_0)_{\alpha=1}^U, \\ R(A), & (P_0)_{\alpha=1}^L \le A \le (P_0)_{\alpha=0}^U. \end{cases}$$

Similarly, we will have the following:

ii. The working vacation period:

$$\mu_{\overline{P1}}(B) = \begin{cases} L(A), & (P_1)_{\alpha=0}^L \le B \le (P_1)_{\alpha=0}^U, \\ 1, & (P_1)_{\alpha=1}^L \le B \le (P_1)_{\alpha=1}^U, \\ R(A), & (P_1)_{\alpha=1}^L \le B \le (P_1)_{\alpha=0}^U. \end{cases}$$

iii. The stationary queue length:

$$\mu_{\overline{E(L)}}(C) = \begin{cases} L(C), & (E(L))_{\alpha=0}^L \le C \le (E(L))_{\alpha=0}^U, \\ 1, & (E(L))_{\alpha=1}^L \le C \le (E(L))_{\alpha=1}^U, \\ R(C), & (E(L))_{\alpha=1}^L \le C \le (E(L))_{\alpha=0}^U. \end{cases}$$

iv. The waiting time:

$$\mu_{\overline{E(W)}}(D) = \begin{cases} L(D), & (E(W))_{\alpha=0}^L \le D \le (E(W))_{\alpha=0}^U, \\ 1, & (E(W))_{\alpha=1}^L \le D \le (E(W))_{\alpha=1}^U, \\ R(D), & (E(W))_{\alpha=1}^L \le D \le (E(W))_{\alpha=0}^U. \end{cases}$$

Example 7.32 Model's Numerical Example

i. The regular busy period:
Suppose the arrival rate $\bar{\lambda}$, the service rate $\bar{\beta}$, working vacation a task is served at a rate $\bar{v}$, and working vacation time $\bar{\theta}$ are assumed to be trapezoidal fuzzy numbers described, respectively, per hour, by:

$$\bar{\lambda} = [41,42,43,44], \bar{\beta} = [51,52,53,54], \bar{v} = [61,62,63,64], \text{ and } \bar{\theta} = [71,72,73,74].$$

Then,

$$\lambda(\alpha) = \min_{w \in S(\lambda)}\{w \in S(\bar{\lambda}), G(x) \geq \alpha\}, \max_{w \in S(\lambda)}\{w \in S(\bar{\lambda}), G(x) \geq \alpha\},$$

where

$$G(x) = \begin{cases} W - 41, & 41 \leq w \leq 42, \\ 1, & 42 \leq w \leq 43, \\ 44 - w, & 43 \leq w \leq 44. \end{cases}$$

That is,

$$\lambda(\alpha) = [41 + \alpha, 44 - \alpha], \mu(\alpha) = [51 + \alpha, 54 - \alpha].$$

$$v(\alpha) = [61 + \alpha, 64 - \alpha], \theta(\alpha) = [71 + \alpha, 74 - \alpha].$$

It is clear that when $w = w_\alpha^U, x = x_\alpha^U, y = y_\alpha^U, z = z_\alpha^U$, A attains its maximum value, and when $w = w_\alpha^L, x = x_\alpha^L, y = y_\alpha^L, z = z_\alpha^L$, A attains its minimum value (Figure 7.16).

From the generated, for the given input value of $\bar{\lambda}, \bar{\mu}, \bar{\theta}$, and $\bar{\beta}$,
i. For fixed values of w, x, and y, A decreases as z increases.
ii. For fixed values of x, y, and s, A decreases as w increases.
iii. For fixed values of y, x, and w, A decreases as x increases.
iv. For fixed values of z, w, and z, A decreases as y increases.

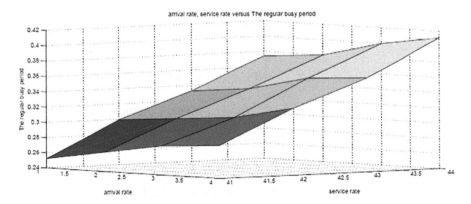

FIGURE 7.16 Arrival rate, service rate versus the regular busy period.

The smallest value of P_0 occurs when w takes its lower bound. That is, $w = 41 + \alpha$ and x, y, and z take their upper bounds given by $x = 54 - \alpha$, $y = 64 - \alpha$, and $z = 74 - \alpha$, respectively. The maximum value of P_0 occurs when $w = 44 - \alpha$, $x = 51 + \alpha$, $y = 61 + \alpha$, and $z = 71 + \alpha$.

If both $\left[|\overline{P_0}| \right]_\alpha^L$ and $\left[\overline{P_0} \right]_U^\alpha$ are invertible with respect to α, then the left shape function $L(A) = \left\{ \left| \overline{P_0} \right|_\alpha^L \right\}^{-1}$ and the right shape function $R(A) = \left\{ \left| \overline{P_0} \right|_\alpha^U \right\}^{-1}$ can be obtained, from which the membership function $\mu_{\overline{P_0}}(A)$ can be constructed as:

$$\mu_{\overline{P_0}}(A) = \begin{cases} L(A), & A_1 \le A \le A_2, \\ 1, & A_2 \le A \le A_3, \\ R(A), & A_3 \le A \le A_4. \end{cases} \tag{7.266}$$

In the same way as before, we obtain the following:

ii. The working vacation period:

$$\mu_{\overline{P_1}}(B) = \begin{cases} L(B), & B_1 \le B \le B_2, \\ 1, & B_2 \le B \le B_3, \\ R(B), & B_3 \le B \le B_4. \end{cases} \tag{7.267}$$

The values of B_1, B_2, B_3, and B_4 are obtained from (7.267) as follows:

$$\mu_{\overline{P_1}}(B) = \begin{cases} L(B), & 1.3203 \le B \le 1.3773, \\ 1, & 1.3773 \le B \le 1.4133, \\ R(B), & 1.4133 \le B \le 1.4628. \end{cases}$$

iii. The stationary queue length:

$$\mu_{\overline{EL}}(C) = \begin{cases} L(C), & C_1 \le C \le C_2, \\ 1, & C_2 \le C \le C_3, \\ R(C), & C_3 \le C \le C_4. \end{cases} \tag{7.268}$$

The values of C_1, C_2, C_3, and C_4 are obtained from (7.269) as follows:

$$\mu_{\overline{E(L)}}(C) = \begin{cases} L(C), & 0.1198 \le C \le 0.1657, \\ 1, & 0.1657 \le C \le 0.2054, \\ R(C), & 0.2054 \le C \le 0.2245. \end{cases}$$

iv. The waiting time:

$$\mu_{\overline{E(W)}}(D) = \begin{cases} L(D), & D_1 \leq D \leq D_2, \\ 1, & D_2 \leq D \leq D_3, \\ R(D), & D_3 \leq D \leq D_4. \end{cases} \tag{7.269}$$

The values of D_1, D_2, D_3, and D_4 are obtained from (7.269) as follows:

$$\mu_{\overline{E(W)}}(D) = \begin{cases} L(D), & 0.0751 \leq D \leq 0.0987, \\ 1, & 0.0987 \leq D \leq 0.1175, \\ R(D), & 0.1175 \leq D \leq 0.1412. \end{cases}$$

Fixing the vacation rate by a crisp value $\bar{\theta} = 72.6$ and $\bar{v} = 61.3$, and taking arrival rate $\bar{\lambda} = [41,42,43,44]$ and service rate $\bar{\beta} = [1,52,53,4]$, both trapezoidal fuzzy numbers, the values of the regular busy period are generated and are plotted in Figure 7.17. From this figure, it can be observed that as $\bar{\lambda}$ increases, the regular busy period increases for the fixed value of the service rate, whereas for fixed value of arrival rate, the regular busy period decreases as the service rate increases.

Similar conclusion can be obtained for the case $\bar{\theta} = 72.6$ and $\bar{v} = 62.3$. Again for fixed values of taking $\bar{\lambda} = [41,42,43,44]$, and service rate $\bar{\beta} = [1,52,53,4]$, the graphs of the working vacation period are drawn in Figures 7.18 and 7.19, respectively. These figures show that as arrival rate increases, the working vacation period also increases, while the working vacation period decreases as the service rate increases in both cases.

It can also be seen from the data generated that the membership value of the regular busy period is 0.41 and the membership value of the working vacation period is 0.46 when the ranges of arrival rate, service rate, and the vacation rate lie in the intervals (41, 42.4), (52, 54.6), and (72.8, 73.4), respectively.

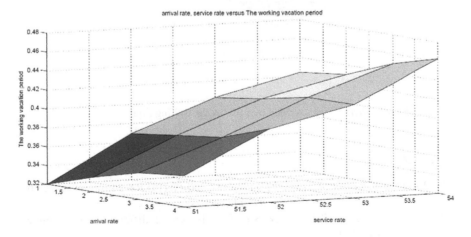

FIGURE 7.17   Arrival rate, service rate versus the working vacation period.

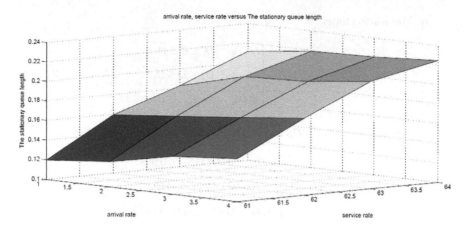

FIGURE 7.18   Arrival rate, service rate versus the stationary queue length.

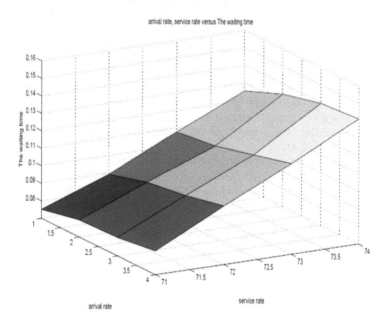

FIGURE 7.19   Arrival rate, service rate versus the waiting time.

EXERCISES

7.1. Find similar relations to (7.212), (7.213), and (7.214) for a finite pure death process and consequently for a B-D process.

7.2. Show that $\Psi_n(s), 0, 1, 2, \ldots, N$, satisfies the set of differential–difference equations (7.230).

7.3. Show the relation (7.241) is true.

7.4. Consider a B-D process where λ_i and μ_i are given as:

$$a. \begin{cases} \lambda_i = 20, & i = 0,1,2,... \\ \mu_i = 25, & i = 1,2,..., \end{cases} \quad b. \begin{cases} \lambda_i = 10, & i = 0,1,2,... \\ \mu_i = 20, & i = 1,2,..., \end{cases}$$

$$c. \begin{cases} \lambda_i = 6, & i = 0,1,2,... \\ \mu_i = 8, & i = 1,2,..., \end{cases} \quad d. \begin{cases} \lambda_i = 10, & i = 0,1,2,... \\ \mu_i = 40, & i = 1,2,..., \end{cases}$$

   i. Find the stationary probability vector.

   ii. Find the stationary probability for state 3.

7.5. Consider a B-D process with Poisson birth and exponential death distributions, where death occurs singly. Denote the birth and death rates as λ and μ, respectively, given as:

$$a. \begin{cases} \lambda = 10, \\ \mu = 11. \end{cases} \quad b. \begin{cases} \lambda = 30, \\ \mu = 50. \end{cases} \quad c. \begin{cases} \lambda = 2, \\ \mu = 8. \end{cases}$$

Let E_k denote the state of the system; that is, population at any time is k. Find the

   i. Stationary probability vector,

   ii. Stationary probability that population is 2,

   iii. Stationary probability that there are no one in the system.

7.6. A small town beauty salon has two stylists, Angel and Barbara, and an additional chair for one waiting customer. If a customer arrives when there are three customers in the salon, she leaves. From the past experience, the average time between arrivals of customers is 30 minutes. Angel completes serving a customer at a rate of 30 minutes, while Barbara does it in an hour. Angel has priority serving the customer since she is faster than Barbara, when there is only one customer in the shop. Further assume that inter-arrival times and service times of customers are independent exponential random variables.

   a. Draw a state transition diagram with possible states and corresponding birth/death rates.

   b. After a long time pass, what is the probability that the shop is empty?

   c. What is the pmf of the number of customers in the shop?

   d. After a long time pass, what is the probability that an arriving customer is turned away?

   e. What is the distribution of the number of customers arriving in the first 3 hours?

   f. What is the probability that no customer arrives in a 3-hour time interval?

   g. What is the mean number of arrivals in 3 hours?

   h. What is the distribution of the time until the fifth customer arrives?

   i. What is the probability that the time until the fifth customer arrives is <2.5 hours?

7.7. Consider a walker who walks on a real line starting at 0 with moving one step forward with probability p and backward with probability q, $p+q=1$. Let X_n describe the position of the walker after n steps.

    a. What is the probability that walker is at the point 0 on the line after two steps?

    b. What is the probability that walker is at the point -1 on the line after three steps?

    c. What is the probability that walker is at the point 3 on the line after three steps?

    d. Suppose the walker is at point 4 after 10 steps, does the probability that it will be at point 8 after 16 steps (6 more steps) depend on how it moves to point 4 within the first 10 steps?

    e. Are $X_{10} - X_4$ and $X_{16} - X_{12}$ independent?

    f. Are $X_{10} - X_4$ and $X_{12} - X_8$ independent?

7.8. Refer to Definition 7.4. Answer the following questions:

    i. What is the probability that the sequence $\{S_n, n \geq 1\}$ contains a term for which a threshold at τ is crossed?

    ii. What is the distribution of the smallest n for which $S_n, \geq \tau$?

7.9. Derive the mean and variance of the population size for the finite pure birth process.

7.10. Show that $\Psi_n(s)$ satisfies the system (7.222).

7.11. Prove (7.233).

7.12. Show that γ_ks given by (7.239) are positive.

7.13. Prove (7.243).

7.14. Prove (7.250).

7.15. Prove (7.251).

7.16. Prove (7.252).

7.17. Prove (7.253).

Appendix

See Tables A.1–A.6.

TABLE A.1
Binomial Probability Distribution

Following are the probabilities for x successes in n independent trials, with the probability of success p

n	x	.010	.050	.100	.150	.200	.250	.300	.350	.400	.450	.500	.550	.600	.650	.700	.750	.800	.850	.900	.950
5	0	.951	.774	.590	.444	.328	.237	.168	.116	.078	.050	.031	.019	.010	.005	.002	.001	.000	.000	.000	.000
	1	.048	.204	.328	.392	.410	.396	.360	.312	.259	.206	.156	.113	.077	.049	.028	.015	.006	.002	.000	.000
	2	.001	.021	.073	.138	.205	.264	.309	.336	.346	.337	.312	.276	.230	.181	.132	.088	.051	.024	.008	.001
	3	.000	.001	.008	.024	.051	.088	.132	.181	.230	.276	.312	.337	.346	.336	.309	.264	.205	.138	.073	.021
	4	.000	.000	.000	.002	.006	.015	.028	.049	.077	.113	.156	.206	.259	.312	.360	.396	.410	.392	.328	.204
	5	.000	.000	.000	.000	.000	.001	.002	.005	.010	.019	.031	.050	.078	.116	.168	.237	.328	.444	.590	.774
10	0	.904	.599	.349	.197	.107	.056	.028	.014	.006	.003	.001	.000	.000	.000	.000	.000	.000	.000	.000	.000
	1	.091	.315	.387	.347	.268	.188	.121	.072	.040	.021	.010	.004	.002	.000	.000	.000	.000	.000	.000	.000
	2	.004	.075	.194	.276	.302	.282	.233	.176	.121	.076	.044	.023	.011	.004	.001	.000	.000	.000	.000	.000
	3	.000	.010	.057	.130	.201	.250	.267	.252	.215	.166	.117	.075	.042	.021	.009	.003	.001	.000	.000	.000
	4	.000	.001	.011	.040	.088	.146	.200	.238	.251	.238	.205	.160	.111	.069	.037	.016	.006	.001	.000	.000
	5	.000	.000	.001	.008	.026	.058	.103	.154	.201	.234	.246	.234	.201	.154	.103	.058	.026	.008	.001	.000
	6	.000	.000	.000	.001	.006	.016	.037	.069	.111	.160	.205	.238	.251	.238	.200	.146	.088	.040	.011	.001
	7	.000	.000	.000	.000	.001	.003	.009	.021	.042	.075	.117	.166	.215	.252	.267	.250	.201	.130	.057	.010
	8	.000	.000	.000	.000	.000	.001	.004	.011	.023	.044	.076	.121	.176	.233	.282	.302	.302	.276	.194	.070
	9	.000	.000	.000	.000	.000	.000	.000	.002	.004	.010	.021	.040	.072	.121	.188	.268	.347	.387	.315
	10	.000	.000	.000	.000	.000	.000	.000	.000	.000	.001	.003	.006	.014	.028	.056	.107	.197	.349	.599
15	0	.860	.463	.206	.087	.035	.013	.005	.002	.000	.000	.000	.000	.000	.000	.000	.000	.000	.000	.000	.000
	1	.130	.366	.343	.231	.132	.067	.031	.013	.005	.002	.000	.000	.000	.000	.000	.000	.000	.000	.000	.000
	2	.009	.135	.267	.286	.231	.156	.092	.048	.022	.009	.003	.001	.000	.000	.000	.000	.000	.000	.000	.000
	3	.000	.031	.129	.218	.250	.225	.170	.111	.063	.032	.014	.005	.002	.000	.000	.000	.000	.000	.000	.000
	4	.000	.005	.043	.116	.188	.225	.219	.179	.127	.078	.042	.019	.007	.002	.001	.000	.000	.000	.000	.000
	5	.000	.001	.010	.045	.103	.165	.206	.212	.186	.140	.092	.051	.024	.010	.003	.001	.000	.000	.000	.000
	6	.000	.000	.002	.013	.043	.092	.147	.191	.207	.191	.153	.105	.061	.030	.012	.003	.001	.000	.000	.000
	7	.000	.000	.000	.003	.014	.039	.081	.132	.177	.201	.196	.165	.118	.071	.035	.013	.003	.001	.000	.000
	8	.000	.000	.000	.001	.003	.013	.035	.071	.118	.165	.196	.201	.177	.132	.081	.039	.014	.003	.000	.000
	9	.000	.000	.000	.000	.001	.003	.012	.030	.061	.105	.153	.191	.207	.191	.147	.092	.043	.013	.002	.000
	10	.000	.000	.000	.000	.000	.001	.003	.010	.024	.051	.092	.140	.186	.212	.206	.165	.103	.045	.010	.001
	11	.000	.000	.000	.000	.000	.000	.001	.002	.007	.019	.042	.078	.127	.179	.219	.225	.188	.116	.043	.005
	12	.000	.000	.000	.000	.000	.000	.000	.000	.002	.005	.014	.032	.063	.111	.170	.225	.250	.218	.129	.031
	13	.000	.000	.000	.000	.000	.000	.000	.000	.000	.001	.003	.009	.022	.048	.092	.156	.231	.286	.267	.135
	14	.000	.000	.000	.000	.000	.000	.000	.000	.000	.000	.000	.002	.005	.013	.031	.067	.132	.231	.343	.366
	15	.000	.000	.000	.000	.000	.000	.000	.000	.000	.000	.000	.000	.002	.005	.013	.035	.087	.206	.463
20	0	.818	.358	.122	.039	.012	.003	.001	.000	.000	.000	.000	.000	.000	.000	.000	.000	.000	.000	.000	.000
	1	.165	.377	.270	.137	.058	.021	.007	.002	.000	.000	.000	.000	.000	.000	.000	.000	.000	.000	.000	.000
	2	.016	.189	.285	.229	.137	.067	.028	.010	.003	.001	.000	.000	.000	.000	.000	.000	.000	.000	.000	.000
	3	.001	.060	.190	.243	.205	.134	.072	.032	.012	.004	.001	.000	.000	.000	.000	.000	.000	.000	.000	.000

(Continued)

TABLE A.1 (*Continued*)
Binomial Probability Distribution

Following are the probabilities for *x* successes in *n* independent trials, with the probability of success *p*

n	x	.010	.050	.100	.150	.200	.250	.300	.350	.400	.450	.500	.550	.600	.650	.700	.750	.800	.850	.900	.950
	4	.000	.013	.090	.182	.218	.190	.130	.074	.035	.014	.005	.001	.000	.000	.000	.000	.000	.000	.000	.000
	5	.000	.002	.032	.103	.175	.202	.179	.127	.075	.036	.015	.005	.001	.000	.000	.000	.000	.000	.000	.000
	6	.000	.000	.009	.045	.109	.169	.192	.171	.124	.075	.037	.015	.005	.001	.000	.000	.000	.000	.000	.000
	7	.000	.000	.002	.016	.055	.112	.164	.184	.166	.122	.074	.037	.015	.005	.001	.000	.000	.000	.000	.000
	8	.000	.000	.000	.005	.022	.061	.114	.161	.180	.162	.120	.073	.035	.014	.004	.001	.000	.000	.000	.000
	9	.000	.000	.000	.001	.007	.027	.065	.116	.160	.177	.160	.119	.071	.034	.012	.003	.000	.000	.000	.000
	10	.000	.000	.000	.000	.002	.010	.031	.069	.117	.159	.176	.159	.117	.069	.031	.010	.002	.000	.000	.000
	11	.000	.000	.000	.000	.000	.003	.012	.034	.071	.119	.160	.177	.160	.116	.065	.027	.007	.001	.000	.000
	12	.000	.000	.000	.000	.000	.001	.004	.014	.035	.073	.120	.162	.180	.161	.114	.061	.022	.005	.000	.000
	13	.000	.000	.000	.000	.000	.000	.001	.005	.015	.037	.074	.122	.166	.184	.164	.112	.055	.016	.002	.000
	14	.000	.000	.000	.000	.000	.000	.000	.001	.005	.015	.037	.075	.124	.171	.192	.169	.109	.045	.009	.000
	15	.000	.000	.000	.000	.000	.000	.000	.000	.001	.005	.015	.036	.075	.127	.179	.202	.175	.103	.032	.002
	16	.000	.000	.000	.000	.000	.000	.000	.000	.000	.001	.005	.014	.035	.074	.130	.190	.218	.182	.090	.013
	17	.000	.000	.000	.000	.000	.000	.000	.000	.000	.000	.001	.004	.012	.032	.072	.134	.205	.243	.190	.060
	18	.000	.000	.000	.000	.000	.000	.000	.000	.000	.000	.000	.001	.003	.010	.028	.067	.137	.229	.285	.189
	19	.000	.000	.000	.000	.000	.000	.000	.000	.000	.000	.000	.000	.002	.007	.021	.058	.137	.270	.377	
	20	.000	.000	.000	.000	.000	.000	.000	.000	.000	.000	.000	.000	.000	.001	.003	.012	.039	.122	.358	

TABLE A.2
Poisson Probability Distribution

Following are the cumulative probabilities for *x* events in a unit time with the rate of events μ

					μ					
x	.01	.01	.02	.03	.04	.05	.06	.07	.08	.09
0	.995	.99	.9802	.9704	.9608	.9512	.9418	.9324	.9231	.9139
1	1.0000	1.0000	.9998	.9996	.9992	.9988	.9983	.9977	.997	.9962
2	1.0000	1.0000	1.0000	1.0000	1.0000	1.0000	1.0000	.9999	.9999	.9999
3	1.0000	1.0000	1.0000	1.0000	1.0000	1.0000	1.0000	1.0000	1.0000	1.0000

					μ					
x	.10	.20	.30	.40	.50	.60	.70	.80	.90	1.00
0	.9048	.8187	.7408	.6703	.6065	.5488	.4966	.4493	.4066	.3679
1	.9953	.9825	.9631	.9384	.9098	.8781	.8442	.8088	.7725	.7358
2	.9998	.9989	.9964	.9921	.9856	.9769	.9659	.9526	.9371	.9197
3	1.0000	.9999	.9997	.9992	.9982	.9966	.9942	.9909	.9865	.981
4	1.0000	1.0000	1.0000	.9999	.9998	.9996	.9992	.9986	.9977	.9963
5	1.0000	1.0000	1.0000	1.0000	1.0000	1.0000	.9999	.9998	.9997	.9994
6	1.0000	1.0000	1.0000	1.0000	1.0000	1.0000	1.0000	1.0000	1.0000	.9999
7	1.0000	1.0000	1.0000	1.0000	1.0000	1.0000	1.0000	1.0000	1.0000	1.0000

TABLE A.3

F-Distribution Table (Critical Values of F at 5% Significance Level)

v_2	v_1														
	1	2	3	4	5	6	7	8	9	10	12	14	16	18	20
1	161.45	199.50	215.71	224.58	230.16	233.99	236.77	238.88	240.54	241.88	243.91	245.36	246.46	247.32	248.01
2	18.51	19.00	19.16	19.25	19.30	19.33	19.35	19.37	19.38	19.40	19.41	19.42	19.43	19.44	19.45
3	10.13	9.55	9.28	9.12	9.01	8.94	8.89	8.85	8.81	8.79	8.74	8.71	8.69	8.67	8.66
4	7.71	6.94	6.59	6.39	6.26	6.16	6.09	6.04	6.00	5.96	5.91	5.87	5.84	5.82	5.80
5	6.61	5.79	5.41	5.19	5.05	4.95	4.88	4.82	4.77	4.74	4.68	4.64	4.60	4.58	4.56
6	5.99	5.14	4.76	4.53	4.39	4.28	4.21	4.15	4.10	4.06	4.00	3.96	3.92	3.90	3.87
7	5.59	4.74	4.35	4.12	3.97	3.87	3.79	3.73	3.68	3.64	3.57	3.53	3.49	3.47	3.44
8	5.32	4.46	4.07	3.84	3.69	3.58	3.50	3.44	3.39	3.35	3.28	3.24	3.20	3.17	3.15
9	5.12	4.26	3.86	3.63	3.48	3.37	3.29	3.23	3.18	3.14	3.07	3.03	2.99	2.96	2.94
10	4.96	4.10	3.71	3.48	3.33	3.22	3.14	3.07	3.02	2.98	2.91	2.86	2.83	2.80	2.77
11	4.84	3.98	3.59	3.36	3.20	3.09	3.01	2.95	2.90	2.85	2.79	2.74	2.70	2.67	2.65
12	4.75	3.89	3.49	3.26	3.11	3.00	2.91	2.85	2.80	2.75	2.69	2.64	2.60	2.57	2.54
13	4.67	3.81	3.41	3.18	3.03	2.92	2.83	2.77	2.71	2.67	2.60	2.55	2.51	2.48	2.46
14	4.60	3.74	3.34	3.11	2.96	2.85	2.76	2.70	2.65	2.60	2.53	2.48	2.44	2.41	2.39
15	4.54	3.68	3.29	3.06	2.90	2.79	2.71	2.64	2.59	2.54	2.48	2.42	2.38	2.35	2.33
16	4.49	3.63	3.24	3.01	2.85	2.74	2.66	2.59	2.54	2.49	2.42	2.37	2.33	2.30	2.28
17	4.45	3.59	3.20	2.96	2.81	2.70	2.61	2.55	2.49	2.45	2.38	2.33	2.29	2.26	2.23
18	4.41	3.55	3.16	2.93	2.77	2.66	2.58	2.51	2.46	2.41	2.34	2.29	2.25	2.22	2.19
19	4.38	3.52	3.13	2.90	2.74	2.63	2.54	2.48	2.42	2.38	2.31	2.26	2.21	2.18	2.16

(Continued)

TABLE A.3 (Continued)
F-Distribution Table (Critical Values of F at 5% Significance Level)

v_2	v_1 1	2	3	4	5	6	7	8	9	10	12	14	16	18	20
20	4.35	3.49	3.10	2.87	2.71	2.60	2.51	2.45	2.39	2.35	2.28	2.22	2.18	2.15	2.12
21	4.32	3.47	3.07	2.84	2.68	2.57	2.49	2.42	2.37	2.32	2.25	2.20	2.16	2.12	2.10
22	4.30	3.44	3.05	2.82	2.66	2.55	2.46	2.40	2.34	2.30	2.23	2.17	2.13	2.10	2.07
23	4.28	3.42	3.03	2.80	2.64	2.53	2.44	2.37	2.32	2.27	2.20	2.15	2.11	2.05	2.05
24	4.26	3.40	3.01	2.78	2.62	2.51	2.42	2.36	2.30	2.25	2.18	2.13	2.09	2.05	2.03
25	4.24	3.39	2.99	2.76	2.60	2.49	2.40	2.34	2.28	2.24	2.16	2.11	2.07	2.04	2.01
26	4.22	3.37	2.98	2.74	2.59	2.47	2.39	2.32	2.27	2.22	2.15	2.09	2.05	2.02	1.99
27	4.21	3.35	2.96	2.73	2.57	2.46	2.37	2.31	2.25	2.20	2.13	2.08	2.04	2.00	1.97
28	4.20	3.34	2.95	2.71	2.56	2.45	2.36	2.29	2.24	2.19	2.12	2.06	2.02	1.99	1.96
29	4.18	3.33	2.93	2.70	2.55	2.43	2.35	2.28	2.22	2.18	2.10	2.05	2.01	1.97	1.94
30	4.17	3.32	2.92	2.69	2.53	2.42	2.33	2.27	2.21	2.16	2.09	2.04	1.99	1.96	1.93
35	4.12	3.27	2.87	2.64	2.49	2.37	2.29	2.22	2.16	2.11	2.04	1.99	1.94	1.91	1.88
40	4.08	3.23	2.84	2.61	2.45	2.34	2.25	2.18	2.12	2.08	2.00	1.95	1.90	1.87	1.84
50	4.03	3.18	2.79	2.56	2.40	2.29	2.20	2.13	2.07	2.03	1.95	1.89	1.85	1.81	1.78
60	4.00	3.15	2.76	2.53	2.37	2.25	2.17	2.10	2.04	1.99	1.92	1.86	1.82	1.78	1.75

(Continued)

TABLE A.3 (Continued)
F-Distribution Table (Critical Values of F at 5% Significance Level)

v_2	v_1 1	2	3	4	5	6	7	8	9	10	12	14	16	18	20
70	3.98	3.13	2.74	2.50	2.35	2.23	2.14	2.07	2.02	1.97	1.89	1.84	1.79	1.75	1.72
80	3.96	3.11	2.72	2.49	2.33	2.21	2.13	2.06	2.00	1.95	1.88	1.82	1.77	1.73	1.70
90	3.95	3.10	2.71	2.47	2.32	2.20	2.11	2.04	1.99	1.94	1.86	1.80	1.76	1.72	1.69
100	3.94	3.09	2.70	2.46	2.31	2.19	2.10	2.03	1.97	1.93	1.85	1.79	1.75	1.71	1.68
120	3.92	3.07	2.68	2.45	2.29	2.18	2.09	2.02	1.96	1.91	1.83	1.78	1.73	1.69	1.66
150	3.90	3.06	2.66	2.43	2.27	2.16	2.07	2.00	1.94	1.89	1.82	1.76	1.71	1.67	1.64
200	3.89	3.04	2.65	2.42	2.26	2.14	2.06	1.98	1.93	1.88	1.80	1.74	1.69	1.66	1.62
250	3.88	3.03	2.64	2.41	2.25	2.13	2.05	1.98	1.92	1.87	1.79	1.73	1.68	1.65	1.61
300	3.87	3.03	2.63	2.40	2.24	2.13	2.04	1.97	1.91	1.86	1.78	1.72	1.68	1.64	1.61
400	3.86	3.02	2.63	2.39	2.24	2.12	2.03	1.96	1.90	1.85	1.78	1.72	1.67	1.63	1.60
500	3.86	3.01	2.62	2.39	2.23	2.12	2.03	1.96	1.90	1.85	1.77	1.71	1.66	1.62	1.59
600	3.86	3.01	2.62	2.39	2.23	2.11	2.02	1.95	1.90	1.85	1.77	1.71	1.66	1.62	1.59
750	3.85	3.01	2.62	2.38	2.23	2.11	2.02	1.95	1.89	1.84	1.77	1.70	1.66	1.62	1.58
1000	3.85	3.00	2.61	2.38	2.22	2.11	2.02	1.95	1.89	1.84	1.76	1.70	1.65	1.61	1.58

TABLE A.4
The Standard Normal Probability Distribution

$$F(z) = P(X \le z) = \frac{1}{\sqrt{2\pi}} \int_{-\infty}^{z} e^{-t^2/2}\, dt$$

$$0 \le P(Z \le z) \le 0.09$$

$P(Z \le z)$

z	.00	.01	.02	.03	.04	.05	.06	.07	.08	.09
-3.4	.000337	.000325	.000313	.000302	.000291	.00028	.00027	.0026	.000251	.000242
-3.3	.000483	.000466	.00045	.000434	.000419	.000404	.00039	.000376	.000362	.000349
-3.2	.000687	.000664	.000641	.000619	.000598	.000577	.000557	.000538	.000519	.000501
-3.1	.000968	.000935	.000904	.000874	.000845	.000816	.000789	.000762	.000736	.000711
-3.0	.00135	.001306	.001264	.001223	.001183	.001144	.001107	.00107	.001035	.001001
-2.9	.001866	.001807	.00175	.001695	.001641	.001589	.001538	.001489	.001441	.001395
-2.8	.002555	.002477	.002401	.002327	.002256	.002186	.002118	.002052	.001988	.001926
-2.7	.003467	.003364	.003264	.003167	.003072	.002980	.002890	.002803	.002718	.002635
-2.6	.004661	.004527	.004396	.004269	.004145	.004025	.003907	.003793	.003681	.003573
-2.5	.006210	.006037	.005868	.005703	.005543	.005386	.005234	.005085	.00494	.004799
-2.4	.008198	.007976	.00776	.007549	.007344	.007143	.006947	.006756	.006569	.006387
-2.3	.010724	.010444	.01017	.009903	.009642	.009387	.009137	.008894	.008656	.008424
-2.2	.013903	.013553	.013209	.012874	.012545	.012224	.011911	.011604	.011304	.011011
-2.1	.017864	.017429	.017003	.016586	.016177	.015778	.015386	.015003	.014629	.014262
-2.0	.022750	.022216	.021692	.021178	.020675	.020182	.019699	.019226	.018763	.018309
-1.9	.028717	.028067	.027429	.026803	.02619	.025588	.024998	.024419	.023852	.023295
-1.8	.035930	.035148	.03438	.033625	.032884	.032157	.031443	.030742	.030054	.029379

(Continued)

TABLE A.4 (Continued)
The Standard Normal Probability Distribution

$$F(z) = P(X \le z) = \frac{1}{\sqrt{2\pi}} \int_{-\infty}^{z} e^{-t^2/2}\, dt$$

$0 \le P(Z \le z) \le 0.09$

$P(Z \le z)$

z	.00	.01	.02	.03	.04	.05	.06	.07	.09
-1.7	.044565	.043633	.042716	.041815	.040059	.039204	.038364	.037538	.036727
-1.6	.054799	.053699	.052616	.051551	.049471	.048457	.04746	.046479	.045514
-1.5	.066807	.065522	.064255	.063008	.060571	.05938	.058208	.057053	.055917
-1.4	.080757	.07927	.077804	.076358	.073529	.072145	.070781	.069437	.068112
-1.3	.096800	.095098	.093418	.091759	.088508	.086915	.085343	.083793	.082264
-1.2	.11507	.113139	.111232	.109349	.10565	.103835	.102042	.100273	.098525
-1.1	.135666	.133500	.131357	.129238	.125072	.123024	.121000	.119000	.117023
-1.0	.158655	.156248	.153864	.151505	.146859	.144572	.142310	.140071	.137857
-.9	.18406	.181411	.178786	.176186	.171056	.168528	.166023	.163543	.161087
-.8	.211855	.20897	.206108	.203269	.197663	.194895	.192150	.18943	.186733
-.7	.241964	.238852	.235762	.232695	.226627	.223627	.22065	.217695	.214764
-.6	.274253	.270931	.267629	.264347	.257846	.254627	.251429	.248252	.245097
-.5	.308538	.305026	.301532	.298056	.29116	.28774	.284339	.280957	.277595
-.4	.344578	.340903	.337243	.333598	.326355	.322758	.319178	.315614	.312067
-.3	.382089	.378280	.374484	.370700	.363169	.359424	.355691	.351973	.348268
-.2	.420740	.416834	.412936	.409046	.401294	.397432	.393580	.389739	.385908
-.1	.460172	.456205	.452242	.448283	.440382	.436441	.432505	.428576	.424655

TABLE A.5
The (Student) *t*-Probability Distribution

$$F(x) = P(X \leq x) = \frac{\Gamma\left(\dfrac{d+1}{2}\right)}{\Gamma\left(\dfrac{d}{2}\right)\sqrt{d\pi}} \int_{-\infty}^{x} \left(1 + \frac{t^2}{d}\right)^{-(d+1)/2} dt$$

$$0.0005 \leq P(X \leq x) \leq 0.30$$

$P(X \leq x)$

d	.30	.20	.10	.05	.025	.01	.005	.001	.0005
1	.726543	1.376382	3.077684	6.313752	12.706205	31.820516	63.656741	318.308839	636.619249
2	.617213	1.060660	1.885618	2.919986	4.302653	6.964557	9.924843	22.327125	31.599055
3	.584390	.978472	1.637744	2.353363	3.182446	4.540703	5.840909	10.214532	12.923979
4	.568649	.940965	1.533206	2.131847	2.776445	3.746947	4.604095	7.173182	8.610302
5	.559430	.919544	1.475884	2.015048	2.570582	3.364930	4.032143	5.893430	6.868827
6	.553381	.905703	1.439756	1.943180	2.446912	3.142668	3.707428	5.207626	5.958816
7	.549110	.896030	1.414924	1.894579	2.364624	2.997952	3.499483	4.785290	5.407883
8	.545934	.888890	1.396815	1.859548	2.306004	2.896459	3.355387	4.500791	5.041305
9	.543480	.883404	1.383029	1.833113	2.262157	2.821438	3.249836	4.296806	4.780913
10	.541528	.879058	1.372184	1.812461	2.228139	2.763769	3.169273	4.143700	4.586894
11	.539938	.875530	1.363430	1.795885	2.200985	2.718079	3.105807	4.024701	4.436979
12	.538618	.872609	1.356217	1.782288	2.178813	2.680998	3.054540	3.929633	4.317791
13	.537504	.870152	1.350171	1.770933	2.160369	2.650309	3.012276	3.851982	4.220832
14	.536552	.868055	1.345030	1.761310	2.144787	2.624494	2.976843	3.787390	4.140454
15	.535729	.866245	1.340606	1.753050	2.131450	2.602480	2.946713	3.732834	4.072765
16	.535010	.864667	1.336757	1.745884	2.119905	2.583487	2.920782	3.686155	4.014996

(*Continued*)

TABLE A.5 (Continued)
The (Student) t-Probability Distribution

$P(X \leq x)$

d	.30	.20	.10	.05	.025	.01	.005	.001	.0005
17	.534377	.863279	1.333379	1.739607	2.109816	2.566934	2.898231	3.645767	3.965126
18	.533816	.862049	1.330391	1.734064	2.100922	2.552380	2.878440	3.610485	3.921646
19	.533314	.860951	1.327728	1.729133	2.093024	2.539483	2.860935	3.579400	3.883406
20	.532863	.859964	1.325341	1.724718	2.085963	2.527977	2.845340	3.551808	3.849516
21	.532455	.859074	1.323188	1.720743	2.079614	2.517648	2.831360	3.527154	3.819277
22	.532085	.858266	1.321237	1.717144	2.073873	2.508325	2.818756	3.504992	3.792131
23	.531747	.857530	1.319460	1.713872	2.068658	2.499867	2.807336	3.484964	3.767627
24	.531438	.856855	1.317836	1.710882	2.063899	2.492159	2.796940	3.466777	3.745399
25	.531154	.856236	1.316345	1.708141	2.059539	2.485107	2.787436	3.450189	3.725144
26	.530892	.855665	1.314972	1.705618	2.055529	2.478630	2.778715	3.434997	3.706612
27	.530649	.855137	1.313703	1.703288	2.051831	2.472660	2.770683	3.421034	3.689592
28	.530424	.854647	1.312527	1.701131	2.048407	2.467140	2.763262	3.408155	3.673906
29	.530214	.854192	1.311434	1.699127	2.045230	2.462021	2.756386	3.396240	3.659405
30	.530019	.853767	1.310415	1.697261	2.042272	2.457262	2.749996	3.385185	3.645959
32	.529665	.852998	1.308573	1.693889	2.036933	2.448678	2.738481	3.365306	3.621802
34	.529353	.852321	1.306952	1.690924	2.032244	2.441150	2.728394	3.347934	3.600716
36	.529076	.851720	1.305514	1.688298	2.028094	2.434494	2.719485	3.332624	3.582150
38	.528828	.851183	1.304230	1.685954	2.024394	2.428568	2.711558	3.319030	3.565678
40	.528606	.850700	1.303077	1.683851	2.021075	2.423257	2.704459	3.306878	3.550966
50	.527760	.848869	1.298714	1.675905	2.008559	2.403272	2.677793	3.261409	3.496013
60	.527198	.847653	1.295821	1.670649	2.000298	2.390119	2.660283	3.231709	3.460200
120	.526797	.846786	1.293763	1.666914	1.994437	2.380807	2.647905	3.210789	3.435015

TABLE A.6A
Critical Values for Chi-squareDistribution

d.f.	.995	.99	.975	.95	.9	.1	.05	.025	.01
1	.00	.00	.00	.00	.02	2.71	3.84	5.02	6.63
2	.01	.02	.05	.10	.21	4.61	5.99	7.38	9.21
3	.07	.11	.22	.35	.58	6.25	7.81	9.35	11.34
4	.21	.30	.48	.71	1.06	7.78	9.49	11.14	13.28
5	.41	.55	.83	1.15	1.61	9.24	11.07	12.83	15.09
6	.68	.87	1.24	1.64	2.20	10.64	12.59	14.45	16.81
7	.99	1.24	1.69	2.17	2.83	12.02	14.07	16.01	18.48
8	1.34	1.65	2.18	2.73	3.49	13.36	15.51	17.53	20.09
9	1.73	2.09	2.7	3.33	4.17	14.68	16.92	19.02	21.67
10	2.16	2.56	3.25	3.94	4.87	15.99	18.31	20.48	23.21
11	2.60	3.05	3.82	4.57	5.58	17.28	19.68	21.92	24.72
12	3.07	3.57	4.40	5.23	6.30	18.55	21.03	23.34	26.22
13	3.57	4.11	5.01	5.89	7.04	19.81	22.36	24.74	27.69
14	4.07	4.66	5.63	6.57	7.79	21.06	23.68	26.12	29.14
15	4.60	5.23	6.26	7.26	8.55	22.31	25.00	27.49	30.58
16	5.14	5.81	6.91	7.96	9.31	23.54	26.30	28.85	32.00
17	5.70	6.41	7.56	8.67	10.09	24.77	27.59	30.19	33.41
18	6.26	7.01	8.23	9.39	10.86	25.99	28.87	31.53	34.81
19	6.84	7.63	8.91	10.12	11.05	27.20	30.14	32.85	36.19
20	7.43	8.26	9.59	10.85	12.44	28.41	31.41	34.17	37.57
22	8.64	9.54	10.98	12.34	14.04	30.81	33.92	36.78	40.29
24	9.89	10.86	12.40	13.85	15.66	33.20	36.42	39.36	42.98
26	11.16	12.20	13.84	15.38	17.29	35.56	38.89	41.92	45.64

(Continued)

TABLE A.6A (*Continued*)
Critical Values for Chi-squareDistribution

d.f.	.995	.99	.975	.95	.9	.1	.05	.025	.01
28	12.46	13.56	15.31	16.93	18.94	37.92	41.34	44.46	48.28
30	13.79	14.95	16.79	18.49	20.60	40.26	43.77	46.98	50.89
32	15.13	16.36	18.29	20.07	22.27	42.58	46.19	49.48	53.49
34	16.50	17.79	19.81	21.66	23.95	44.90	48.60	51.97	56.06
38	19.29	20.69	22.88	24.88	27.34	49.51	53.38	56.90	61.16
42	22.14	23.65	26.00	28.14	30.77	54.09	58.12	61.78	66.21
46	25.04	26.66	29.16	31.44	34.22	58.64	62.83	66.62	71.20
50	27.99	29.71	32.36	34.76	37.69	63.17	67.50	71.42	76.15
55	31.73	33.57	36.40	38.96	42.06	68.80	73.31	77.38	82.29
60	35.53	37.48	40.48	43.19	46.46	74.4	79.03	83.3	88.38
65	39.38	41.44	44.60	47.45	50.88	79.97	84.82	89.18	94.42
70	43.28	45.44	48.76	51.74	55.33	85.53	90.53	95.02	100.43
75	47.21	49.48	52.94	56.05	59.79	91.06	96.22	100.84	106.39
80	51.17	53.54	57.15	60.39	64.28	96.58	101.88	106.63	112.33
85	55.17	57.63	61.39	4.75	68.78	102.08	107.52	112.39	118.24
90	50.20	61.75	65.65	69.13	73.29	107.57	113.15	118.14	124.12
95	63.25	65.90	69.92	73.52	77.82	113.04	118.75	123.86	129.97
100	67.33	70.06	74.22	77.93	82.36	113.5	124.34	129.56	135.81

TABLE A.6B
Critical Values for Sign Test

	One-Tailed				
	$\alpha = 0.005$	$\alpha = 0.01$	$\alpha = 0.025$	$\alpha = 0.05$	$\alpha = 0.010$
			Two-Tailed		
n	$\alpha = 0.005$	$\alpha = 0.005$	$\alpha = 0.005$	$\alpha = 0.005$	$\alpha = 0.020$
1	–	–	–	–	–
2	–	–	–	–	–
3	–	–	–	–	–
4	–	–	–	–	0
5	–	–	–	0	0
6	–	–	0	0	0
7	–	0	0	0	1
8	0	0	0	1	1
9	0	0	1	1	2
10	0	0	1	1	2
11	0	1	1	2	2
12	1	1	2	2	3
13	1	1	2	3	3
14	1	2	2	3	4
15	2	2	3	3	4
16	2	2	3	4	4
17	2	3	4	4	5
18	3	3	4	5	5
19	3	4	4	5	6
20	3	4	5	5	6
21	4	4	5	6	7
22	4	5	5	6	7
23	4	5	6	7	7
24	5	5	6	7	8
25	5	6	7	7	8

TABLE A.6C
Critical Values for Mann–Whitney Test

Two-Tailed $\alpha = 0.05$ (One-Tailed $\alpha = 0.25$)

n_1/n_2	1	2	3	4	5	6	7	8	9	10	11	12	13	14	15	16	17	18	19	20
1	—	—	—	—	—	—	—	—	—	—	—	—	—	—	—	—	—	—	—	—
2	—	—	—	—	—	—	—	—	0	0	0	1	1	1	1	1	2	2	2	2
3	—	—	—	—	0	1	1	2	2	3	3	4	4	5	5	6	6	7	7	8
4	—	—	—	0	1	2	3	4	4	5	6	7	8	9	10	11	11	12	13	13
5	—	—	0	1	2	3	5	6	7	8	9	11	12	13	14	15	17	18	19	20
6	—	—	1	2	3	5	6	8	10	11	13	14	16	17	19	21	22	24	25	27
7	—	—	1	3	5	6	8	10	12	14	16	18	20	22	24	26	28	30	32	34
8	—	0	2	4	6	8	10	13	15	17	19	22	24	26	29	31	34	36	38	41
9	—	0	2	4	7	10	12	15	17	21	23	26	28	31	34	37	39	42	45	48
10	—	0	3	5	8	11	14	17	21	23	26	29	33	36	39	42	45	48	52	55
11	—	0	3	6	9	12	16	19	23	26	29	33	37	40	44	47	51	55	58	62
12	—	1	4	7	11	14	18	22	26	29	33	37	41	45	49	53	57	61	65	69
13	—	1	4	8	12	16	20	24	28	33	37	41	45	50	54	59	63	67	72	76
14	—	1	5	9	13	17	22	26	31	36	40	45	50	55	59	64	67	74	78	83
15	—	1	5	10	14	19	24	29	34	39	44	49	54	59	64	70	75	80	85	90
16	—	1	6	11	15	21	26	31	37	42	47	53	59	64	70	75	81	86	92	98
17	—	2	6	11	17	22	28	34	39	45	51	57	63	67	75	81	87	93	99	105
18	—	2	7	12	18	24	30	36	42	48	55	61	67	74	80	86	93	99	106	112
19	—	2	7	13	19	25	32	38	45	52	58	65	72	78	85	92	99	106	113	119
20	—	2	8	14	20	27	34	41	48	55	62	69	76	83	90	98	105	112	119	127

(Continued)

TABLE A.6C (Continued)
Critical Values for Mann–Whitney Test

Two-Tailed $\alpha = 0.01$ (One-Tailed $\alpha = 0.005$)

n_1/n_2	1	2	3	4	5	6	7	8	9	10	11	12	13	14	15	16	17	18	19	20
1	—	—	—	—	—	—	—	—	—	—	—	—	—	—	—	—	—	—	—	—
2	—	—	—	—	—	—	—	—	—	—	—	—	—	—	—	—	—	—	0	0
3	—	—	—	—	—	—	—	—	0	0	0	1	1	1	2	2	2	2	3	3
4	—	—	—	—	—	0	0	1	1	2	2	3	3	4	5	5	6	6	7	8
5	—	—	—	—	0	1	1	2	3	4	5	6	7	7	8	9	10	11	12	13
6	—	—	—	0	1	2	3	4	5	6	7	9	10	11	12	13	15	16	17	18
7	—	—	—	0	1	3	4	6	7	9	10	12	13	15	16	18	19	21	22	24
8	—	—	—	1	2	4	6	7	9	11	13	15	17	18	20	22	24	26	28	30
9	—	—	0	1	3	5	7	9	11	13	16	18	20	22	24	27	29	31	33	36
10	—	—	0	2	4	6	9	11	13	16	18	21	24	26	29	31	34	37	39	42
11	—	—	0	2	5	7	10	13	16	18	21	24	27	30	33	36	39	42	45	46
12	—	—	1	3	6	9	12	15	18	21	24	27	31	34	37	41	44	47	51	54
13	—	—	1	3	7	10	13	17	20	24	27	31	34	38	42	45	49	53	56	60
14	—	—	1	4	7	11	15	18	22	26	30	34	38	42	46	50	54	58	63	67
15	—	—	2	5	8	12	16	20	24	29	33	37	42	46	51	55	60	64	69	73
16	—	—	2	5	9	13	18	22	27	31	36	41	45	50	55	60	65	70	74	79
17	—	—	2	6	10	15	19	24	29	34	39	44	49	54	60	65	70	75	81	86
18	—	—	2	6	11	16	21	26	31	37	42	47	53	58	64	70	75	81	87	92
19	—	0	3	7	12	17	22	28	33	39	45	51	56	63	69	74	81	87	93	99
20	—	0	3	8	13	18	24	30	36	42	46	54	60	67	73	79	86	92	99	105

TABLE A.6D
Critical Values for Wilcoxon Signed-Rank Test

	Two-Tailed		One-Tailed	
n	$\alpha = 0.05$	$\alpha = 0.01$	$\alpha = 0.05$	$\alpha = 0.01$
5	–	–	0	–
6	0	–	2	–
7	2	–	3	0
8	3	0	5	1
9	5	1	8	3
10	8	3	10	5
11	10	5	13	7
12	13	7	17	9
13	17	9	21	12
14	21	12	25	15
15	25	15	30	19
16	29	19	35	23
17	34	23	41	27
18	40	27	47	32
19	46	32	53	37
20	52	37	60	43
21	58	42	67	49
22	65	48	75	55
23	73	54	83	62
24	81	61	91	69
25	89	68	100	76
26	98	75	110	84
27	107	83	119	92
28	116	91	130	101
29	126	100	140	110
30	137	109	151	120

Bibliography

A

Abate, J. and Whitt, W. (1988). Approximations for the $M/M/1$ busy-period distribution, in: O.J. Boxma, R. Syski (Eds.), *Queueing Theory and its Applications, Liber Amicorum Professor J.W. Cohen*, North-Holland, Amsterdam, pp. 149–191.

Abramowitz, M. and Stegun, I. A., (Eds.) (1972). *Modified Bessel Functions 1 and k, §9.6 in Handbook of Mathematical Functions with Formulas, Graphs, and Mathematical Tables*, 9th printing, Dover, New York, pp. 374–377.

Ahn, C. (1998). An evaluation of phase I cancer clinical trial designs, *Statistics in Medicine* 17, p. 15371549. [PubMed: 9699228].

Aho, A. V. and Ullman, J. D. (1973). *The Theory of Parsing, Translation and Compiling*, Prentice-Hall, Englewood Cliffs, NJ.

Ammar, S. I., Helan, M. M., Al Amri, and Faizah, T. (2013). The busy period of an $M/M/1$ queue with balking and reneging, *Applied Mathematical Modelling* 37, pp. 9223–9229.

Anderson, W.J. (1991). *Continuous-Time Markov Chains, An Applications-Oriented Approach*. Springer-Verlag, New York.

Andrews, L. C. (1992). *Special Functions of Mathematics for Engineers*, 2nd edition, McGraw-Hill, Inc., New York.

Apostol, T. M. (1969). *Calculus, Volume II*, 2nd edition, John Wiley & Sons, Hoboken, NJ.

Arbuthnot, J. (1710). An argument for divine providence, taken from the constant regularity observed in the births of both sexes, *Philosophical Transactions* 27, pp. 186–190.

Arfken, G. (1985). *Mathematical Methods for Physicists*, 3rd edition, Academic Press, Orlando, FL, pp. 307–308.

Arnold, V. I. (1993). On A. N. Kolmogorov, in S. Zdravkovska, P. A. Duren (Eds.), *Golden Years of Moscow Mathematics (Providence R.I.)*, American Mathematical Society, London Mathematical Society, London, pp. 129–153.

B

Babb, J., Rogatko, A., and Zacks, S. (1963). Cancer phase I clinical trials: Efficient dose escalation with overdose control, *Statistics in Medicine* 17, pp. 1103–1120.

Bartko, J. J. (1966). The intraclass correlation coefficient as a measure of reliability, *Psychological Reports* 19(1), pp. 3–11.

Bean, N. and Latouche, G. (2010). Approximations to quasi-birth-and-death processes with infinite blocks, *Advances in Applied Probability* 42, pp. 1102–1125.

Bell, J. L. (2008, revised 2015). The Axiom of Choice, Stanford Encyclopedia of Philosophy. (The latest version, Spring 2019 Edition, Edward N. Zalta (Ed.).

Belyaev, Yu. K. (1967). Elements of the general theory of point processes (Appendix to Russian translation of: H. Cramér, M. Leadbetter, Stationary and related stochastic processes, Wiley).

Black, M. (1963). Reasoning with Loose concepts', *Dialogue* 2, pp. 1–12.

Blyth, C. R. and Pathak, P. K. (1986). A note on easy proofs of Stirling's formula, *American Mathematical Monthly* 93, pp. 376–379.

Borkakaty, B., Agarwal, M., and Sen, K. (2010) Lattice path approach for busy period density of $GI^a/G^b/1$ queues using \$ Coxian distributions, *Applied Mathematical Modelling* 34, pp. 1597–1614.

Boxma, O. J., Konheim, and Alan, G. (1981). Approximate analysis of exponential queueing systems with blocking, *Acta Informatica* 15, pp. 19–66.

Brockmeyer, E., Halstrom, H. L., and Jensen, A. (1948). The life and works of A.K. Erlang, Transactions of the Danish Academy of Technical Science 2.

Brualdi, R. A. (2010). *Introductory Combinatorics*, 5th edition, Prentice-Hall, Upper Saddle River, NJ, ISBN 978-0-13-602040-0.

Brush, S. G. (1968). A history of random processes, *Archive for History of Exact Sciences* 5(1), p. 25. doi:10.1007/BF00328110. ISSN 0003-9519.

Burke, P. J. (1956). The output of a queueing system, *Operations Research* 4, pp. 699–704.

Burke, P. J. (1972). The Output Processes and Tandem queues. *Symposiuim on Computer-Communications Networks and Teletrafic*, Polytechnic Press of the Polytechnic Institute of Brooklyn.

Burke, C. J., Estes, W. K., and Hellyer, S. (1954). Rate of verbal conditioning in relation to stimulus variability, *Journal of Experimental Psychology* 48, pp. 153–161.

C

Casella, G., Berger, R. (1990). *Statistical Inference (The Wadsworth & Brooks/Cole Statistics/Probability Series)*, Duxbury Press, Florence, KY.

Cheney, E. W and Kincaid, D. R. (1994). *Numerical Mathematics and Computing*, Brooks Cole, California, CA.

Publishing Co., Pacific Grove, CA, 3rd edition, 1994.

Clenshaw, C. W. (1962). Mathematical Tables, Vol. 5: Chebyshev Series for Mathematical Functions. Department of Scientific and Industrial Research, American Mathematical Society. https://www.jstor.org/stable/2003739.

Conolly, B.W (1971). The busy period for the infinite capacity service system *M/G/*1, in studii diprobabilit^a statistica e Ricerca opertiva in onore di G. Pompilj. *Instituto di calcolo delle probabilit^a Universita di Roma, Oderisi Gubbio, Roma*, pp. 128–130.

Conolly, B. W. (1974). The generalized state dependent Erlangian queue: the busy period, *J. Appl. Prob.* 11, pp. 618–623.

Coolidge, J. L. (1949). The story of the Binomial theorem, *The American Mathematical Monthly* 56(3), pp. 147–157. doi:10.2307/2305028.

Cox, D. R. and Isham, V. (1980). *Point Processes*, Chapman & Hall, London.

D

David, F. N. (1962). *Games, Gods & Gambling: A History of Probability and Statistical Ideas*, Charles Griffin Co. Ltd., London (Reprinted Dover Publications, 1998).

David, F. N. (1970). Dicing and gaming (a note on the history of probability), *Studies in the History of Statistics and Probability* 1, pp. 1–17.

Derman, C. (1957). Nonparametric up and down experimentation, *Annals of Mathematical Statistics* 28, pp. 795–798.

Disney, R. L. (1962). Some multichannel queueing problems with order entry. *Journal of Ind ustrial and Engineering* 13, pp. 46–48.

Disney, R. L. (1963). Some results of multichannel queueing problems with ordered entry - An application to conveyor theory, *Journal of Industrial and Engineering* XIV(2), pp. 105–108.

Disney, R. L. and Kiessler, P. C. (1987). *Traffic Processes in Queueing Networks: A Markov Renewal Approach*. The John Hopkins University Press, Baltimore, MD.

Disney, R. L. and Kögin, D. (1958). Queuing networks: A survey of their random processes, *SIAM Review* 27(3), pp. 335–400.

Doob, J. (1953). *Stochastic Processes*, John Wiley & Sons, Hoboken, NJ. ISBN 0-471-52369-0.

Doob, J. (1984). *Classical Potential Theory and Its Probabilistic Counterpart*, Springer Science & Business Media, Berlin.

Durham, S. D., Montazer-Haghighi, A., Trueblood, R. (1989). A Queueing Network for A Two- Processor System with Task Splitting and Feedback. *Proceeding of the Twentieth Annual Pittsburgh Conference on Modeling and Simulation*, part 3, Computers, Computer Architecture and Networks, Modeling and Simulation, pp. 1189–1193.

Durham, S., Flournoy, N., and Haghighi, A. M. (1993). Up-and-down designs, *Computing and Statistics: Interface* 25, pp. 375–384.

E

Edwards, A. W. F. (1987). *Pascal's Arithmetic Triangle*, Griffin, London.

Erlahg, A. K. (1917). Solution of some problems in the theory of probabilities of significance in automatic telephone exchanges, *The Post Office Electrical Engineer's Journal* 10, pp. 189–197.

Erlander, S. (1965). The remaining busy period for a single server queue with Poisson input, *Operations Research* 14, pp. 444–459.

F

Fayolle, G. (1979). Méthodes Analytiques pour les filess d'attente couplées. Theses, Université Paris VI.

Feller, W. (1968). *Stirling's Formula. §2.9 in An Introduction to Probability Theory and Its Applications*, Vol. 1, 3rd edition, Wiley, New York, pp. 50–53.

Feller, W. (1971). *An Introduction to Probability Theory and Its Applications*, Vol. 2, 3rd edition, Wiley, New York, p. 45.

Fillmore, C. J. (1969). Toward a modern theory of case, in: Reibel and Schane (eds.), *Modern Studies in English*, Prentice-Hall, Toronto, ON, pp. 361–375.

Fisher, R. A. and Tippett, L. H. C. (1928). Limiting forms of the frequency distribution of the largest and smallest member of a sample, *Proceedings of the Cambridge Philosophical Society* 24(2), pp. 180–190.

Foley, R. D. and Disney, R. L. (1983). Queues with delayed feedback. *Advances in Applied Probability.* 15, pp. 162–182.

Fraenkel, A. A., Bar-Hillel, Y., and Levy, A. (1973). *Foundations of Set Theory*, Elsevier, pp. 156–157. ISBN 978-0-08-088705-0.

G

Gao, S. and Wang, J. (2014). Performance and reliability analysis of an $M/G/1$-G retrial queue with orbital search and non-persistent customers, *European Journal of Operational Research* 236, pp. 561–572.

Gibson, D. and Seneta, E. (1987). Augmented truncations of infinite stochastic matrices, *Journal of Applied Probability* 24, pp. 600–608.

Glaisher, J. W. L. (July 1871). On a class of definite integrals, *London, Edinburgh, and Dublin Philosophical Magazine and Journal of Science.* 4. Taylor & Francis. 42(277), pp. 294–302. Retrieved 6 December 2017.

Goguen, J. A. (1969). 'The logic of inexact concepts', *Synthese* 19, pp. 325–373.

Goguen, Jr., J. A. (1974). Concept representation in natural and artificial languages: Axioms, Extensions and Applications for Fuzzy Sets, *International Journal of Man-Machine Studies* 6, pp. 513–561.

Grassmann, W. K. (1977a). Transient solutions in Markovian queues: An algorithm for finding them and determining their waiting-time distributions. *European Journal of Operational Research* 1, pp. 396–402.

Grassmann, W. K. (1977b). Transient solutions in Markovian queueing systems. *Computers & Operations Research* 4, pp. 47–53.

Grassmann, W. K. (1990). Computational methods in probability theory, Chapter 5. *Handbooks in OR & MS*, Vol. 2, p. 200. Elsevier, North-Holland.

Graham, R. L., Knuth, D. E., and Patashnik, O. (1994). *Concrete Mathematics: A Foundation for Computer Science*, 2nd edition, Addison-Wesley, Reading, MA, p. 162.

Green, M. L., Krinik, A., Mortensen, C., Rubino, G. and Swift, R. J. (2003). Transient Probability Functions: A Sample Path Approach, *Discrete Mathematics and Theoretical Computer Science* AC, pp. 127–136.

Gumbel, E. J. (1958). *Statistics of Extremes*, Columbia University Press, New York.

H

Haghighi, A. M. (1981). A many-server queueing system with feedback, *Bulletin of Iranian Mathematical Society* 9(I) (serial No. 16), pp. 65–74.

Haghighi, A. M. and Mishev, D. P. (2007). A tandem queueing system with task-splitting, feedback and blocking, *International Journal of Operational Research (IJOR)* 2(2), pp. 208–230, 2007. http://www.inderscience.com/info/inarticletoc.php?jcode=ijor&year=2007&vol=2&issue=2.

Haghighi, A. M. and Mishev, D. P. (2009). Analysis of a two-node task- splitting feedback tandem queue with infinite buffers by functional equation, *International Journal of Mathematics in Operations Research* 1(1/2), pp. 246–277. http://www.inderscience.com/info/inarticletoc.php?jcode=ijmor&year=2009&vol=1&issue=1/2.

Haghighi, A. M. and Mishev, D. P. (2013a). *Difference and Differential Equations with their Applications in Queuing Theory*, John Wiley & Sons Inc., Hoboken, NJ. http://www.wiley.com/WileyCDA/Section/id-WILEY2_SEARCH_RESULT.html?query=haghighi.

Haghighi, A. M. and Mishev, D. P. (2013b). Stochastic three-stage hiring model as a tandem queueing process with bulk arrivals and Erlang phase-type selection, $M^X/M^{(k,K)}/1-M^Y/E_J/1-\infty$, *International Journal of Mathematics in Operations Research (IJMOR)* 5(5), pp. 571–603.

Haghighi, A. M. and Mishev, D. P. (2014). *Queuing Models in Industry and Business*, 2nd Edition, Nova Science Publishers, Inc., New York. https://www.nova-publishers.com/catalog/product_info.php?products_id=42055&osCsid=7ed271dac267e38ea22b93987f939075.

Haghighi, A. M. and Mishev, D. P. (2016a). Busy period of a single-server Poisson queueing system with splitting and batch delayed-feedback, *International Journal of Mathematics in Operational Research* 8(2), pp. 239–257.

Haghighi, A. M. and Mishev, D. P. (2016b). Stepwise explicit solution for the joint distribution of queue length of a *MAP* single-server service queueing system with splitting and varying batch size delayed-feedback, *International Journal of Mathematics in Operational Research* 9(1), pp. 39–64.

Haghighi, A. M. and Mishev, D. P. (2018). Chapter 7: Stochastic modeling in indusry and management, in: M. Ram and J. P. Davim (Eds.), *A Modeling and Simulation in Industrial Engineering*, Springer, Switzerland. https://link.springer.com/chapter/10.1007/978-3-319-60432-9_7.

Haghighi, A. M. and Trueblood, R. (1997). Busy Period Analysis of a Parallel Multi-Processor System with Task Split and Feedback. The Advances in System Science and Applications (ASSA), Special Issue.

Haghighi, A. M. and Mishev, D. P. (2006). A parallel priority queueing system with finite buffers. Journal of Parallel and Distributed Computing, 66, pp. 379–392, http://www.sciencedirect.com/science/journal/07437315/66/3.

Haghighi, A. M., Chukova, S., and Mishev, D. P. (2007). Busy Period of a Delayed-Service Single-Server Poisson Queue. American Institute of Physics (AIP), http://www.aip.org, *Conference Proceedings*, Volume 946, online October 18, 2007. Applications of Mathematics in Engineering and Economics, *33rd International Conference*, Sozopol (Bulgaria), 8–14 June 2007. ISBN: 978-0-7354-0460-1. Editor: Michail D. Todorov, Technical University of Sofia.

Haghighi, A. M., Chukova, S., and Mishev, D. P. (2008). A two station tandem queueing system with delayed-service, *International Journal of Operational Research* 3(4), pp. 239–257. http://www.inderscience.com/info/inarticletoc.php?jcode=ijor&year=2008&vol=3&issue=4.

Haghighi, A. M., Lian, J., and Mishev, D. P. (2011a). *Advanced Mathematics for Engineers with Applications in Stochastic Processes*, Nova Science Publishers, New York.

Haghighi, A. M., Chukova, S., and Mishev, D. P. (2011b). Single-Server Poisson queueing system with delayed-feedback: Part 1, *International Journal of Mathematics in Operational Research (IJMOR)* 3(1), p. 363.

Hahn, G. J. and Meeker, W. Q. (1987). *Statistical Intervals: A Guide for Practitioners*, John Wiley & Sons, Inc. Computer programs for obtaining 2m are also available [e.g., ANORIN in IMSL (1987)].

Hald, A. (1990). *A History of Probability and Statistics and Their Applications before 1750*. Wiley, New York.

Halmos, P. R. (1960). *Naive Set Theory*, Van Nostrand, New York.

Hannibalsson, I. and Disney, R. L. (1977). An M/M/1 queues with delayed feedback, *Naval Res. Logist, Quart.* 24, pp. 281–291.

Haridass, M. and Arumuganathan, R. (2011). Analysis of a batch arrival general bulk service queueing system with variant threshold policy for secondary jobs, *International Journal of Mathematics in Operational Research* 3(1), pp. 56–77.

Havil, J. (2003). *Gamma: Exploring Euler's Constant*, Princeton University Press, Princeton, NJ, p. 157.

Hempel, C. G. (1952). Fundamentals of concept formation in empirical science, in *International Encyclopedia of Unified Science*, vol. 2, The University of Chicago Press, Chicago, IL.

Hintikka, J. and Suppes, P. (eds.), (1966). *Aspects of Inductive Logic*, North-Holland Publ. Co., Amsterdam.

Hintikka, J., Moravcsik, J. and Suppes, P. (Eds.), (1973). *Approaches to Natural Language*, D. Reidel Publ. Co., Dordrecht.

Hogg, R. V., Tanis, E. A. (1993). *Formats and Editions of Probability and Statistical Inference*, 4th Edition, Macmillan, New York.

Hotelling, H. and Pabst, M. R. (1936). Rank correlation and tests of significance involving no assumption of normality, *The Annals of Mathematical Statistics* 7, pp. 29–43.

I

Indrajitsingha, S. K., Samanta, P. N., and Misra, U. K. (2019). A fuzzy two-warehouse inventory model for single deteriorating item with selling-price-dependent demand and shortage under partial-backlogged condition, *Applications and Applied Mathematics: An International Journal (AAM)*, 14(1), pp. 511–536.

Irvine, A. D, Deutsch, H. (2014). Russell's Paradox, in: E. N. Zalta of philosophy, Sanford University, CA, pp. 1–11.

Ivanova, A., Montazer-Haghighi, A., Mohanty, S. G. and Durham, S. D. (2003). Improved up-and-down designs for phase I trials, *Statistics in Medicine* 22, p. 6982. doi: 10.1002/sim.1336, John Wiley & Sons, Ltd.

J

Jacobson, M. (1982). Statistical Analysis of Counting Processes, Lecturer notes in statistics, 12, Springer.

Jain, M. and Upadhyaya, S. (2012). Optimal repairable $M^X/G/1$ queue with Bernoulli feedback and setup, *International Journal of Mathematics in Operational Research* 4(6), pp. 679–702.

James, B. M. and Yexiao, J. X. (1995). A generalization of the beta distribution with applications, *Journal of Econometrics* 66(1–2), pp. 133–152. https://www.sciencedirect.com/science/article/pii/0304407694016124.

Jean-Claude, M., Wilson, S. S., Gernet, J., and Dhombres, J. (1987). *A History of Chinese Mathematics*, Springer, Switzerland.

Johnson, P. (1972). *A History of Set Theory*, Weber & Schmidt, Prindle, ISBN 0-87150-154-6.

Johnson, N. L. and Kotz, S. (1995). *Continuous Univariate Distributions*, Vol. 2, 2nd edition, Chapter 21, Wiley, Hoboken, NJ. ISBN 978-0-471-58494-0.

K

Kannadasan, G. and Sathiyamoorth, N. (2018). The analysis of $M/M/1$ queue with working vacation in fuzzy environment, *Applications and Applied Mathematics: An International Journal (AAM)*, 13(2), pp. 566–577.

Kárász, P. (2013). Equilibrium distribution for bulk-arrival $M/G/1$ system with vacation, *Journal of Mathematical Sciences* 191(4), pp. 480–484.

Karlin, S. and McGregor, J. (1958). Many server queueing processes with Poisson input and exponential service times, *Pacific Journal of Mathematics* 8, pp. 87–118.

Karlin, S. and Taylor, H. M. (1975). *A First Course in Stochastic Processes*, 2nd edition, Academic Press, New York.

Kaufmann, A. (1972). *Theory of Fuzzy Sets*, Masson, Paris.

Kendall, M. (1953). The Analysis of Economic Time Series, Part I, Prices. *Journal of the Royal Statistical Society*, 96.

Kendall, M. G. (1956). The beginnings of a probability calculus, *Biometrika* 43, pp. 1–14.

Kerstan, J., Matthes, K., and Mecke, J. (1978). *Infinitely divisible point processes*, Wiley, Hoboken, NJ (Translated from German).

Khinchin, A. Y (1960). *Mathematical Methods in the Theory of Queueing*, Griffin (Translated from Russian), Hardcover, London.

Khinchine, A. Y. (1932). Mathematical theory of a stationary queue, *Mathematics Sb.* 39(4), pp. 73–84 and 103–144 (in Russian).

Khinchine, A. Y. (1969). *Mathematical Methods in the Theory of a Queueing*, 2nd edition, Griffin's statistical monographs & courses, no. 7, London.

Kim, S-H. and Whitt, W. (2013). Statistical analysis with Little's law, *Operations Research* 61(4), pp. 1030–1045.

Kingman, J. F. C. (1961). Tv10 similar queues in parallel, *Annals of Mathematical Statistics* 32, pp. 1314–1323.

Kleinrock, L. and Gail, R. (1996). *Queueing Systems, Problems and Solutions*, John Wiley & Sons, Inc, Hoboken, NJ.

Klinerock, L. (1975). *Queueing Systems, Volume I: Theory*, John Wiley & Sons, Hoboken, NJ.

Klir, G. J. and Yuan, B. (1995). *Fuzzy sets and fuzzy logic: Theory and applications*, Prentice Hall , Upper Saddle River, NJ.

Knuth, D. (1968). Semantics of context-free languages, *Mathematical Systems Theory* 2, pp. 127–145.

Kolmogorov, A. (1933). Grundbegriffe der Wahrscheinlichkeitsrechnung (in German), Julius Springer, Berlin.

Kolmogorov, A. (1950). *Foundations of the Theory of Probability*, Chelsea, New York (2nd ed. 1956, ISBN 0-8284-0023-7. Retrieved 2016-02-17. Translation. http://uf.catalog.fcla.edu/uf.jsp?st=UF024777441&ix=pm&I=0&V=D&pm=1).

L

Lakoff, G. (1971). Linguistic and natural logic, in: D. Davidson and G. Harman (eds.), *Semantics of Natural Languages*, D. Reidel Publishing Co., Dordrecht.

Lakoff, G. (1973). Hedges: A study of meaning criteria and the logic of fuzzy concepts', *Journal of Philosophical Logic* 2, pp. 458–508.

Larsen, R. J. and Morris, L. M. (1998). *An Introduction to Probability and Its Applications*, Prentice Hall, Upper Saddle River, NJ.

Lebesgue, H. (1904). *Leçons sur l'intégration et la recherche des fonctions primitives.* Gauthier-Villars, Paris.

Lether, F. G. (1993). Elementary approximation for erf(X), *Journal of Quantitative Spectroscopy and Radiative Transfer* 49(5), pp. 573–577. doi: 10.1016/0022-4073(93)90068-S, Elsevier.

Levy, J. D. (2006). *The State After Statism: New State Activities in the Age of Liberalization*, Harvard University Press, Cambridge, MA, p. 469. ISBN 978-0-674-02276-8.

Liptser, R. S. and Shiryaev, A. N. (1978). *Statistics of Random Processes, II. Applications*, Springer (Translated from Russian), New York.

Łukasiewicz, J. (1951). *Aristotle's Syllogistic*, Clarendon Press, Oxford.

M

Maistrov, L. E. (1974). *Probability Theory - A Historical Sketch*, Academic Press, New York.

Martin, W. A. (1973). Translation of English into MAPL Using Winogard's Syntax, State Transition Networks, and a Semantic Case Grammar', MIT APG Internal Memo 11, April.

McDonald, J. B. and Xub, Y. J. (1995). A generalization of the beta distribution with applications. *Journal of Econometrics* 66, 133–152. Corr: 1995;69, 427–428.

Mohanty, S. G., Montazer Haghighi, A., and Trueblood, R. (1993). On the transient behavior of a finite birth-death process with an application. *Computers and Operations Research* 20(3), pp. 239–248.

Montazer-Haghighi, A. (1973). *Axiomatic System and Axiomatic Set Theory*, Institute of Statistics and Informatics, Tehran, Iran.

Montazer-Haghighi, A. (1976). Axiomatic Systems and Axiomatic Set Theory, Publication of Institute of Statistics and Informatic, No. 52. Written in Farsi (Persian language) with a special preface by Professor Mohsen Hashtroodi.

Montazer Haghighi, A. and Mohanty, S. G. (2019). In Honor and Memory of Professor Emeritus Lajos Takács, Book Series: Developments in Mathematics, Editors: George E. Andrews, Christian Krattenthaler, and Alan Krinik, Springer International Publishing, Print ISBN: 978-3-030-11101-4, Electronic ISBN: 978-3-030-11102-1 https://www.amazon.com/gp/product/3030111016/ref=ppx_od_dt_b_asin_title_ s00?ie=UTF 8&psc=1.

Montazer Haghighi, A., Medhi, J., and Mohanty, S. G. (1986). On a multi-server Markovian queueing system with balking and reneging. *Computers and Operations Research* 13, pp. 421–425.

Moore, C. N. (2007). Random walks, *Ramanujan Mathematical Society Mathematics Newsletter* 17(3), pp. 78–84.

Murofushi, T. and Sugeno, M. (1989). An interpretation of fuzzy measures and the Choquet integral as an integral with respect to a fuzzy measure, *Fuzzy Sets and Systems* 29(2), pp. 201–227.

Murphy, J. A. and O'Donohoe, M. R. (1975). Some properties of continued fractions with applications in Markov processes, *IMS Journal of Applied Mathematics*, 16 (1), pp. 57–71.

N

Nakade, K. (2000) New bounds for expected cycle time in tandem queues with blocking, *European Journal of Operational Research* 125, pp. 84–92.

Nakamura, G. (1971). A feedback queueing model for an interactive computer system, *AFIPS Proceedings of the Fall Joint Conference*.

Nance, R. E., Bhat, V. N., Billy, G. C., and Claybrook, V. N. (1972). Busy period analysis of a time- sharing system: Transform inversion, *Journal of the ACM* 19, pp. 453–462.

Neuts, M. M. F. (1989). *Structured Stochastic Matrices of M/G/1 Type and Their Application*, Marcel Dekker, Inc., New York.

Neyman, J. (1937). Outline of a theory of statistical estimation based on the classical theory of probability, *Philosophical Transactions of the Royal Society of London* 236, pp. 333–380.

O

Olofsson, P. and Andersson, M. (2012). *Probability, Statistics and Stochastic Processes*, 2nd edition, John Wiley & Sons, Inc., Hoboken, NJ.

P

Parlett, B. N. (1980). *The Symmetric Eigenvalue Problem*, Prentice Hall, Englewood Cliff, NJ.

Pearson, K. (1895). Contributions to the mathematical theory of evolution. II. Skew variation in homogeneous material. *Philosophical Transactions of the Royal Society A: Mathematical, Physical and Engineering Sciences* 186, pp. 343–414.

Pearson, K. (1905). The problem of the random walk, *Nature* 72(1865), p. 294.

Pearson, E. S. and Kendall, M. G. (Eds.) (1970). *Studies in the History of Statistics and Probability, Volume 1*, Hafner, Darien, Connecticut.

Pearson and Kendall (1970). Kendall, M. G. and Plackett, R. L. (Eds.) (1977). *Studies in the History of Statistics and Probability, Volume 2*, Macmillan, New York.

Pólya, G. (2004). How to Solve It: A New Aspect of Mathematical Method (Princeton Science Library) Princeton Science Li Edition with new foreword by John H. Conway. Prior printings: 1945, 1957, 1971, 1973, 1985, and 1988.

Porter, T. (1986). *The Rise of Statistical Thinking, 1820–1900*, Princeton University Press, Princeton, NJ.

Prabhu, N. (2007). *Stochastic Processes: Basic Theory and its Applications*, World Scientific Publishing Co. Ltd, Singapore.

Prats'ovytyi, M. V. (1996). Singularity of distributions of random variables given by distributions of elements of the corresponding continued fraction, *Ukrainian Mathematical Journal* 48(8), pp. 1229–1240, Springer. https://rd.springer.com/article/10.1007%2FBF02383869.

Q

Quine, W. V. (1953). *From a Logical Point of View*, Harvard Univrsity Press, Cambridge.

R

Ramakrishna Rao, K. V. (2007). Date of Pingala - Origin of Binary Computation, All Empires History Community, February 17. http://www.allempires.com/forum/forum_posts.asp?TID=17915.

Ramaswami, V. (1988a). A stable recursion for the steady state vector in Markov chains of *M/G/*1 type, *Stochastic Models* 4(1), pp. 183–188.

Ramaswami, V. (1988b). Nonlinear matrix equations in applied probability – Solution techniques and open problems, *SIAM Review* 30(2), pp. 256–263.

Rescher, N. (1969). *Many-Valued Logic*, McGraw-Hill, New York.

Rice, J. A. (2007). *Mathematical Statistics and Data Analysis*, 3rd Edition, Thomson, Brooks/Cole.

Rice, S. O. (1962) Single server systems–II. Busy period, *Bell System Technology Journal* 41, pp. 279–310.

Russell, B. A. W. (1919). *Introduction to Mathematical Philosophy*, Routledge, London.

S

Schank, R. C. (1972). Conceptual dependency: A theory of natural language understanding, *Cognitive Psychology* 3, pp. 552–631.

Schneider, I. (1988). The market place and games of chance in the fifteenth and sixteenth centuries, in: C. Hay (Ed.), *Mathematics from Manuscript to print*, Clarendon Press; Oxford University Press, New York.

Servi, L.D. and Finn, S.G. (2002). *M/M/1* queue with working vacations *(M/M/1/WV)*, *Performance Evaluation* 50, pp. 41–52.

Shafer, G. (1978). Non-additive probabilities in the work of Bernoulli and Lambert, *Archive for History of Exact Sciences* 19, pp. 309–370.

Shafer, G. (1982). Bayes's two arguments for the rule of conditioning, *Annals of Statistics* 10, pp. 1075–1089.

Shafer, G. 1988. The St. Petersburg paradox, *Encyclopedia of Statistical Sciences* 8, pp. 865–870. Samuel Kotz and Norman L. Johnson, editors. New York (Wiley).

Shanbhag, D. N. (1966). On infinite server queues with batch arrivals, *Journal Applied Probability* 3, pp. 274–279.

Simon, H. A. and Siklossy, L. (Eds.), (1972). *Representation and Meaning Experiments with Information Processing Systems*, Prentice-Hall, Englewood Cliffs, NJ.

Singh, C. J., Jain, M., and Kumar, B. (2012). $M^X/G/1$ queuing model with state dependent arrival and Second Optional Vacation, *International Journal of Mathematics in Operational Research* 4(1), pp. 78–96.

Smith, D. E. (1951). *A History of Mathematics*, Dover Publications, Inc., New York.

Smith, D. E. (1959). *A Source Book in Mathematics*, Dover Publications, Inc., New York.

Smith, C. A. B. (1966). Graphs and composite games, *Journal of Combinatoral Theory* 1, 51–81, reprinted in slightly modified form in: A Seminar on Graph Theory (F. Harary, ed.), Holt, Rinehart and Winston, New York, NY, 1967.

Srivastava, H. M. and Kashyap, B. R. K. (1982). *Special Functions in Queueing Theory and Related Stochastic Processes*, Academic Press, New York.

T

Täcklind, S. (1944). Elementare Behandlung vom Erneuerungsproblem, *Skand. Aktuarietidsk* 27, pp. 1–15.

Takács, L. (1958). On a probability problem in the theory of counters, *Annals of Mathematical Statistics*. 29, pp. 1257–1263.

Takács, L. (1962). *Introduction to the Theory of Queues*, Oxford University Press, Oxford.

Takács, L. F. (1963a). A single-server queue with feedback, *The Bell System Technical Journal* 42(2), pp. 505–519.

Takács, L. F. (1963b). Delay distributions for one line with Poisson input, general holding times and various orders of services, *The Bell System Technical Journal* 42, pp. 487–503.

Takács, L. (1967). *Combinational Methods in the Theory of Stochastic Processes*, John Wiley and Sons, New York.

Takács, L. (1974). A single-server queue with limited virtual waiting time, *Journal Applied Probability* 11, 612–617.

Takács, L. F. (1975). *Some Remarks on a Counter Process*, Case Western Reserve University, Cleveland, OH.

Takács, L. (1997). Ballot problems, in: N. Balakrishnan (Ed.), *Advances in Combinatorial Methods and Applications to Probability and Statistics*, Birkhäuser, Boston, MA, pp. 97–114.

Takács, L. (1999). The distribution of the sojourn time of the Brownian excursion, *Methodology and Computing in Applied Probability* 1(1), pp. 7–28.

Tarski, A. (1956). *Logic, Semantics, Metamathematics*, Clarendon Press, Oxford.

Tijms, H. C. (2003). *A First Course in Stochastic Models*, John Wiley & Sons Ltd, West Sussex, England.

Tukey, J. W. (1977). *Exploratory Data Analysis*, 1st edition, Pearson, London. ISBN 0-201-07616-0.

V

van Fraassen, B. C. (1969). Presuppositions, Supervaluations and Free Logic, in K. Lambert (Ed.), *The Logical Way of Doing Things*, Yale University Press, New Haven.

Vijaya Laxmi, P. and Yesuf, O. M. (2011). Renewal input infinite buffer batch service queue with single exponential working vacation and accessibility to batches, *International Journal of Mathematics in Operational Research* 3(2), pp. 219–243.

Ville, J. (1939). Étude critique de la notion de collectif. *Bulletin of the American Mathematical Society. Monographies des Probabilités (in French). Paris* 3(11), pp. 824–825. doi: 10.1090/S0002-9904-1939-07089-4. Zbl 0021.14601. Review by Doob, Vlastos, Gregory (1972). Zeno of Elea, in The Encyclopedia of Philosophy (Paul Edwards, ed.), New York.

Vlastos, G. (1972). Zeno of Elea, in: P. Edwards (Ed.), *The Encyclopedia of Philosophy*, The Macmillan Co. and The Free Press, New York.

W

Wang, Z. and Klir, G. J. (1972). *Fuzzy Measure Theory*, Plenum Press, New York.

Ward, M. D. and Gundiach, E. (2016). *Introduction to Probability*, W.H Freeman & Company, New York.

Weibull, W. (1951). A statistical distribution function of wide applicability, *Journal of Applie d Mechanics, Transactions ASME* 18(3), pp. 293–297.

Whittaker, E. T. and Robinson, G. (1967). *Stirling's Approximation to the Factorial. §70 in The Calculus of Observations: A Treatise on Numerical Mathematics*, 4th edition, Dover, New York, pp. 138–140.

Whitlock, T. S. (1973). Modeling computer systems with time-varying Markov-chains. Ph. D. thesis, University of North Carolina.

Wilkinson, J. H. (1965). Tile Algebraic *Eigenvalue Problem*. Clarendon Press, Oxford.

Wilks, S. S. (1962). *Mathematical Statistics*, 1st Edition, Wiley Publication in Mathematical Statistics, Wiley; ASIN: B011SK1WX4.

Wolfowitz, J. (1942). Additive partition functions and a class of statistical hypotheses, *The Annals of Mathematical Statistics*, 13, pp. 247–279.

X

Xie, M. (1989). On the solution of renewal-type integral Equationso, *Communications in Statistics, B*. 18, pp. 281–293.

Z

Zadeh, L. A. (1965). Fuzzy sets, *Information and Control* 8(3), pp. 338–353.

Zadeh, L. A. (1968). Fuzzy algorithms, *Information and Control* 12(2), pp. 94–102. doi: 10.1016/S0019-9958(68)90211-8.

Zadeh, L. A. (1971). Quantitative fuzzy semantics', *Information Sciences* 3, pp. 159–176.

Zadeh, L. A. (1972). A fuzzy-set-theoretic interpretation of linguistic hedges, *Journal of Cybernetics* 2, pp. 4–34.

Zadeh, L. A. (1973). The Concept of a Linguistic Variable and Its Application to Approximate Reasoning', Memorandum No. ERL-M411, Electronics Research Lab., Univ. of Calif., Berkeley, Calif., October.

Zadeh, L. A. (1974). 'A Fuzzy-Algorithmic Approach to the Definition of Complex or Imprecise Concepts', Memorandum No. ERL-M474, Electronics Research Lab., Univ. of Calif., Berkeley, Calif.

Zadeh, L. A. (1978). Fuzzy sets as a basis for a theory of probability, *Fuzzy Sets and Systems* 1, pp. 3–28.

Index

9 780367 500863